Element Speciation in
Bioinorganic Chemistry

CHEMICAL ANALYSIS

A SERIES OF MONOGRAPHS ON ANALYTICAL CHEMISTRY AND ITS APPLICATIONS

Editor
J. D. WINEFORDNER

VOLUME 135

A WILEY-INTERSCIENCE PUBLICATION

JOHN WILEY & SONS, INC.

New York / Chichester / Brisbane / Toronto / Singapore

Element Speciation in Bioinorganic Chemistry

Edited by

SERGIO CAROLI

Istituto Superiore di Sanità
Rome, Italy

A WILEY-INTERSCIENCE PUBLICATION

JOHN WILEY & SONS, INC.

New York / Chichester / Brisbane / Toronto / Singapore

This text is printed on acid-free paper.

Copyright © 1996 by John Wiley & Sons, Inc.

All rights reserved. Published simultaneously in Canada.

Reproduction or translation of any part of this work beyond that permitted by Section 107 or 108 of the 1976 United States Copyright Act without the permission of the copyright owner is unlawful. Requests for permission or further information should be addressed to the Permissions Department, John Wiley & Sons, Inc., 605 Third Avenue, New York, NY 10158-0012.

Library of Congress Cataloging in Publication Data:

Element speciation in bioinorganic chemistry / edited by Sergio Caroli.
    p. cm.—(Chemical analysis : v. 135)
    "A Wiley-Interscience publication."
    Includes index.
    ISBN 0-471-57641-7 (alk. paper)
    1. Bioinorganic chemistry—Methodology. 2. Speciation (Chemistry)
I. Caroli, Sergio. II. Series.
QP531.E44   1996
574.19'214—dc20   95-11167

Printed in the United States of America

10 9 8 7 6 5 4 3 2 1

CONTENTS

CONTRIBUTORS		xvii
FOREWORD		xix
PREFACE		xxi
CUMULATIVE LISTING OF VOLUMES IN SERIES		xxiii

CHAPTER 1 CHEMICAL SPECIATION: A DECADE OF PROGRESS 1
S. Caroli

1.1. A Leap Ahead	1
1.2. A Composite Picture	3
1.2.1. Some General Considerations	3
1.2.2. Selected Examples	8
1.2.2.1. General Aspects	8
1.2.2.2. Instrumental Developments	9
1.2.2.3. Environmental Applications	11
1.2.2.4. Biological Applications	16
1.3. A Look into the Future	18
References	18

CHAPTER 2 DIRECT METHODS OF SPECIATION OF HEAVY METALS IN NATURAL WATERS 21
A. M. Mota and M. L. Simões Gonçalves

2.1. The Ecological Importance of Speciation	21
2.1.1. Different Types of Metal Species	22
2.1.2. Environmental Factors Affecting Speciation	23
2.2. Constituents of Natural Waters	25
2.2.1. Inorganic Constituents	25
2.2.2. Organic Matter	25

CONTENTS

2.2.3. Colloids and Particulate Matter	27
2.3. Methodology for Speciation	28
2.3.1. Various Approaches	28
2.3.2. Kinetic Aspects	30
2.4. Direct Methods of Speciation	32
2.4.1. Introduction	32
2.4.2. Physical Techniques	34
2.4.2.1. Centrifugation	34
2.4.2.2. Filtration	35
2.4.2.3. Ultrafiltration	37
2.4.2.4. Dialysis	39
2.4.2.5. Gel Filtration Chromatography	40
2.4.3. Chemical Techniques	40
2.4.3.1. Oxidative Destruction of Organic Matter	40
2.4.3.2. Liquid–Liquid Extraction	41
2.4.3.3. Ionic Exchange and Adsorbent Resins	42
2.4.3.4. Voltammetric Techniques	44
2.4.3.4.1. Metals Determined by Voltammetric Techniques	47
2.4.3.4.2. Major Factors Influencing the Voltammetric Signal of Natural Waters	48
2.4.3.4.3. Some Applications to Natural Waters	51
2.4.4. Species-Specific Techniques	54
2.4.4.1. Potentiometry with Ion-Specific Electrodes	54
2.4.4.2. Gas Chromatography and/or Hydride Generation	57
2.4.4.3. High-Performance Liquid Chromatography	59
2.4.4.4. Other Chemical Methods	60
2.4.5. Bioassays	61
2.4.5.1. Determination of Free Element Concentration and Toxicity	61
2.4.5.2. Comparison Between Biochemical and Chemical Processes	62
2.4.6. Comprehensive Speciation Schemes	63

	2.5. Fundamentals of Voltammetric Methods	68
	2.5.1. General Comments	68
	2.5.2. Speciation in Noncomplexing Media	70
	2.5.3. Speciation in Complexing Media	70
	2.5.3.1. Inert Complexes	71
	2.5.3.1.1. Speciation of M and ML	71
	2.5.3.1.2. Complexing Titration Curves	73
	2.5.3.2. Labile Complexes	74
	2.5.3.2.1. Speciation from i_m Measurements	75
	2.5.3.2.2. Speciation from ΔE Measurements	75
	2.5.3.2.3. Complexing Titration Curves	77
	2.5.3.3. Quasi-labile Complexes	79
	2.5.4. Complexation of Heterogeneous Ligands	83
	References	87
CHAPTER 3	**NONCHROMATOGRAPHIC METHODS OF ELEMENT SPECIATION BY ATOMIC SPECTROMETRY**	**97**
	M. de la Guardia	
	3.1. Introduction	97
	3.2. Selective Atomization of Different Chemical Forms of the Same Element	100
	3.3. Speciation Based on Selective Extractions	102
	3.4. Speciation Based on Derivatization Procedures	105
	3.5. Speciation Based on Selective Volatilization	110
	3.6. Miscellaneous Methods	112
	3.7. Concluding Remarks	114
	References	115
CHAPTER 4	**DEVELOPMENT OF NEW METHODS OF SPECIATION ANALYSIS**	**121**
	I. T. Urasa	
	4.1. Introduction	121
	4.2. Factors Affecting Element Speciation	122

	4.2.1. Sample Treatment and Processing	122
	4.2.2. Sample Acidification	123
4.3.	Analytical Aspects	125
	4.3.1. Measurement Methods	125
	4.3.2. Element-Selective Detectors	129
	4.3.3. Coupling of IC with DCP–AES	132
	4.3.4. Detection Power	133
	4.3.5. Influence of the Mobile Phase on Detector Performance	134
	4.3.6. Influence of the Chemistry of the Analyte and Sample Matrix	135
4.4.	Applications	140
	4.4.1. Chromium Speciation	140
	4.4.2. Monitoring of the Element Transformation Rate	143
	4.4.3. Application to Clinical and Related Measurements	144
	4.4.4. Solution Chemistry of Phosphorus	147
	4.4.5. Determination of Trace Heavy Metals in Water	147
	4.4.6. Determination of the Binding Capacity of Natural Waters	149
4.5.	Conclusions	150
	References	151

CHAPTER 5 NEUTRON ACTIVATION ANALYSIS AND RADIOTRACER METHODS IN BIOINORGANIC CHEMISTRY — 155

H. A. Das and J. R. W. Woittiez

5.1.	Introduction	155
	5.1.1. Historic Background	155
	5.1.2. Definitions and Limitations	157
	5.1.3. General Aspects	157
5.2.	Outline of Neutron Activation Analysis	161

5.2.1. Principles	161
5.2.2. Apparatus	164
5.2.3. Irradiation Facilities for NAA	166
5.2.3.1. Nuclear Reactors	166
5.2.4. Practical Aspects of NAA of Biological Materials	168
5.2.4.1. NAA with Short-Lived Radionuclides at ECN	168
5.2.4.2. NAA with Long-Lived Radionuclides at ECN	169
5.2.4.3. Standards	171
5.2.4.4. Sampling and Sample Preparation	172
5.2.4.5. The Error Budget	173
5.2.4.5.1. Statistical Errors	173
5.2.4.5.2. Systematic Errors	174
5.3. Separation of Proteins and Its Control by Radiotracers and Neutron Activation	174
5.3.1. General Remarks	174
5.3.2. Gel Filtration Chromatography	175
5.3.3. Gel–Element Interaction of Gel Filtration Materials	175
5.3.4. Radiotracer Experiments	178
5.3.5. Use of NAA to Detect Protein-Bound Elements	179
5.4. An Experiment Involving NAA of Desalted Human Serum	181
5.4.1. Purpose	181
5.4.2. Samples	181
5.4.3. The Procedure Using NAA	182
5.4.3.1. NAA with Short-Lived Radionuclides	182
5.4.3.2. NAA with Long-Lived Radionuclides	182
5.4.4. Results	182
5.4.4.1. Desalting on Bio P-2	183
5.4.4.1.1. Short-Lived Radionuclides	183
5.4.4.1.2. Long-Lived Radionuclides	185

	5.5. Elements in Human Serum Proteins or Other Proteins and Related Radiotracer Experiments	187
	5.5.1. General Remarks	187
	5.5.2. Distribution of Trace Elements	188
	5.5.3. Some Applications of Radiotracers	189
	5.6. Conclusions	191
	References	192
CHAPTER 6	QUALITY CONTROL OF RESULTS OF SPECIATION ANALYSIS	195
	Ph. Quevauviller, E. A. Maier, and B. Griepink	
	6.1. Introduction	195
	6.2. The Need for Development of New Analytical Techniques in Speciation Analysis	196
	6.3. Overview of Quality Assurance Principles	197
	6.3.1. General Remarks	197
	6.3.2. Statistical Control	199
	6.3.3. Comparison with Results of Other Methods	201
	6.3.4. Use of Certified Reference Materials	201
	6.3.5. Intercomparisons	202
	6.4. Potential Sources of Error in Speciation Analysis	203
	6.4.1. Extraction	204
	6.4.2. Derivatization	205
	6.4.3. Separation	207
	6.4.4. Detection	209
	6.4.5. Primary Calibrants and Internal Standards	209
	6.4.6. Further Efforts to Improve the State of the Art and Method Validation	211
	6.4.6.1. Determination of As(III), As(V), and Organo-As Compounds	214
	6.4.6.2. Determination of Butyl- and Phenyl-Sn Compounds	215
	6.4.6.3. Determination of Methyl-Hg	216

	6.5. Errors in Trace Element Speciation in Sediments and Soils	217
	6.6. Conclusions	219
	References	220
CHAPTER 7	**ALUMINUM AND SILICON SPECIATION IN BIOLOGICAL MATERIALS OF CLINICAL RELEVANCE**	**223**

A. Sanz-Medel, B. Fairman, and Katarzyna Wróbel

7.1. Introduction	223
7.2. Total Aluminum and Silicon Analyses in Biological Materials	225
7.2.1. Total Aluminum Determination	227
7.2.2. Total Silicon Determination	229
7.3. Aluminum Speciation Techniques	232
7.3.1. Analytical Techniques for Speciation of Aluminum	232
7.3.1.1. Nonchromatographic Techniques: Determination of Ultrafiltrable Serum Aluminum	232
7.3.1.2. Chromatographic Techniques: Aluminum–Protein Speciation	234
7.3.1.3. Low-Molecular-Weight Aluminum Species: Aluminum–Citrate	238
7.3.1.4. Aluminum–Complexing Drugs: Speciation of Aluminum–Desferrioxamine	239
7.4. The Role of Aluminum and Silicon in Neurological Disorders: The Quest for Speciation	243
7.4.1. Dialysis Encephalopathy and Other Diseases	243
7.4.2. Alzheimer's Disease	243
7.4.3. Species Favoring Aluminum and Silicon Transport to the Brain	244
7.5. Does Silicic Acid Prevent Aluminum Toxicity via Formation of Aluminosilicates?	246
7.6. Conclusions	246
References	248

CHAPTER 8 SPECIATION OF TRACE ELEMENTS IN MILK BY HIGH-PERFORMANCE LIQUID CHROMATOGRAPHY COMBINED WITH INDUCTIVELY COUPLED PLASMA ATOMIC EMISSION SPECTROMETRY 255

E. Coni, A. Alimonti, A. Bocca, F. La Torre, D. Pizzuti, and S. Caroli

8.1. Introduction	255
8.2. Experimental Work	258
8.2.1. General Aspects	258
8.2.2. Sampling Procedure	259
8.2.3. Determination of the Total Element Concentrations	260
8.2.3.1. Sample Treatment	260
8.2.3.2. Analysis	261
8.2.4. Determination of Chemical Species	263
8.2.4.1. Sample Treatment	263
8.2.4.2. Analysis	264
8.3. Evaluation of Data	265
8.3.1. Total Concentrations of Elements	265
8.3.2. Concentration of Element Species	267
8.3.3. Relative Distribution of Trace Elements	273
8.4. Conclusions	282
References	283

CHAPTER 9 ORGANOTIN COMPOUNDS IN MARINE ORGANISMS 287

S. Chiavarini, C. Cremisini, and R. Morabito

9.1. Introduction	287
9.2. Toxicity	289
9.3. Metabolism and Accumulation	297
9.4. Legal Provisions	306
9.4.1. France	306
9.4.2. United Kingdom	307
9.4.3. United States	307
9.4.4. Italy	308
9.4.5. Other Countries	308

9.5.	Regional Cases	308
	9.5.1. France	308
	9.5.2. United Kingdom	309
	9.5.3. United States	310
	9.5.4. Italy	315
	9.5.5. Other Countries	319
9.6.	Conclusions	320
	References	321

CHAPTER 10 TIN SPECIATION MONITORING IN ESTUARINE AND COASTAL ENVIRONMENTS 331

Ph. Quevauviller and O. F. X. Donard

10.1.	Introduction	331
10.2.	The Development of Environmental Concerns Regarding Sn Compounds	333
10.3.	Geochemical Pathways and Transformations	335
	10.3.1. Methylation Reactions	335
	10.3.2. Volatilization Reactions	336
	10.3.3. Degradation Reactions	337
10.4.	Sources of Environmental Variability	339
	10.4.1. Temporal Variability	340
	10.4.2. Spatial Variability	341
	10.4.3. Partitioning in Marine Waters and Sediments	342
10.5.	Strategies for Organotin Monitoring	347
	10.5.1. Sampling Program Design	347
	10.5.2. Choice of Samples	348
	10.5.3. Water Samples	348
	10.5.4. Suspended Matter Samples	353
	10.5.5. Sediment Samples	354
	10.5.6. Sample Storage	354
	10.5.7. Determination	355
10.6.	Conclusions	356
	References	357

CHAPTER 11 THE ANODIC STRIPPING VOLTAMMETRIC TITRATION PROCEDURE FOR STUDY OF TRACE METAL COMPLEXATION IN SEAWATER — 363

G. Scarponi, G. Capodaglio, C. Barbante, and P. Cescon

- 11.1. Introduction — 363
- 11.2. Methodology: Theoretical Aspects — 365
  - 11.2.1. Single 1:1 Complex Formation — 366
  - 11.2.2. Ligand Competition: Two 1:1 Complexes of One Metal with Two Ligands — 373
    - 11.2.2.1. The van den Berg Approach — 379
    - 11.2.2.2. The Ruzic Approach — 380
    - 11.2.2.3. Ligand Competition: Generalization — 383
  - 11.2.3. Metal Competition: Two 1:1 Complexes of One Ligand with Two Metals — 383
  - 11.2.4. Other than 1:1 Complex Formation — 392
- 11.3. Methodology: The Instrumental Approach — 393
  - 11.3.1. DPASV Instrumentation — 394
  - 11.3.2. Preparation of the Thin-Mercury Film Electrode — 394
  - 11.3.3. Total Metal Concentration — 395
  - 11.3.4. Metal Speciation (Voltammetric Titration of Ligands) — 398
  - 11.3.5. Defining the DPASV-Measured Species in a Seawater Sample — 399
- 11.4. Contamination Control — 402
  - 11.4.1. Laboratory and Chemicals — 403
  - 11.4.2. Sampling — 405
  - 11.4.3. Filtration — 406
  - 11.4.4. Storage — 407
- 11.5. Review of the Literature — 408
  - 11.5.1. Copper — 408
  - 11.5.2. Zinc and Cobalt — 411
  - 11.5.3. Lead and Cadmium — 412

	11.6. Conclusions	413
	References	414

CHAPTER 12 PROBLEMS OF SPECIATION OF ELEMENTS IN NATURAL WATERS: THE CASE OF CHROMIUM AND SELENIUM   **419**

L. Campanella

	12.1. Introduction	419
	12.2. Chemical and Biological Aspects	420
	12.3. The Case of Chromium	423
	12.4. The Case of Selenium	432
	12.4.1. Procedures	436
	12.4.2. Application of the Speciation Model	436
	12.5. Conclusions	441
	References	442

CHAPTER 13 ARSENIC SPECIATION AND HEALTH ASPECTS   **445**

S. Caroli, F. La Torre, F. Petrucci, and N. Violante

	13.1. Introduction	445
	13.2. Experimental Approaches	446
	13.3. Ongoing Activities	451
	13.4. Conclusions	459
	References	461

INDEX   **465**

CONTRIBUTORS

A. Alimonti, Department of Applied Toxicology, Istituto Superiore di Sanità, Rome, Italy

C. Barbante, Department of Environmental Sciences, University of Venice, and Center for Studies on Environmental Chemistry and Technology, CNR, Venice, Italy

A. Bocca, Department of Food Chemistry, Istituto Superiore di Sanità, Rome, Italy

L. Campanella, Department of Chemistry, University "La Sapienza," Rome, Italy

G. Capodaglio, Department of Environmental Sciences, University of Venice, and Center for Studies on Environmental Chemistry and Technology, CNR, Venice, Italy

S. Caroli, Department of Applied Toxicology, Istituto Superiore di Sanità, Rome, Italy

P. Cescon, Department of Environmental Sciences, University of Venice, and Center for Studies on Environmental Chemistry and Technology, CNR, Venice, Italy

S. Chiavarini, Department of Analysis and Monitoring of the Environment, ENEA, Environment, Energy and Health Sector, Rome, Italy

E. Coni, Department of Food Chemistry, Istituto Superiore di Sanità, Rome, Italy

C. Cremisini, Department of Analysis and Monitoring of the Environment, ENEA, Environment, Energy and Health Sector, Rome, Italy

H. A. Das, Netherlands Energy Research Foundation, ECN, Petten, The Netherlands

M. de la Guardia, Department of Analytical Chemistry, Faculty of Chemistry, University of Valencia, Burjasot, Valencia, Spain

O. F. X. Donard, Laboratory of Photophysics and Molecular Photochemistry, University of Bordeaux I, Talence, France

B. Fairman, Department of Physics and Analytical Chemistry, University of Oviedo, Oviedo, Spain

B. Griepink, European Commission, Standards, Measurements and Testing Programme, Brussels, Belgium

F. La Torre, Department of Pharmaceutical Chemistry, Istituto Superiore di Sanità, Rome, Italy

E. A. Maier, European Commission, Standards, Measurements and Testing Programme, Brussels, Belgium

R. Morabito, Department of Analysis and Monitoring of the Environment, ENEA, Environment, Energy and Health Sector, Rome, Italy

A. M. Mota, Centro de Química Estrutural, Instituto Superior Técnico, Lisbon, Portugal

F. Petrucci, Department of Applied Toxicology, Istituto Superiore di Sanità, Rome, Italy

D. Pizzuti, Department of Food Chemistry, Istituto Superiore di Sanità, Rome, Italy

Ph. Quevauviller, European Commission, Standards, Measurements and Testing Programme, Brussels Belgium

A. Sanz-Medel, Department of Physics and Analytical Chemistry, University of Oviedo, Oviedo, Spain

G. Scarponi, Department of Environmental Sciences, University of Venice, Venice, Italy

M. L. Simões Gonçalves, Centro de Química Estrutural, Instituto Superior Técnico, Lisbon, Portugal

I. T. Urasa, Department of Chemistry, University of Hampton, Hampton, Virginia

N. Violante, Department of Applied Toxicology, Istituto Superiore di Sanità, Rome, Italy

J. R. W. Woittiez, Netherlands Energy Research Foundation, ECN, Petten, The Netherlands

Katarzyna Wróbel, Institute of Chemistry, Białystok, Poland

FOREWORD

The concept of *metal speciation* in the sense of distinguishing fractions of the total element concentration according to some peculiar properties such as being "bioavailable" or "algae assimilable" grew slowly in the late 1950s and early 1960s. It seems that, as fate would have it, investigations into the cycling of radioisotopes released by atomic weapons test explosions initially triggered the interest of some scientists around the globe in this rather exotic subject matter.

It was in early 1968 that I first encountered the subject because of a quite innocent request by a colleague from the Biology Division at the Joint Research Centre, Ispra, Italy, who wished me to determine the physical-chemical fractions of cobalt in the waters of Lake Maggiore, with particular attention to vitamin B_{12} (cobalamin). Needless to say, the request remained without a satisfactory answer owing to the insufficient analytical means available to us in those years. Nevertheless, the seed of a lifelong interest was planted, and I soon discovered that elsewhere some other laboratories were engaged in this type of research activity, especially in water-related projects: first of all the groups around the late Hans Wolfgang Nürnberg in Jülich, Germany; Marco Branica and Znonimir Pucar in Zagreb, Yugoslavia (now Croatia), E. Duursma in Monaco; and T. M. Florence in Australia.

The rapid development of trace element analytical chemistry in the 1960s, the advent of the Massman furnace and the Kemula electrode, which opened hitherto prohibitive concentration ranges of elements in water to every analyst, promoted the development of operationally defined procedures for analyte fractionation in water.

The generous support of the NATO Advanced Studies Institute allowed us to realize the dream of gathering together master scientists as well as scholars at Nervi, Italy, in 1980 for an unforgettable week of discussion that opened new horizons in methodological development and application.

During the previous decade, the speciation concept had been rapidly applied to other areas of the environment. In the mid-1970s, G. Müller and U. Förstner (Heidelberg), R. Chester (Liverpool), and A. Tessier (Sainte-Foy, Québec, Canada) had pioneered the development of methodologies for the discrimination of mobile and less mobile element species in sediments. Then, in the early 1980s, the Commission of the European Communities (CEC) issued for the first time sewage sludge and soil reference materials yielding, besides the total element concentration, data on the concentrations of analyte fractionation by the aqua regia method. This was followed some years later by the sediment reference material series, which offered data obtained by a four-step metal fractionation method. This approach developed further through a series of collaborative studies and workshops into the CEC Common Procedure, for which we prepared a series of sediment reference materials that were under study by the Standards, Measurements and Testing (SMT) Programme of the European Union [formerly the Community Bureau of Reference (BCR)] attempting the certification of element fraction concentrations.

Single chemical species analyses for tributyltin, methylmercury, and arsenobetaine, to mention a few, were first achieved in the 1980s owing to the development and refinement of hyphenated analytical techniques applied to sediment and biota analysis.

Currently, a series of candidate reference materials prepared at Ispra over the last few years is undergoing certification analysis by the SMT Programme. We anticipate that the advent of these materials on the market will provide a further impetus in an area where more accurate and precise data are urgently needed. The present monograph will no doubt help in this respect.

HERBERT MUNTAU

Environment Institute
Joint Research Centre of the European Union
Ispra Establishment
Ispra, Italy

PREFACE

Some time ago, I became aware of the frequency with which the term *speciation* appears in the literature. The concept was certainly not new to me; my laboratory had been actively involved in speciation for a number of years. What I did perceive, however, was how quickly the scientific community had responded to this rather revolutionary approach crossing many chemistry-related disciplines. It also became apparent to me that such a proliferation of studies could be misguiding and deceptive insofar as it could obscure, through its very abundance, the overall fabric in which such investigations should take place. It seemed to me that the time had come to make a stand and account for the advances attained to date. This book will focus on the two major issues brought about by chemical speciation—the scientific rationale behind it, and its applicability to priority themes for the protection of human health and the environment. I have consulted with internationally renowned experts, many of whom contributed a chapter centered on one of the two criteria mentioned above. The project was submitted for approval to the Editor of the Chemical Analysis Series, Professor J. D. Winefordner, with whom the final layout of the book was agreed upon and to whom I am gratefully indebted for constructive criticisms and strong encouragement.

The book is divided into two main sections: in the first, fundamental and instrumental aspects of speciation are treated; in the second, applications of speciation-oriented techniques to current public health and ecological challenges are discussed. Chapters 1 through 6 examine the following topics: the wealth of information published on chemical speciation (S. Caroli); the theoretical features of and consolidated approaches to the analysis of metal species in aqueous media (A. M. Mota and M. L. Simões Gonçalves); the capabilities of atomic absorption spectrometry for the direct identification of different element forms (M. de la Guardia); the development of novel hyphenated techniques (I. T. Urasa); the analytical potential of neutron activation analysis and radiotracer methods for assessment of the role of elements in protein chemistry (H. A. Das and J. R. W. Woittiez); and the parameters that may impair data credibility and comparability in speciation analysis (Ph. Quevauviller, E. A. Maier, and B. Griepink). The topics of the remaining seven chapters are clearly aimed at the solution of practical problems and are inextricably interwoven with the aspects highlighted in the preceding chapters.

These include the following: the performance of chemical speciation for aluminum and silicon in biological fluids (A. Sanz-Medel, B. Fairman, and K. Wróbel); the speciation of element species in milk (E. Coni, A. Alimonti, A. Bocca, F. La Torre, D. Pizzuti, and S. Caroli); the metabolism and accumulation of organotin compounds in the marine environment, as well as legal provisions pertaining to them (S. Chiavarini, C. Cremisini, and R. Morabito); the monitoring of tin species in estuarine and coastal milieus (Ph. Quevauviller and O. F. X. Donard); the study of trace metal complexation in seawater (G. Scarponi, G. Capodaglio, C. Barbante, and P. Cescon); the speciation of chromium and selenium in natural waters (L. Campanella); and the identification in seafood of arsenic compounds noxious to humans (S. Caroli, F. La Torre, F. Petrucci, and N. Violante). The scenario that emerges will help promote the opening of substantially new avenues in bioinorganic chemistry and the life sciences. It is hoped that interested readers will find this book an incentive to further expand an already promising horizon.

SERGIO CAROLI

Rome, Italy
March 1996

ACKNOWLEDGMENT

The skills and forbearance of Mr. Massimo Delle Femmine in dealing with the various drafts of this book are gratefully acknowledged.

CHEMICAL ANALYSIS

A SERIES OF MONOGRAPHS ON ANALYTICAL CHEMISTRY AND ITS APPLICATIONS

J. D. Winefordner, *Series Editor*

Vol.	1.	**The Analytical Chemistry of Industrial Poisons, Hazards, and Solvents.** *Second Edition.* By the late Morris B. Jacobs
Vol.	2.	**Chromatographic Adsorption Analysis.** By Harold H. Strain (*out of print*)
Vol.	3.	**Photometric Determination of Traces of Metals.** *Fourth Edition* Part 1: General Aspects. By E. B. Sandell and Hiroshi Onishi Part IIA: Individual Metals, Aluminum to Lithium. By Hiroshi Onishi Part IIB: Individual Metals, Magnesium to Zirconium. By Hiroshi Onishi
Vol.	4.	**Organic Reagents Used in Gravimetric and Volumetric Analysis.** By John F. Flagg (*out of print*)
Vol.	5.	**Aquametry: A Treatise on Methods for the Determination of Water.** *Second Edition* (*in three parts*). By John Mitchell, Jr. and Donald Milton Smith
Vol.	6.	**Analysis of Insecticides and Acaricides.** By Francis A. Gunther and Roger C. Blinn (*out of print*)
Vol.	7.	**Chemical Analysis of Industrial Solvents.** By the late Morris B. Jacobs and Leopold Schetlan
Vol.	8.	**Colorimetric Determination of Nonmetals.** *Second Edition.* By the late David F. Boltz and James A. Howell
Vol.	9.	**Analytical Chemistry of Titanium Metals and Compounds.** By Maurice Codell
Vol.	10.	**The Chemical Analysis of Air Pollutants.** By the late Morris B. Jacobs
Vol.	11.	**X-Ray Spectrochemical Analysis.** *Second Edition.* By L. S. Birks
Vol.	12.	**Systematic Analysis of Surface-Active Agents.** *Second Edition.* By Milton J. Rosen and Henry A. Goldsmith
Vol.	13.	**Alternating Current Polarography and Tensammetry.** By B. Breyer and H. H. Bauer
Vol.	14.	**Flame Photometry.** By R. Herrmann and J. Alkemade
Vol.	15.	**The Titration of Organic Compounds** (*in two parts*). By M. R. F. Ashworth
Vol.	16.	**Complexation in Analytical Chemistry: A Guide for the Critical Selection of Analytical Methods Based on Complexation Reactions.** By the late Anders Ringbom
Vol.	17.	**Electron Probe Microanalysis.** *Second Edition.* By L. S. Birks
Vol.	18.	**Organic Complexing Reagents: Structure, Behavior, and Application to Inorganic Analysis.** By D. D. Perrin

xxiv   CHEMICAL ANALYSIS: A SERIES OF MONOGRAPHS

Vol. 19. **Thermal Analysis.** *Third Edition.* By Wesley Wm. Wendlandt
Vol. 20. **Amperometric Titrations.** By John T. Stock
Vol. 21. **Reflectance Spectroscopy.** By Wesley Wm. Wendlandt and Harry G. Hecht
Vol. 22. **The Analytical Toxicology of Industrial Inorganic Poisons.** By the late Morris B. Jacobs
Vol. 23. **The Formation and Properties of Precipitates.** By Alan G. Walton
Vol. 24. **Kinetics in Analytical Chemistry.** By Harry B. Mark, Jr. and Garry A. Rechnitz
Vol. 25. **Atomic Absorption Spectroscopy.** *Second Edition.* By Morris Slavin
Vol. 26. **Characterization of Organometallic Compounds** (*in two parts*). Edited by Minoru Tsutsui
Vol. 27. **Rock and Mineral Analysis.** *Second Edition.* By Wesley M. Johnson and John A. Maxwell
Vol. 28. **The Analytical Chemistry of Nitrogen and Its Compounds** (*in two parts*). Edited by C. A. Streuli and Philip R. Averell
Vol. 29. **The Analytical Chemistry of Sulfur and Its Compounds** (*in three parts*). By J. H. Karchmer
Vol. 30. **Ultramicro Elemental Analysis.** By Günther Tölg
Vol. 31. **Photometric Organic Analysis** (*in two parts*). By Eugene Sawicki
Vol. 32. **Determination of Organic Compounds: Methods and Procedures.** By Frederick T. Weiss
Vol. 33. **Masking and Demasking of Chemical Reactions.** By D. D. Perrin
Vol. 34. **Neutron Activation Analysis.** By D. De Soete, R. Gijbels, and J. Hoste
Vol. 35. **Laser Raman Spectroscopy.** By Marvin C. Tobin
Vol. 36. **Emission Spectrochemical Analysis.** By Morris Slavin
Vol. 37. **Analytical Chemistry of Phosphorus Compounds.** Edited by M. Halmann
Vol. 38. **Luminescence Spectrometry in Analytical Chemistry.** By J. D. Winefordner, S. G. Schulman and T. C. O'Haver
Vol. 39. **Activation Analysis with Neutron Generators.** By Sam S. Nargolwalla and Edwin P. Przybylowicz
Vol. 40. **Determination of Gaseous Elements in Metals.** Edited by Lynn L. Lewis, Laben M. Melnick, and Ben D. Holt
Vol. 41. **Analysis of Silicones.** Edited by A. Lee Smith
Vol. 42. **Foundations of Ultracentrifugal Analysis.** By H. Fujita
Vol. 43. **Chemical Infrared Fourier Transform Spectroscopy.** By Peter R. Griffiths
Vol. 44. **Microscale Manipulations in Chemistry.** By T. S. Ma and V. Horak
Vol. 45. **Thermometric Titrations.** By J. Barthel
Vol. 46. **Trace Analysis: Spectroscopic Methods for Elements.** Edited by J. D. Winefordner
Vol. 47. **Contamination Control in Trace Element Analysis.** By Morris Zief and James W. Mitchell
Vol. 48. **Analytical Applications of NMR.** By D. E. Leyden and R. H. Cox
Vol. 49. **Measurement of Dissolved Oxygen.** By Michael L. Hitchman
Vol. 50. **Analytical Laser Spectroscopy.** Edited by Nicolò Omenetto
Vol. 51. **Trace Element Analysis of Geological Materials.** By Roger D. Reeves and Robert R. Brooks

Vol.	52.	**Chemical Analysis by Microwave Rotational Spectroscopy.** By Ravi Varma and Lawrence W. Hrubesh
Vol.	53.	**Information Theory As Applied to Chemical Analysis.** By Karel Eckschlager and Vladimir Štěpánek
Vol.	54.	**Applied Infrared Spectroscopy: Fundamentals, Techniques, and Analytical Problem-solving.** By A. Lee Smith
Vol.	55.	**Archaeological Chemistry.** By Zvi Goffer
Vol.	56.	**Immobilized Enzymes in Analytical and Clinical Chemistry.** By P. W. Carr and L. D. Bowers
Vol.	57.	**Photoacoustics and Photoacoustic Spectroscopy.** By Allan Rosencwaig
Vol.	58.	**Analysis of Pesticide Residues.** Edited by H. Anson Moye
Vol.	59.	**Affinity Chromatography.** By William H. Scouten
Vol.	60.	**Quality Control in Analytical Chemistry.** *Second Edition.* By G. Kateman and L. Buydens
Vol.	61.	**Direct Characterization of Fineparticles.** By Brian H. Kaye
Vol.	62.	**Flow Injection Analysis.** By J. Ruzicka and E. H. Hansen
Vol.	63.	**Applied Electron Spectroscopy for Chemical Analysis.** Edited by Hassan Windawi and Floyd Ho
Vol.	64.	**Analytical Aspects of Environmental Chemistry.** Edited by David F. S. Natusch and Philip K. Hopke
Vol.	65.	**The Interpretation of Analytical Chemical Data by the Use of Cluster Analysis.** By D. Luc Massart and Leonard Kaufman
Vol.	66.	**Solid Phase Biochemistry: Analytical and Synthetic Aspects.** Edited by William H. Scouten
Vol.	67.	**An Introduction to Photoelectron Spectroscopy.** By Pradip K. Ghosh
Vol.	68.	**Room Temperature Phosphorimetry for Chemical Analysis.** By Tuan Vo-Dinh
Vol.	69.	**Potentiometry and Potentiometric Titrations.** By E. P. Serjeant
Vol.	70.	**Design and Application of Process Analyzer Systems.** By Paul E. Mix
Vol.	71.	**Analysis of Organic and Biological Surfaces.** Edited by Patrick Echlin
Vol.	72.	**Small Bore Liquid Chromatography Columns: Their Properties and Uses.** Edited by Raymond P. W. Scott
Vol.	73.	**Modern Methods of Particle Size Analysis.** Edited by Howard G. Barth
Vol.	74.	**Auger Electron Spectroscopy.** By Michael Thompson, M. D. Baker, Alec Christie, and J. F. Tyson
Vol.	75.	**Spot Test Analysis: Clinical, Environmental, Forensic and Geochemical Applications.** By Ervin Jungreis
Vol.	76.	**Receptor Modeling in Environmental Chemistry.** By Philip K. Hopke
Vol.	77.	**Molecular Luminescence Spectroscopy: Methods and Applications** (*in three parts*). Edited by Stephen G. Schulman
Vol.	78.	**Inorganic Chromatographic Analysis.** By John C. MacDonald
Vol.	79.	**Analytical Solution Calorimetry.** Edited by J. K. Grime
Vol.	80.	**Selected Methods of Trace Metal Analysis: Biological and Environmental Samples.** By Jon C. Van Loon
Vol.	81.	**The Analysis of Extraterrestrial Materials.** By Isidore Adler

Vol.	82.	**Chemometrics.** By Muhammad A. Sharaf, Deborah L. Illman, and Bruce R. Kowalski
Vol.	83.	**Fourier Transform Infrared Spectrometry.** By Peter R. Griffiths and James A. de Haseth
Vol.	84.	**Trace Analysis: Spectroscopic Methods for Molecules.** Edited by Gary Christian and James B. Callis
Vol.	85.	**Ultratrace Analysis of Pharmaceuticals and Other Compounds of Interest.** By S. Ahuja
Vol.	86.	**Secondary Ion Mass Spectrometry: Basic Concepts, Instrumental Aspects, Applications and Trends.** By A. Benninghoven, F. G. Rüdenauer, and H. W. Werner
Vol.	87.	**Analytical Applications of Lasers.** Edited by Edward H. Piepmeier
Vol.	88.	**Applied Geochemical Analysis.** By C. O. Ingamells and F. F. Pitard
Vol.	89.	**Detectors for Liquid Chromatography.** Edited by Edward S. Yeung
Vol.	90.	**Inductively Coupled Plasma Emission Spectroscopy: Part I: Methodology, Instrumentation, and Performance; Part II: Applications and Fundamentals.** Edited by J. M. Boumans
Vol.	91.	**Applications of New Mass Spectrometry Techniques in Pesticide Chemistry.** Edited by Joseph Rosen
Vol.	92.	**X-Ray Absorption: Principles, Applications, Techniques of EXAFS, SEXAFS, and XANES.** Edited by D. C. Konnigsberger
Vol.	93.	**Quantitative Structure-Chromatographic Retention Relationships.** Edited by Roman Kaliszan
Vol.	94.	**Laser Remote Chemical Analysis.** Edited by Raymond M. Measures
Vol.	95.	**Inorganic Mass Spectrometry.** Edited by F. Adams, R. Gijbels, and R. Van Grieken
Vol.	96.	**Kinetic Aspects of Analytical Chemistry.** By Horacio A. Mottola
Vol.	97.	**Two-Dimensional NMR Spectroscopy.** By Jan Schraml and Jon M. Bellama
Vol.	98.	**High Performance Liquid Chromatography.** Edited by Phyllis R. Brown and Richard A. Hartwick
Vol.	99.	**X-Ray Fluorescence Spectrometry.** By Ron Jenkins
Vol.	100.	**Analytical Aspects of Drug Testing.** Edited by Dale G. Deutsch
Vol.	101.	**Chemical Analysis of Polycyclic Aromatic Compounds.** Edited by Tuan Vo-Dinh
Vol.	102.	**Quadrupole Storage Mass Spectrometry.** By Raymond E. March and Richard J. Hughes
Vol.	103.	**Determination of Molecular Weight.** Edited by Anthony R. Cooper
Vol.	104.	**Selectivity and Detectability Optimizations in HPLC.** By Satinder Ahuja
Vol.	105.	**Laser Microanalysis.** By Lieselotte Moenke-Blankenburg
Vol.	106.	**Clinical Chemistry.** Edited by E. Howard Taylor
Vol.	107.	**Multielement Detection Systems for Spectrochemical Analysis.** By Kenneth W. Busch and Marianna A. Busch
Vol.	108.	**Planar Chromatography in the Life Sciences.** Edited by Joseph C. Touchstone
Vol.	109.	**Fluorometric Analysis in Biomedical Chemistry: Trends and Techniques Including HPLC Applications.** By Norio Ichinose, George Schwedt, Frank Michael Schnepel, and Kyoko Adochi
Vol.	110.	**An Introduction to Laboratory Automation.** By Victor Cerdá and Guillermo Ramis

CHEMICAL ANALYSIS: A SERIES OF MONOGRAPHS

Vol. 111. **Gas Chromatography: Biochemical, Biomedical, and Clinical Applications.** Edited by Ray E. Clement
Vol. 112. **The Analytical Chemistry of Silicones.** Edited by A. Lee Smith
Vol. 113. **Modern Methods of Polymer Characterization.** Edited by Howard G. Barth and Jimmy W. Mays
Vol. 114. **Analytical Raman Spectroscopy.** Edited by Jeannette Graselli and Bernard J. Bulkin
Vol. 115. **Trace and Ultratrace Analysis by HPLC.** By Satinder Ahuja
Vol. 116. **Radiochemistry and Nuclear Methods of Analysis.** By William D. Ehmann and Diane E. Vance
Vol. 117. **Applications of Fluorescence in Immunoassays.** By Ilkka Hemmila
Vol. 118. **Principles and Practice of Spectroscopic Calibration.** By Howard Mark
Vol. 119. **Activation Spectrometry in Chemical Analysis.** By S. J. Parry
Vol. 120. **Remote Sensing by Fourier Transform Spectrometry.** By Reinhard Beer
Vol. 121. **Detectors for Capillary Chromatography.** Edited by Herbert H. Hill and Dennis McMinn
Vol. 122. **Photochemical Vapor Deposition.** By J. G. Eden
Vol. 123. **Statistical Methods in Analytical Chemistry.** By Peter C. Meier and Richard Zund
Vol. 124. **Laser Ionization Mass Analysis.** Edited by Akos Vertes, Renaat Gijbels, and Fred Adams
Vol. 125. **Physics and Chemistry of Solid State Sensor Devices.** By Andreas Mandelis and Constantinos Christofides
Vol. 126. **Electroanalytical Stripping Methods.** By Khjena Z. Brainina and E. Neyman
Vol. 127. **Air Monitoring by Spectroscopic Techniques.** Edited by Markus W. Sigrist
Vol. 128. **Information Theory in Analytical Chemistry.** By Karel Eckschlager and Klaus Danzer
Vol. 129. **Flame Chemiluminescence Analysis by Molecular Emission Cavity Detection.** Edited by David Stiles, Anthony Calokerinos, and Alan Townshend
Vol. 130. **Hydride Generation Atomic Absorption Spectrometry.** Edited by Jiri Dedina and Dimiter L. Tsalev
Vol. 131. **Selective Detectors: Environmental, Industrial, and Biomedical Applications.** Edited by Robert E. Sievers
Vol. 132. **High-Speed Countercurrent Chromatography.** Edited by Yoichiro Ito and Walter D. Conway
Vol. 133. **Particle-Induced X-Ray Emission Spectrometry.** By Sven A. E. Johansson, John L. Campbell, and Klass G. Malmqvist
Vol. 134. **Photothermal Spectroscopy Methods for Chemical Analysis.** By Stephen Bialkowski
Vol. 135. **Element Speciation in Bioinorganic Chemistry.** Edited by Sergio Caroli

Element Speciation in
Bioinorganic Chemistry

CHAPTER

1

CHEMICAL SPECIATION: A DECADE OF PROGRESS

S. CAROLI

Istituto Superiore di Sanità, 00161 Rome, Italy

1.1. A LEAP AHEAD

The life sciences are benefiting more and more from the development of innovative analytical methodologies. At present, this is especially so as regards the identification, separation, and quantification of the various forms under which a chemical substance may occur in matrices as diverse as biological fluids, tissues and organs, foodstuffs; fresh-, sea-, and wastewaters, plant materials, soils and sediments, among many others. Although entirely satisfactory definitions for the different aspects covered by this overall approach are still lacking and consequently full agreement on terminology is as yet far from being achieved, the word *speciation* has become extremely popular and is widely used in current applications focused on trace elements (1–5). Probably the most comprehensive definition is the one worked out by the International Union of Pure and Applied Chemistry (IUPAC) in an attempt to reconcile various overly specialized views. It states that speciation is "the process yielding evidence of the atomic or molecular form of an analyte." The validity of this statement lies in that it can accommodate both organic compounds and inorganic substances. In the latter case, moreover, it holds for the determination of both specific molecular combinations of an element and the different states of oxidation of an atom.

In this connection it should also be stressed that there is at present much debate as to whether the term speciation should be extended to include processes that allow fractions of a given element to be measured on the basis of their degree of leachability from complex matrices, e.g., soils or sediments. This is obviously a matter of convention and therefore arbitrary to a certain extent; however, this broader meaning would appear justified when in the material being tested distinct compounds are found to which the element under

Element Speciation in Bioinorganic Chemistry, edited by Sergio Caroli.
Chemical Analysis Series, Vol. 135.
ISBN 0-471-57641-7 © 1996 John Wiley & Sons, Inc.

consideration is bound with different degrees of·strength. The net outcome is, again, that various chemical species, or subsets thereof, can be unequivocally determined, exactly as happens in all other instances of speciation (6–8).

What is certainly common to all possible interpretations of the concept of speciation is that there is an attempt, on the one hand, to establish which forms of a chemical element possess the highest potential for noxious effects on organisms and ecosystems and, on the other hand, to clarify which species of an essential element are actually bioavailable (9, 10). Both aspects are obviously crucial to the further development of bioinorganic chemistry as a discipline substantially contributing to the preservation and improvement of human welfare.

It is not by chance that a wealth of information has been accruing in this field over the past few years: the dramatic growth in the number of sophisticated instrumental techniques that can be usefully combined to achieve speciation has in recent times allowed for the proliferation of investigations that have now produced an impressive body of evidence. From this point of view highly effective separation methods and precise, accurate, and sensitive detection systems are both needed and must be effectively combined. Progress made so far in analytical equipment is thoroughly illustrated in several chapters in this book, together with representative applications. Therefore, this aspect will not be dealt with in any detail here except for an outline of some basic features. In this regard, it is first of all worth stressing that most often the analytes of interest are at the trace or ultratrace level and that, consequently, the concentrations of the different chemical species of an element are even more elusive. Secondly, the equilibria governing the formation of such species are quite complex and sensitive to any external interference. This leads to the conclusion that an appropriate analytical approach to the speciation challenge must inevitably rely on superior separation and detection capabilities while affecting as little as possible the original distribution of the element among matrix components.

Furthermore, it goes without saying that accuracy and precision of measurements should be guaranteed by all possible means. For this reason, hyphenated separation–quantification systems are to be preferred, as manipulation of samples and run time are thereby minimized. It is not surprising, then, that those analytical methods that are best suited to be combined on line to carry out speciation studies are also those that generally have been the most successful and have undergone the quickest development (11, 12). For the sake of completeness, however, it should be noted that there are some interesting alternative routes that deserve our particular attention, e.g., attempts to determine the sought-for species directly *in situ* or to model their behavior by computer simulation. The state of the art as regards speciation modes may be traced back to three main experimental categories, based on physical, chemical, and biochemical principles, as set forth in Table 1.1

Table 1.1. Principal Classes of Speciation Procedures

Methods centered on physical properties: Adsorption phenomena; centrifugation; some chromatographic techniques; cold trapping; dialysis; distillation; element-specific detection; filtration

Methods centered on chemical properties: some chromatographic techniques; complexation; derivatization; element-specific detection; extraction; ion-exchange processes; redox reactions

Methods centered on biochemical properties: enzymatic reactions; immunoassay measurements

In this context, it is self-evident that complementary techniques can be advantageously combined within each of these three subsets and among them as well, in order to attain at the same time both isolation of the species of interest and their quantification (13, 14). The borderline between these two general tasks is sometimes not easily recognizable, if not seemingly absent. In fact, it may occur under some circumstances that an analytical method is highly specific for a given chemical species. In instances like this the separation step is not visible only because it is intrinsic to the detection system, which is able to sense selectively the form at hand. Many examples of this built-in specificity are possible: among others, enzymatic reactions, immunoassay tests, and colorimetry trials. In all cases the sought-for chemical species of the element triggers a highly distinctive chemical, physical, or biological process that leads to its unequivocal measurement. From a general point of view, however, it is the proper combination of separation and detection procedures that sets the stage for successful speciation studies. In this regard such valuable techniques as gel filtration chromatography, supercritical fluid extraction, and bioassays in the former group and atomic absorption spectrometry (AAS), inductively coupled plasma mass spectrometry (ICP–MS) and anodic stripping voltammetry (ASV) in the latter group show much promise. Past achievements and future prospects clearly support this view, as will be abundantly described in later chapters of this book.

1.2. A COMPOSITE PICTURE

1.2.1. Some General Considerations

The mounting awareness that speciation studies are contributing significantly to the progress of the life sciences is well reflected by the number of papers, reports, books, and other scientific publications that have become available in recent years. To gauge the real dimensions of this phenomenon as well as to estimate in a more quantitative and impartial fashion well-established trends and emerging tendencies, we carried out a survey of the literature in this field that is retrievable through relevant documentation services. Rather arbitrarily, our scan took into consideration only what appeared from 1983 up to 1994, because we felt that before that period the number of works on speciation were still too scanty and pioneering in nature to provide any useful indication as to the subsequent evolution. The considerations which follow are, therefore, based on some 12 years of progress in speciation research and applications, i.e., a time span from which realistic insights can be gained. In this period some 2060 publications could be identified as eligible for inclusion in the roster of those works mainly focused on element speciation.

Four different yet complementary criteria were chosen by which we sought to classify the wealth of information thus collected, namely, the *topic* dealt with by the study considered, the *country* where the study was carried out, the *language* in which it was published, and the *type of document* used to disseminate it. The general picture that becomes manifest is certainly intriguing, as illustrated in detail in Figs. 1.1 through 1.4, although some particular facets could be easily anticipated. Several interesting conclusions can be drawn from the data reported therein.

As regards the subject matter of investigations (Fig. 1.1), many of them (more then 40%) appear to be devoted to either general arguments (e.g., pertaining to fundamental and theoretical aspects, reviews, or instrumentation and the like) or sets of elements grouped according to a specific characteristic, pattern, or exigency. On the other hand, as regards single elements, those more inclined to form complexes, adducts, and associations of any kind with other substances are quite understandably more often taken into account. Groups Ib, IIb, IIIb, IVb, Vb, VIa, VIb, and VIII of the periodic table as well as actinides are from this point of view investigated much more often than the remaining ones, with elements such as Al, As, Cd, Cr, Cu, Fe, Hg, Pb, and Zn ranking highest of all. Just the opposite holds true for Groups Ia, IIa, IIIa, IVa, and Va and lanthanides, which rather understandably have stimulated the curiosity of scientists in this field only to a limited extent.

In terms of the geographic location of the institutions where the studies were conducted (Fig. 1.2), almost half of them belong to European states, including those in the east. In this context it worth noting that the highest ranking goes in order to the United Kingdom, Germany, Russia, France, and Italy. Nearly at the same level are laboratories in North America. Moreover, the active role played by some Asian countries, especially Japan and China, should not be overlooked.

With respect to the languages used to convey research findings to the readers (Fig. 1.3), it is not surprising that an overwhelming portion of the

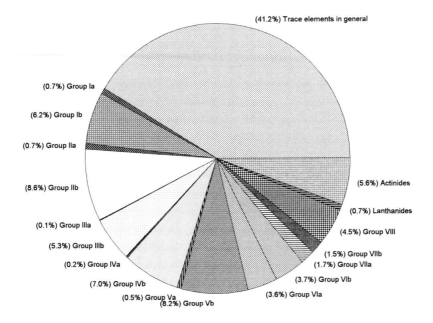

```
Group Ia   (Li, Na, K, Cs)           Group Ib   (Cu, Ag, Au)
Group IIa  (Be, Mg, Ca, Sr, Ba)      Group IIb  (Zn, Cd, Hg)
Group IIIa (Y)                       Group IIIb (B, Al, Ga, Tl)
Group IVa  (Ti, Zr, Hf)              Group IVb  (C, Si, Ge, Sn, Pb)
Group Va   (V, Nb)                   Group Vb   (N, P, As, Sb)
Group VIa  (Cr, Mo, W)               Group VIb  (S, Se, Te, Po)
Group VIIa (Mn, Tc)                  Group VIIb (F, Cl, Br, I)
Group VIII (Fe, Co, Ni, Ru, Pd, Ir, Pt)
```

Figure 1.1. Percentage of publications devoted to speciation of chemical elements according to analyte type over the period 1983–1994 (2060 publications were scanned). Elements in parentheses for each group are those more frequently studied. "Trace elements in general" stands for works focused on general aspects or not specifically centered on one element.

studies was published in English. Less expected was that, although at a distance of an order of magnitude from English, Chinese and Russian were utilized more frequently than other languages. Immediately after them in frequency came Japanese and German.

Finally, as regards the publications preferred to disseminate speciation studies (Fig. 1.4), North American and European journals each comprise

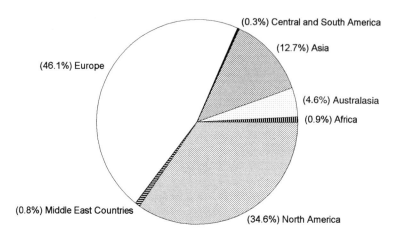

Figure 1.2. Percentage of publications devoted to speciation of chemical elements according to the region where each study was done over the period 1983–1994 (2060 publications were scanned).

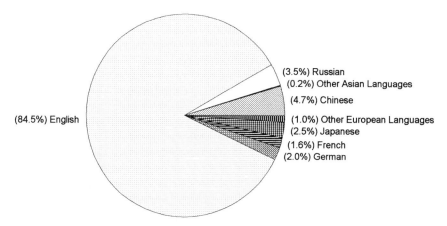

Figure 1.3. Percentage of publications devoted to speciation of chemical elements according to the language in which each study was presented over the period 1983–1994 (2060 publications were scanned).

nearly a third of the entire mass of published material. Among these, the predominant ones are periodicals such as *Analytica Chimica Acta, Analytical Chemistry*, and *Fresenius' Journal of Analytical Chemistry* [prior to Vol. 339 (1991) called *Fresenius' Zeitschrift für Analytische Chemie*]. Nor should the rather high percentage of information flow be neglected that finds its way through conference proceedings, university dissertations, books, and reports,

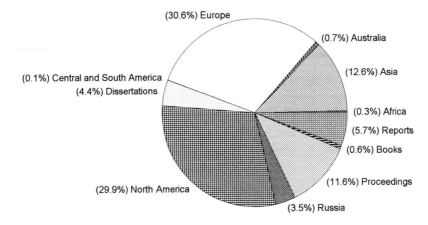

Figure 1.4. Percentage of publications devoted to speciation of chemical elements according to publication type over the period 1983–1994 (2060 publications were scanned).

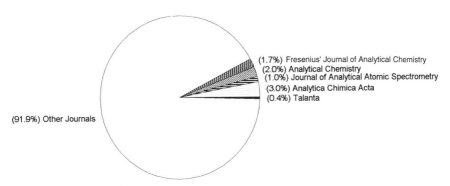

Figure 1.5. Percentage of papers published in specific journals (2060 publications were scanned over the period 1983–1994).

an unequivocal sign of the creativity that earmarks this research area. Figure 1.5 highlights the percentage of works appearing in these journals vs. the total of journals considered.

It would obviously be impracticable to cite even briefly all the contributions on speciation that have enriched the relevant literature to date. Nonetheless, it seems appropriate to mention some representative papers that have opened new avenues for the development of this discipline, and these are briefly discussed below.

1.2.2. Selected Examples

1.2.2.1. General Aspects

In a multiauthored volume devoted to analytical approaches in trace element speciation, current challenges and strategies were presented (2). The chapters contained in this book examined the following topics in order: various aspects concerned with the proper collection of samples; preparation and storage of samples for speciation analysis; the role of speciation analysis in assessing bioavailability and toxicity in natural waters; physicochemical separation methods for speciation studies in aquatic samples; the relevance of electrochemical techniques for trace element speciation in diverse waters; the development and applicability of high-performance liquid chromatography (HPLC) as well as gas chromatography (GC) in trace element speciation; and specific sets of problems posed by trace element speciation in both sediments and biological systems. Especially designed to assist the analytical chemist, the book proves of value in guiding the reader to a correct choice of methods amid the intricate abundance of sophisticated analytical instrumentation available today.

In a 1992 paper by Behne, the need for speciation in the various fields of biology and medicine was highlighted with particular reference to the methods available and those being developed (15). The author pointed out how often toxic effects are exerted by elements bound to small molecules, whereas their presence in high-molecular-weight substances have clear relevance to essential biochemical mechanisms. Precautions were suggested to minimize the adverse effects of preanalytical steps (e.g., homogenization, fractionation of tissues and cells, cell disruption, and fractionation of subcellular compounds). In this connection it is of the highest importance that the biological structures of the components to be separated remain intact. The keen exigency for fully reliable quality control schemes in speciation analysis and the production of dedicated certified reference materials were also touched upon.

An overview of the state of the art of element speciation as undertaken by the Standards, Measurements and Testing (SMT) Programme of the European Union [formerly the Community Bureau of Reference (BCR)] was published, reflecting in particular the discussions held at a workshop centered on improvement of speciation analysis in environmental matrices held in Arcachon, France, in 1990 (4). Activities under the auspices of the SMT Programme were shown to involve some 30 institutions in the European Union and to cover topics as diverse as derivatization, organometallic synthesis and chemistry, separation techniques, matrix digestion, sample pretreatment, extraction, cleanup, recovery, and detection. Several ongoing projects launched by the SMT Programme were also illustrated in detail, namely, the determination of extractable trace element contents in soils and sediments of total and mobile forms of Al; Cr(III) and Cr(VI); Se(IV), Se(VI), and Se organic forms; As(III), As(V), and As organic compounds; trimethyl- and triethyl-Pb compounds; methyl-Hg; and butyl- and phenyl-Sn compounds. Emphasis was placed upon the stringent requirements for high accuracy, reproducibility, and traceability that characterize such collaborative efforts and underlie the philosophy of the SMT Programme as well as the ensuing intercomparisons aimed at checking and improving the quality of speciation analyses made by participants.

1.2.2.2. Instrumental Developments

The pros and cons of element speciation techniques based on the combination of chromatographic systems with atomic detectors were discussed by Hill and Ebdon (16). From this standpoint GC has the obvious advantage of conveying the sample to the detector directly in the gaseous form, thus avoiding problems of inefficiency associated with nebulization. This approach is however limited to volatile compounds, whereas in most cases metal speciation concerns non-volatile species. Some practical solutions to frequent problems inherent in these and other combinations of techniques are described, in particular as regards electrothermal atomization AAS (ETA–AAS), inductively coupled plasma atomic emission spectrometry (ICP–AES), direct current plasma AES (DCP–AES), and atomic fluorescence spectrometry (AFS). The continuous mode of operation was clearly indicated as an ideal approach to deal with sample streams. This approach is greatly facilitated by the commercial availability of many components that can be easily assembled and readily decoupled.

A critical review of the various techniques for the determination of elemental species was presented by Lund (17). The different branches of spectrometry, electrochemistry, chromatography, nuclear magnetic resonance spectrometry, and mass spectrometry were considered. Particular attention was devoted to the application of such methodologies to the analysis of natural waters, air, soil, and biological materials. The importance of chemical speciation to bioavailability was stressed, together with the need for operationally defined speciation schemes that can achieve separation of chemical forms (or groups thereof) according to size, ionic charge, polarity, redox state, redox properties, and binding strength.

The miniaturization of HPLC systems was proposed by Ishii et al. (18). At that time the possibilities offered by microchromatographic systems also showed analytical potential for speciation studies. Column volume was thought to be reducible to 1–2% of the normal value without any deleterious effect on chromatographic performance. Theoretical aspects were also taken into account.

An interface for coupling a microcolumn liquid chromatographic system with a cold-vapor AAS detector was described and applied to the speciation of Hg compounds in wastewater by Munaf et al. (19). Preconcentration of the analyte was achieved by injecting the water samples into a Develosil-ODS (30 μm) column and subsequently by eluting with cysteine–acetic acid through the STR-ODS-H (5 μm) separation column. The Hg species thus fractionated were first mixed with $K_2S_2O_8$ and $CuSO_4$ to convert organic Hg compounds into Hg(II) ions and then added with $SnCl_2$ and NaOH to achieve reduction to elemental Hg, which was finally swept to the AAS detector. This approach enabled the successful separation of $HgCl_2$, CH_3HgCl and C_2H_5HgCl to be achieved with a detection power in the order of magnitude of tens to hundreds of micrograms per liter.

The separation of several groups of cations and anions was accomplished by Bushee et al. using ion-paired reversed-phase HPLC (20). Coupling of this chromatographic technique with conventional refractive index or ICP–AES detection enabled them to achieve the subsequent speciation of elements within each group and of various oxyanions. Matrix and salt interferences were found to have little impact, whereas the spectrometric moiety of the system turned out to be fully compatible with mixed aqueous organic eluents in the gradient mode, e.g., in the presence of methanol, ethanol, or isopropanol. Complete element speciation profiles are thus attainable, as is the case for the separation of Cu^{2+} from Cu^+ and of arsenate from arsenite and dimethylarsenate.

Using a similar experimental approach, the same team reported the separation and speciation of Cd, Hg, and Zn by extraction chromatography with reversed-phase HPLC, followed by the aforementioned detection modes (21). In terms of reproducibility of overall retention times with both refractive index and ICP–AES detection, standard deviation was always better than $\pm 5\%$. On the other hand, corrosion of metal components of the chromatographic equipment due to contact with the mobile phase created major difficulties by causing severe backpressure in the column.

Conventional paired-ion reversed-phase HPLC in combination with either refractive index or ICP–AES detection was reported to be an excellent means to separate and determine Cr(III) and Cr(VI) ions (22). The merits of tetrabutylammonium hydroxide and sodium n-alkylsulfonate as counterions in the mobile phase were ascertained and discussed. The use of the two reagents in a sequential mode allow first the cation Cr^{3+} and then the anion CrO_4^{2-} to be distinguished after reinjection, as each counterion can retain only one of the two species while the other appears in the solvent front. Total time of analysis did not exceed 10 min. The major limitation of this approach was that the achievable detection power was about 2 orders of magnitude higher than the concentrations normally encountered in foods, beverages, water bodies, and environmental matrices. Finally, possible ways of improving the performance of this method were outlined.

Differences in volatilization behavior from the matrix of As compounds were utilized to obtain information on the chemical states of this element (23). A device consisting of a quartz heating tube and a Teflon collecting vessel was set up and connected to a microwave-induced plasma AES (MIP–AES) or AAS detector. The system was applied to the determination of the total As content of NIST [National Institute of Standards and Technology (U.S.)] standard reference materials (orchard leaves, tomato leaves, oak leaves) with good recovery yields. One major disadvantage of the plasma detector was the instability caused by the presence of other volatile compounds.

The suitability of DCP–AES to detect the element content of fractions emerging from an ion chromatography (IC) system was experimentally ascertained with particular regard to the performance of speciation studies (24). A comparison of the conventional conductivity detector with the aforementioned on-line combination was also carried out both by bypassing the conductivity detector itself and by introducing the chromatographic effluent into it first and then into the spectrometric system in a tandem mode. Plasma stability effects were identified as a potential source of disturbance for baseline reproducibility and noise level. On the other hand, mobile phase flow characteristics must be carefully optimized in order to avoid a significant decrease (by a factor of 2–4) of the analytical signal due to sample dilution effects. Low mobile phase flow rates with large sample introduction loops or a preconcentration step were consequently recommended. The IC–DCP–AES system was tested in the case of the separation and measurement of the species As(III), As(V), Cr(III), Cr(VI), Fe(II), Fe(III), Mn(II), Mn(VII), Pt(IV), V(IV), and V(V), also taking into account all possible transformations induced by changes in pH.

The suitability of MIP–AES for on-line coupling with GC was thoroughly described by Bulska, who showed this approach to have high potential for speciation studies (25). To this end, particularly attractive results were obtained with the Beenakker-type cylindrical TM_{010} cavity, as it can generate and sustain a microwave plasma at atmospheric pressure with both argon and helium. A number of applications of such systems for the speciation of As-, Hg-, Pb-, and Se-containing compounds was reviewed and critically evaluated.

1.2.2.3. *Environmental Applications*

Frimmel and Gremm reported on the role of element speciation in water analysis, a theme still largely neglected (26). A number of physical, chemical, and biological approaches for species determination in aquatic systems were given that were of potential value for updating legislation pertaining to natural water sources and wastewater. From this standpoint oxidation states, coordination forms, and electron configurations were examined in detail, together with the major aspects related to species formation in water media, i.e., toxicity, bioavailability, mobility, eutrophication, flocculation, and adsorption ability. These concepts were illustrated with As(III), As(V), Cr(III), Cr(VI), Fe(II), Fe(III), Mn(II), Mn(IV), S($-$II), S(0), S(VI), N($-$III), N(0), N(V), Hg(II), methyl-Hg, Sn(II), Sn(IV), and alkyl-Sn compounds, among others. In addition to this, the influence of complexation reactions and change of oxidation states on the solubility and transport of pollutants is illustrated.

An exhaustive review on the analysis of total As and As forms in marine environmental specimens, covering more than 190 references, was prepared by Francesconi et al. (27). The discussion focused first on the properties of As species of general relevance to man and the environment (AsO_2^-, AsO_4^{3-}, $AsCl_3$, $CH_3AsO(OH)_2$, $(CH_3)_2AsO(OH)$, CH_3AsO, AsH_3, CH_3AsH_2, $(CH_3)_2AsH$, $(CH_3)_3As$, $(CH_3)_4As^+$, arsenocholine, arsenobetaine, and dimethylarsinyl-ribosides). Storage and treatment aspects necessary to put samples into a form suitable for As determination were subsequently considered for seawater, sediments, interstitial water, and biological materials. Moreover, methodologies for the molecular characteriztion of As species were critically assessed together with the specific features of detection techniques more suited for their quantification, in particular molecular absorption spectrophotometry, neutron activation analysis (NAA), ASV, hydride generation flame AAS, ETA–AAS, atomic emission spectrometry with glow discharge, MIP–AES and ICP–AES excitation sources, ICP–MS, and hydride generation AFS. The authors provide information on commercially available certified reference materials (CRMs) for total As, while stressing the need for CRMs for individual As forms as well.

The separation of different dissolved Al species was achieved in acidic water samples of the Belgian Campine bogs (28). These are small natural bodies of water generally not larger than a few thousand square meters with a depth of 1–2 m. Their content in total Al is in the milligram per liter range. The analytical approach consisted in sampling of the water, double-step filtration to remove the larger particles first and then the smaller ones, separation by means of ion-exchange chromatography, and final detection by ICP–AES of the element combined with humic acid as well as of the fluoride-complexed, hydrolyzed, and free Al. The global Al content was instead assessed by instrumental NAA. The metal concentration in bogs was ascertained to be controlled both by the acidity of soil and groundwater and by the level of humic acid. The former parameter triggers the release of Al from soil, whereby at low pH values the hydrolyzed species $Al(OH)^{2+}$ and $Al(OH)_2^+$ occur. The latter factor, in turn, tends to increase the total buffer capacity of soil and Al solubility by forming rather stable complexes. As regards fluoride, mainly originating from atmospheric deposition, the observed concentrations allow only a small aliquot of Al to be complexed, AlF^{2+} being the only species detected. The overall result of such multiple equilibria is that the formation of the more toxic hydrolyzed species can take place only between pH 4 and 6 and that their concentration is substantially reduced by the aforementioned complexation reaction.

The performance of automated and manual flow injection-based methods for the speciation of Al in natural waters was compared by Berdén et al. (29). Two major problems are associated with this kind of determination, namely, sampling and sample storage for these steps may easily perturb the dynamic equilibria in water and species detection, as the mutual equilibria existing in solution are altered by any reaction used for this purpose. Noninvasive techniques would therefore be ideal if it were not for their generally low sensitivity and specificity. The only way to circumvent these drawbacks is to resort to a flow injection approach to exploit differences in reaction kinetics, i.e., the reactions applied to detect a specific form must be much faster than those responsible for the reestablishment of equilibria. Storage, in turn, was found to be more sensitive for solutions with a lower content of labile Al, larger amounts of organic matter, and pH values in the range of 5–6. Detection limits, although not entirely satisfactory (ca. $10\,\mu g\,L^{-1}$), appear consistent with those provided by other methods quoted in the literature.

It is well known that humic substances in natural waters can complex heavy metals, thus forming species of environmental interest that must be separated, identified, and quantified. To this aim, Cd, Cu, and Pb complexed by humic and fulvic acids in freshwater were filtered after 1:1 dilution and selectively and quantitatively adsorbed on a column loaded with diethylaminoethyl-Sephadex A-25, a macroreticular weak-base anion exchanger (30). While simple metal cations are not adsorbed, the species retained by the column can be eluted with HNO_3 and then conveyed to an ETA–AAS detector. The column performance was tested at different concentrations of the two acids and at various solution volumes and pH values. Shape and diameter of the anion exchanger particles were not affected by repeated treatments. Additional experiments were necessary to ascertain the extent of positive and negative errors that might occur and go unnoticed because of partial mutual compensation. In the case of pond water, 30 mL samples were sufficient to complete the analysis. The conclusions was that complexed Cd, Cu, and Pb were in the range of ca. 0.01, 0.58–1.15, and 0.14–0.17 $\mu g\,L^{-1}$, respectively, with humic substances at ca. $100\,\mu g\,L^{-1}$.

Separation of nontoxic from toxic element species in road runoff was undertaken by Morrison and Florence, as the latter species can significantly pollute nearby environmental "compartments" (i.e., environmental subdivisions) (31). A combination of medium exchange and sample acidification techniques was thus developed to allow for the calculation of the individual effects of chemicals (complex-forming agents and surfactants) on the ASV deposition and stripping steps for Cd, Cu, and Pb. Some of the problems arising from ASV lability measurements were discussed. The differential pulse mode gave much higher values for the toxic fraction of the metals considered than that achievable by direct ASV.

An analytical methodology was developed by Stein and Schwedt to provide information about the physical-chemical form, toxicity, and ecochemical behavior of Cr in wastewater (32). Results show that Cr(VI) is unstable whereas most of Cr(III) after centrifugation is associated with macromolecular particles, either complexed or colloidal. Only a small part of total Cr was found to be ionized and therefore potentially toxic.

An attempt to contribute to the understanding of the chemistry of Cu in river, lake, and pond waters was made by Miwa et al. (33). A microcolumn packed with a strong-acid cation-exchange resin, AG50W-X12, or a chelating resin, Chelex-100, was used to concentrate selectively free Cu^{2+} and easily dissociable Cu complexes. AAS was then resorted to for the analysis of the metal. The sorption of Cu in the absence or presence of particulate species, such as negatively charged colloidal aggregates, as well as with or without CO_2, was throughly examined. Both experimental data and theoretical predictions converged in that the dominant forms turned out to be Cu^{2+} and $CuCO_3^0$.

Miyake and Suzuki postulated that the formation of intermediate polymolecular organic compounds of Hg occurred in seawater as a consequence of the decomposition of various constituents of marine plankton (34). This element was separated from western North Pacific seawater samples immediately after collection with nonmetallic devices, followed by filtration through 0.45 μm Millipore septa. The procedure enabled contamination and loss phenomena to be minimized. Preliminary measurements revealed that the total content of Hg dissolved in seawater was between 8 and 24 ng L^{-1}, with higher values in surface and coastal waters due to anthropic activities. After being trapped on XAD-2 resin, Hg was eluted and converted to inorganic forms prior to the determination by cold-vapor AAS. While particulate Hg ranged from 0.5 to 0.9 ng L^{-1} (3–10% of the total), the dissolved organic forms were in the interval 4–11 ng L^{-1}, i.e., from one-third to one-half of the total. Furthermore, the vertical distributions of total and organic Hg were ascertained at three different stations. Data showed that in the temperate zone the metal content increased along the water column below 6000 m, whereas in the subtropical region no increment was observed as measurements approached the bottom.

The movement, behavior, and fate of the different species of Hg in aquatic ecosystems are primary factors when the environmental impact of this element is assessed. A short review was presented on this topic by Schroeder, who summarized present knowledge on the total Hg concentration level expected in open ocean, in coastal seawater, in freshwater rivers and lakes, and in rainwater (35). The advantages and drawbacks of computational methods were set forth, focusing in particular on parameters such as ligand concentration, pH, temperature, redox potential, and known dissociation and formation constants. As regards experimental techniques Schroeder emphasized the need for operationally defined methods, stressing that the unambigous determination of individual chemical forms was hampered by the lack of sensitive and selective approaches. Two examples were given to illustrate such operational schemes, one reported by Lindqvist and Rhode (36) based on differences in species volatility and reactivity, and one discussed by Tessier et al. (7) using sequential extraction. Finally, it was emphasized that an ideal method for rigorous chemical speciation of Hg in natural waters or sediments should be characterized by high detection power (below $1\,\text{ng}\,\text{L}^{-1}$), good selectivity, broad range of applicability, absence of sampling artifacts, minimal preanalytical treatment, and availability of suited CRMs.

The simultaneous presence of the two pollutants Pb and the trisodium salt of nitrilotriacetic acid (NTA) in several environmental media may lead to the formation of complexes of the former with the latter. Those of the type $Pb(NTA)_2$ were detected by direct current (dc) and differential pulse polarography in natural waters (37). Computer simulations were devised to simulate the influence of NTA on Pb speciation. The presence of Ca^{2+} and Mg^{2+} hindered the formation of the NTA complexes unless the concentration of these complexes was higher than 10^{-4}–10^{-3} M. The stability constant of the $Pb(NTA)_2$ species was calculated as $\log \beta = 17.0 \pm 0.2$ in $0.1\,\text{M}\,KNO_3$. It was stressed that the formation of such complexes may cause a higher solubility of the metal when particles with high levels of NTA due to detergents are present in water bodies.

The mobility of selected elements (Al, Cd, Fe, Pb, and Ti) in soil and the noxious consequences for plants were discussed by Smirnová et al. (38). An elaborated speciation scheme was worked out on the basis of a number of extraction, filtration, and precipitation steps followed by either flame AAS or ETA–AAS for the final detection of the metal content of each fraction. This methodology was applied to minerals such as chlorite, kaolin, tourmaline, calcite, dolomite, and illite.

A fundamental contribution to the standardization of the sequential extraction procedure of particulate trace elements from sediments and soils was made in 1979 by Tessier et al. (7). This now classical study fully developed a systematic approach that has since been widely applied worldwide with only minor variations. The merits of the procedure were ascertained in the case of Cd, Co, Cu, Fe, Mn, Ni, Pb, and Zn in bottom sediments. Fractions likely to be affected by environmental conditions were chosen. These were as follows, in order: the exchangeable fraction, i.e., the one accounting for sorption–desorption processes of trace elements as a consequence of changes in water ionic composition; the fraction bound to carbonates, as significant trace metal concentrations can be associated with these compounds, strongly dependent on pH variations; the fraction bound to Fe and Mn oxides, as these may exist as modules or concretions, are excellent scavengers, and may be unstable under anoxic conditions; the fraction bound to organic matter, mainly humic and fulvic acids; the residual solid, made up notably of primary and secondary minerals with trace metals within their crystal structure and therefore not liable to be released in solution. To simulate the natural mechanism of dissolution each fraction was treated with extracting reagents of increasing power, starting with sodium acetate (or $MgCl_2$) and ending with an HF–$HClO_4$ mixture. Total trace metal concentrations were also measured, along with inorganic and organic carbon levels. Flame AAS was always utilized to quantify trace elements. Precision and accuracy were more than satisfactory, with due account being given to the inherent inhomogeneity of sediments.

The suitability of the extraction–derivatization–GC–AAS methodology for quantification of ionic organo-Pb species in fresh grass and leaf samples was investigated by Van Cleuvenbergen et al. (39). Two different pretreatment approaches were developed, both based on extraction with a 25% aqueous $(CH_3)_4NOH$ solution, but differing as to the complexity of the filtration and extraction steps. With the second, more complex treatment, recoveries for diethyl-, dimethyl-, triethyl-, and trimethyl-Pb were significantly better than with the simplified treatment (some 84–104%). This study offered a useful insight into the accumulation mechanism of pollutants in plants, with significant effects on human health considered, especially as regards the food chain.

1.2.2.4. Biological Applications

Violante et al. have described a novel approach for the separation and determination of six biologically and environmentally significant As compounds using HPLC combined on-line with ultraviolet (UV) decomposition and a hydride generation system with ICP–AES as the detector (40). Inorganic AsO_2^- and AsO_4^{3-} and organic (monomethylarsonic acid, dimethylarsinic acid, arsenobetaine, and arsenocholine) species were separated on a strong anion-exchange column connected with NH_2-based column using a phosphate buffer as the mobile phase. Conversion into hydrides was required to increase detectability of the six compounds by ICP–AES and thus match their expected levels in natural matrices. UV irradiation therefore became necessary because of the resistance of arsenobetaine and arsenocholine to the hydride generation process. Caroli et al. reported further advances in this area, achieved by employing an HPLC-combined ICP–MS system (41).

The metallothionein-bound content of Cd, Cu, and Zn was assayed by Heilmeier et al. in human and rat tissues (namely, brain, heart, small intestine, kidneys, liver, lungs, muscle, pancreas, and spleen) (42). After homogenization, tissues were centrifuged at $18,000\,g$ and the supernatants were analyzed by ICP–AES and AAS. Depending on organ type and origin, Zn concentrations (in micrograms per gram) as high as 55 in human liver and 35 in rat lungs were detected, while the highest levels of Cu were 6 in human liver and 8.5 in rat kidneys and below 0.2 for all rat organs. These findings reinforced the view that the presence of Cd in human tissues may induce the formation of metallothioneins. This may have as a consequence not only an increase in intracellular metal binding sites but also a disturbance in the homeostatic mechanism of both endogenous and potentially toxic elements.

Guy et al. differentiated free and protein-bound Cd in lobster digestive gland extracts by coupling an HPLC system with flame AAS detection (43). Four types of separation media were investigated, the best being Fractogel HW40F because of its resistance to compression. In the sample extraction process it was deemed mandatory to make the final solution 0.01 M in toluenesulfonyl fluoride in order to prevent protein degradation by protease enzymes present in the lobster digestive gland extracts and thus alteration of the originally bound Cd. Ion–gel interactions were also minimized by proper choice of the element and by not allowing the duration of the chromatographic runs to exceed 10 min. The same investigators also studied the behavior of Cu and Zn. Detection limits achievable for the three metals Cd, Cu, and Zn were, respectively, 0.80, 1.6, and $0.80\,\mu g\,g^{-1}$. The procedure developed by these authors should be applicable to other systems where free elemental species are to be distinguished from bound forms, provided that these last are not labile.

Gardiner et al. investigated the distribution of Co, Cu, Fe, and Zn in human blood serum, milk serum, and seminal fluid as well as in synthetic aqueous solutions by on-line coupled size-exclusion HPLC–ICP–AES (44). From an operative point of view it was noted that a large dead volume between the chromatographic column and the plasma detector adversely affects both the chromatographic resolution and the detection power. This problem could be alleviated in part by using small-diameter tubes to connect the two analytical systems. Sample volumes in the range $250-1000\,\mu L$ were found to be adequate in most cases. Although rather preliminary, this study indicated that the constituents of the various biological fluids associated with the aforementioned elements can be rather well characterized, with potential applications in disease diagnosis.

Ferritin plays a key role in the metabolism of Fe and can be stored in a soluble, nontoxic form for the synthesis of heme and related proteins (45). In light of this, the separation and analysis of the various Fe-bound fractions of the substance was achieved in our laboratory by coupling on-line HPLC with ICP–AES, resulting in a powerful tool for both quantitative and qualitative studies. The method showed a high potential for detecting possible adulterations of commercial products containing mixtures of Fe-proteins other than ferritin. Examples were given in this report to illustrate the applicability of the technique for testing and characterizing ferritin drugs supplied by various producers.

1.3. A LOOK INTO THE FUTURE

The dramatic increase in recent years in the number of studies in the field of element speciation provides clear evidence of the push forward that this investigative area is beginning to impart to many experimental disciplines, bioinorganic chemistry in particular. A full coming of age of this research field, however, still depends on the resolution of three issues of major importance: (i) the need for harmonized approaches, at both the national and international level, that can facilitate the systematic planning of new projects while avoiding any useless duplication of effort and ineffective exploitation of the resources available; (ii) the need for ready availability of CRMs with known contents of given chemical forms for a number of essential and potentially toxic elements in a variety of matrices in order to allow for better accuracy and comparability of experimental data; and (iii) the need for closer cooperation between speciation-oriented scientists and suppliers of instrumentation so that new equipment can be developed that will fulfill the exacting requirements imposed upon separation and detection techniques when they are combined for speciation measurements. Once these three goals are achieved, speciation might indeed be considered to be emerging out of its infancy.

REFERENCES

1. I. S. Krull, *Trace Metal Analysis and Speciation*. Elsevier, Amsterdam and Paris, 1991.
2. G. E. Batley, *Trace Elements Speciation: Analytical Methods and Problems*, 350 pp. CRC Press, Boca Raton, FL, 1989.
3. S. Caroli, *Microchem. J.* **51**, 64 (1995).
4. Ph. Quevauviller, O. F. X. Donard, E. A. Maier, and B. Griepink, *Mikrochim. Acta* **109**, 169 (1992).
5. S. Hetland, I. Martinsen, B. Radzuk, and Y. Thomassen, *Anal. Sci.* **7**, 7 (1991).

6. A. Tessier, P. G. C. Campbell, and N. J. Bisson, *J. Geochem. Explor.* **16**, 77 (1982).
7. A. Tessier, P. G. C. Campbell, and N. J. Bisson, *Anal. Chem.* **51**, 844 (1979).
8. C. Kheboian and C. F. Bauer, *Anal. Chem.* **59**, 1417 (1987).
9. M. Bernhard, F. E. Brinckman, and P. J. Sadler eds., *The Importance of Chemical Speciation in Environmental Processes*, Report of the Dahlem Workshop, Berlin, 1984, *Life Sci. Res. Rep. No.* **33**. Springer-Verlag, Berlin, 1986.
10. J. O. Nriagu, ed., *Changing Metal Cycles and Human Health*, Report of the Dahlem Workshop, Berlin, 1993, *Life Sci. Res. Rep. No.* **28**. Springer-Verlag, Berlin, 1994.
11. R. Harrison and S. Rapsomanikis, eds., *Environmental Analysis Using Chromatography Interfaced with Atomic Spectroscopy*, p. 370. Ellis Horwood, Chichester, UK, 1988.
12. R. J. A. Van Cleuvenbergen, W. M. R. Dirkx, Ph. Quevauviller, and F. C. Adams, *Int. J. Environ. Anal. Chem.* **47**, 21 (1992).
13. T. M. Florence, *Talanta* **29**, 345 (1982).
14. T. M. Florence and G. E. Batley, *CRC Crit. Rev. Anal. Chem.* **9**, 219 (1980).
15. D. Behne, *Analyst (London)* **117**, 555 (1992).
16. S. Hill and L. Ebdon, *Eur. Spectrosc. News* **58**, 21 (1985).
17. W. Lund, *Fresenius' Z. Anal. Chem.* **337**, 557 (1990).
18. D. Ishii, K. Asai, T. Hibi, T. Jonokuchi, and M. Nagaya, *J. Chromatogr.* **144**, 157 (1977).
19. E. Munaf, H. Haraguchi, D. Ishii, T. Takeuchi, and M. Goto, *Anal. Chim. Acta* **235**, 399 (1990).
20. D. Bushee, I. S. Krull, R. N. Savage, and S. B. Smith, Jr., *J. Liq. Chromatogr.* **5**, 463 (1982).
21. D. Bushee, D. Young, I. S. Krull, R. N. Savage, and S. B. Smith, Jr., *J. Liq. Chromatogr.* **5**, 693 (1982).
22. I. S. Krull, D. Bushee, R. N. Savage, R. G. Schleicher, and S. B. Smith, Jr., *J. Liq. Chromatogr.* **5**, 267 (1982).
23. B. Sarx and K. Bächmann, *Trace Elem. Anal. Chem. Med. Bio.* **2**, 713 (1983).
24. V. D. Lewis, S. H. Nam, and I. T. Urasa, *J. Chromatogr. Sci.* **27**, 468 (1989).
25. E. Bulska, *J. Anal. At. Spectrom.* **7**, 201 (1992).
26. F. H. Frimmel and T. Gremm, *Fresenius' J. Anal. Chem.* **350**, 7 (1994).
27. K. A. Francesconi, J. S. Edmonds, and M. Morita, in *Arsenic in the Environment. Part I. Cycling and Characterization* (J. O. Nriagu, ed.), p. 189. Wiley, New York, 1994.
28. E. Courtijn, C. Vandecasteele, and R. Dams, *Sci. Total Environ.* **90**, 191 (1990).
29. M. Berdén, N. Clarke, L. G. Danielsson, and A. Sparén, *Water, Air, Soil Pollut.* **72**, 213 (1994).
30. M. Hiraide, S. Prasan Tillekeratne, K. Otsuka, and A. Mizuika, *Anal. Chim. Acta* **172**, 215 (1985).
31. G. M. P. Morrison and T. M. Florence, *Sci. Total Environ.* **93**, 481 (1990).

32. K. Stein and G. Schwedt, *Fresenius' J. Anal. Chem.* **350**, 38 (1994).
33. T. Miwa, M. Murakami, and A. Mizuike, *Anal. Chim. Acta* **219**, 1 (1989).
34. Y. Miyake and Y. Suzuki, *Deep-Sea Res.* **30**, 615 (1983).
35. W. H. Schroeder, *Trends Anal. Chem.* **8**, 339 (1989).
36. O. Lindqvist and H. Rhode, *Tellus* **37B**, 136 (1985).
37. A. M. Mota and M. L. Gonçalves, *Water Res.* **24**, 587 (1990).
38. L. Smirnová, M. Vačková, and M. Žfiemberyová, *Chem. Listy* **85**, 1235 (1991).
39. R. Van Cleuvenbergen, D. Chakraborti, and F. Adams, *Anal. Chim. Acta* **228**, 72 (1990).
40. N. Violante, F. Petrucci, F. La Torre, and S. Caroli, *Spectroscopy* **7**, 36 (1992).
41. S. Caroli, F. La Torre, F. Petrucci, and N. Violante, *Environ. Sci. Pollut. Res.* **1**, 205 (1994).
42. H. E. Heilmeier, P. Schramel, G. A. Drasch, E. Kretschmer, and K. H. Summer, *Trace Elem. Anal. Chem. Med. Biol.* **4**, 495–500 (1987).
43. R. D. Guy, C. L. Chou, and J. F. Uthe, *Anal. Chim. Acta* **174**, 269 (1985).
44. Ph. E. Gardiner, P. Brätter, B. Gercken, and A. Tomiak, *J. Anal. At. Spectrom.* **2**, 375 (1987).
45. S. Caroli, F. La Torre, F. Petrucci, O. Senofonte, and N. Violante, *Spectroscopy* **8**, 46 (1993).

CHAPTER

2

DIRECT METHODS OF SPECIATION OF HEAVY METALS IN NATURAL WATERS

A. M. MOTA AND M. L. SIMÕES GONÇALVES

Centro de Química Estrutural, Instituto Superior Técnico, Lisbon, Portugal

2.1. THE ECOLOGICAL IMPORTANCE OF SPECIATION

Knowledge of the several chemical forms (speciation) that an element can have in natural waters, as well as of their stability and kinetics of dissociation, is of primary importance because their toxicity, bioavailability, bioaccumulation, mobility, and biodegradability depend on the chemical species. So, speciation studies are of interest to chemists doing research on the toxicity and chemical treatment of natural waters, to biologists inquiring about the influence of the chemical species on animals and plants, and to geochemists investigating the possibilities of transport of the elements in the environment.

In terms of environmental contamination it is worth noting that although the levels of pollutants are usually small in natural waters, the values can be higher in certain zones due, for example, to an output of industrial and domestic wastes, to complex mixture processes, or to adsorption. On the other hand, it is known that aquatic plants and animals, in particular some mollusks (bivalves), can accumulate high levels of toxic elements with concentration factors of 10^3 or 10^4. High levels of heavy metals in different chemical species can therefore be introduced into the food chain of other animals, and man in particular, with important ecological consequences.

We should point out that in toxicity studies test results showing acute toxicity are not the only important finding; also requiring consideration are lethal or chronic effects on growth and attainment of the adult state, as well as interference with movement, choice of habitat, selection of food, or even the self-protection reaction. In this context the parameter $^{96}LC_{50}$, defined as the concentration killing 50% of exposed organisms within 96 h, is important

---

Element Speciation in Bioinorganic Chemistry, edited by Sergio Caroli.
Chemical Analysis Series, Vol. 135.
ISBN 0-471-57641-7 © 1996 John Wiley & Sons, Inc.

especially for fish and macroinvertebrates. For smaller organisms a similar parameter is used but with shorter times.

The various types of metal species in natural waters differ in their toxicity and bioavailability, their speciation being influenced by environmental factors (1, 2).

2.1.1. Different Types of Metal Species

a. Free Metal Ions (e.g., Cu_{aq}^{2+}, Pb_{aq}^{2+}). It is well known that heavy metals in the free form are very toxic to living organisms. Free ions like Cu^{2+} and Cd^{2+} are the most toxic species to unicellular algae for instance, probably because they form complexes with the ligands at the surface of the membrane as the first step of assimilation, it being possible that a change of their coordination chemistry takes place inside the cell.

b. Inorganic Complexes (e.g., $CdCl_2$, $PbCO_3$). It is known that in general hard water, usually alkaline (major components: Ca^{2+}, Mg^{2+}, and CO_3^{2-}), reduces toxicity of heavy metals to the microbiota probably due to carbonate complexation. Sometimes it is not clear from the literature whether there is a special type of complex that is preferentially toxic, since its concentration is proportional to the free metal if constant ligand excess is used in all experiments ([ML] = K[M][L], where K is the formation constant, [ML] is the concentration of the complex, and [M] and [L] are, respectively, the free metal and free ligand concentrations).

c. Metal Organic Compounds with Low Molecular Weight (e.g., MSR, MOOCR, copper complexes with amino acids, alkylmetal compounds). The presence of organics in natural waters may sometimes completely change speciation. For example, organic exudates from bioorganisms can form strong complexes with heavy metals, probably being involved in minimizing the toxicity of metals in natural ecosystems, as has been noticed with the alga *Selenastrum capricornutum* Printz. In fact, in biological models contaminated with Cd^{2+}, under optimal conditions of temperature, light, and nutrients, the alga releases more organics forming Cd complexes with smaller rate constants of dissociation, the uptake of this heavy metal simultaneously decreasing (3). With Pb contamination it has also been noticed that this alga only releases a significant amount of organics with chelating groups for Pb in the presence of this heavy metal (4).

Alkylmetal complexes are very toxic probably because they can pass directly through the biological membrane since they are lipid soluble. For example, aerosols that are emitted by automobiles and contain alkylated forms of Pb can enter the lungs and be absorbed directly into the bloodstream.

d. Organic Complexes with Macromolecular Weight (e.g., M-lipids, M-humic acid polymers, M-pigments, M-polysaccharides. Complexation of heavy metals with large organic molecules generally reduces the toxicity of the metals to microbiota because these complexes cannot pass easily through the biological membrane.

e. Metal Species in the Form of Highly Dispersed Colloids [e.g., $(FeO)OH$, $Fe(OH)_3$, MnO_2, Ag_2S, and Cu_2S]. In general solubility products of different pecies in natural waters are not attained. However, Fe and Mn in oxic waters exist mainly in the form of hydroxides and oxides, and in anoxic media Ag(I) and Cu(I) can exist as sulfides.

f. Metal Species Adsorbed on Colloids, Particulate Matter, and Sediments, or Incorporated in Organisms. Colloids and particles together with sediments generally have the highest levels of heavy metals in natural waters. Clay minerals (montmorillonite, kaolinite, illite, etc.) posses surfaces that are predominantly negative and can adsorb metal ions at the pH of most natural ecosystems. The same happens to hydrous metal oxides that can exchange protons and hydroxyl ions for cations and anions, respectively, thereby decreasing their potential uptake by microbiota; these characteristics of particulate matter depend on several abiotic factors including pH and type and concentration of ligands. However, it must be taken into consideration that the heavy metals adsorbed on solid surfaces can be redissolved due, for example, to acid precipitation or complexation with ligands from wastewater and to the filter action of feeders such as mollusks that can ingest solid particles. The presence of organic ligands can increase or decrease adsorption of heavy metals on solid surfaces. So, for example, glutamic acid increases adsorption of Cu(II) on Fe oxide particles since adsorption occurs through a carboxylate group, and an aminocarboxylic group is still available to be bound to Cu. On the other hand, adsorption of picolinic acid on Fe oxides prevents adsorption of Cu, as chelating groups are no longer available for further complexation. With adsorption of fulvic acid (a fraction of humic substances) on Fe oxide surfaces, the following possibilities can occur: one end of the molecule can bind the oxide surface or the other may bind either to the surface or to a competing cation (5).

2.1.2. Environmental Factors Affecting Speciation

a. Redox Conditions. Under anoxic conditions, e.g., those found in some interstitial waters of sediments, deep waters, sewage, and closed basins as some fjords, speciation can change completely. For example, reducing conditions may lead to microbial conversion of sulfate to sulfide, with the subsequent precipitation of sulfide salts of metals (e.g., HgS, CdS), which greatly reduces the bioavailability and toxicity of these metals to microbiota. In fact, the only metals that show significant binding to soluble organic ligands in a sewage effluent are Co and Ni, since the reactivity of the other metals is limited due to the extreme insolubility of their sulfides. At higher redox potentials most metals tend to be dissolved (1).

The redox potential of the environment also determines the oxidation state of some metal ions. For example, Fe(III) can be reduced to Fe(II), Mn(IV) to Mn(II), Cu(II) to Cu(I), Cr(VI) to Cr(III), and As(V) to As(III), which greatly influence the bioavailability and toxicity of the element. In fact, Fe(II) and Mn(II) are soluble in natural waters with a deficiency of oxygen at normal pH values, but precipitate at higher oxidation states. Copper(II) is a transition element, but Cu(I) is a "b"-type cation, with more affinity for N and S groups. Chromium(VI) is, in general, more toxic than Cr(III), although the essentiality and toxicity of an element is always a question of biological species and concentration. Anoxic conditions lead to the formation of thiocomplexes and organic ligands with sulfur sites, which have more affinity for "b"-type cations.

In sediments of natural waters the main reductant is the settling biomass, which—by microbial mediation—consumes oxygen and reduces Mn salts, Fe hydroxides, and sulfate. The products of this reduction, especially Fe(II) and sulfide, can interact with As(V) and Cr(VI) producing As(III) and Cr(III), the oxidation of reduced species by oxygen being in many cases a slow process.

Iron speciation in fogwaters strongly depends on the presence of light: in the daytime the main reaction is photoreduction of Fe(III) species to Fe(II), which may be reoxidized to Fe(III). During the night oxidation of Fe(II) is expected to be complete, unless in the presence of sulfite ion or other organic reductants that may reduce Fe(III) in the absence of light. The concentrations of dissolved Fe(II) in fogwater are much higher than those measured in oxic surface waters, and so atmospheric deposition may increase the availability of Fe to aquatic biota (6, 7).

b. *Macroconstituents and Ionic Strength.* Alkaline earth cations that exist in many natural waters in a large percentage can compete with heavy metals for the inorganic and/or organic groups. However, one should be aware that this competition is not so drastic since the alkaline earth cations are of the "a" type, with more affinity for oxygen donors, contrary to what happens with those of "b" type, with more affinity for N and S atoms, the transition cations having an intermediate behavior.

From the concentrations of the macroconstituents the ionic strength can be calculated, this parameter directly influencing the activity coefficient of the species in solution and therefore the values of the stoichiometric stability constants, defined in terms of concentrations.

c. *pH.* The values of pH can influence speciation either by shifting the acid–base equilibrium and redox reactions or by competition between protons and metal ions for ligand groups and/or for the surface groups in adsorption processes.

d. Temperature and Pressure. These parameters affect stability and rate constants as well as specific processes such as growth and photosynthesis. Temperature and pressure, which can change, respectively, from about 25 to 9 °C and 1 to 1000 atm in the ocean, drastically affect the type of bioorganisms present, differing from the surface to the bottom due to the necessary biological adaptation.

e. Synergistic and Antagonistic Effects. The presence of several heavy metals that can compete for the same or different sites of proteins in a living organism might affect the toxicity and bioavailability of these metals, it being known that Cu and Zn ions together are about 2.5 more toxic to fish (e.g., the guppy *Poecilia reticulata*) than separately (5).

2.2. CONSTITUENTS OF NATURAL WATERS

Three types of constituents—inorganic, organic, and particulate matter including colloids—are considered in the following subsections.

2.2.1. Inorganic Constituents

The concentration of inorganic macroconstituents is quite different in the various types of natural waters such as rivers, freshwater lakes and/or rainwaters (Table 2.1) (1, 8), seawater being the only natural water whose composition can be considered practically constant all over the world (1). From this point of view seawater, which is in a stationary state with the same input (especially from the rivers) and output (owing to sedimentation), is indeed a good medium to be studied, since the results can easily be generalized, the only exceptions being some closed seas with very high salinity due to evaporation.

2.2.2. Organic Matter

Organic matter in natural waters is due to living organisms, in particular to exudates of the algae (mainly polypeptides and polysaccharides) and also to pollution. In seawater there exist, in general, levels of dissolved organic matter of about $0.1-1\,\mathrm{mg\,L^{-1}}$ C ($1\,\mathrm{g\,L^{-1}}$ corresponding to 1–4 mM of complexing sites) and in freshwaters higher levels, in many cases of about $10\,\mathrm{mg\,L^{-1}}$ C. However, interstitial waters of sediments, some coastal waters where there is a bloom of phytoplankton, and very polluted waters can have levels of about

Table 2.1. Inorganic Macroconstituents of Some Natural Waters

	Rain[a,e]	Rivers[b,e]	Lakes		Oceans[e]
			Thirlmere[c,e]	Lake Kivu[d,e]	
Na^+	0.083	0.274	0.139	5.29	485
Mg^{2+}	0.008	0.169	0.013	3.58	55.2
Ca^{2+}	0.008	0.374	0.09	0.12	10.6
K^+	0.005	0.059	0.007	2.49	10.6
Cl^-	0.093	0.220	0.151	1.55	566
SO_4^{2-}	0.033	0.117	0.060	0.25	29.3
HCO_3^-	0.007	0.958	0.061	12.5	2.4
pH	4.5	7.0	6.7	—	7.0
Ionic strength	0.21	2.08	0.51	18.8	700

[a,b] Mean values are given for rainwaters and river waters, respectively, with considerable variation from sample to sample.
[c] United Kingdom.
[d] East Africa, on the border between Zaire and Rwanda.
[e] Concentrations in millimoles per liter.

100–150 mg L^{-1} or sometimes higher, the formation of organic complexes being probably very important in these media.

The release by living organisms of peptides, polyphenols, amino acids, nucleic bases, and macrocyclic compounds such as ferrichromes, porphyrins (vitamin B$_{12}$, chlorophylls), pigments, and siderophores of the hydroxamic acid type, can explain the high degree of complexation of certain natural media, especially of Fe(III) and Cu(II), the strongest ligands probably having complexing groups with N and S atoms or oxygen donors in an adequate conformation. It should be emphasized that organic matter is, in general, metastable, but it can persist for quite a long time in natural waters, being involved in acid–base and complexation reactions. Iron (III)/Fe(II) and/or Mn(IV)/Mn(II) systems, particularly in the presence of light, may catalyze the oxidation of some organics such as those that contain hydroxy carboxylic functional groups, i.e., gallic, tannic, humic, and fulvic acids, under natural water conditions (1).

Among the organic matter found in natural waters, humic substances play an important role. Humic matter, due to the decomposition of organisms, is a mixture of organic polymers with hydrophobic and charged hydrophilic groups, having polyelectrolytic and polyfunctional characteristics. Operationally, it can be divided into three different fractions corresponding to a decrease in the molecular weight and complexity: humine (the nonsoluble fraction), humic acids (which precipitate at acid pH values), and fulvic acids (the soluble fraction). In solution these substances can exist free or associated, e.g., with amino acids such as glycine, aspartic acid, alanine, serine, and glutamic acid, at least partially due to decomposition of organisms (9). The main hydrophilic groups of humic and fulvic acids are —COOH, —NH_2, —OH, and —SH, it being noted that even when they are separated from the hydrophobic groups only by a methyl group they do not hinder adsorption of these compounds on neutral particles (10). The origin of humic substances found in the ocean is the organic matter, obtained from the decomposition of plankton and sea organisms, and pedogenic matter, carried by soil runoff into rivers and then to the sea. Marine organic matter produced in situ has less aromaticity and lower molecular weight than pedogenic matter, where aromaticity is in part due to lignin. For instance, the analysis of fulvic acids extracted from the sea might show an aquatic fraction with a molecular weight of about 1000 and a pedogenic fraction about 10 times as high. Another difference between these two fractions is the higher content of N and S in organic matter, which can influence the sedimentation of trace metals through complexation. A large fraction of pedogenic matter, especially the high-molecular-weight fraction, is removed in estuaries when the river water mixes with seawater, due mainly to neutralization of the negatively charged groups with Ca^{2+} and Mg^{2+} and their coprecipitation with Fe(III) and Al(III) oxohydroxycolloids that coagulate with the increase of ionic strength.

2.2.3. Colloids and Particulate Matter

Adsorption of cation and anions on metal oxides is considered to be one type of coordination reaction (surface reaction), protons and hydroxyl ions being released from the surface. Both the electrostatic forces and the chemical aspect play an important role in this process, as well as the competition of the solvent and other molecules in solution for the surface.

The electrical charge on each particle's surface, forming a double layer, prevents the flocculation of colloidal particles. As the electric charge decreases with the increase of salinity (which compresses the double layer and neutralizes the charge), the level of colloids in seawater is smaller than in freshwater, drastically decreasing in estuaries where aggregation occurs. Certain substances such as macromolecules, polyelectrolytes, and ions of higher charge [e.g., Fe(III) and Al(III), metals of type "a"] have a high affinity for oxygen donor surface groups and consequently a strong tendency to be accumulated on the particle/water interface, destabilizing the colloids and promoting their coagulation or flocculation (1). On the other hand, colloids are very often stabilized by organic films, as is the case for the Fe oxide colloids that are generally covered by humic substances in both freshwater and seawater (pseudocolloids).

2.3. METHODOLOGY FOR SPECIATION

The speciation of trace elements in natural waters is a difficult task because they exist in very low concentrations in a medium with several macroconstituents, often in a nonequilibrium situation, and so various approaches should be used in order to increase the information about the system. It is important to emphasize that all approaches have some advantages and limitations, the best knowledge of the medium being obtained when all the information is put together. Another important point is that the operations involved in the measurements might well alter the original conditions; thus, full details of the analytical procedures used are relevant to any estimate of the alterations done.

2.3.1. Various Approaches

a. Determination in situ. This is the method least disturbing to the composition of the system under study, but generally little information can be obtained due to the enormous complexity of natural systems and/or the low concentration range of trace elements.

b. Determination in a Natural Pretreated Sample. This approach is used in order to have a simpler system, eventually with more concentrated species; it provides more detailed information, but conclusions might be distorted due to the absence of some constituents and the change of concentration.

c. Chemical Models. In this approach a very simple system is used in the laboratory as a first approximation to the real sample, followed by the addition of other constituents in order to study their influence on the previous model. It is important to mention that these studies should be made under well-controlled conditions of temperature, pressure, and ionic strength, since these variables strongly influence the values of the constants and consequently the speciation results. In particular, the carbonate system, one of the most important buffers in natural water, is very much affected by those parameters (11). In spite of the considerable interest of these studies for the understanding of the chemical reactions in natural media (thermodynamic and kinetic parameters can be determined and used in computer simulations), their application as "models" of complex natural mixtures cannot be made without significant adaptations.

d. Biological Models. Another approach to the study of speciation and the influence of the organics released by the organisms under diverse conditions—i.e., different nutrients, temperature, light, and contaminants—is the use of biological models, although it is difficult to extrapolate the results directly to field situations. These models also make it possible to study the uptake of contaminants (in particular heavy metals) by organisms, to know whether defense mechanisms are present, and to relate speciation in solution to toxicity.

e. Fine Speciation. Fine speciation is the determination of the percentage of each species by means of computer programs, generally assuming equilibrium conditions. For this calculation it is necessary to know the total concentration of metals, inorganic and organic ligands, colloids, and particulate matter as well as the stability constants of the complexes formed in solution and at the surface (adsorption constants), all of whose values strongly influence speciation.

Computer simulations show the mutual influence on speciation of metals and ligands present in solution. On the other hand, these calculations can also predict the disturbance of the system due, for instance, to some pH variation (say, acid rain) or to an increase in the concentration of one or more species (such as a local input of wastes). A simulated experiment has shown that when a 10^{-6} M total concentration of natural organics (a concentration value that can be found in natural waters) is added to freshwater with only inorganics, the concentrations of free Cu, Zn, and Cd (with concentrations generally lower than 10^{-8} M) decrease respectively by a factor of about 160, 2, and 1.4, which can greatly influence toxicity to bioorganisms (12). From computer simulations it can also be seen that, in fresh- and seawaters with organic matter, Cu exists in the form of organic complexes in a significant concentration, contrary to other elements that exist mainly in the free form or complexed to inorganic ligands (1).

Fine speciation generally assumes equilibrium in natural systems, which might exist in some localized zones, or to which natural media tend, as the real solutions always do. However, in natural waters as a whole, there are always transient and spatial gradients, because some reactions do not attain equilibrium during the residence time of the water. The type of reactions that mainly cause a disturbance of equilibrium are photosynthesis, erosion of rocks, processes of diffusion and mixture, sedimentation, and assimilation by organisms. Generally the rate constants of the following types of reaction decrease in this order: acid–base reactions; hydration; formation of ionic pairs; formation and dissociation of chelates; reactions between different phases (adsorption, precipitation, assimilation by organisms); certain redox reactions; sedimentation and mixtures. If kinetics is being considered in fine speciation, it is only necessary to consider the kinetic parameters of the slowest reaction, since this is the rate-determining step; if a chemical species enters several reactions with similar rate constants of formation and dissociation, a stationary state is attained (13).

A deviation from equilibrium in natural waters is more probable when organic matter is present because, although metastable, the organics can be adsorbed on particles and colloids and form complexes with dissociation rate constants that are small compared with the residence time of the element. The residence time of most elements is usually controlled by sedimentation (due to uptake by organisms, adsorption, coprecipitation, and ionic exchange), although a few elements can precipitate directly in natural waters, such as MnO_2, SiO_2 and (FeO)OH.

2.3.2. Kinetic Aspects

In order to distinguish diverse groups of species with different kinetics of dissociation, it is important to know the time scale of the analytical technique used and to compare it with the dissociation rate of the complex. In this context two different categories of complexes, inert and labile, can be defined:

a. Inert Complex. The kinetics of dissociation is too slow compared with the time of the analytical measurement, i.e., during the measuring time of the technique there is no time for the dissociation reaction to occur. This situation has been observed in complexes adsorbed on particles, incorporated into living organisms, and with metal ions strongly bound to large molecules (14, 15).

b. Labile Complex. The kinetics of dissociation is fast enough so that a thermodynamic equilibrium between the complex and the free ion (ML $\Leftrightarrow$ M + L) is always reached during the measuring time.

It is important to bear in mind that this distinction between inert and labile is not an absolute concept for each compound but depends on the time scale of the analytical technique, i.e., a complex that behaves as if it is inert with a certain technique might behave as if it is labile with a slower one. Such complexes with an intermediate behavior are called quasi-labile.

In this context Davison estimated theoretically whether a complex can be experimentally detected as a labile species by voltammetric techniques, considering that the kinetic current (i_k), which depends on the time scale of the technique, is negligible compared with diffusion current (i_d) when $i_k/i_d \leqslant 0.10$ (see Section 2.5.3.3) (16). Based on this assumption it was estimated that Cu complexes with tartarate, glycine, and citrate should be labile in terms of anodic stripping voltammetry (ASV), contrary to Cu complexes with nitrilotriacetic acid (NTA) and ethylenediaminetetraacetic acid (EDTA), which should be inert, in agreement with experimental results (16, 17). However, in this sort of study several reaction mechanisms should be considered. For example, for EDTA, two mechanisms must be taken into account, even at basic pH values:

$$Cu^{2+} + EDTA \rightarrow Cu\,EDTA$$

and

$$Cu^{2+} + H\,EDTA \rightarrow Cu\,EDTA + H^+$$

Figura and McDuffie defined curves of x (the fraction of ML dissociated in the time t) vs. log k_d (the dissociation rate constant of the complex) for the time scale of ASV measurement ($t = 2$ ms) and Ca-Chelex column ($t = 7$ s), in the presence of an excess of ligand, according to the expression (18):

$$1/(1-x) = \exp(k_d t) \tag{1}$$

From those curves it could be anticipated that the complex Cd NTA, with $k_d = 1.6\,\text{s}^{-1}$, is 100% labile in terms of the Ca-Chelex column and not labile in ASV terms, as experimentally determined.

A rate constant of about $0.1\,\text{s}^{-1}$ was anticipated for the Cu(II) humic acid complex, since the complex was not detected by ASV at pH = 7.8 and was 54% determined by Ca-Chelex-100, which agrees with the literature, although the heterogeneous character of the ligand was not considered (19).

Figura and McDuffie also used the Ca-Chelex batch technique, which has large time scales ($\geqslant 1$ h), and assumed that the labile species detected by this method (but not labile in terms of the Ca-Chelex column) probably include relatively stable complexes such as Cd, Pb, and Zn EDTA and possibly metals bound to colloids. With the batch technique, where there is no local excess of resin, the uptake of the metal depends not only on the kinetics of dissociation of the complexes but also on some experimental variables that influence the contact degree between metal ion and resin, such as the resin-to-volume ratio of solution and stirring rate. In this case:

$$\ln([ML]_0/[ML]_t) = k_b[CaR]t/K_{ML}[L] \tag{2}$$

where $[ML]_0$ and $[ML]_t$ are, respectively, the initial concentration of the complex and its value at the instant t; K_{ML} is the stability constant of the complex; R represents the groups of the resin; and k_b is the rate constant of the reaction

$$CaR + M \xrightarrow{k_b} MR + Ca$$

The value $k_b[CaR]$ can be estimated to be $5 \times 10^{-3}\,\text{s}^{-1}$ from an experiment without ligand, under the same experimental conditions, but values of K_{ML}

and [L] are necessary to estimate the kinetics of the system. If $K_{ML} = k_f/k_d$ is substituted in Equation 1, an expression of the same type as the one obtained for the Ca-Chelex column can be written:

$$1/(1-x) = \exp(k'k_d t) \tag{3}$$

where $k' = 5 \times 10^{-3}/(k_f[L])$, k_f and k_d being, respectively, the rate formation and rate dissociation constants of the complex. So, in this case, the boundary condition depends not only on the dissociation rate constant of the complex but also on $k_f[L]$. The limit value $0.1\,s^{-1}$ chosen by the authors for $k_f[L]$ seems reasonable since the Cd EDTA^{2-} complex, which is only 14% labile in terms of the Ca-Chelex column (20), was determined to 100% labile by the batch method (18).

For the component i in a mixture of ML_i complexes, the following expression is valid:

$$X_i/f_i = 1 - \exp(-k_i t) \tag{4}$$

where $\sum_i X_i = X$ and $\sum_i f_i = 1$.

2.4. DIRECT METHODS OF SPECIATION

2.4.1. Introduction

Many analytical methods determine only the total concentration of a particular element in the sample, a pretreatment being generally required. However, it is much more important to know in which chemical form the element is present and the concentration of each species, since these are the parameters that determine the effect of the element on life, as well as its biodegradability and mobility.

Direct methods of speciation can be used: (a) to divide organic compounds in the sample into homologous groups and analyze them (however, organic speciation is beyond the scope of this chapter, since only element speciation will be discussed); (b) to separate and/or determine the concentration of different element complexes with inorganic and/or organic ligands. In the latter context the separation of organic matter aims at determining the element content associated with each fraction.

Table 2.2 presents the most important "direct methods of speciation" (21). the first three groups are based on physical and chemical interactions of the technique for the species under study, and bioassays are based on the interaction of the element species with living organisms. From the combinations of different direct methods of analysis, comprehensive speciation

Table 2.2. Direct Methods of Speciation

1. *Physical techniques*—based on size, density, or charge of the species	• Centrifugation • Membrane filtration • Ultrafiltration • Dialysis • Gel filtration chromatography
2. *Chemical techniques*—based on redox, complexation, and/or adsorption properties	• Oxidative destruction of organic matter • Liquid extraction • Ionic exchange and adsorbent resins • Voltammetry
3. *Species-specific techniques*—based on techniques that are specific to certain species	• Potentiometry with specific electrodes • Gas chromatography and/or hydride generation • High-performance liquid chromatography • Other chemical methods
4. *Bioassays*—influence of the metal ion on the growth (or inhibition) of organisms	
5. *Comprehensive speciation schemes*—combinations of different direct methods of speciation	

schemes can be outlined, some of which are subsequently presented in the text as examples.

The first two types of methods, with the exception of voltammetry, allow for the separation of the sample into different fractions of homologous compounds. In physical techniques the separation is based on size fraction, and in chemical techniques it is based on chemical affinity. After the application of the separative method, the element content of each fraction has to be determined by the application of an analytical technique that allows for the determination of the total element concentration, such as atomic absorption spectrometry (AAS), neutron activation analysis (NAA), or X-ray fluorescence spectrometry (XFS).

It is important to be aware that speciation of trace elements might be altered during the application of the technique involved due to the dissociation of the complex under the concentration gradients created, if the complex is not inert within the time scale of the method. In this case the determination is affected by the dissociation process, and a distinction, should be made between element-associated labile and nonlabile species for the method used (some comprehensive speciation schemes outlined in section 2.4.4 are based on this differentiation).

Some of the analytical techniques presented in Table 2.2 allow the determination of very low concentrations of trace elements to be carried out in the same range of concentrations found in natural systems, avoiding the use of preliminary preconcentration procedures. In this context voltammetric methods are among the most powerful used owing to their high detection power and the wide information that can be obtained from the experimental results, but their interpretation is not always easy. In fact, a correct interpretation of voltammetric signals requires a detailed understanding of the theory of the method, taking into account various factors that affect experimental results. For this reason the fundamentals of voltammetric methods will be more thoroughly discussed in Section 2.5.

Before presentation of the different "direct methods of speciation," we should emphasize the importance of sampling and storage method (22–25).

Teflon or polyethylene sampling and storage containers are recommended and can be decontaminated by soaking in approximately 10% HNO_3 for at least 48 h, then rinsed with distilled water and also with the sample. Conditioning the analytical cell with a solution containing the major elements of the sample, e.g., seawater, practically eliminates adsorption losses. It has been noticed that the storage of estuarine waters with low salinity in polyethylene containers for 24 h leads to a loss of 30% of Pb, suggesting appreciable adsorption by the container walls (26). Acidification, freeze-drying, and freezing can induce irreversible changes of trace element species, and samples should be stored at 4 °C, it having been noted that this can be done for periods as long as 3 weeks with no apparent deleterious effects (27, 28). A clean room or at least a special room with filtered air supply is necessary for work at ultratrace levels.

2.4.2. Physical Techniques

2.4.2.1. Centrifugation

Centrifugation, as well as filtration (Section 2.4.2.2), can be used to remove particles from a suspension in natural waters, the rotation rate and time necessary for the preferential sedimentation of the particles being carefully selected in the former technique. So, both techniques can be used to separate metal ions dissolved in natural waters from the fraction associated with large molecules or colloid particles. For example, all particles with diameters ⩾ 190 nm and specific density about 2.5 were separated from river and lake waters by 30 min of centrifugation at 3000 rpm, conditions that did not remove the humic matter (23, 29).

In the centrifugation technique separation depends not only on the size but also on the density, since small dense particles can settle during centrifugation, contrary to what happens to larger and less dense ones. In fact, the amount of centrifuged matter is generally higher than the fraction retained by conventional filtration (minimal pore size 0.4–0.5 μm), especially when small dense particles of metals (e.g., Al, Fe, Mn, or Ti particles) are present in solution (30, 31).

For suspended particles with diameters > 15 nm (below which ultrafiltration is necessary) filtration is in general preferable to centrifugation, since density effects are eliminated and size fractionation can be more successfully achieved. However, filtration also has problems, as can be seen in the next subsection, and centrifugation is generally faster and requires less attention from the operator.

2.4.2.2. Filtration

Chemical species in natural waters are in general operationally divided into "dissolved" (filtrable or soluble) and "particulate" (nonfiltrable), depending upon their ability to pass through a filter with a nominal pore size in the range of 0.4 to 0.5 μm (32). However, the transition between the two fractions takes place in a continual way, polymers and colloids (with diameters $\leqslant 1\,\mu$m) passing to the filtrate side (12). Table 2.3 shows the radius of various species in natural waters.

The applicability of filters to trace speciation depends on their ability to efficiently separate two specific size fractions, rather than to simply remove particulate material from suspension. This is related to pore size selectivity and susceptibility to contamination and adsorption, which is a function of the composition of the membrane (33). In fact, filter pores should be uniform in size and remain constant throughout the filtration, while the stated nominal pore size should be close to the effective pore size. This value can be determined from retention curves (percentage of retention vs. particle diam-

Table 2.3. Typical Sizes of Species in Natural Waters

Species	Estimated Diameter (μm)
Suspended particles	1–100
Colloids	5×10^{-3}–1
Humic substances	2×10^{-3}–6×10^{-3}
Aquometal ions	3×10^{-4}–8×10^{-4}
H$_2$O molecule	2×10^{-4}

eter), the slope of the curve reflecting the degree of uniformity of the pores (steeper slopes lead to more selective filters).

Filtration processes present some problems due to contamination of the filters and losses of trace compounds by adsorption, which is very significant in environmental samples owing to the extremely low trace element concentration (34–37). In order to avoid contamination with heavy metals, filters should be rinsed with diluted HNO_3. To overcome adsorption of trace compounds on the filter, both filter and equipment may be conditioned with large volumes of samples or preconditioned with 0.1 M $CaNO_3$, since Ca^{2+} forms surface complexes and saturates the groups of apparatus and filters (38–41). Other filtration problems that can occur if storage and filtration periods are too long arise from the degradation of water samples and/or from the decrease of the flux through the membrane, which is due to changes in osmotic pressure, interactions between different solutes and the solute with the membrane, variations of hydraulic resistance, and particle coagulation. Coagulation occurs when the particle concentration on the retention side of the membrane becomes larger than in the bulk solution (concentration polarization), leading to gel formation followed by the clogging of the membrane. On the other hand, colloids whose diameter is smaller than the pore size of the filter can be retained owing to adsorption due to specific interaction between the colloids and the membrane surface, which essentially depends on the chemical nature of the surfaces involved, in particular of the membrane.

The ratio of the particle concentration at the membrane surface (C_0) to that in the bulk solution (C) can be represented by the following approximate equation (13, 42, 43):

$$C_0/C = \exp(v\sigma\delta/D) \qquad (5)$$

where v is the flow rate per unit of area (cm s^{-1}); s is the retention coefficient ($\sigma = 1$, fully retained; $\sigma = 0$, not retained); δ is the diffusion layer thickness (cm); and D is the diffusion coefficient (cm^2 s^{-1}).

In syringe filtration a positive pressure is applied to the sample by the piston in order to decrease filtration time, flow rates being of the order of 100–300 mL min^{-1}. From the previous equation it can be estimated that, with this technique, the final concentration C_0 is much higher than C if particle size is between 0.04 and 0.3 mm (44). With such a high particle concentration, coagulation becomes very rapid, producing agglomerates that are deposited and may form a clogging gel independent of the chemical nature of the membrane.

In order to minimize coagulation all parameters must be selected in such a way that the C_0/C ratio is as small as possible, using: (a) very small flow rates, v (however, if large volumes are involved, this will imply longer filtration time, increasing the risk of adsorption and contamination); (b) efficient stirring at the membrane, which increases the diffusion flux of the particle back into solution, reducing filter clogging and allowing for faster filtration and better size discrimination (however, turbulence produces the simultaneous adverse effect of increasing the rate of orthokinetic coagulation of colloids in solution); (c) small values of the concentration factor V_0/V_f (V_0 and V_f are the initial and final volumes in the filtration cell, respectively), so that bulk concentration in the cell does not increase excessively; and (d) washing filtration, which consists of adding solvent to the cell during the filtration so that the volume to be filtered is maintained constant. This technique is more time consuming than conventional filtration (also called concentration filtration), but gives more reproducible results because the concentration of the species in the cell remains approximately constant.

In order to decrease the time of filtration processes, both positive or negative pressure can be applied, although large pressure differentials should be avoided to prevent cell rupture. Positive pressure filtration with an inert gas is generally preferred, since it reduces the risk of airborne contamination and eliminates pH fluctuation associated with CO_2 evolution during vacuum filtration (negative pressure), which, in low-buffered waters, can cause a change of 2 pH units through the membrane, accompanied by displacement of solubility equilibrium (45–47).

In conclusion, in speciation studies filtration must be performed at low controlled flow rates of approximately $10\,cm\,h^{-1}$, preferably with sequential filters with increasing pore size (the cascade system) and stirring the solution under N_2 atmosphere, in order to divide the chemical species into different size fractions with small concentration factors.

It is worth noting the need to develop new methods specifically oriented toward analysis of natural colloids, taking care to minimize the problems induced by currently employed techniques, although such studies are outside the scope of the present chapter [see Perret et al. (42)]. In fact, an efficient separation between colloids and dissolved molecules is not always achieved, since colloidal particles of Fe, Mn, and clays, coated with humic acid, sometimes can pass through the 0.45 μm filter. This fraction is very important in speciation of heavy metals, because in general the highest amount of metals is the one specifically adsorbed on colloidal humic acid, especially in freshwater.

2.4.2.3. Ultrafiltration

This technique is similar to filtration, but the membranes have a nominal pore size smaller than 20 nm and can be used to further discriminate the size continuum of materials in natural waters (33). Ultrafilters are available with

Table 2.4. Pore Size of Ultrafiltration Filters

Amicon Filters	Pore Size (nm)
XM 300	14
XM 100 A	6.0
XM 50	3.5
PM 30	2.4
PM 10	1.8
UM 2	1.2
UM 05	1.0

nominal pore sizes ranging from 1.0 to 20 nm (Table 2.4) and can separate species into dissolved and nondissolved fractions, the problems being similar to those of the filter technique just described (48). These problems are, however, generally more important in ultrafiltration, since smaller concentrations of groups of species are involved.

Visking membranes with a molecular weight cutoff (MWCO) of 14,000 are the most easily available ultrafilters, but in speciation studies the Spectrapor 6000 MWCO is also widely used. The MWCO is the retentive capacity of an ultrafilter, a quantity referring to the molecular weight of a solute that is 90% retained, i.e., with pore sizes on the order of 1–20 nm, the MWCO range is 500–300,000. These values are only nominal because fractionation is achieved by molecular size rather than weight. So, for spherical molecules (e.g., globular proteins), a reasonable agreement is obtained; contrary to what happens for linear compounds (e.g., polyethylene glycol molecules), which can exhibit diffusibility through membranes (49).

Buffle et al. noticed that factors such as electrolyte concentration, pressure, and even pH do not seem to have a significant effect on ultrafiltration results, except in extreme conditions (50). However, the results greatly depend on interactions of organic matter with dissolved molecules or interactions of colloidal particles with the membrane and with each other, forming aggregates.

Washing and concentration ultrafiltration procedures were examined by Buffle et al. in order to optimize the fractionation of organic material (50). The authors recommended a sequential ultrafiltration (cascade system) with a MWCO of 10,000 down to 5000 and then 500 in order to reduce the adsorptive effect of high-molecular-weight compounds observed when samples are filtered through membranes with very small pore sizes in comparison with the radius of the organics.

Hoffman et al. used the concentration technique associated with sequential ultrafiltration, where large concentration gradients were avoided by reducing the volume only by 50% at each step (51). A mass balance formalism was developed to calculate the concentration in each molecular-weight range.

2.4.2.4. Dialysis

The dialysis technique, where a sample solution is placed in contact with a blank through a membrane permeable to dissolved species, has also been used to separate colloids from molecules in solution. Ideally, the dialysis membrane should be permeable only to species in true solution, but in practice some high-molecular-weight material, and even colloids, can pass through membranes with pore sizes of from 1 to 5 nm, i.e., 1000–5000 MWCO. However, as with filtration and ultrafiltration, the size discrimination of the dialysis membrane depends on the type used.

When dialysis equilibration is attained, dissolved neutral species should have the same concentration in both diffused and retained solutions (33). However, if the retained solution has ionic species that do not pass through the membrane, different concentrations of dissolved ionic species will be found on both sides of the membrane once equilibrium has been attained, according to Donnan theory (52).

Two major disadvantages of the dialysis technique are the long time required for equilibrium to be attained (ca. 24 h) and the fact that the membranes often have a negative charge, interfering with diffusion rates of charged molecules. Under such conditions negative species experience a smaller effective pore size owing to repulsion of charges of the same type and require a longer time to equilibrate. This phenomenon is used with good advantage in Donnan dialysis, where charged species may be separated by using membranes impermeable to anionic complexes (53). On the other hand, dialysis can be carried on without assistance, presenting no concentration effects as filtration and ultrafiltration do, since the volumes of the retained and diffused solutions are constant during the separation.

Guy and Chakrabarti compared ultrafiltration with pore sizes of 1.2, 1.8, 2.4, 3.5, and 6.0 nm and dialysis, both associated with AAS, to determine total element concentration in each fraction, and concluded that ultrafiltration is preferable because of the inconvenience of long equilibration times and the Donnan effect of dialysis (54).

The long equilibration time of dialysis (the high timescale of the technique) increases the possibility of dissociation of metal complexes at the membrane surface, causing shifts of the equilibrium.

The recovery efficiency in dialysis decreases as the ionic strength of the sample increases, but this effect can be counteracted by increasing the ionic strength of the diffused solution (54).

Trace element contamination, one of the problems common to dialysis, filtration, and ultrafiltration, can be at least partially avoided if the membrane is decontaminated by soaking in mineral acids (55–59).

2.4.2.5. Gel Filtration Chromatography

Gel filtration chromatography (GFC), also referred to as gel chromatography, gel filtration, gel permeation chromatography, exclusion chromatography, and molecular sieve chromatography, enables a complete size continuum of trace element species to be performed. Details of the technique can be found in several reviews and textbooks (60–63). In this technique porous polymer beads are swollen with a solvent and vertically packed into a chromatographic column. Distilled water or, more commonly, a dilute solution of Groups I and II metal cations, have been used as an eluent (64–67).

When the sample is applied to the column and the eluent is added, size fractionation is achieved depending on the diffusion rate of a molecule into and out of the interstitial cavities of the beads (33). So, large molecules are excluded from the pores and eluted first, followed by other molecules in a decreasing order of size. The volume range of molecules can be altered by using beads of different pore size.

GFC has an advantage over ultrafiltration in terms of speciation, because element concentration can be determined over a continuum size spectrum, rather than at discrete size ranges. However, this technique is limited to high element concentrations due to the relatively small sample amount and large eluent volume, which leads to high dilution factors and large blank values. In this context GFC is an attractive method for separating element species into molecular size fractions in sewage effluents and in contaminated sample solutions of soils (65, 68, 69).

GFC must be carried out with careful control of the eluent with respect to pH, ionic strength, composition, and temperature, in order to limit the dissociation of complexes and adsorption effects. Preconcentration procedures prior to GFC must be avoided, and this must be a general rule in all speciation schemes, because of the unknown effects they might have on the physicochemical speciation of trace elements.

2.4.3. Chemical Techniques

2.4.3.1. Oxidative Destruction of Organic Matter

In order to determine the fraction of elements bound to organics, wet chemical techniques including digestion with perchloric, nitric, and peroxydisulfuric acids, photooxidation with UV radiation, and ozonolysis have been used to destroy organic matter.

Ultraviolet (UV) irradiation causes not only a temperature increase but also a pH rise due to loss of dissolved CO_2, which leads to precipitation of Fe(III) hydroxides. Even if acidification at a pH of about 5 (with 0.1 M $HClO_4$) prior to UV irradiation is used, this does not necessarily prevent deposition of oxyhydroxides. On the other hand, UV irradiation does not completely destroy the organics in solution in natural waters, although it presents the advantage of being a technique that avoids problems of contamination, contrary to what happens with wet oxidation. Hence, the photooxidation procedure may be applicable only with natural waters with low levels of organics and exhibiting low Fe concentrations (21).

Ozone treatment (ozonolysis) is used to remove trace element contaminants from industrial effluents (70). Although this technique also avoids significant problems of contamination and the procedure solubilizes elements associated with organic material, it apparently causes the oxidative precipitation of several metals including Fe, Mn, and Pb, scavenging other elements in solution. So, ozonolysis is a possible oxidative procedure for speciation studies, but only when the precipitation of metal oxyhydroxides is unlikely to occur.

2.4.3.2. Liquid–Liquid Extraction

Lipid-soluble compounds, known to be highly toxic, can be at least roughly estimated by extraction with organic solvents such as chloroform and hexane–butanol, which have dielectric constants similar to those of biological membranes (71). The metal content associated with these fractions is determined, after extraction, with a total element concentration technique, as usual.

Dissolved elements in natural waters have been extracted with ammonium pyrrolidinedithiocarbamate (APDC), dithizone and APDC, and methyl isobutyl ketone (e.g., Zn; Zn and Cu; Cd and Pb), due to the exceptionally strong complexes that dithizone, diethyldithiocarbamate, and their derivatives form with heavy metals (72, 73). So, all naturally occurring soluble complexes with most heavy metals in natural waters are likely to be dissociated and extracted by these compounds if there is no kinetic complication, the nonextractable fractions being constituted most probably by trace elements adsorbed or occluded in organic and inorganic particles and colloids.

Note, however, that it is always difficult to characterize the fractions of heavy metals extracted by organic solvents, since neutral species may not be completely recovered and, on the other hand, heavy metals associated with organic colloids may interfere.

2.4.3.3. Ionic Exchange and Adsorbent Resins

In this method the sample is passed through a resin inside a column, the resin being previously conditioned with a buffer at the same pH as that of the sample, so that the pH of the solution is maintained constant during its passage through the column; the different fractions are then eluted with adequate solutions and subsequently analyzed together with the unretained aliquot.

The time scale of the technique depends on the column flow rate, the Chelex-100 column being considered as a transient technique with a time scale of about 6–9 s; this is considerably longer than the dissociation time of the labile species measured by ASV, which is of the order of milliseconds, but much faster than the contact time of a batch technique, where the sample reacts over a rather long period (e.g., 3 days) with the resin in a vessel (18). The application of ion-exchange resins to speciation studies relies upon the presence of labile and nonlabile species for the resins, since the complexes with slow dissociation rates relative to the solution/resin contact time are not, in principle, retained (20, 74). On the other hand, for adsorbent resins molecules larger than the pore size of the resins themselves are not retained since they are quantitatively rejected. So, trace elements associated with colloidal particles such as colloidal Fe(III) and Fe-humic acid colloids are not retained by most ion-exchange resins owing to the resin pore size of 1–2 nm, which is too small to allow for the entrance of colloids into the resin network (28).

Cation exchange resins, or more commonly resins where chelating functional groups have been incorporated into the polymeric macroporous network, are used in environmental analysis. From this point of view Chelex-100 of the type Dowex A1 with an iminodiacetate group (IDA) has been the most widely used chelating resin, due to its ability to extract several elements simultaneously and owing to its commercial availability in quite a pure and inexpensive form. Before the sample is passed into the column it is important to convert Chelex-100 into the Na^+, Ca^+, or NH_4^+ form, since in the H^+ form its ion-exchange capacity is poorer due to the weak acidity of the chelating groups. The retention of trace elements associated with complexes with stability constants larger than those with functional groups of chelating resin is unlikely to occur, since functional groups with high affinities or transition elements have been chosen in order to attain maximal recovery from solution. However, siderophores in natural waters may compete for the resin groups, since they sequester significant concentrations of, e.g., Cu and Fe (75, 76).

Copper and/or other elements have been determined in natural waters after passage through the resin. The resin can fix the ion either directly, by a cation-exchange mechanism, or indirectly, if the resin fixes organic matter complexed with metal ions. Some examples will now be given. (a) Zorken et al.

(77) and Sweilel et al. (78) analyzed Cu(II) in seawater after passage of the sample through a column with a cation-exchange resin. Assuming that none of the Cu complexes with organics and bound to colloidal matter bore a positive charge, the authors concluded that the fraction retained by the column corresponded only to inorganic species. Having determined the concentration of these compounds and knowing the concentration of inorganic ligands from the pH and salinity values, the investigators could estimate the speciation of Cu(II). (b) Cu(II) complexes with humic and fulvic acids can be separated from fresh-water, either due to their anionic properties, by using an anion-exchange resin (79), or due to the hydrophobic properties of humic compounds, by using a nonionic macroreticular resin (79, 80). However, since in the latter case significant amounts of simple metal cations can also be sorbed on the resin, the cation exchange sites have been saturated with Ir(III) with more affinity for the groups than univalent and bivalent ions. The only causes of interference in this study are organic species other than humic complexes that can also be sorbed on the Ir resin, depending on the aromaticity and size of the hydrophobic groups. (c) Speciation of Cu(II), Cd(II), and Zn(II) in natural water samples has been done by on-line flame AAS, after sequential passage through a Chelex-100 resin and AGMP-1 macroporous anion resin. The Chelex-100 retained hydrated metal ions and cations released from labile species, including complexes in solution and possibly metals loosely associated with colloidal matter. Nonlabile complexes and elements strongly associated with colloids are not retained by Chelex-100 and are therefore passed through an AGMP-1 resin with a large pore size that can retain negatively charged complexes and elements associated with negatively charged polymers such as humic material. Elements strongly associated with very large colloidal particles and neutral nonlabile complexes may not be retained owing to the exclusion limit of AGMP-1 resin [molecular weight (MW) ca. 75,000], although retention due to hydrophobic interactions is possible (81); (d) C-18 SEP-PAK prepacked columns containing octadecyl carbon moieties chemically bound to a silica gel support have been used in order to isolate and concentrate dissolved organic matter and organic hydrophobic complexes of metal ions from different types of natural waters, such as estuaries or oceans (82–87). This approach has the advantage that, because weak interactive forces are involved in this technique, the possibility of denaturing the isolated organic matter and of disrupting the complexes with metal ions is minimized. Although the retention mechanism on SEP-PAK C-18 resin has not been completely formulated, the process is most probably of weak adsorption in the nonpolar stationary phase C-18. After lake water was passed through C-18 SEP-PAK columns, it was noted that 55–68% of total Zn was not retained, 6–10% was retained by the resin and eluted with methanol-water solution, and 22–37% was eluted by 0.6 M HCl. The fraction eluted by methanol–water represents nonpolar hydropho-

bic complexes, and the one eluted with 0.6 M HCl may include nonpolar complexes with some hydrophilic character as well as possibly Zn fixed by unreacted silanol groups. It was also noted that in addition to Zn bound to hydrophobic organic complexes there exist Zn complexes in the fraction eluted with 0.6 M HCl and in the nonretained solution.

Finally, some research work has placed emphasis on chelation using sulfur, rather than oxygen- and nitrogen-binding sites, since thiol resins are more suitable to simulate the biological reactivity of an element in a water sample (71). In fact, the toxicity of a dissolved element species to aquatic organisms depends on its ability to react with the biological membrane, where the carrier proteins may be similar to metallothioneines, which react with metals mainly via sulfhydryl groups; once inside the cell, the element can be bound by glutathione, a similar cytoplasmic sulfhydryl compound (71). Bio-Rad SM2 resin has been used as a ligand model with thiol groups, having few leachable element impurities and a molecular weight exclusion limit of 14,000. Sugimura et al. used Amberlite XAD-2 resin, similar to Bio-Rad SM2, to separate organically bound Fe and other elements from seawater (88). The adsorbed elements were then determined after elution with methanol; however, if there was also adsorption of inorganic forms, they might not have been eluted with this solvent. Florence noted that thiol and oxine porous glass resins were generally more efficient than Chelex-100 in removing metal ions (71). This author also obseved that, in order to determine organically associated trace elements, SM2 resin is more convenient than conventional organic solvents, such as hexane or chloroform, and that the very small blanks associated with this resin allow trace elements to be measured at low concentrations. Metal–lipid soluble compounds, which cannot be detected by voltammetric methods or Chelex-100 resin if the complexes are not labile for these techniques, may be detected with SM2 resins.

In conclusion, ion-exchange resin is an attractive technique because separation can be carried out with little manipulation and low contamination (in general, the resin has a high level of purity). Furthermore, when compared, say, with liquid–liquid extraction, it involves less contamination with element impurities from solvents.

2.4.3.4. *Voltammetric Techniques*

In voltammetry a potential is applied to the working electrode connected to the reference and to the auxiliary electrodes, all of them being immersed in the sample solution. If a redox reaction occurs between the first interface and the species in solution, the concentrations of these species decrease at the electrode surface. Thus, a diffusion gradient between this interface and the bulk solution is created, the diffusion current being directly related to the bulk concentration of the redox species involved in the process. The variation of the species concentration during the redox process does not disturb the global concentration, since only weak currents (a few milliamperes) are developed between the working and auxiliary electrodes in voltammetric techniques due to the small dimension of the former, i.e., the bulk concentration remains constant during the measurement (89, 90). The working electrode more commonly used in single-step voltammetric techniques is the dropping mercury electrode (DME). It has an easy renewal capability and an excellent reproducibility, whereas a large range of negative potentials can be scanned in aqueous solution due to the high H^+ reduction overpotential in mercury. The scan mode of potential vs. time limits the detection power of voltammetric techniques (Table 2.5).

Voltammetric measurements should be performed without the interference of oxygen, which presents irreversible reduction waves for potentials more negative than 0 V. For this reason oxygenated solutions should be deaerated, or a technique with very small time scale should be used, such as rapid square wave polarography. In this technique, with a detection limit of the order of pulse methods, a symmetrical square wave is superimposed in a staircase and the entire polarogram is obtained in one drop of the DME (91).

The voltammetric signal (current vs. potential) is influenced not only by the concentration of redox species and their thermodynamic equilibria but also by kinetic factors, and so information can be obtained about kinetic and complexing parameters of the medium. Besides these factors, adsorption of neutral or electroreactive species can also influence the voltammetric signal. In this context, all the factors that affect diffusion and kinetic currents should be rigorously known and controlled in order to enable the voltammogram obtained to be correctly interpreted; in this connection, the presence and absence of ligands and their heterogeneity will be discussed in Section 2.5.

The application of voltammetric methods to speciation of heavy metals in natural waters has been developed mainly since the 1970s. The use of voltammetric techniques for complexation studies of metal ions and macromolecular ligands in natural aquatic media has been critically reviewed by van Leeuwen et al. (92). These methods are of great interest for speciation studies in natural waters, allowing for (a) determination of free ion concentration in the absence of ligands, or in their presence within certain restrictions; (b) detection of labile or quasi-labile complexes within the time scale of the technique and determination of inert complexes, if their potential of reduction is in the potential range of voltammetry; and (c) determination of complexing and mass transport parameters (i.e., complexing capacity of the medium, stability constant and rate dissociation constant of the complexes formed, and mean diffusion coefficient if only labile species are present).

Table 2.5. Detection Limit (DL) and Signal of Various Polarographic Techniques[a]

Method	$E(t)$	Measurements of i	i vs. E
Direct current (dc) polarography DL: 10^{-5} M		For each instantaneous t	
Sampled DL: 10^{-6} M		During the last 50 ms of each drop	
Normal pulse polarography (NPP) DL: 10^{-7} M		During Δt of each drop	i_l, i_d, i_r at $E_{1/2}$
Differential pulse polarography (DPP) DL: 10^{-8} M		$i_{t2} - i_{t1}$ for each drop	i_p at E_p

[a] The following symbols are adopted: t_d, drop time; $E_{1/2}$, half-wave potential; E_p, potential of the peak; i_l and i_r, residual current ($C_0 = C$) and limiting current ($C_0 = 0$), where C_0 and C represent, respectively, the solution concentration at the electrode surface and that in the bulk solution; i_d, diffusion current; i_p, peak current. Scan rate = 2–5 mV/s; pulse amplitude in DPP = 25–50 mV; pulse time = 50 ms; $\Delta t \leqslant$ pulse time; t_d = 1–5 min.

2.4.3.4.1. Metals Determined by Voltammetric Techniques.

These techniques are applied to some transition elements of the first row (some of them vital to human life), as well as some "b" type metals, often toxic (e.g., Cd, Pb, and Zn) (93). The former metal ions are mainly determined by differential pulse polarography (DPP), with a detection limit of 10^{-8} M, whereas the latter, with a very low concentration in natural waters, are determined by stripping techniques, with a higher detection power due to a preconcentration step. Stripping techniques in dc mode have a detection limit of about 10^{-9} M if the hanging mercury drop electrode (HMDE) is used, or of about 10^{-10} M for the mercury film electrode (MFE), which has a higher sensitivity due to its higher area/volume ratio. The detection limit can still decrease by about 1 order of magnitude if the potential is scanned in the differential pulse mode (Table 2.6). During the preconcentration step of stripping techniques the metal ion can be concentrated "into" or "on" the electrode, respectively, by reduction of the metal ion in the amalgam form or by adsorption of the metal in the complex form. This step is followed, respectively, by a stripping process of oxidation (ASV), or reduction [cathodic stripping voltammetry (CSV)], where the signal measured is proportional to the concentration of the preconcentrated metal. ASV has been used in the determination of metals soluble in Hg (e.g., Cd, Cu, Pb, and Zn) (72, 94–100), whereas the adsorption technique (CSV) has been applied to some Hg-insoluble metals (e.g., Co, Mo, Ni, S, Se, and Ti) (101–105) and also to some soluble metals such as Cu (106–109), Al (110), and Zn (98, 111).

Stripping techniques have recently been coupled with mercury microelectrodes (electrodes with diameters in the range 0.1–50 μm). These electrodes, obtained by plating mercury onto a proper substrate with a suitable geometry (112–116), have some important advantages over electrodes of conventional size, namely, their use in media of low ionic strength (≈ 0.001 M), application

Table 2.6. Different Modes of Stripping Techniques

Mode of the Stripping Step	Electrode	i vs. E (in the Stripping Step)
Direct current mode DL: 10^{-8} M	HMDE	
Differential pulse mode DL: 10^{-9} M	HMDE	
Differential pulse mode DL: 10^{-11} M	MFE	

to stripping techniques without the need to stir the solution during the preconcentration step, and determination *in situ* in interstitial waters of sediments and/or in microsamples (117–119).

2.4.3.4.2. Major Factors Influencing the Voltammetric Signal of Natural Waters. In natural waters small organic and inorganic ligands as well as macromolecules and surface groups (e.g., polysaccharides, nucleic acids, proteins, humic matter, and oxyhydroxides) can complex heavy metals, the voltammetric signal obtained being dependent on several factors, as discussed below.

The simplest situation is when all the complexes are electrochemically inert within the time scale of the technique, which may happen in fractions of inorganic and organic colloids. The voltammetric signal is then directly related to the free metal concentration in the bulk solution (Section 2.5.3.1). The other limiting case is when all the complexes are labile, i.e., when there is equilibrium between the free metal ion and the complexes formed even at the electrode surface. The global process is then controlled by diffusion of all metal species to the electrode, and a mean diffusion coefficient should be defined (Section 2.5.3.2). The intermediate situation is when complexes are quasi-labile (Section 2.5.3.3); the interpretation of the voltammetric signal is much more complicated in this case, especially when the diffusion coefficients of complexes are different from those of metal ions as happen for macromolecules [a complete interpretation of this situation is still being formulated (120, 121)].

Besides being influenced by the degree of lability, the voltammetric signal also depends on the reversibility of the redox reaction, which may be affected by the external conditions of the medium. Other practical problems associated with ASV applied to natural waters, e.g., direct reduction of the inert complex at the applied potential and surface effect, have been discussed elsewhere (122, 123).

The natural sample under study should have an ionic strength sufficiently high to minimize the migration current ($I > 100\,C_M$, where I represents the ionic strength and C_M the total metal concentration) and the ohmic overpotential iR (i representing the current intensity and R the resistance). On the other hand, the sample should have a buffer capacity and an excess of natural complexing agents so that the concentrations of the proton and of the free ligand are similar near the electrode and in the bulk solution.

The heterogeneity of natural samples, due to different complexing sites present in solution, might be responsible for a broadening of the peak width or for a spread over the potential axis in the polarographic wave, which may influence the results obtained (124). On the other hand, owing to the heterogeneity of the sample, the first sites to be complexed might be characterized by an inert behavior, the following sites becoming probably less inert or even labile with the decrease of their complexing affinity. In this regard, mixture effects, due to the presence of different type of ligands in solution, can be decreased by a prior separation into homologous groups. However, humic and fulvic acids that can be found in natural waters show a heterogeneous behavior due to their own composition, with a large charge density and different complexing sites (polyelectrostatic and polyfunctional effects, respectively). It should be pointed out that, although a large number of voltammetric studies on the interaction of metal ions with fulvic acids has been reported, the macromolecular nature of these ligands is often not considered owing to their high complexity. The necessity of considering this aspect has been critically reviewed (13). Heterogeneity can also be found in pure inorganic oxides and hydroxides present due to the polyelectrostatic effect, although these particles have only one type of functional group (125). In the presence of macromolecules aggregation or gel formation may occur, mainly due to attractive forces between species with different charges and/or neutralization of charges of the same signal. This can be avoided if a previous separation between dissolved molecules and colloids is done; however, even with 0.45 μm filters this separation is not complete since some colloids can pass through.

One of the most common problems in voltammetric techniques is the interference caused by the adsorption of organic constituents of the sample on the electrode. This effect may dramatically affect the voltammetric signal, since it increases the organic concentration at the electrode surface and thus interferes with the redox process, in the presence or absence of complexation, as discussed for DPP (126–129) and normal pulse polarography (NPP) (10, 130, 131). In ASV adsorption may affect both preconcentration and stripping steps, the contribution in each becoming more difficult to interpret (132–136). Moreover, adsorption–desorption processes of organic compounds can lead to voltammetric peaks that can interfere with or be mistaken for faradic peaks. Various paths have been tried and/or developed in order to overcome adsorption effects. These are enumerated below:

a. In order to elucidate the kinetics and measure the extent of adsorption of organics in the absence of metals, studies are required under the same experimental conditions as in speciation studies, i.e., at the same pH, ionic strength, range of concentrations, stirring rate, and potential. For this purpose alternating current (ac) voltammetry is a good choice, since the capacity of the double layer at the electrode surface as a function of time and potential can easily be determined from current measurements, in the absence of redox processes (137). From these values the degree of coverage of the electrode by adsorbed molecules vs. time and potential can be found (138). Analysis of these curves provides information about equilibrium parameters (adsorption constant and maximum coverage) and about the kinetics of the adsorption process, where two steps should be considered: (i) diffusion of the adsorbed molecules to the electrode, and (ii) adsorption kinetics at the interface. However, so far, mathematical treatments have considered only the global process controlled by one of these steps or at equilibrium. For large molecules with fast adsorption kinetics at the interface, the limiting step is the diffusion process (139–143). Competing with the diffusion step, a slow adsorption step might occur for some macromolecules due to the possible defolding and reorientation of the adsorbed molecule at the interface, under electrostatic and chemical affinity effects. In the literature some adsorption studies of natural dissolved organic matter by ac voltammetry can be found, but further investigation of the kinetics of the global process is required (144–148).

b. The problems of adsorption in speciation studies can sometimes be minimized by using normal and reverse pulse polarography, if an initial potential where adsorption does not occur can be applied and if during the application of the pulse (≈ 50 ms) the adsorption can be neglected.

c. To minimize adsorption effects in ASV measurements, the deposition potential should be chosen in the range where adsorption of natural organic matter is negligible, if possible; adsorption of humic matter depends on its origin, but generally adsorption becomes much weaker for potentials more negative than -1.0 V vs. SCE (saturated calomel electrode) (149, 150). Extrapolation to zero contact time between the electrode and the solution has also been suggested to decrease adsorption effects (132).

d. Coating of the working electrode with selective membranes has been used as a means of circumventing the organic interferences in ASV, thereby preventing some organic interferents from reaching the voltammetric interface. Obviously, a compromise between exclusion of organic matter and the unhindered transport of the metal ion must be sought.

In this last context the membrane used most is Nafion, a non-cross-linked copolymer of tetrafluoroethylene and vinylsulfonic acid that acts as a cation-exchange resin. The coating procedure is performed by applying a solution of the polymer to the electrode surface, in general glassy carbon. Subsequently mercury can be plated onto the electrode. The thickness of the polymer film is such that it does not interfere with the mass transport of the species under analysis, although it can discriminate especially against negatively charged organic species, due to repulsion of fluoride ions and sulfonic groups of Nafion. The polymer is chemically inert, nonelectroactive, hydrophobic, and insoluble in water, and the modification procedure is convenient and fast. The extent of peak depression at the Nafion-coated thin mercury film electrode caused by detergents is in the following order: cationic > nonionic > anionic. Indeed positive species easily penetrate a negatively charged membrane, since Nafion has a remarkably high affinity for hydrophobic cations, with ion exchange selectivity coefficients vs. Na^+ ranging from about 1×10^4 to 6×10^6, the coefficient increasing with hydrophobicity of the species (151, 152). When compared with conventional mercury film electrodes, modified electrodes show major advantages: improved resistance to interference from surface-active compounds, increased sensitivity, and better mechanical stability of the mercury film. Quite small detection limits for Pb and Ag, on the order of 5×10^{-10} M and 2×10^{-12} M, respectively, have been attained by ASV with a glassy carbon electrode coated with a film of Nafion in which the crown ether dicyclohexyl-18-crown-6 (DC18 C6) has been incorporated. The Pb complex with DC18 C6 is strongly adsorbed on the electrode surface by the Nafion cation-exchange film, owing to the strong interaction with large organic cations. If the total concentrations and stability constants of the species are known, a speciation model can be established (153, 154). Nafion can also be deposited on microelectrodes to prevent adsorption problems.

Cellulose acetate–coated mercury film electrodes have also been used in analyses of trace metals by ASV. The coating provides an effective barrier on the mercury surface, thus eliminating the effects of various organic surfactants such as gelatin, it having been noted by Wang and Hutchins-Kumar (155) that $100 \mu g\, g^{-1}$ of this compound does not alter the voltammetric response. The controlled permeability can be achieved by hydrolyzing the polymeric coating in alkaline media, which increases the porosity of cellulose acetate by breaking its chain into small fragments. This leads to an increase of selectivity mainly based on size-minimizing interferences such as adsorption of organic surfactants and overlapping stripping peaks (155). However, this electrode modification procedure requires at least 70 min, which is inconvenient for practical ASV work. A new bilayer polymeric coating with a cellulose acetate layer on the top of a Nafion film can be used to differentiate between various organic cations (156).

2.4.3.4.3. Some Applications to Natural Waters. Differential pulse anodic stripping voltammetry (DPASV) and differential pulse cathodic stripping voltammetry (DPCSV) have been used to determine total trace concentrations of several heavy metals in different types of natural waters, as well as free, inorganic, and organically bound metal ions, thereby allowing the stability constants of organic complexes to be calculated.

Speciation studies of heavy metals are ecologically very important since, as previously pointed out, complexation of these ions can trigger the input of toxic metals into marine food webs and the marine sedimentation cycle.

When considering complexation in natural waters, one should be aware that organic matter in these media can present diverse binding sites. Since the sites with higher affinity for the metal ion will be the first ones to be complexed, the conditional stability constant (K_{cond}) should decrease with the increase of the metal ion in the titration of a natural sample up to saturation of the ligand ($K_{cond} = [ML]/[M'][L']$, where L' and M' represent, respectively, all L and M species different from ML). In this context K_{cond} represents an average conditional parameter $\bar{K}_{cond}$, which takes into account several sites complexed with the metal. Both graphic methods and/or nonlinear regression analysis have been used to determine $\bar{K}_{cond}$ and C_L simultaneously in natural waters, for inert or labile systems, based on discrete binding site models, where a small number of discrete sites are assumed (157, 158). Experimental data of natural waters are often fitted to one or two 1:1 metal–organic complexes (97, 98, 159), but a higher spectrum of sites may be present due to the heterogeneity of natural samples, as observed in CSV experiments with several different detection windows (160, 161). An effort has been made to describe the complexation of organic substances present in natural waters by discrete models with more than two binding sites (162, 163). Some authors claim that in natural samples and/or in the presence of humic or fulvic substances a continuous variation of complexing parameters should be assumed due to their high degree of heterogeneity, and therefore distribution functions should be used instead of conventional complexing parameters (Section 2.5.4).

A few examples will suffice to illustrate how speciation is carried out in natural waters. In samples of offshore Pacific Ocean waters, where the levels of heavy metals are very low, an unexpectedly high affinity of Zn and Cu has been observed, the organic complexes behaving as inert in voltammetric measurements. This might be due to the low C_M/C_L ratio and because the metal ions go first to the groups with higher chemical affinities, which together with electrostatic and conformational effects lead to high stability constants. In fact, for a total Zn concentration of 0.1–0.2 nM and a ligand concentration 10 times higher, 95% of total Zn is organically complexed in the euphotic zone with an average value for the conditional stability constant of log $\bar{K}_{cond} = 10.8$ (98). The total concentration of Cu found changes from 0.6 to 1.8 nM between the surface and 1000 m depth, but free Cu varies from 1.6×10^{-5} nM to 0.01 nM, i.e., there is a variation of 3 orders of magnitude, with significant implications concerning the toxicity and bioavailability of Cu in open ocean waters. This is due to the presence of extremely strong sites with log $\bar{K}_{cond} = 11.5$ at a concentration of about 1.8 nM, which is predominant when Cu complexation occurs in surface water, and a weaker class of ligands (log $K_{cond} \approx 8.5$) at higher concentrations (8–10 nM) (97). On the other hand, it was found that, in general, more than 99% of dissolved Cu is organically complexed in estuarine waters, with calculated pCu^{2+} values from 11.1 to 12.8. The results strongly suggest that Cu is transported by rivers to the ocean mainly in the form of organic complexes (108).

It should be mentioned that, although in the foregoing examples the complexes are inert, higher C_M/C_L ratios may indicate quasi-labile or even labile behavior within the time scale of the voltammetric measurements, due to the heterogeneity of the sample (Section 2.5.4). In this respect, reported values based on assumptions that inert organic complexes are present without a previous check of the degree of lability should be regarded with some degree of skepticism.

Van den Berg et al. determined in Atlantic Ocean seawater two classes of ligands for Zn (C_L = 26 and 62 nM), with log $\bar{K}_{cond}$ = 8.4 and 7.5, for a total Zn concentration of 45 ± 15 nM (159). Water from a Swiss lake was analyzed, and the results obtained compared quite well with the previous ones, i.e., two types of ligands (C_L = 10 and 90 nM), with log $\bar{K}_{cond}$ = 8.5 and 8.0, were found for a similar total Zn concentration. The agreement obtained for Zn speciation in different types of natural waters points to the fact that among the most important inputs of organic complexing matter in these media are the exudates of organisms. On the other hand, the higher value obtained for the average conditional stability constant in waters of the Pacific Ocean (log $\bar{K}_{cond}$ = 10.8) can be explained by the lower Zn concentration found in these samples and by the higher affinity of heterogeneous ligands involved in complexation, as noted earlier (98).

The determination of Cu in a eutrophic lake reveals a seasonal pattern similar to that of the productivity of algae, which makes it probable that Cu(II) is complexed with biologically produced ligands (109). Significant amounts of Cu-complexing capacity (100–300 nM) have been measured by DPASV during an extended bloom of the dinoflagellate *Gymnodinium sanguineum* (100). This value increases linearly with cell density. The conditional stability constant found decreases from log $\bar{K}_{cond}$ = 7.8 to 6.6 with the increase in the total ligand concentration, the changes in $\bar{K}_{cond}$ pointing to the heterogeneity of the ligands involved. Trace metal speciation of dissolved species may thus be altered by exudates of some dinoflagellates as well as diatoms, and toxicity may be ameliorated.

Other applications of voltammetric methods to natural waters can be found in the determination *in situ* of labile trace metals such as Cd(II), Cu(II), Pb(II), and Zn(II), carried out directly in oxygen-saturated seawater by means of square wave stripping voltammetry with a mercury film electrode and without previous degassing with nitrogen. Indeed, square wave voltammetry can be used in the speciation of metal ions in the presence of oxygen, owing to the high speed of its scanning process. For this study a submersible probe with a flow-through cell was used with an MFE or HMDE. Owing to the design of the cell, there is a well-controlled hydrodynamic flux around the drop with small perturbation of the voltammetric signal (111).

Mercury microelectrodes have been applied to the determination of heavy metals in seawater and natural waters of low conductivity, using DPASV (164), and *in situ* in the presence of dissolved oxygen, when coupled with square wave voltammetry (165).

ASV and DPP have also been adapted to an on-line analytical system to automatically measure trace metals (e.g., Cd, Cr, Cu, Ni, Pb, and Zn) in a wastewater stream for several days (166).

2.4.4. Species-Specific Techniques

2.4.4.1. *Potentiometry with Ion-Specific Electrodes*

In this technique the difference of potential between the working and the reference electrodes, immersed in the sample solution, is measured under conditions of zero current ($i \leqslant 10^{-12}$ A), i.e., the species concentration remains constant everywhere in solution. The potential of the working electrode (E_W) is directly related to the logarithm of the free ion concentration in solution, according to the following equation:

$$E_W = Q \pm S \ln([M^{n\pm}]) \qquad (6)$$

where Q is a constant term if the ionic strength is kept constant with an appropriate electrolyte and if during the measurements the potential of asymmetry is constant within the experimental errors; S is the Nernstian slope [$S = RT/(nF)$, where R, T, and F are respectively the gas constant, the absolute temperature, and the Faraday constant]; and [$M^{n\pm}$] is the concentration of the free ion.

Free ions can be directly determined from Equation 6 if the Q and S parameters have been previously determined from a calibration curve, this calibration requiring a medium without complexants but representative of the sample (with the same pH, ionic strength, and temperature), since the Q value is much influenced by the matrix of the sample.

Ion-selective electrodes (ISEs), with a membrane that responds to the ion to be measured, enable determinations to be performed of free metal ions for concentrations generally higher than 10^{-6} M (167, 168). Depending on the type of the membrane, ISEs can be classified as detailed hereafter (Table 2.7).

a. Solid crystalline membranes. These are membranes with "holes" in the crystalline net, of appropriate dimensions and charge relative to a specific ion, with a reference electrode (Ag/AgCl) immersed in the internal solution.

Table 2.7. Ion-Selective Electrodes (ISEs) with Various Membranes[a]

Type of Membrane	Ion Measured	Concentration Range (M)	Interferences
Glass: depending on the composition of the glass, different ion exchanges can occur	H^+	1×10^{-14}–1.0	Na^+, Ag^+
	Na^+	1×10^{-8}–10	$H^+, K^+, Li^+, Ag^+, Cs^+, Tl^+$
Crystalline Solid: "Permeable" to specific ions	Cu^{2+}	1×10^{-8}–0.1	$Ag^+, Hg^{2+}, Br^-, Fe^{2+}$
	Pb^{2+}	1×10^{-6}–0.1	Ag^+, Cu^{2+}, Hg^{2+}
	Cd^{2+}	1×10^{-7}–0.1	Ag^+, Cu^{2+}, Hg^{2+}
	F^-	1×10^{-6}–sat.	OH^-
	Cl^-	5×10^{-5}–1.0	$OH^-, Br^-, I^-, S^{2-}, CN^-$
	Br^-	5×10^{-6}–1.0	I^-, S^{2-}, CN^-
	I^-	5×10^{-8}–1.0	$NH_3, S^{2-}, CN^-, S_2O_3^{2-}$
	S^{2-}/Ag^+	1×10^{-7}–1.0	Hg^{2+}
	SCN^-	5×10^{-6}–1.0	$Br^-, I^-, Cl^-, S^{2-}, CN^-, OH^-, S_2O_3^{2-}, NH_3$
	CN^-	8×10^{-6}–0.01	S^{2-}, Br^-, I^-, Cl^-
Liquid: with a hydrophobic ligand that can complex the M ion	Ca^{2+}	5×10^{-7}–1.0	$Zn^{2+}, Fe^{2+}, Cu^{2+}, Pb^{2+}, Hg^{2+}, Sr^{2+}, Ni^{2+}, Ba^{2+}, Mg^{2+}, H^+, Na^+, K^+, Li^+, NH_4^+$
	K^+	1×10^{-6}–1.0	$Na^+, NH_4^+, H^+, Ag^+, Li^+, Tl^+, Cs^+$
	NO_3^-	7×10^{-6}–1.0	$Br^-, F^-, I^-, Cl^-, OAc^-, ClO_4^-, HS^-, CO_3^{2-}, HCO_3^-, PO_4^{3-}, HPO_4^{2-}, NO_2^-, SO_4^{2-}, CN^-$
	BF_4^-	7×10^{-6}–1.0	$NO_3^-, SO_4^{2-}, F^-, Cl^-, I^-, OAc^-, HCO_3^-, OH^-$
	ClO_4^-	7×10^{-6}–1.0	$Br^-, F^-, I^-, Cl^-, OAc^-, CN^-, ClO_4^-, ClO_3^-, CO_3^{2-}, HCO_3^-, HPO_4^{2-}, PO_4^{3-}, H_2PO_4^-, NO_3^-, NO_2^-, SO_4^{2-}$

Source: *Ion-Selective Catalog and Guide to Ion Analysis* (168).

[a] ISEs from different commercial sources may display different interferences and concentration ranges.

b. Glass membranes. These are membranes at whose surface an ion exchange can occur (in all other aspects, this electrode is similar to the previous one).

c. Liquid membranes. Here an internal solution with a polymeric hydrophobic ligand specific for the metal ion under study is separated from the external solution by a porous support impregnated with this solution; a reference electrode is immersed in the internal solution.

Major elements essential to life (e.g., Ca, Cl, F, K, Mg, and Na) can be determined using ISEs, their concentration in most natural waters lying in the concentration range of these electrodes (13). Some applications of ISEs to aquatic media have been treated exhaustively in the literature (93, 169–171).

It is important to mention that the use of electrodes with selective membranes is not necessarily limited to the analysis of inorganic ions but has been extended to the determination of biochemical compounds and dissolved gases (CO_2, NH_3, SO_2, NO/NO_2, and H_2S), these applications being beyond the scope of the present chapter (93, 172–174). Critical reviews of the more significant contributions to the understanding and use of ISEs have also been published (175, 176).

Another electrode type, the amalgam electrode, has been developed mainly for determinations of metal ions for which membrane electrodes cannot be used (e.g., Bi, Cd, Pb, Sn, and Zn), based on their characteristic of forming amalgam with Hg. The concentration of the metal in the amalgam can be measured directly by chronopotentiometric oxidation or by AAS after dissolution in nitric acid. Until now such electrodes have been applied mainly to concentrated solutions in the presence or absence of synthetic ligands, only a few applications having been tried in aquatic media (13).

Some limitations should be taken into account when the determination of free metal ions is being made in natural samples by the potentiometric method:

a. Selectivity. Interfering ions can be an important source of errors, generally leading to an excessive ion concentration.

b. Adsorption. Adsorption of organic matter on the membrane of the ISE modifies the potential of asymmetry and may interact with the electrode mechanism, leading to variations in parameter Q during the measurement (177, 178); the presence of fulvic acids with concentrations lower than 100 mg L^{-1} seems to have little effect on ISE membranes (179).

c. Limit of detection. The lower concentration that can be determined by ISEs is caused by the dissolution of the membrane and various aforementioned adsorption properties.

It should be emphasized that the detection limits of ISEs (Table 2.7) are too high for the determination of trace element concentrations in a nonpolluted water ($C_M < 10^{-8}$ M). On the other hand, in polluted zones, potentiometry can be used only if organic matter is not significantly adsorbed on the electrode membrane.

Owing to these limitations ISEs are mainly used for the determination of the major ions present in natural samples, which are in general of primary importance for life, or for the determination of minor elements, after separation and preconcentration.

2.4.4.2. Gas Chromatography and/or Hydride Generation

A chromatographic technique coupled with a sensitive element detector system provides a powerful and versatile tool to perform trace element speciation in natural waters, since concentrations of specific element complexes are determined separately. Since AAS detectors have high detection power, efforts have been made to directly couple such detectors with gas chromatography (GC) (180, 181). Other types of detector that can be used are electron capture, flame photometry, flame ionization, mass spectrometry, and plasma emission spectroscopy detectors (182, 183).

Hydride generation has been used in speciation studies, sometimes coupled with GC, sometimes not, as an initial extraction step. This procedure is frequently found in the determination of volatile species such as Hg and Pb alkyl compounds, or in the determination of organic complexes that can form volatile hydrides, such as alkyl complexes of As, Pb, Se, and Sn. These metals may also be reduced directly to hydride metals, depending on their oxidation state, pH, and ligands in solution, and so selective reduction steps can be used in their speciation (31, 184–194).

In hydride generation the elements or organic complexes are reduced with $NaBH_4$. Applications of this technique to environmental analysis have been reviewed by Braman (195). The volatile species formed are removed from solution with the flow of an inert gas. If a hydride is formed, the element is directly determined by AAS. In the presence of mixtures of metallorganic hydrides a separation method is frequently applied using GC.

If the organic compound is volatile (e.g., alkyl compounds of Pb), a previous separation can be done, the species being removed from solution with an inert gas (184, 196, 197). The resulting flow then enters a trap cooled with liquid nitrogen, where the reduction with $NaBH_4$ is performed.

Alkyl Sn compounds, used in paints as stabilizers, biocides, and antifouling agents and in bactericides, can produce serious problems through their impact on biota (notably in certain shellfish and mollusks), although it is worth noting that environmental methylation can also contribute to the existence of this type of compound in natural waters.

A nonconventional GC was developed to perform speciation studies of alkyl compounds, either coupled or not with hydride generation, where the liquid sample is trapped in a chromosorb or silanized chromatographic quality glass wool column. Subsequently the trap is heated and the separation of elemental species is achieved mainly due to the difference in their boiling points the elements being finally determined by a specific detector coupled with GC.

Some applications of these techniques to natural waters are as follows. Dissolved As and Sb in natural waters exist mainly in the trivalent and pentavalent oxidation states, the atomization signal for both elements with the hydride technique depending on their oxidation rate and pH values. Simultaneous determination of As(III) and Sb(III) can be done by hydride generation at a pH of about 7 without interference from As(V) and Sb(V). The same holds for As(V) and Sb(V) at a pH of about 1 (184, 191). Selective hydride generation coupled to liquid-nitrogen-cooled trapping and a GC–photoionization detector has detection limits of 10 pM for As and 3.3 pM for Sb (156). Low concentrations of Se can interfere with the determination of As. This interference can be eliminated by preventing Se hydride from entering the atomizer, e.g., by adding a sufficient quantity of Cu to the solution under analysis to prevent the formation of Se hydride.

Selenium(IV) can be determined by hydride generation, in contrast to Se(VI), which cannot be reduced by $NaBH_4$ (92). Thus, Se(VI) must be reduced to Se(IV) prior to its determination. Sodium iodide can be used as a reducing agent in very low amounts; otherwise the reduction proceeds to elemental Se. For analysis of surface water, Se(IV) is first determined in a slightly acidified sample and then total Se is quantified after reduction with hydrochloric acid. By means of the hydride technique detection, even better limits than those directly achieved by graphite furnace AAS can be obtained. Chau et al. determined volatile dimethylselenide and dimethyldiselenide produced by biological activity in the atmospheres of seawater sedimentary systems by a combination of GC and AAS (198). Holen et al. deposited Se electrolytically onto a Pt wire and volatilized the element in an argon/hydrogen flame (199).

Van Loon reported some applications of GC coupled with AAS, where volatile compounds of Pb, Hg, and As have been determined (200). In some instances a conventional GC separation has been utilized with or without an initial extraction step, e.g., in the analysis of alkylated Pb complexes in water, sediment, and fish samples (201–204). Dimethyl-, diethyl-, and dibutyl-Hg have also been analyzed in fish and sediments by GC (205).

Arsenate, arsenite, mono-, di-, and trimethyl arsine, monomethylarsonic and dimethylarsinic acids (MMAA and DMAA), and trimethylarsine oxide have been analyzed in natural waters with detection limits of several nanograms per liter. The arsines are volatilized from the sample by gas stripping, and the other species are then selectively reduced to the corresponding arsines and volatilized. The arsines are collected in a cold trap cooled with liquid nitrogen and then separated by slow warming of the trap or by GC. Analyses have been done in seawater and freshwater (from rivers, lakes, and rain), and the formed species As(III), As(V), MMAA, and DMAA have been determined (193, 206–208). The uptake of arsenate from seawater, the biosynthesis of organoarsenic compounds, and the release of arsenite, MMAA, and DMAA compounds have been studied in pure cultures of marine phytoplankton species. All species released substantial amounts of MMAA and DMAA into the environment, the production of arsenate also being common, which can explain the occurrence of these As compounds in the aquatic environment (209). Arsenite can be produced directly from arsenate by many marine organisms, although it may also occur as an intermediate in the transformation of organoarsenic compounds back into arsenate. No evidence for reductive biomethylation has been observed in the marine environment, and no volatile organoarsenic compounds have been found in seawater. On the other hand, the biosynthesis of organoarsenic compounds by marine planktonic algae and subsequent reactions are quite common in the marine biochemistry of As. The concentration of arsenate in surface seawater approaches that of the nutrient phosphate in many regions of the oceans. Probably as a result of this, marine algae possess mechanisms to transform arsenate into less toxic forms such as large organoarsenic compounds (193). Arsenic speciation can also be done by using ionic exchange chromatography (210, 211).

2.4.4.3. *High-Performance Liquid Chromatography*

Advances in high-performance liquid chromatography (HPLC) coupled with an AAS detector have led to enhanced resolution and speed of analysis, as reviewed by Fernandez (212) and Horlick (213).

In HPLC coupled with AAS the eluent from the column is normally pumped directly into the flame, thereby allowing flow rates to be controlled by the HPLC program, the detection limits in the flame being solvent dependent.

Organo-Cr, tetramethyl-Pb, and tetraethyl-Pb as well as alkyl-Sn complexes have been determined by this technique (214–216). Organo-Sn compounds ($R_m Sn^{(4-m)+}$, with $m = 2$ or 3 and R = alkylaryl or alicyclic), which are of relevance in natural waters due to the painting of ships, have been speciated in trace quantities by graphite furnace AAS coupled with HPLC, employing a commercial bounded-phase strong cation-exchange (SCEX) column (217).

Higher sensitivity should in principle be possible with graphite furnace AAS, but generally noncontinuous chromatograms are obtained, it being necessary to stop the eluent flow for each analysis (218–220). Arsenic speciation in drinking waters has been investigated by HPLC coupled with on-line hydride generation, but the sensitivity is not as good as with gas–liquid chromatography (GLC) coupled with hydride techniques (216, 218, 219, 221).

2.4.4.4. *Other Chemical Methods*

Some other analytical methods can be used to determine specific forms, such as optical methods [spectrophotometry, electron paramagnetic resonance (EPR), or fluorescence]. Concentrations of certain trace metal complexes can be directly determined by spectrophotometric procedures, as has happened, for example, for Al and Fe (68). Copper(I) has been spectrophotometrically determined using neocuproine complexant, and it has been noted that it constitutes 5–10% of the total Cu at various locations of the upper marine water column in the Atlantic Ocean and Gulf of Mexico, although thermodynamically it should not exist in oxygenated waters (222). In marine systems Cu(I) may be produced by reducing agents such as some organic compounds in the presence of light or biological systems and stabilized by chloride (223). Copper has also been measured in fogwater by the bathocuproine method, and concentrations in the range 0.1–1 mM were found, representing anything between 4% and 95% of the total Cu that was measured by AAS (224). In fact, Cu(I), produced in the presence of various reductants such as organic compounds and radicals that are either involved or not in photochemical reactions, may be complexed by sulfite when the latter is present in excess in fogwaters, which prevents its rapid oxidation. In experiments using concentration ranges of Cu and sulfite close to those in fogwaters, it has been observed that the reduction of Cu(II) to Cu(I) by sulfite is pH dependent and occurs rapidly at pH > 6 (224). Measurements of Fe(II) concentration up to 200 mM in fogwaters (pH 3–7) have been colorimetrically performed with ferrozyne and total Fe determined by AAS (6, 7). The lability of natural organic–metal complex has been studied by spectrophotometry, adding a colorimetric ligand A to the natural sample and following the rate formation of the complex MA vs. time. For example, the kinetic spectrum of Cu dissociation in estuarine waters was studied using the colorimetric reagent 4-(2-pyridylazo)-resorcinol (225). To improve the detection power of this technique, rather poor for trace metal ion concentrations in natural waters, laser thermal lensing has been proposed (226).

Electron spin resonance (ESR) can give indications as to the structural characteristics of organic complexes. In fact, ESR spectra of humic matter in natural waters shows that there is complexation with Fe(III), Cu(II), and Mn(II) in most of the samples (227). It has been suggested that Fe^{3+} and Cu^{2+} are strongly bound and protected in inner-sphere complexes with ligand sites arranged in a square-planar (distorted octahedral) configuration around the central ion, it being probable that there exists more than one class of binding sites for Cu. Manganese(II) species have been determined quantitatively in seawater and freshwater by ESR. With this technique Mn(III) and Mn(IV) cannot be detected (228, 229).

Fluorescence spectroscopy has been applied to speciation studies of fulvic acids, since heavy metals produce a considerable decrease in fluorescence emission of these acids by accelerating the rate of intersystem crossing (230). In addition to this, paramagnetic ions accelerate nonradiative processes, also causing a reduction in fluorescence intensity (231). In fact, a close relationship has been found to exist between fluorescence quenching of fulvic acid in the presence of Cu^{2+} and the free metal concentration measured by ion-selective electrode potentiometry (232). The difference of intensity of fluorescence at 450 nm after excitation at 350 nm at the natural pH and after acidification can give an estimate of the heavy metals bound.

Flow injection analysis (FIA) has also been used in speciation studies, in particular in natural waters with various types of detectors (electrochemical, such as potentiometry with specific electrodes or voltammetry, and optical, such as UV/visible spectrophotometry, AAS, or ICP–AES). This type of procedure has been used, e.g., in groundwaters of polluted soils and for distinguishing Fe(II) from Fe(III) species (233–235).

2.4.5. Bioassays

2.4.5.1. Determination of Free Element Concentration and Toxicity

The frequent observation that metal ion concentration is toxic to aquatic biota and that a much more consistent pattern of toxicity can be described on the basis of free element concentration rather than total element concentrations, suggests the possible application of bioassay data to determine free metal ion concentration. This technique, which is employed especially when chemical speciation methods are not available, usually requires a previous calibration curve of toxicity vs. free element concentration in the absence of any complexant but under the same experimental conditions as those of the sample, in particular pH, temperatures, and hardness of water (236). However, Borgmann has carried out determinations in natural waters before and after addition of weak complexing agents such as amino acids without a previous calibration curve (237). He then has calculated free element concentration assuming that the difference in toxicity before and after addition of the ligand was entirely due to complexation with the added organic, the amino complex presenting no toxicity. This procedure is particularly useful for bioassays where unknown inert complexing agents cannot be removed.

Increased toxicity at higher pH values can be explained by the assumption that element surface complexation—and consequently uptake by organisms—increases with pH owing to less competition of hydrogen ions for complexing sites. If, on the other hand, toxicity increases at smaller pH, this may be because the ligands in solution have more affinity for the protons than the surface groups, more free element being released that can bind the groups at the surface of the cell. It should also be emphasized that speciation changes with pH and that overly acidic or alkaline solutions can damage the cell membrane. Water hardness (i.e., Ca and Mg) also frequently affects free element toxicity, and this can obscure the pH effect when toxicity is studied in waters where hardness increases simultaneously with pH.

Bioassay experiments are important because from them one can estimate the free element and simultaneously its toxicity to the organism being used, which can be expressed in terms of $^{96}LC_{50}$. However, all of these conclusions are reached based on the assumption that the free element is the most toxic species due to the formation of surface complexes at the cell membrane. This is not always the case, since some complexes are more toxic, in particular lipid-soluble species; moreover, it is known that filter feeders appear to accumulate larger amounts of some elements if they are in the particulate rather than in the ionic form.

Direct toxicity measurements in natural waters can be done using bioassays with algae, fish, or other animals, which in principle should provide the best speciation measurement. In this context unicellular algae should be a useful monitor in chemical speciation, since algae cultures are easy to maintain, are susceptible to low concentrations of elements, the effects being detected in a period of some days, and provide sufficient statistical data for analysis since in a typical experiment 10^3–10^6 cells are available per flask (238).

It should be noted, however, that bioassays are slow, have poor precision, provide toxicity results only on one biological species demonstrating only growth effects, and have a narrow detection window outside which the growth inhibition is either too severe or not detectable. Also, laboratory results cannot be directly extrapolated to the field owing to the existence of many involved variables.

Chemical speciation schemes (Section 2.4.6), on the other hand, are more amenable to standardization and can yield information on the various chemical forms in water (71).

2.4.5.2. *Comparison Between Biochemical and Chemical Processes*

The uptake of an element by living organisms involves the following steps: (i) transport of the free and/or complexed element to the biological cell due to concentration gradients; (ii) dissociation of the complex at the surface of the cell; and (iii) uptake of the element by the cell.

This mechanism can be compared, in an approximate way, to voltammetric or potentiometric processes, where the electrode replaces the biological cell and redox reactions replaces the assimilation. Whitfield and Turner concluded that the determination of the available fraction by ASV is similar to the biological uptake if the first two steps control the global mechanism, i.e., if transport and/or dissociation of the complexes are much slower than assimilation (redox reaction in electrochemical method) (239, 240). In this situation electrochemical availability is similar to bioavailability. On the other hand, if uptake (redox reaction in electrochemical method) is the rate-determining step of the global mechanism, biological and voltammetric results cannot be compared. In that case, since there is always equilibrium at the surface of the cell, potentiometry with specific electrodes would be more representative of the biological cell if the method has adequate detection power. Another technique that can simulate what happens biologically if the uptake is controlled by biochemical reactions inside the cell, rather than by transport in the diffusion layer, is the batch resin method, which has a sufficiently high time scale (on the order of hours).

In those cases where the global process is partially controlled by transport in the diffusion layer, the element-uptake behavior of an organism can be chemically simulated by voltammetric techniques with applied potentials that are not in the limiting plateau, i.e., without complete depletion of the element at the surface.

Problems of organic adsorption and irreversibility at the Hg interface may occur when natural samples are being studied; if so, the electrode mechanism would not be similar to the uptake by the biological cell, even in the absence of lipid-soluble species. It can also happen that organisms change the chemistry of the reactional zones, segregating complexing agents that can bind elements vital to them (e.g., siderophores) or releasing complexing agents that can decrease the toxicity of element contamination, probably being involved in mechanisms of detoxification (3, 4).

2.4.6. Comprehensive Speciation Schemes

Various combinations of the direct methods discussed above have been proposed for several speciation schemes. Batley and Florence separated several groups of species of Cd, Pb, and Cu in seawater in terms of degree of lability, applying ASV to UV-irradiated samples, to untreated samples, and to UV-irradiated and untreated samples passed through a chelating resin column (241). The resin used first was Chelex-100, but later Florence used a thiol resin, in principle more suitable for determining reactive elements in a water sample (71). Conventional anion-exchange resins have also been used in such speciation studies (242). Laxen and Harrison combined ultrafiltration with ASV and Chelex-100 measurements to develop a detailed scheme mainly based on size fractionation of the element species (39). The same authors proposed a speciation scheme using filters and ultrafilters with porosity from 12 μm to 1.8 nm and analyzed the various fractions by AAS and ASV with and without previous UV irradiation. They also determined the labile fraction in terms of Chelex-100 resin in column (243). Further physicochemical speciation developments include the use of other resins to complement Chelex labile measurements.

Nielsen and Lund proposed a speciation scheme combining filtration and acidification (244). From 0.45 μm pore filtration followed by acidification at $pH = 1$, they determined what they called the dissolved labile element (a). By filtration of the initial sample and strong digestion with concentrated HNO_3, they determined the total concentration of the dissolved element (b), the fraction ($b - a$) thus being the dissolved nonlabile element. The several water samples analyzed showed significant amounts of dissolved nonlabile element, indicating that acidification at $pH = 1$ does not release all the element present in the dissolved fraction. On the other hand, from acidification of the initial sample at $pH = 1$ followed by filtration, the authors could estimate fraction c, the value ($c - a$) being the particulate acid-labile element, i.e., elements weakly bound to large particles. Finally, after a strong digestion without filtration, the total concentration of element d in the nonfiltered sample was determined, the value $[(d - b) - (c - a)]$ representing the elements that are retained by the filter and not released by acidification at $pH = 1$, i.e., strongly bound to larger particles.

Hart and Davies used a scheme based on a 0.40 μm membrane filter, the dissolved fraction being analyzed in terms of dialysis, either combined or not with ion exchange and a Chelex batch procedure (242, 245). They developed a batch technique with equilibration times of 16–168 h to overcome possible effects of slow kinetics of cation chelation with the groups of the resin. The advantage of this method lies in the fact that its applicability is not limited to those elements that can be determined by ASV.

Figura and McDuffie, using ASV and Ca-Chelex resin in column and in batch techniques, developed a speciation method according to the kinetics of the species, where the order of magnitude of the dissociation rate constants of the complexes could also be estimated (18). The authors first performed a 0.40 μm filtration of the sample, inert complexes virtually corresponding to all particulate forms retained on the filter (246). After this step the filtrate was divided into two parts with the same volume: one part was analyzed by ASV at $pH = 4.0$ with acetate (in order to be well buffered); the other was passed through a Ca-Chelex column with a flux of 2.7 mL min^{-1}. The heavy metals retained in the column were eluted with HNO_3 partially neutralized with NH_4OH, then analyzed by ASV. The species not retained by the column were stirred in a Ca-Chelex batch for 3 days, then filtered and analyzed by ASV after digestion with a mixture of $HNO_3 + HClO_4$. The species retained by the batch technique were eluted and analyzed following the same procedure as used for the fraction retained by the column, and the total concentration of this fraction was determined after digestion with $HNO_3 + HClO_4$. Using this methodology the authors were able to establish a speciation scheme for Cu, Pb, Cd, and Zn in samples of rivers, estuaries, and wastes. The time scale of the techniques in the experimental conditions used was 2 ms for ASV, 7 s for the column resin technique, and 3 days for the batch technique. With such a wide range of time scales it was possible to divide the species into groups according to marked differentiation of dissociation rate constants, which is very significant in terms of bioavailability. It was noted by the authors that Cd and Zn generally predominate in forms comparatively more labile than Cu and Pb, which seems reasonable if one considers only the thermodynamic aspects of the chemistry of these elements and the type of organic groups commonly found in natural waters.

Another scheme is based on seawater samples reacting continuously with 10 g of Chelex-100 resin in a batch technique. At selected time intervals the stirring is stopped to allow the resin to settle and Zn to be analyzed by AAS (246, 247). The labile element corresponds to the amount removed during the initial 0.5 h of the experiment and consists mostly of inorganic and organic complexes in solution. Moderately labile Zn might denote a first-order dissociating organic complex present in quite variable concentrations and with a characteristic $k_d = 0.36 \pm 0.08\,h^{-1}$. Inert Zn complexes virtually correspond to all particulate forms of Zn retained on the 0.40 μm filter.

Boussemart *et al.* have used DPASV and fluorescence quenching to measure the complexing capacities and conditional stability constants of complexes in natural waters. In this approach the measured electrochemical signal reflects the presence of labile Cu in solution, while spectrofluorimetry yields an estimate of the uncomplexed fluorescent ligands. Speciation in solution is calculated using these values (248).

Stiff proposed a speciation scheme for the determination of Cu in polluted freshwaters, where this cation can be associated with suspended solids and/or give rise to different soluble chemical forms (249). The most likely soluble species are free Cu^{2+} as well as Cu complexed with carbonate, cyanide, humic matter, amino acids, and polypeptides. Analytical results showed that much of the Cu present in river waters was associated with suspended solids and that soluble Cu consisted almost entirely of complexed forms. To separate suspended solids from soluble species, filtration under pressure through 0.45 μm pores was performed, the soluble fractions being analyzed as follows. (i) Total soluble Cu—determination by spectrophotometry at 454 nm with neocuproine, after nitric acid digestion. (ii) Copper "cyanide complexes"—calculation after subtraction from total soluble Cu of the fraction determined spectrophotometrically at 433 nm with 3-propyl-5-hydroxy-5-D-arabinotetrahydroxybutyl-3-thiazilidine-2-thione (PHTTT). The value experimentally obtained does not include the fraction of Cu complexed with cyanide or with other complexants with similar stability constants ($\log \beta \approx 22$), since PHTTT is not competitive enough for Cu as compared with these ligands. (iii) Copper–humic complexes—extraction with ethanol and determination of Cu associated with it by the neocuproine method. (iv) Free Cu and carbonate complex—determination of free Cu by ISEs, with a detection limit of 10^{-6} M. Copper carbonate was evaluated from Cu^{2+} and CO_3^{2-} free concentrations, knowing the pH and the ionic strength of the sample as well as the stability constant of the complex formation. (v) Copper complexed by amino acids and polypeptides—this fraction is estimated by subtracting the free and complexed forms determined as described above from the total soluble Cu.

Speciation schemes for elements that can have more than one oxidation number in natural waters—for example, As, Cr, Fe, Mn, and Sb—are also illustrated.

Under anoxic conditions Fe and Mn are present in solution mainly in the soluble forms Fe(II) and Mn(II), while under oxygenated conditions they are present as Fe(III) and Mn(IV) hydrolyzed forms (particulate matter), although in special cases some Fe(III) (in general less than 1%) can be stabilized in soluble form by natural complexing agents. Manganese (II) and Fe(II) can also be present in oxygenated waters at low levels ($\leqslant 1 \mu M$) due either to anthropogenic input or to photochemical reduction of higher oxidation states. Manganese(II) and Fe(II) in anoxic conditions are ideally determined by voltammetric methods because there is no need to remove unwanted oxygen signals. Under favorable conditions measurements may be performed directly on untreated waters, yielding information about speciation. Since these elements are abundant in nature they can usually be measured by direct polarographic analyses (e.g., DPP) instead of stripping techniques. In fact, Mn(II) can exist in anoxic waters at concentrations of 0.1 mM, and Fe(II) can exist in such waters at higher levels. The hydrolyzed forms of Fe(III) and Mn(IV) in particulate form generally do not interfere with such determinations (250, 251). Polarographic measurements of Fe(II) and Mn(II) were performed either directly in the field or in the laboratory after sampling and storage for comparison. In field analyses lake water samples are directly pumped into an especially designed polarographic cell through which a continuous flow of N_2 is maintained during the filling. Electrochemical reduction of Mn(II) (at $E \approx -1.5$ V vs. SCE) can suffer from the interference of Cr(III) in anoxic waters since it is reduced at a similar potential (252, 253). In this case the composition of the medium must be modified to resolve the signals (254). Iron(II) is usually determined by its irreversible reduction to the metal form by DPP at -1.4 V vs. SCE in natural waters (251–253). In principle it is also possible to measure Fe(II) by its oxidation to Fe(III), but the potential for this reaction very much depends on pH and the presence of complexing agents. Furthermore, in untreated natural waters the reaction occurs at potentials that are too positive to be measured by a mercury electrode. In the presence of S($-$II) and Fe(II) a peak may be observed at $E \approx -1.0$ V vs. SCE that is due to the reduction to Fe(0) of FeS adsorbed on the electrode. Under oxygenated conditions ASV has been used to determine hydrolyzed products of Fe(III) directly on a mercury/graphite electrode or on the HMDE (255, 256). Dissolved Fe(III) often reveals a small differential pulse (DP) signal at -0.35 to -0.40 V vs. SCE. Polarographic measurements of Fe(III) at $E \approx -0.20$ V vs. SCE have also been performed on natural waters with high concentrations of humic substances (257). Manganese oxides (MnOx) in unfiltered samples can be determined using three different approaches, as follows (44). (i) The concentration of particulate Mn(IV) is computed by subtracting the polarographically measured concentration from the total concentration determined by AAS; this value is correct if there are no particulate forms of Mn(II) and all the soluble Mn(II) species are polarographically labile with the same diffusion coefficient. (ii) Total Mn is polarographically measured after reduction of MnOx by sodium bisulfite at pH = 3; manganese(II) is also polarographically determined *in situ*, and the concentration of MnOx is computed by subtraction. (iii) Manganese(IV) is measured directly by colorimetry (258).

When the electrochemical properties of Fe(II) were used and this technique combined with spectrophotometric methods and filtration, a scheme to differentiate among the several forms of Fe and S as a function of their size, their oxidation states, and the nature of the species formed was implemented in a lake where seasonal changes occur (259). The total concentrations of Fe(II) and of Fe(II) + Fe(III) were obtained after digestion of the sample (without filtration) by hydrochloric acid, and then colorimetrically determined with *o*-phenanthroline and by AAS, respectively. The concentrations of dissolved Fe(III) and total Fe were determined by the same methods after filtration through a 0.45 μm filter. The concentrations of Fe(II) and S($-$II) labile species were determined by Davison and colleagues directly without filtration by using DPP (252, 253, 260). The nonlabile dissolved species were assumed to be in a colloidal form; the difference between total and dissolved species gives the particulate fraction. Iron and Mn transformation at the O_2/S($-$II) transition layer in a eutrophic lake was also analyzed by DeVitre et al. using voltammetric and spectrophotometric methods (44).

As regards inorganic As(III), As(V), Sb(III), and Sb(V), these species can be determined by AAS with hydride generation. At neutral pH, in fact, As(III) and

Sb(III) can be determined by hydride generation alone since they are selectively reduced (184, 191, 261). In order to determine the concentration of As(V) and Sb(V) at neutral pH, a pre-reduction should be performed before determination of the lower oxidation state, KI in hydrochloric acid solution being the most suitable reductant.

Determination of dissolved Cr(VI) and Cr(III) is carried out by passing the sample through anion and cation-exchange columns, since the first species exists as an anion and the second as a cation (262). The total metal content is then determined by graphite furnace AAS. Speciation of Cr in seawater can also be done after preconcentration of Cr(III) with hydrated Fe(III) oxide and of the chromate with Tris–pyrrolidine thiocarbamate Co(III), the particulate matter being subsequently analyzed by XFS (263). It has been observed that Cr(III) formed under anoxic conditions is strongly bound to particles and thus retained in the sediment (264).

2.5. FUNDAMENTALS OF VOLTAMMETRIC METHODS

2.5.1. General Comments

In these techniques a scan of potential vs. time is applied to the working electrode connected to a reference electrode and to an auxiliary interface, all of them immersed in the sample solution. A low current between the working and auxiliary electrodes is produced due to the small dimensions of the former (89, 90). The interface more commonly used in voltammetric techniques is the DME, owing to its renewal capability, its excellent reproducibility, and the wide range of negative potentials that can be scanned in aqueous solution due to the high H^+ reduction overpotential on Hg. Two factors contribute to the total current: (i) *capacitive current*, due to the charge of the double layer at the electrode interface; and (ii) *faradic current*, due to redox reactions that can occur at the electrode surface. The decrease of the species concentration near to the electrode creates a gradient of concentration between the surface and the bulk solution, which causes the diffusion of species toward the electrode. If the diffusion process is solely responsible for the limiting current, observed when the solution concentration at the interface, C_0, is null, the diffusion current (i_d) can be derived from Fick's laws, giving the well-known Cottrell expression:

$$i_d = QD^{1/2}C \qquad (7)$$

where D and C are, respectively, the diffusion coefficient and the concentration of the redox species in the bulk solution; and Q is equal to $60.7 \, n \cdot m^{2/3} \cdot t_d^{1/6}$ for the DME with a linear variation of potential vs. time (dc polarography). In the latter expression n is the number of electrons involved in the redox process, m is the flow rate of Hg, and t_d is the drop time; the constant 60.7 implies the following units: i_d (A), m (g s^{-1}), t_d (s), D (cm^2 s^{-1}), and C (M).

When a pulse mode is used, the current i_d for NPP or i_p for DPP (Table 2.5) is directly related to concentration by an expression similar to Equation 7. In these analytical methods Q has a different expression, but it is still dependent on the geometry of the electrode, on the number of electrons involved in the redox process, and on the time scale of the technique.

From Table 2.5 it can also be seen that (i) depending on the pulse mode, different detection limits are obtained, always lower than in dc polarography; (ii) there is another characteristic parameter of the signal measured, the half-wave potential ($E_{1/2}$) for NPP or dc polarography, and the peak potential (E_p) for DPP. In this discussion i_d and $E_{1/2}$ for NPP and dc polarography or i_p and E_p for DPP will be represented in a more general way by i_m and E_m, respectively.

In order to enhance the detection power of these techniques a voltammetric method with preconcentration should be used, followed by a stripping step, where the signal i vs. E is measured. The detection limit can be still further decreased if electrodes with a higher area/volume ratio are employed, such as the MFE (Table 2.6). Depending on the type of the preconcentration, two different stripping techniques are possible:

a. ASV. A constant potential is applied to the electrode in the limiting current of the redox reaction of the metal, in order to reduce it to the amalgamated form (M(Hg)), during a constant time t_d and in a well-controlled stirred medium. This step is followed by anodic stripping, where the redissolution of the amalgamated species on the electrode occurs.

b. CSV. A ligand L' (e.g., dimethylglyoxime) is added to the sample solution and a potential value is applied to the electrode during a constant time t_d and in a well-controlled stirred medium, so that the complex formed in solution with L' is partly adsorbed on the electrode. At the applied potential the metal ion should be electrochemically inert. This step is followed by a cathodic stripping, where the metal adsorbed in the complex form is reduced. CSV can also be done without addition of auxiliary ligands, when the metal ion is in a form that can be directly adsorbed on the electrode. This is the case of, say, the insoluble hydroxides of Fe(III) and Mn(IV) electrochemically produced at the electrode surface and deposited on it in a stirred medium. During the stripping step these species are reduces to soluble forms.

Both preconcentration processes increase the metal concentration "into" or "onto" the electrode compared with the bulk solution, increasing in this way the sensitivity of the method. In fact, the peak height of the voltammetric signal in ASV (or in CSV) is proportional to the concentration of the reduced metal [or adsorbed complex (ML'_{ads})] directly related to the bulk concentration of the metal ion [or dissolved complex (ML')] in the absence of any other complexants. In stripping techniques an equation similar to Equation 7 is applied, where the Q parameter depends not only on the variables described before but also on the stirring of the medium, on the deposition (or adsorption) time, and on the area of the electrode, these experimental conditions being very well controlled during the preconcentration step. Equation 7 can be written in the general form:

$$i_m = QD^r C \qquad (8)$$

where the parameter r, equal to 1/2 for dc polarography, NPP or DPP, in stripping techniques has an experimental value between 2/3 and 1/2, owing to the dependence of the diffusion layer thickness on the stirring.

Three different situations will be outlined in the following subsections: (i) there is no ligand present in solution; (ii) L is a simple ligand, with a diffusion coefficient equal to or lower than the aquo-metal ion; (iii) L is a macromolecule with different functional groups, and a diffusion coefficient lower than the aquo-metal ion.

2.5.2. Speciation in Noncomplexing Media

In a noncomplexing medium the $E_{1/2}$ value in NPP or dc polarography, or the E_p value in DPP or stripping techniques, have, for each technique, a constant characteristic value for the redox reaction of the metal ion regardless of its concentration but dependent on the medium. The element concentration can be directly determined from the maximum intensity of the current, after a previous calibration to determine the slope (QD^r) of Equation 8.

2.5.3. Speciation in Complexing Media

Let us consider the basic scheme of the global chemical and electrochemical reactions that involve the electroactive metal ion M and the ligand L:

$$M + L \underset{k_d}{\overset{k_f}{\rightleftarrows}} ML$$
$$+ne \updownarrow$$
$$(MHg)$$

where ML stands for the complex; (MHg) for the amalgam of the reduced metal ion; and k_f and k_d for the formation and dissociation rate constants of ML ($K = k_f/k_d$, where K is the thermodynamic stability constant of ML).

The species complexed with metal ions are involved in the following processes: (i) diffusion of M and ML to the electrode surface; (ii) dissociation of ML [the characteristic lifetime of the complex ($1/k_d$ for a first order or pseudo-first-order reaction) can be compared with the time scale of the electrochemical technique (t_m), the complex being labile if $k_d(t_m/k_f[L])^{1/2} \gg 1$ for $D_{ML} \approx D_M$, i.e., if the complex dissociation in the diffusion layer is very fast compared to diffusion (265); the relative importance of the complexation kinetics depends on the value of t_m, which is different for each technique and can change for the same analytical method (t_m is the drop time in dc polarography, the pulse duration in pulse polarography, and the stirring rate in stripping techniques)]; (iii) charge transfer of M, and of ML if the complex is inert but electroactive during the time scale of the technique (the system will be reversible if the charge transfer is very fast compared to diffusion and/or complex dissociation). Here it is always assumed that the global process is controlled by diffusion and/or chemical reactions.

2.5.3.1. Inert Complexes

2.5.3.1.1. Speciation of M and ML. As referred to earlier, inert complexes are those for which there is no time for dissociation in the diffusion layer during the time scale of the technique, i.e., M and ML act as individual species. In this case two peaks (or waves) may be observed in polarographic measurements, corresponding respectively to the reduction of the free metal ion and to the direct reduction of the complex, if it is electroactive. The measured currents i_m^M and i_m^{ML} are directly related to M and ML concentrations, respectively, and the potentials E_m^M and E_m^{ML} have constant values regardless of the concentration of M and/or ML (Fig. 2.1):

$$i_m^M = QD_M^r[M] \tag{9}$$

$$i_m^{ML} = QD_{ML}^r[ML] \tag{10}$$

where D_M and D_{ML} stand for the diffusion coefficients of the free metal ion and of the ML complex, respectively, and r can be substituted by its experimental value.

In most systems ML is not an electroactive species in the potential range of voltammetry, since E_m^{ML} is generally more negative than the discharge potential of H^+, and so i_m^{ML} and E_m^{ML} cannot be measured. Equation 9 is only valid in ASV if the potential applied during the deposition step is sufficiently

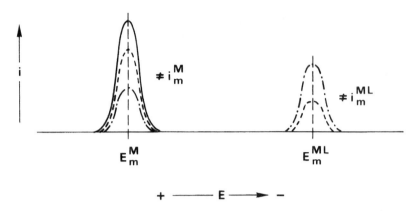

Figure 2.1. DP polarograms for an inert electroactive complex. Different ligand concentrations are used: $C_L = 0$ (—); C_{L_1} (---) $< C_{L_2}$ (-·-). Two peaks are observed, respectively, at E_m^M and E_m^{ML}; i_m^M and i_m^{ML} are directly related, respectively, to [M] and [ML]. If $D_{ML} = D_M$, then $(i_m^M)_{C_L = 0} = (i_m^M + i_m^{ML})_{C_L \neq 0}$.

negative to reduce the metal ion but not so negative that the complex is also reduced.

The concentration of M can be found directly from Equation 9, after a previous calibration with a stock solution of metal ion so that (QD_M^r) can be determined. The concentration of ML is obtained from $(C_M - [M])$, where C_M represents the total concentration of metal. If the complex is electroactive, ML concentration can also be determined from Equation 10 for direct redox measurements, QD_{ML}^r being calculated from the calibration curve for the complex in the presence of an excess of ligand, so that $[ML] \approx C_M$. From the ratio of the slopes of Equations 9 and 10 and the D_M value, found in appropriate tables (266), D_{ML} can be calculated.

An easy way to find out what types of complex are present in solution consists in acidifying the sample. For inert complexes the potential of the polarographic peak E_m^M should remain constant, but i_m^M should increase due to the larger dissociation of inert complexes in the bulk solution at lower pH values. If the peak current increases but the potential of the signal shifts toward more positive values with the decrease of pH, then the complex is labile or quasi-labile, as can be seen below. If i_m^M and E_m^M remain constant when the solution is acidified, then the metal ion is not complexed. The discussion of lability of the complex dissociation at the interface is valid in terms of potential for polarographic measurements, but not for stripping techniques; for ASV, the potential of the signal is related to the lability of the complex formation during the stripping process, and so no information about the lability of the system during the deposition step can be obtained from the potential shift.

2.5.3.1.2. Complexing Titration Curves. Complexing titration curves, obtained by addition of successive aliquots of a metal stock solution to the sample, give the *complexing capacity* (C_c) of the medium, defined as the ligand sites available for the metal ion (Fig. 2.2). For inert complexes, $i_m^M = 0$ if C_M is sufficiently low so that all the metal is complexed. For C_M values so large that all ligands are in the complexed form, $C_L = [ML]$ and therefore $[M] = C_M - C_L$. From Equation 9, we have

$$i_m^M = QD_M^r((C_M)_s + (C_M)_{ad} - C_L) \qquad (11)$$

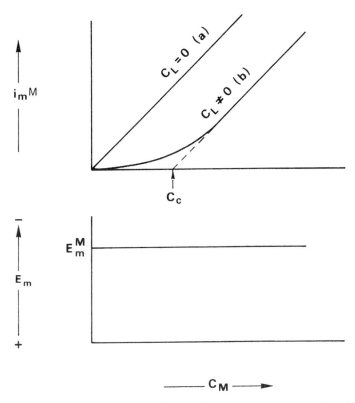

Figure 2.2. Titration curves i_m vs. C_M and E_m vs. C_M for inert complexes. After saturation, curve (b) tends to become parallel to curve (a). Current i_m and potential E_m are measured at the potential E_m^M; C_M, total concentration of metal added; C_c, complexing capacity of the sample for metal M. Curves: (a) uncomplexing medium; (b) complexing medium.

where the indexes s and ad stand, respectively, for the sample and for the metal added. The relation i_m^M vs. $(C_M)_{ad}$ is a straight line with the same slope as that obtained in the absence of a ligand, and with an intercept at $i_m^M = 0$ from which the complexing capacity can be obtained. For intermediate C_M values the equilibrium $M + L \rightleftarrows ML$ is not shifted for the complex formation since there is neither excess of ligand nor of metal. It is important to note that the range where $i_m^M = 0$, i.e., $C_L = [ML]$, depends not only on the C_L/C_M ratio but also on the K_{ML} value; lower values of K_{ML}, for the same C_L, decrease the ranges where $C_L = [ML]$.

In Fig. 2.2 the inertness of the complex can be checked from the consistency of E_m values for various additions of C_M, followed by polarographic measurements.

2.5.3.2. Labile Complexes

Labile complexes are those where a thermodynamic equilibrium between M and ML is established during the measuring time of the technique even at the interface. Let us consider the more general case where $D_L \approx D_{ML} < D_M$. In this case a mean diffusion coefficient $\bar{D}$ should be defined, taking into account the diffusion of M and ML:

$$\bar{D} = \frac{[M]}{C_M} D_M + \frac{[ML]}{C_M} D_{ML} \tag{12}$$

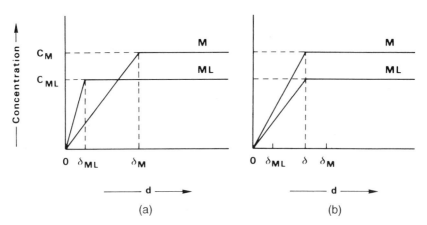

Figure 2.3. Concentration profiles vs. distance to the electrode surface under limiting conditions, if $D_{ML} < D_M$: (a) inert complex, where δ_M and δ_{ML} are, respectively, the thickness of the diffusion layer of M and ML; (b) labile complex, with a common diffusion layer δ for M and ML.

with $C_M = [M] + [ML]$. Equation 12 shows that M and ML do not act separately, as is true for inert complexes, but have a simultaneous diffusion with a mean diffusion coefficient $\bar{D}$, since they are in instantaneous equilibrium with each other (Fig. 2.3).

2.5.3.2.1. Speciation from i_m Measurements.
If $D_{ML} \approx D_M$, then, from Equation 12, we would have $\bar{D} \approx D_M$, and i_m would be similar in the presence or absence of the ligand, the value of this parameter not allowing for speciation studies.

On the other hand, if $D_{ML} < D_M$, Equation 8 should be written as (92)

$$i_m^{M+L} = Q\bar{D}^r C_M \tag{13}$$

The parameter $\bar{D}$ can be determined from Equation 13 if we replace i_m^{M+L} by its experimental value. Q and r should be determined from a previous calibration with a stock solution of metal ion, in the absence of ligand. From the $\bar{D}$ value and Equation 12, [M] and [ML] can be determined if the diffusion coefficients D_M and D_{ML} are known. The value of D_M can be found in appropriate tables (266), and that of D_{ML} can be obtained from experiments with excess of ligand, so that $\bar{D} = D_{ML}$ (Section 2.5.3.2.3). It should be emphasized that i_m^{M+L} is larger than the value obtained by the summation of the separate diffusion of M and ML, due to the contribution of the mutual interactions between M and ML.

In order to apply Equation 12 to the determination of [M] and [ML], $\bar{D}$ should be constant in the diffusion layer for the same C_M value, i.e., an excess of ligand should be present so that $[L] \approx C_L$ within an error of 10%. However, this excess should not be so large that $C_M \approx [ML]$, since in this case a large error in [M] would be obtained.

In natural waters, where organic and inorganic ligands are present, and in the absence of inorganic particles and/or colloids, M should be substituted by M' in Equation 12, where M' represents the inorganic fraction.

2.5.3.2.2. Speciation from ΔE Measurements.
Since M and ML act as one interconvertible species, the voltammogram i vs. E has only one wave (or peak) for labile complexes, with a potential (E_m^{M+L}) more negative than that observed in the absence of the ligand (E_m^M) (Fig. 2.4). The shift between E_m^M and E_m^{M+L} is due to the additional energy necessary for the dissociation of the complex, which is expressed mathematically by the DeFord–Hume expression (267):

$$\Delta E = E_m^M - E_m^{M+L} = RT/(nF) \ln((\bar{D}/D_M)^r (1 + K[L]^\circ)) \tag{14}$$

where the index "°" stands for the concentration at the electrode surface, and

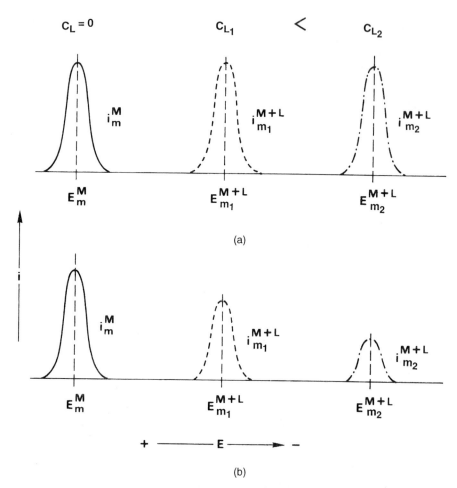

Figure 2.4. DP polarograms obtained for labile complexes. Different ligand concentrations are used: $C_L = 0$ (—); C_{L_1} (---) $< C_{L_2}$ (-·-). Only one peak is observed, and E_m^{M+L} is shifted to more negative values with the increase of C_L: (a) $D_{ML} = D_M \rightarrow i_m^{M+L}$ is a constant; (b) $D_{ML} < D_M \rightarrow i_m^{M+L}$ decreases with the increase of C_L.

$(1 + K[L]°)$ is, in terms of metal concentration, equal to $(C_M/[M])°$. This expression is usually resorted to in the presence of an excess of ligand $(C_L/C_M > 20$ for direct redox measurements, i.e., with one step involved) so that $[L]° \approx [L] \approx C_L$. In fact, in the absence of such excess the increase in L° concentration due to the additional dissociation of the complex at the interface during the reduction of the metal ion is not negligible, leading to a distortion of the voltammogram (268). In the stripping step of ASV, C_M° becomes larger than C_M, and so the condition of a large excess of ligand should be replaced by $C_L/C_M^\circ > 20$, so that $[L]^\circ \approx C_L$. For usual stirring rates and drop radius, it was found that $C_M^\circ \approx 15 C_M t_d$ (min) (269). The ratio C_L/C_M (or $C_L/C_M^\circ > 20$) can be lowered for weaker complexes.

From Equation 14 the concentration of M can be calculated in the presence of a large excess of ligand, $(1 + K[L]^\circ)$ being replaced by $C_M/[M]$ and $(\bar{D}/D_M)^r$ by its experimental value obtained from the current values (Equation 13). On the other hand, the stability constant K can be directly determined from this equation, if $[L]^\circ \approx C_L$.

For $D_{ML} \ll D_M$, a large error is associated with the determination of $C_M/[M]$ from potential values if $\Delta E - (RT/nF)\ln(\bar{D}/D_M)^r$ is within the experimental error; in this case speciation from i_m measurements should give more accurate values than from ΔE measurements. If $D_{ML} \approx D_M$, then $i_m^{M+L} \approx i_m^M$ and the error in [M] and [ML] will be much lower when determined from the DeFord–Hume equation.

If in solution there exist other metal ions besides M competing for the ligand L and other ligands besides L competing for M, and if the free concentrations of these ions and ligands are constant, then a conditional stability constant, K_{cond}, can be defined: $K_{cond} = [ML]/([M'][L'])$, where L' and M' represent, respectively, all L species not bound to M and M species not bound to L. Such is the case of a buffered natural water where the other metal ions and ligands are macroconstituents.

If different complexes of 1:1 type are formed with different ligands L_i, then $K[L]$ in Equation 14 should be replaced by $\Sigma K_i[L_i]$. In the general case different labile complexes with different coordination numbers j might be formed between M and L, and so $\Sigma K_{ij}[L_i]^j$ should replace $K[L]$, the parameter $\bar{D}$ in this equation taking into account the diffusion of all labile species present.

It should be noted that stability constants (conditional or not) can be determined either from potential or from current values whenever speciation can be obtained from these values and C_M and C_L are known parameters.

2.5.3.2.3. Complexing Titration Curves. If successive aliquots of a metal stock solution of known concentration are added to a labile complexing solution, i_m vs. C_M will be different in the presence and absence of the complexant only if $D_{ML} < D_M$ (Fig. 2.5). In the presence of an excess of ligand so that $C_M = [ML]$, Equation 13 will represent a linear plot i_m^{M+L} vs. C_M, if $[M]D_M \ll [ML]D_{ML}$:

$$i_m^{M+L} = Q D_{ML}^r C_M \qquad (15)$$

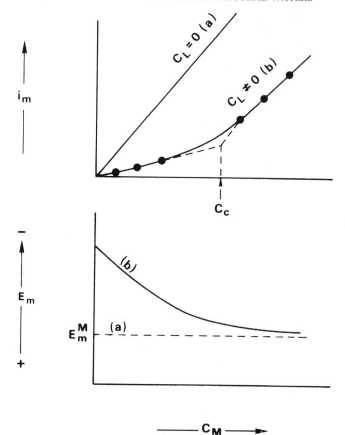

Figure 2.5. Titration curves i_m vs. C_M and E_m vs. C_M for labile complexes, if $D_{ML} < D_M$. If $D_{ML} = D_M$, then i_m vs. C_M will be the same for both curves (a) and (b). After saturation and for high C_M values the slope of (b) tends asymptotically to the slope of (a). C_M, total concentration of metal added; C_c, complexation capacity of the sample for metal M. Curves: (a) uncomplexing medium, with current i_m and potential E_m measured at E_m^M; (b) complexing medium, with current i_m and potential E_m measured at E_m^{M+L}.

from which D_{ML} can be determined. In the presence of a large excess of metal so that $C_L \approx [ML]$, Equations 12 and 13 will give

$$i_m^{M+L} = Q(D_M - (C_L/C_M)(D_M - D_{ML}))^r C_M \qquad (16)$$

where the slope of i_m^{M+L} tends asymptotically to the slope of i_m^M obtained in the absence of the ligand. Figure 2.5 presents a sudden change in the slope of the curve when saturation of the ligand is attained, the complexing capacity being determined from this point.

For the first additions of the metal ion ($C_M \ll C_L$), the value of i_m might be close to zero if $C_M \approx$ [ML] and $D_{ML} \to 0$. The same behavior ($i_m \approx 0$) is often observed for inert complexes if the complexes are sufficiently stable so that [M] ≈ 0.

In spite of the apparent similarity between the curves i_m vs. C_M in Figs. 2.2 and 2.5 they should not be mistaken for each other, and in order to distinguish them the plots of E_m vs. C_M for direct redox measurements should be analyzed. If the complexes are labile, E_m (more negative than E_m^M) has a constant value in the presence of an excess of ligand; in its absence E_m is shifted to more positive values with the increases of the metal content up to attainment of the potential of the metal in noncomplexing medium (curve a of Fig. 2.5). However, for inert complexes E_m always has a constant value (Fig. 2.2).

2.5.3.3. Quasi-labile Complexes

For quasi-labile complexes the limiting current is controlled by diffusional transport and by chemical kinetics, since during the timescale of the technique and under limiting conditions the ML complex is only partially dissociated at the electrode surface. The extent of dissociation depends on the dissociation rate constant of the complex and on the timescale of the technique, increasing with the increase of both factors (270–272).

In the presence of quasi-labile complexes the polarogram obtained with excess of ligand (Fig. 2.6) shows one peak due to the metal reduction derived from the partial dissociation of the complex at the electrode surface, with a potential value depending on the ligand concentration and on the diffusion

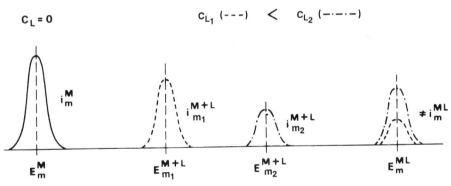

Figure 2.6. DP polarograms of a quasi-labile complex, for an electroactive complex ML. Different ligand concentrations are used: $C_L = 0$ (—); C_{L_1} (---) $< (C_{L_2}$ (-··-). Two peaks are observed respectively at E_m^{ML} and E_m^{M+L}. E_m^{ML} is constant, and E_m^{M+L} is shifted to more negative values with the increase of C_L; i_m^{M+L} decreases with the increase of C_L. If $D_{ML} = D_M$, then $(i_m^M)_{C_L=0} = (i_m^{M+L} + i_m^{ML})_{C_L \neq 0}$.

coefficients of M and ML. If ML is an electroactive species in the potential range of voltammetry, a second peak due to the direct reduction of the complex, with a constant potential value, is also obtained. The relative heights of the two peaks depend on the dissociated fraction of ML at the electrode surface and on the diffusion coefficient of M and ML.

The measured kinetic current, i_k, is given by (273)

$$\frac{i_k}{i'_m} = \frac{qD'\dfrac{k_d t_m^{1/2}}{(k_f[L])^{1/2}}}{1 + qD'\dfrac{k_d t_m^{1/2}}{(k_f[L])^{1/2}}} \qquad (17)$$

where i'_m is the hypothetical current controlled by diffusion ($i'_m = Q\bar{D}^r C_M$); D' depends on the diffusion coefficients ($D' = (\bar{D}/D_{ML})(D_M/D_{ML})^r$); t_m is the measuring time; q is $10^{-3}nFA/(Qt_m^{1/2})$, with n being the number of electrons, F the Faraday constant, and A the area of the electrode; and the other parameters have their usual meaning. The parameter q depends on the polarographic technique used, being close to 1 and 2, respectively, for dc polarography and NPP. For DPP it depends on the pulse amplitude (ΔE) and on the number of electrons involved in the redox process (n), being close to 10 for the typical value of $n\Delta E = 50\,\text{mV}$.

Equation 17 is only valid if [L] is constant in the diffusion and reactional layer, i.e., if there is an excess of ligand, so that $[L] \approx C_L$.

Two distinct situations can be outlined from Equation 17: (i) if the dissociation rate is much higher than the rate of diffusion, then the second term of the denominator in Equation 17 is much higher than unity, and so $i_k = i'_m$, i.e., the limiting current is controlled by diffusion; (ii) if, on the contrary, the dissociation rate is slow enough so that

$$D'k_d(t_m/(k_f[L]))^{1/2} \ll 1 \qquad (18)$$

then the global process is controlled purely kinetically:

$$\frac{i_k}{i'_m} = \frac{qD'k_d t_m^{1/2}}{(k_f[L])^{1/2}} \qquad (19)$$

In practice it is assumed that the kinetic current i_k is negligible when $i_k/i'_m \leqslant 0.1$ (16). If (KC_L) is large enough (K is the complex stability constant), so that $\bar{D} = D_{ML}$, the expression for D' becomes $(D_M/D_{ML})^r$, a simplification assumed in some reports dealing with macromolecules (13, 121). If the rate constant of formation k_f is not known, a value of $10^9 - 10^{10}\,\text{M}^{-1}\,\text{s}^{-1}$ can be expected as the

FUNDAMENTALS OF VOLTAMMETRIC METHODS   81

highest limit, assuming that the slowest step in the complex formation is the dehydratation of the cation, which is the usual case (274). One of the best ways to determine the dissociation rate constant of quasi-labile complexes is by cyclic voltammetry or square wave voltammetry, due to their wide range of timescales of techniques (275).

If the process is kinetically controlled and in the presence of an excess of ligand, Equation 20 can be deduced (273):

$$\Delta E = E_m^M - E_m^{M+L} = RT/(nF) \ln ((\bar{D}/D_M)^r (1 + K[L]^0)(i_k/i'_m)) \qquad (20)$$

Koryta proposed a solution for a set of complexes ML, ML$_2$,..., ML$_n$, assuming that: (i) there is only one slow step corresponding to the conversion ML$_m \to$ ML$_{m-1}$ at the electrode surface, the complexes of higher and lower order being in equilibrium with the other species; (ii) the ligand is in excess; (iii) the free cation as well as the complexes from ML to ML$_{m-1}$ are electroactive, whereas the other complexes are electroinactive; and (iv) all the species have similar diffusion coefficients (276). Under these conditions the following expression can be obtained:

$$\frac{i_k}{i'_m - i_k} = \frac{q(k_d \beta_m [L]^m \sum_{j=0}^{m-1} (\beta_j [L]^j) t_m)^{1/2}}{\sum_{j=m}^{n} (\beta_j [L]^j)} \qquad (21)$$

A general mathematical treatment, valid for any set of rate constants and diffusion coefficients, cannot be done without numerical simulations; such work is in progress (121).

In ASV technique it is also possible to forecast whether a complex is labile or not in terms of dissociation (deposition step). Davison has shown that, for free metal ions and complexed species with the same diffusion coefficient D, the current i_k is given by (16)

$$i_k = nFADC_M \left(\delta + \frac{K^{3/2}[L]D^{1/2}}{k_f^{1/2}(1 + K[L])^{1/2}} \right)^{-1} \qquad (22)$$

where A is the area of the electrode; δ is the thickness of the the diffusion layer; and the other parameters have their usual meaning. The value of δ can be experimentally determined from the electrolysis on the same electrode of a known concentration of a metal ion solution, in a reproducible stirred medium. In this case, i_{el}, the measured current, has a constant value during the electrolysis and δ is given by

$$\delta = nFADC_M/i_{el} \qquad (23)$$

The parameter δ is related to the timescale t_m of ASV by the following expression:

$$\delta = (\pi D t_m)^{1/2} \qquad (24)$$

The timescale of ASV can also be changed using rotating disk electrodes where a film of Hg is previously deposited. Timescales of 0.001–0.3 s are found according to the expression $(2\pi f)^{-1}$, where f is the frequency of the rotating disk (89). When the rotating speed of the electrode w is low enough for the complex to be completely labile, the disk current will be given by Equation 25 according to the Levich theory, which assumes the same diffusion coefficient D for all the species (277):

$$(i_m^{M+L})_{\text{labile}} = \frac{nFAD^{2/3}w^{1/2}C_M}{1.61v^{1/6}} \qquad (25)$$

where v is the kinematic viscosity and the other symbols have their usual meaning. If, on the other hand, the speed of rotation of the electrode is so high that the complex is completely inert, we have

$$(i_m^M)_{\text{inert}} = \frac{nFAD^{2/3}w^{1/2}K'C_M}{1.61v^{1/6}(K'+1)} \qquad (26)$$

where $K' = [M]/[ML]$ is the ratio of the free metal to that of the complex concentrations.

For intermediate values of w, if K' is small and if the sum of the rate constants of dissociation and formation of the complex is high, the measured current is i_k and we can have, according to Levich theory (278),

$$1/i_k = \frac{D^{1/6}(1+K')^{1/2}K'^{-1/2}k_d^{-1/2} + 1.61w^{-1/2}}{nFAD^{2/3}C_M} \qquad (27)$$

From Equations 25, 26, and 27 we can determine, respectively: (i) the D value, from $(i_m^{M+L})_{\text{labile}}$ vs. $w^{1/2}$ for small w values; (ii) the parameter K', from $(i_m^M)_{\text{inert}}$ vs. $w^{1/2}$ for high frequencies; and (iii) the parameter k_d, from $(1/i_k)$ vs. $w^{-1/2}$, if K' is known.

By this technique Shuman and Michael determined k_d to be ca. $2\,s^{-1}$ for Cu complexes in seawater samples as an average value of the rate constant of dissociation of Cu complexes with natural organic ligands, it being possible by this method to determine rate constants between 0.1 and $400\,s^{-1}$ (279). This method can also be used to determine stability constants, as can the other voltammetric methods in general.

2.5.4. Complexation of Heterogeneous Ligands

Heterogeneous ligands undergo polyfunctional and polyelectrostatic effects, due, respectively, to different complexing sites and to a high charge density. Each complexing site (with a specific chemical group and/or specific surroundings) acts differently in the complexation of the metal. The complexation is also affected by the charge of the ligand and the ionic strength of the medium. In this context it is assumed that different ML_i complexes are being formed during titration of the ligand with a stock solution of metal, L_i standing for each site. If two or more groups are bound to a metal ion, all these groups are taken as one coordination site, and so only complexes of the 1:1 type are considered. For these ligands the stability constant K should be replaced by the average complexing parameter $\bar{K}$, defined as

$$\bar{K} = \frac{\sum_i [ML_i]}{[M] \sum_i [L_i]} \tag{28}$$

where the term $\sum [L_i]$ accounts for the multiple sites of the heterogeneous ligand, with a large number of different groups in one molecule. On the other hand,

$$\sum_i [L_i] = C_L - \sum_i [ML_i] \tag{29}$$

If L interacts with other ions M' besides M, $\sum[L_i]$ should be replaced by $\sum_i[L_i']$, where L_i' stands for all the ligands not complexed to the metal M. In this case $\bar{K}$ represents a conditional parameter, $\bar{K}_{cond}$, but only if $[M']/[M]$ and $[L']/[L]$ are constant values, as commented in Section 2.5.3.2.

The parameter $\bar{K}$ is calculated from Equation 28, since $[M]$ and $\sum[ML_i]$ can be experimentally determined for each addition of metal ion from i_m or E_m measurements if the complexes are labile (Section 2.5.3.2) or from i_m values if they are inert (Section 2.5.3.1). The diffusion coefficients of all complexes formed in solution are considered similar although smaller than D_M. For the determination of $\sum[L_i]$ the concentration of C_L should be known in molarity, where C_L represents the concentration of the total complexing sites.

The complexing parameters $\bar{K}$ and k_d (k_d stands for the dissociation rate constant of the complex), whose values vary from point to point in a complexometric titration of the heterogeneous ligand with the metal ion or vice versa, exhibit the following behavior: (i) the strongest complexing sites are first saturated, $\bar{K}$ decreasing with the increase of the metal content; (ii) since, as a first approximation, the dissociation rate constant varies inversely to K ($K = k_f/k_d$, where k_f is assumed to be constant because the slow step in the formation process is generally the dehydratation of M; k_d should increase with the increase of C_M). Owing to the heterogeneity of the ligand the complex might exhibit an inert behavior for $C_L \gg C_M$, becoming more labile with the increase of C_M.

The diffusion coefficient D_{ML} is similar to D_L, being more or less independent of C_M. However, small differences in its value may be observed due to the charge of the molecule, an increase in C_M leading generally to a decrease in its charge, which might affect the D_{ML} value if the molecule becomes less unfolded (different conformation).

To study the heterogeneity of fulvic acids, Cabaniss and Shuman used an empirical N-site model (where N represents the number of binding sites) to fit the results obtained from a set of different titration curves with Cu, and found $N = 5$ (162). Another approach used to study the heterogeneity of natural samples is an N-site electrostatic model, where the major factor affecting $\log \bar{K}$ is the net charge on the ligand (163).

When complexing parameters have continuous variations, as may happen in natural waters or biological systems, distribution functions, which represent the probability of finding sites with an affinity constant K, should be used instead of discrete models (280–284). Two types of semianalytical methods, assuming a continuous distribution, have been published in the literature on the basis of different definitions of the binding parameter for heterogeneous samples: the differential equilibrium parameter (124, 280, 284–286) and the *local binding parameter* (281–283, 287–289).

The concept of the differential equilibrium parameter (K_{DEF}), which is the complexing parameter obtained for a small increase of [L] and of the corresponding complex ML, has been developed in particular by Gamble et al. (280). Later Buffle deduced an analytical expression for K_{DEF} using the same sort of mathematical treatment (284).

More recently, the polarographic curve i vs. E for NPP and dc polarography, based on the heterogeneity of the ligand, has been deduced by Filella et al. (124). The theoretical asymmetric sigmoidal curves presented in Fig. 2.7, valid for labile complexes, are qualitatively explained on the basis that, at the beginning of the rising part of the wave the metal bound to the weaker sites is reduced, and the remaining complexes, being more stable, require more negative potential values, which spreads the wave observed for fully labile complexes. From this curve the authors determined the parameters D_{ML} and Γ, where Γ is related to the heterogeneity of the ligand (124). The Γ value obtained for natural media is in the range 0.3–0.7, depending on the ligand and on the metal ($\Gamma = 1$ when there are no heterogeneous ligands, decreasing to zero with the increase of heterogeneity).

The value of K_{DEF} can be found from the titration of the heterogeneous ligand with a metal solution. In this case C_L has a constant value and, in the

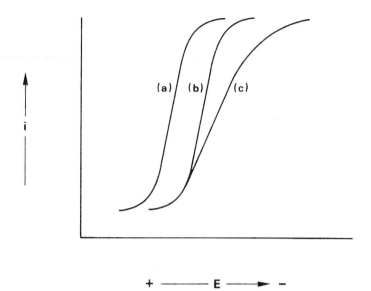

Figure 2.7. Direct current (dc) polarographic curves for the reduction of M (124). Only labile complexes are considered in the presence of the ligand: (a) noncomplexing medium; (b) simple ligand [curve (b) is parallel to curve (a)]; (c) heterogeneous ligand ($\Gamma = 0.3$):

presence of an excess of ligand, K_{DEF} is defined by (124, 284)

$$K_{DEF} = -\frac{\alpha^2}{C_M}\left(\frac{1}{1+(\alpha-1)d\ln(C_M)/d\ln\alpha}\right) \tag{30}$$

where $\alpha = C_M/[M]$. This parameter can be determined from the experimental values i_m and/or E_m obtained from the titration of the ligand, if the complexes are labile, and from i_m, if they are inert. The derivative presented in Equation 30 is responsible for the narrow range of sites considered in the determination of K_{DEF} at each experimental point. The plot of $\log K_{DEF}$ vs. $\log \Theta_t$ characterizes the heterogeneity of the sample (Θ_t represents the overall degree of sites occupied and is experimentally given by $[ML]/C_L$), and the probability density function is defined by

$$f(\log K_{DEF}) = -\frac{d\Theta_t}{d\log K_{DEF}} \tag{31}$$

For heterogeneous ligands with a continuous distribution of complexing sites the parameter K_{DEF} should be used instead of $\bar{K}$, defined in Equation 28, owing to the following facts. (i) The parameter K determined at each point of the titration involves all the sites complexed up to this point, and so its value is, in a certain sense, a mean value for all the sites titrated. The parameter K_{DEF} has a more specific meaning, corresponding to a small range of ligand concentration, which is actually complexing the metal. (ii) For the determination of $\bar{K}$ it is necessary to know the ligand concentration in molarity, as already pointed out, but not for the calculation of K_{DEF}; so, if C_L has a constant value during the titration, K_{DEF} does not depend explicitly on the value of C_L (Equation 30), and if C_L is not constant, only a parameter proportional to C_L is required to determine K_{DEF}, for example, the concentration of dissolved organic carbon.

The local binding parameter assumes that each site i is defined by a binding constant K_i; the degree of sites type i occupied by M, when M is the only species reacting with L_i, is given by

$$\Theta_i = \frac{[ML_i]}{C_{L_i}} = \frac{K_i[M]}{1 + K_i[M]} \tag{32}$$

If the sample has a continuous spectrum of K_i values, Θ_t will be given by the integral equation

$$\Theta_t = \Theta \int_D f(\log K) d\log K \tag{33}$$

where Θ is the local binding function, and D is the range of available K values. The probability density function $f(\log K)$ is obtained by inverting the integral of Equation 33 if Θ is expressed by a function of K and $[M]$ (Equation 32), and Θ_t by its experimental value ($[ML]/C_L$).

Some methods have been presented for analytically solving $f(\log K)$, based on approximations of the local binding function. The affinity spectrum (AS) is a well-known example of this group of methods (281). More recently, LOGA, a new approximation for the local binding isotherm, has been developed (283).

It should be noted that $f(\log K)$ determined either from the local binding function solved analytically or from the differential equilibrium parameter (Equation 31) is always an approximation to the true distribution function. Some distribution functions based on K_{DEF} and on various local binding isotherms have been compared with true distribution functions for two different models of mixtures with individual known stability constants (288, 289). In the absence of experimental errors it was concluded that LOGA is generally a better approximation to the true distribution. Which of these mathematical treatments is the more sensitive to experimental errors still remains to be ascertained.

Although continuous representations are the most realistic ways to interpret a continuous distribution, they present greater difficulties than discrete models in the calculation of species concentration. So far, very few applications have considered the heterogeneity of the sample using voltammetric measurements and continuous distribution (15, 290). No doubt this aspect will require further investigation in the years to come.

REFERENCES

1. W. Stumm and J. Morgan, *Aquatic Chemistry*, 2nd ed. Wiley, New York, 1981.
2. D. R. Turner, M. Whitfield, and A. G. Dickson, *Geochim. Cosmochim. Acta*, **45**, 855 (1981).
3. M. L. Simões Gonçalves, M. F. Vilhena, and M. A. Sampayo, *Water Res.* **22**, 1429 (1988).
4. M. L. Simões Gonçalves, M. F. Vilhena, J. M. F. Sollis, J. M. C. Romero, and M. A. Sampayo, *Talanta* **38**, 1111 (1991).
5. B. T. Hart, *Trace Met. Nat. Waters* **49**, 260 (1982).
6. L. L. Stookey, *Anal. Chem.* **42**, 779 (1970).
7. P. Behra and L. Sigg, *Nature (London)* **344**, 419 (1990).
8. M. Whitfield, in *Activity Coefficients in Electrolyte Solutions* (R. M. Pytkowicz, ed), Vol. 2, Chapter 3. CRC, Boca Raton, FL, 1989.
9. P. H. Nienhuis, in *Marine Organic Chemistry* (E. K. Duursma and R. Dawson, eds), Chapter 3. Elsevier, Amsterdam, 1981.
10. J. Buffle, A. M. Mota, and M. L. Simões Gonçalves, *J. Electroanal. Chem.* **223**, 235 (1987).
11. J. N. Butler, *Carbon Dioxide Equilibria and their Application*. Addison-Wesley, Reading, MA, 1982.
12. W. Stumm and H. Bilinsky, *Adv. Water Pollut. Res. Proc. Int. Conf., 6th, Jerusalem, 1972*, 39 (1973).
13. J. Buffle, *Complexation Reactions in Aquatic Systems*. Ellis Horwood/Wiley (Halsted Press), New York, 1988.
14. M. L. Simões Gonçalves, L. Sigg, and W. Stumm, *Environ. Sci. Technol.* **19**, 141 (1985).
15. M. L. Simões Gonçalves and A. L. Conceiçao, *Sci. Total Environ.* **103**, 185 (1991).
16. W. Davison, *J. Electroanal. Chem.* **87**, 395 (1978).
17. Y. K. Chau, R. Gachter, and K. L. S. Chan, *J. Fish. Res. Board Can.* **31**, 1515 (1974).
18. P. Figura and B. McDuffie, *Anal. Chem.* **52**, 1433 (1980).
19. M. Shuman and J. C. Cromer, *Environ. Sci. Technol.* **13**, 543 (1979).

20. P. Figura and B. McDuffie, *Anal. Chem.* **51**, 120 (1979).
21. S. J. De Mora and R. M. Harrison, *Hazard Assess. Chem.: Curr. Dev.* **3**, 1 (1984).
22. G. E. Batley and D. Gardner, *Water Res.* **11**, 745 (1977).
23. P. Benes and E. Steines, *Water Res.* **9**, 741 (1975).
24. D. P. H. Laxen and R. M. Harrison, *Anal. Chem.* **53**, 345 (1981).
25. L. Mart, *Fresenius' Z. Anal. Chem.* **296**, 350 (1979).
26. J. A. Campbell, M. J. Gardner, and A. M. Gunn, *Anal. Chim. Acta* **176**, 193 (1985).
27. G. E. Batley and D. Gardner, *Estuarine Coastal Mar. Sci.* **7**, 59 (1978).
28. T. M. Florence, *Water Res.* **11**, 681 (1977).
29. P. Benes, E. T. Gjessing, and E. Steinnes, *Water Res.* **10**, 711 (1976).
30. J. C. Duinker, R. F. Nolting, and H. A. van der Sloot, *Neth. J. Sea Res.* **13**, 282 (1979).
31. D. R. Roden and D. E. Tallman, *Anal. Chem.* **54**, 307 (1982).
32. E. D. Goldberg, M. Baker, and D. L. Fox, *J. Mar. Res.* **11**, 197 (1952).
33. S. J. De Mora and R. M. Harrison, *Water Res.* **17**, 723 (1983).
34. J. P. Riley, in *Chemical Oceanography* (J. P. Riley and G. Skirrow, eds.), 2nd ed., Vol. 3, p. 211. Academic Press, London, 1975.
35. K. T. Marvin, R. R. Proctor, and R. A. Neal, *Limnol. Oceanogr.* **17**, 777 (1972).
36. R. Wageman and B. Graham, *Water Res.* **8**, 407 (1974).
37. D. O. Coony, *Anal. Chem.* **52**, 1068 (1980).
38. D. W. Spencer and P. C. Brewer, *Geochim. Cosmochim. Acta* **33**, 325 (1965).
39. D. P. H. Laxen and R. M. Harrison, *Sci. Total Environ.* **19**, 59 (1981).
40. H. W. Nürnberg, P. Valenta, L. Mart, B. Raspor, and L. Sipos, *Fresenius' Z. Anal. Chem.* **282**, 357 (1976).
41. R. Salim and B. G. Cooksey, *J. Electroanal. Chem.* **106**, 251 (1980).
42. D. Perret, R. de Vitre, G. G. Leppard, and J. Buffle, in *Large Lakes, Ecological Structure and Function* (M. M. Tilzer and C. Serruga, eds.). Chapter 12. Springer-Verlag, Berlin, 1990.
43. P. Meares, in *Membrane Separation Processes* (P. Meares, ed.), Chapter 1. Elsevier, Amsterdam, 1976.
44. R. R. DeVitre, J. Buffle, D. Perret, and R. Baudat, *Geochim. Cosmochim. Acta* **52**, 1601 (1988).
45. K. Grasshoff, *Methods of Seawater Analysis*, p. 137. Verlag Chemie, Weinheim, 1976.
46. J. Lecomte, P. Mericam, and M. Astruc, *Heavy Met. Environ., Int. Conf., 3rd*, Amsterdam, *1981*, 678 (1981).
47. D. W. Spencer and P. C. Brewer, *Geochim. Cosmochim. Acta* **33**, 325 (1969).
48. J. C. T. Kwak, R. W. P. Nelson, and D. S. Gamble, *Geochim. Cosmochim. Acta* **41**, 993 (1977).

49. W. F. Blatt, B. G. Hudson, S. M. Robinson, and E. M. Zipilivan, *Nature (London)* **216**, 511 (1967).
50. J. Buffle, P. Deladoey, and W. Haerdi, *Anal. Chim. Acta* **101**, 339 (1978).
51. M. R. Hoffman, E. C. Yost, S. J. Eisenreich, and W. J. Maier, *Environ. Sci. Technol.* **15**, 655 (1981).
52. C. Tanford, *Physical Chemistry of Macromolecules*, p. 224. Wiley, New York, 1961.
53. R. D. Guy and C. Bourque, *Heavy Met. Environ., Int. Conf. 3rd*, Amsterdam, *1981*, 557 (1981).
54. R. D. Guy and C. L. Chakrabarti, *Int. Conf. Heavy Met Environ.* [*Symp. Proc*], *1st*, Toronto, *1975*, 275 (1977).
55. T. M. Florence and G. E. Batley, *Talanta* **24**, 151 (1977).
56. P. Benes and E. Steinnes, *Water Res.* **8**, 947 (1974).
57. R. D. Guy and C. L. Chakrabarti, *Can. J. Chem.* **54**, 2600 (1976).
58. B. T. Hart and S. H. R. Davies, *Aust. J. Mar. Freshwater Res.* **32**, 175 (1981).
59. R. E. Truitt and J. H. Weber, *Anal. Chem.* **53**, 337 (1981).
60. K. H. Altgelt, *Adv. Chromatogr. (N.Y.)* **7**, 1 (1968).
61. H. Determan, *Adv. Chromatogr. (N.Y.)* **8**, 1 (1969).
62. B. J. Hunt and S. R. Holding, *Size Exclusion Chromatography*. Chapman & Hall, New York, 1989.
63. T. Kremmer and L. Boross, *Gel Chromatography*. Wiley, Chichester, 1979.
64. C. Steinberg, *Water Res.* **14**, 1239 (1980).
65. S. F. Sugai and M. L. Healy, *Mar. Chem.* **6**, 291 (1978).
66. R. M. Sterrit and J. N. Lester, *Environ. Pollut., Ser., A* **27**, 37 (1982).
67. F. E. Butterworth and B. J. Alloway, *Heavy Met. Environ., Int. Conf., 3rd*, Amsterdam, *1981*, 713 (1981).
68. T. M. Florence, *Talanta* **29**, 345 (1982).
69. A. Acher, Y. Pistol, and B. Yaron, in *Developments in Arid Ecology and Environmental Quality* (H. Shuvai, ed.), pp. 211–220. Balbaban Int. Sci. Serv., Philadelphia, 1981.
70. R. L. Shambaugh and P. B. Melnyk, *J. Water. Pollut. Control Fed.* **50**, 113 (1978).
71. T. M. Florence, *Anal. Chim. Acta* **141**, 73 (1982).
72. J. C. Duinker and C. J. M. Kramer, *Mar. Chem.* **5**, 207 (1977).
73. A. E. Martell and R. M. Smith, *Critical Stability Constants*. Plenum, New York and London, 1974.
74. P. Figura and B. McDuffie, *Anal. Chem.* **49**, 1950 (1977).
75. D. M. McKnight and F. M. M. Morel, *Limnol. Oceanogr.* **24**, 823 (1979).
76. D. M. McKnight and F. M. M. Morel, *Limnol. Oceanogr.* **25**, 62 (1980).
77. N. G. Zorkon, E. V. Grill, and A. G. Lewis, *Anal. Chim. Acta* **183**, 163 (1986).
78. J. A. Sweileh, D. Lucyk, D. L. B. Kratochvil, and F. F. Cantwell, *Anal. Chem.* **59**, 586 (1987).

79. M. Hiraide, S. P. Tillekeratne, K. Otsuka, and A. Mizuike, *Anal. Chim. Acta* **172**, 215 (1985).
80. M. Hiraide, Y. Arima, and A. Mizuike, *Anal. Chim. Acta* **200**, 171 (1987).
81. Y. Liu and J. D. Ingle, Jr., *Anal. Chem.* **61**, 525 (1989).
82. G. L. Mills and J. G. Quinn, *Mar. Chem.* **10**, 93 (1981).
83. G. L. Mills, A. K. Hanson, Jr., J. G. Quinn, W. R. Lammela, and N. D. Charteen, *Mar. Chem.* **11**, 355 (1982).
84. G. L. Mills and J. G. Quinn, *Mar. Chem.* **15**, 151 (1984).
85. G. L. Mills, G. S. Douglas, and J. G. Quinn, *Mar. Chem.* **26**, 277 (1989).
86. J. R. Donat, P. J. Statham, and K. W. Bruland, *Mar. Chem.* **18**, 85 (1986).
87. K. H. Coale and K. W. Bruland, *Limnol. Oceanogr.* **33**, 1084 (1988).
88. Y. Sugimura, Y. Suzuki, and Y. Miyake, *Proc. Nucl. Energy Agency Semin. Mar. Radioecol.*, 3rd, Tokyo, 131 (1979).
89. A. J. Bard and L. R. Faulkner, *Electrochemical Methods*. Wiley, New York, 1980.
90. A. M. Bond, *Modern Polarographic Methods in Analytical Chemistry*. Dekker, New York, 1980.
91. J. Osteryoung and J. J. O'Dea, in *Electroanalytical Chemistry* (A. J. Bard, ed.), Vol. 14, p. 209. Dekker, New York, 1986.
92. H. van Leeuwen, R. Cleven and J. Buffle, *Pure Appl. Chem.* **61**, 255 (1989).
93. M. Whitfield, in *Chemical Oceanography* (J. P. Riley and G. Skirrow, eds.), 2nd ed., Vol. 4, Chapter 20. Academic Press, London, 1975.
94. A. Zirino, M. Whitfield, and D. Jagner, in *Marine Electrochemistry* (M. Whitfield and D. Jagner, eds.), Chapter 10. Wiley, New York, 1981.
95. H. W. Nürnberg, *Pure Appl. Chem.* **54**, 853 (1982).
96. H. W. Nürnberg, *Sci. Total Environ.* **37**, 9 (1984).
97. K. H. Coale and K. W. Bruland, *Limnol. Oceanog.* **33**, 1084 (1988).
98. J. R. Donat and K. W. Bruland, *Mar. Chem.* **28**, 301 (1990).
99. G. Capodaglio, K. H. Coale, and K. W. Bruland, *Mar. Chem.* **29**, 221 (1990).
100. M. G. Robinson and L. N. Brown, *Mar. Chem.* **33**, 105 (1991).
101. H. Li and C. M. G. van den Berg, *Anal. Chim. Acta* **221**, 269 (1989).
102. S. H. Khan and C. M. G. van den Berg, *Mar. Chem.* **27**, 31 (1989).
103. H. Zhang, C. M. G. van den Berg, and R. Wollast, *Mar. Chem.* **28**, 285 (1990).
104. C. M. G. van den Berg and S. H. Khan, *Anal. Chim. Acta* **231**, 221 (1990).
105. K. Yokoi and C. M. G. van den Berg, *Anal. Chim. Acta* **245**, 167 (1991).
106. M. Whitfield and D. R. Turner, *Nature (London)* **278**, 132 (1979).
107. C. M. G. van den Berg, *Mar. Chem.* **15**, 1 (1984).
108. S. C. Apte, M. J. Gardner, and J. E. Ravenscroft, *Mar. Chem.* **29**, 63 (1990).
109. H. Xue and L. Sigg, *Limnol. Oceanogr.* **38**, 1200 (1993).
110. C. M. G. van den Berg, K. Murphy, and J. P. Riley, *Anal. Chim. Acta* **188**, 177 (1986).

111. M. L. Tercier, J. Buffle, A. Zirino, and R. R. de Vitre, *Anal. Chim. Acta* **237**, 429 (1990).
112. R. M. Wightman and D. D. Wipf, *Electroanal. Chem.* **15**, 267, (1989).
113. M. Fleischmann, S. Pons, D. R. Rolison, and P. Schmidt, eds., *Ultramicroelectrodes*. Datatech Science, Morganton, NC, 1987.
114. C. Wechter and J. Osteryoung, *Anal. Chim. Acta* **234**, 275 (1990).
115. S. P. Kounaves and W. Deng, *J. Electroanal. Chem.* **301**, 77 (1991).
116. R. R. de Vitre, M. L. Tercier, M. Isacopoulos, and J. Buffle, *Anal. Chim. Acta* **249**, 419 (1991).
117. J. Peng and W. Jin, *Anal. Chim. Acta* **264**, 213 (1992).
118. M. Wojciechowski and J. Balcerzak, *Anal. Chim. Acta* **273**, 127 (1990).
119. P. R. Unwin and A. J. Bard, *Anal. Chem.* **64**, 113 (1992).
120. H. G. de Jong, H. P. van Leeuwen, and K. Holub, *J. Electroanal. Chem.* **234**, 1 (1987).
121. H. G. de Jong and H. P. van Leeuwen, *J. Electroanal. Chem.* **234**, 17 (1987).
122. C. J. M. Kramer, Yu Guo-hui, and J. C. Duinker, *Fresenius' Z. Anal. Chem.* **317**, 383 (1984).
123. J. Buffle, A. Tessier, and W. Haerdi, in *Complexation of Trace Metals in Natural Waters*, (C. J. M. Kramer and J. C. Duinker, eds.), p. 301. Martinus Nijhoff/Jung Publishers, The Hague, 1984.
124. M. Filella, J. Buffle, and H. P. van Leeuwen, *Anal. Chim. Acta* **232**, 209 (1990).
125. P. W. Schindler and W. Stumm, in *Aquatic Surface Chemistry*, (W. Stumm, ed.), Chapter 4. Wiley, New York, 1987.
126. J. Buffle, F. L. Greter, and W. Haerdi, *J. Electroanal. Chem.* **101**, 211 (1979).
127. J. Buffle and F. L. Greter, *J. Electroanal. Chem.* **101**, 231 (1979).
128. Z. Kozarac, B. Cosovic, and V. Vojvodic, *Water Res.* **20**, 295 (1986).
129. H. P. van Leeuwen, J. Buffle, and M. Lovric, *Pure Appl. Chem.* **64**, 1015 (1992).
130. H. P. van Leeuwen, in *Hygiene Environmental, Clinical and Pharmaceutical Chemistry* (W. F. Smyth, ed.), p. 382. Elsevier, Amsterdam, 1980.
131. H. P. van Leeuwen, *J. Electroanal. Chem.* **162**, 67 (1984).
132. J. Buffle, J. J. Vuilleumier, M. L. Tercier, and N. Parthasarathy, *Sci. Total Environ.* **60**, 75 (1987).
133. J. E. Gregor and H. K. J. Powel, *Anal. Chim. Acta* **211**, 141 (1988).
134. M. Plavsic, B. Cosovic, and S. Miletic, *Anal. Chim. Acta* **255**, 15 (1991).
135. M. Plavsic and B. Cosovic, *Mar. Chem.* **36**, 39 (1991).
136. G. M. P. Morrison, T. M. Florence, and J. L. Stauber, *Electroanalysis* **2**, 9 (1990).
137. M. Sluyters-Rehbach and J. H. Sluyters, in *Electroanalytical Chemistry* (A. J. Bard, ed.), Vol. 4, Chapter 1, p. 1. Dekker, New York, 1982.
138. J. Heyrovsky and J. Kuta, in *Principles of Polarography* (R. Brdička, ed.), Chapter XIV. Publ. House Czech. Acad. Sci., Prague, 1965.

139. S. Sathyanaryana and K. G. Baikerikar, *J. Electroanal. Chem.* **25**, 209 (1970).
140. T. Yoshida, T. Ohsaka, and S. Nomoto, *Bull. Chem. Soc. Jpn.* **45**, 1585 (1972).
141. S. Nakamura and M. Nakamura, *Electrochim. Acta* **27**, 67 (1982).
142. R. Guidelli and M. R. Moncelli, *J. Electroanal. Chem.* **89**, 261 (1978).
143. J. Buffle and A. Cominoli, *J. Electroanal. Chem.* **121**, 273 (1981).
144. B. Cosovic and V. Vojvodic, *Limnol. Oceanogr.* **27**, 361 (1982).
145. B. Cosovic, V. Vojvodic, and T. Plese, *Water Res.* **19**, 175 (1985).
146. M. Plavsic, V. Vojvodic, and B. Cosovic, *Anal. Chim. Acta* **232**, 131 (1990).
147. B. Raspor, *Sci. Total Environ.* **81**, 319 (1989).
148. B. Raspor, *J. Electroanal. Chem.* **316**, 223 (1991).
149. B. Raspor, H. W. Nürnbeg, P. Valenta, and M. Branica, *Mar. Chem.* **15**, 217 (1984).
150. B. Raspor, H. W. Nürnberg, P. Valenta, and M. Branica, *Mar. Chem.* **15**, 231 (1984).
151. B. Hoyer, T. M. Florence, and G. A. Batley, *Anal. Chem.* **59**, 1608 (1987).
152. M. N. Szentirmay and C. R. Martin, *Anal. Chem.* **56**, 1898 (1984).
153. S. Dong and Y. Wang, *Talanta* **35**, 819 (1988).
154. S. Dong and Y. Wang, *Anal. Chim. Acta* **212**, 341 (1988).
155. J. Wang and L. D. Hutchins-Kumar, *Anal. Chem.* **58**, 402 (1986).
156. J. Wang and P. Tuzhi, *Anal. Chem.* **58**, 3257 (1986).
157. S. C. Apte, M. J. Gardner, and J. E. Ravenscroft, *Anal. Chim. Acta* **212**, 1 (1988).
158. D. R. Turner, M. S. Varney, M. Whitfield, R. F. C. Mantoura, and J. P. Riley, *Sci. Total Environ.* **60**, 1 / (1987).
159. C. M. G. van den Berg and S. Dharmvanij, *Limnol. Oceanogr.* **29**, 1025 (1984).
160. S. C. Apte, M. J. Gardner, J. E. Ravenscroft, and J. A. Turrell, *Anal. Chim. Acta* **235**, 287 (1990).
161. C. M. G. van den Berg, M. Nimmo, P. Daly, and D. R. Turner, *Anal. Chim. Acta* **232**, 149 (1990).
162. S. E. Cabaniss and M. S. Shuman, *Geochim. Cosmochim. Acta* **52**, 185 (1988).
163. E. Tipping and M. A. Hurley, *Geochim. Cosmochim. Acta* **56**, 3627 (1992).
164. S. Daniele, M. A. Baldo, P. Ugo, and G. A. Mazzocchin, *Anal. Chim. Acta* **219**, 19 (1989).
165. R. R. de Vitre, M. L. Tercier, and J. Buffle, *Anal. Proc. (London)* **28**, 74 (1991).
166. B. R. Clark, D. W. De Paoli, D. R. McTaggart, and B. D. Patton, *Anal. Chim. Acta* **215**, 13 (1988).
167. J. Koryta and K. Stulík, *Ion Selective Electrodes*, 2nd ed. Cambridge Univ. Press, Cambridge, UK, 1983.
168. *Ion-Selective Catalog and Guide to Ion Analysis*. Orion Research Incorporated, Boston, 1992.
169. C. H. Culberson, in *Marine Electrochemistry* (M. Whitfield and D. Jagner, eds.), Chapter 6. Wiley, New York, 1981.

170. T. Midorikawa, E. Tanoue, and J. Sugimura, *Anal. Chem.* **62**, 1737 (1990).
171. J. Gardiner, *Water Res.* **8**, 23 (1974).
172. A. P. Turner, I. Karube, and G. S. Wilson, *Biosensors: Fundamentals and Application*, Oxford University Press, New York, 1987.
173. J. Buffle and M. Spoerri, *J. Electroanal. Chem.* **129**, 67 (1981).
174. M. Mascini and C. Cremisini, *Anal. Chim. Acta* **97**, 237 (1978).
175. R. L. Solsky, *Anal. Chem.* **60**, 106R (1988).
176. J. Koryta, *Anal. Chim. Acta* **233**, 1 (1990).
177. J. Anfolt and D. Jagner, *Anal. Chim. Acta* **50**, 23 (1970).
178. G. K. Pagenkopf and K. A. Buell, *Water Res.* **8**, 375 (1974).
179. J. Buffle, F. L. Greter, and W. Haerdi, *Anal. Chem.* **49**, 216 (1977).
180. R. S. Braman, D. L. Johnson, C. C. Foreback, J. M. Ammons, and J. L. Bricker, *Anal. Chem.* **49**, 621 (1977).
181. C. Feldman and D. A. Batistoni, *Anal. Chem.* **49**, 2215 (1977).
182. P. C. Uden, in *Environmental Speciation and Monitoring Needs for Trace Metal-Containing Substances from Energy Related Processes* (F. E. Brinckman and R. H. Fish, eds.), NBS Spec. Publ. No. 618. U.S. Govt. Printing Office, Washington, DC, 1981.
183. J. W. Carnahan, K. J. Mulligan, and J. A. Caruso, *Anal. Chim. Acta* **130**, 227 (1981).
184. M. O. Andreae, *Anal. Chem.* **49**, 820 (1977).
185. R. B. Cruz, C. Loruso, S. George, Y. Thomassen, J. D. Kinrade, L. K. P. Butler, J. Lye, and J. C. van Loon, *Spectrochim. Acta* **35B**, 775 (1980).
186. R. S. Braman and M. A. Tompkins, *Anal. Chem.* **51**, 12 (1978).
187. V. F. Hodge, S. L. Seidel, and E. D. Goldberg, *Anal. Chem.* **51**, 1256 (1979).
188. O. F. X. Donard, S. Rapsomanikis, and J. H. Weber, *Anal. Chem.* **58**, 772 (1986).
189. M. O. Andreae and J. T. Byrd, *Anal. Chim. Acta* **156**, 147 (1984).
190. S. Nakashima, *Analyst* **104**, 172 (1979).
191. M. O. Andreae, J. Asmodé, P. Foster, and L. van't Dack, *Anal. Chem.* **53**, 1766 (1981).
192. J. D. Burton, W. A. Maher, C. I. Measures, and P. J. Statham, *Thalassia Jugosl.* **16**, 155 (1980).
193. M. O. Andreae, *Anal. Chem.* **49**, 820 (1977).
194. B. Welz and M. Melcher, *Spectrochimica Acta* **36B**, 439 (1981).
195. R. S. Braman, *Chem. Anal. (N.Y.)* **64**, 1 (1983).
196. D. T. Coker, *Anal. Chem.* **47**, 386 (1975).
197. Y. K. Chau, P. T. S. Wong, and P. D. Goulden, *Int. Conf. Heavy Met. Environ. [Symp. Prov.], 1st* Toronto, *1975*, 295 (1977).
198. Y. K. Chau, P. T. S. Wong, and P. P. Goulden, *Anal. Chem.* **47**, 2279 (1975).
199. B. Holen, R. Bye, and W. Lund, *Anal. Chim. Acta* **130**, 257 (1981).

200. J. C. van Loon, *Anal. Chem.* **51**, 1139A (1979).
201. W. de Jonghe, D. Chakrabarti, and F. Adams, *Anal. Chim. Acta* **115**, 89 (1980).
202. C. I. Measures and J. D. Burton, *Anal. Chim. Acta* **120**, 177 (1980).
203. D. A. Segar, *Anal. Lett.* **7**, 89 (1974).
204. Y. K. Chau, P. T. S. Wong, G. A. Bengert, and O. Kramar, *Anal. Chem.* **51**, 186 (1979).
205. J. E. Longbottom, *Anal. Chem.* **44**, 1111 (1972).
206. A. M. M. Bettencourt, *J. Sea Res.* **22**, 205 (1988).
207. M. O. Andreae, *Deep-Sea Res.* **25**, 391 (1978).
208. M. O. Andreae, *Limnol. Oceanogr.* **24**, 440 (1979).
209. M. O. Andreae and D. Klumpp, *Environ. Sci. Technol.* **13**, 738 (1979).
210. F. T. Henry, T. O. Kirch, and T. M. Thorpe, *Anal. Chem.* **51**, 215 (1979).
211. F. T. Henry and T. M. Thorpe, *Anal. Chem.* **52**, 80 (1980).
212. F. J. Fernandez, *At. Absorpt. Newsl.* **16**, 33 (1977).
213. G. Horlick, *Anal. Chem.* **54**, 276R (1982).
214. D. R. Jones and S. E. Manahan, *Anal. Lett.* **8**, 569 (1975).
215. C. Botré, F. Cacace, and R. Cozzani, *Anal. Lett.* **9**, 825 (1976).
216. D. Thorburn-Burns, F. Glocking, and M. Harnot, *Analyst* (*London*) **106**, 921 (1981).
217. K. C. Jewett and F. E. Brinckman, *Proc. 178th Nat. Meet. Am. Chem. Soc.*, Washington, DC, 1 (1979).
218. F. E. Brinckman, K. L. Jewett, W. P. Iverson, K. J. Irgolic, K. C. Ehrhardt, and R. A. Stockton, *J. Chromatogr.* **191**, 31 (1980).
219. E. Woolson and N. Aharoson, *J. Assoc. Off. Anal. Chem.* **63**, 523 (1980).
220. A. Y. Centillo and D. A. Segar, *Int. Conf. Heavy Met. Environ.* [*Symp. Proc.*], *1st*, Toronto, *1975*, 183 (1977).
221. R. Iadevaic, N. Aharonson, and E. Woolson, *J. Assoc. Off. Anal. Chem.* **63**, 742 (1980).
222. J. W. Moffett and R. G. Zika, *Geochim. Cosmochim. Acta* **52**, 1849 (1988).
223. M. L. Simões Gonçalves and M. M. Correia dos Santos, *J. Electroanal. Chem.* **143**, 397 (1983).
224. H. Xue, M. L. Simões Gonçalves, M. Reutlinger, L. Sigg, and W. Stumm, *Environ. Sci. Technol.* **25**, 1716 (1991).
225. D. L. Olson and M. S. Shuman, *Geochim. Cosmochim. Acta* **49**, 1371 (1985).
226. C. H. Langford and D. W. Gutzman, *Anal. Chim. Acta* **256**, 183 (1992).
227. N. Senesi, *Anal. Chim. Acta* **232**, 51 (1990).
228. B. A. Burgess, N. D. Chasteen, and H. E. Gaudette, *Environ. Geol.* **1**, 171 (1975).
229. R. Carpenter, *Geochim. Cosmochim. Acta* **47**, 875 (1983).
230. N. Senesi, *Anal. Chim. Acta* **232**, 77 (1990).
231. W. R. Seitz, *Trends Anal. Chem.* **1**, 79 (1981).

232. R. A. Saar and J. H. Weber, *Anal. Chem.* **52**, 2095 (1980).
233. P. O. Scokart, K. Keeus-Verdinne, and R. de Borges, *J. Environ. Anal. Chem.* **29**, 305 (1987).
234. E. B. Milosavljevic, L. Solujic, J. H. Nelson, and J. L. Hendrix, *Fresenius' Z. Anal. Chem.* **330**, 614 (1988).
235. R. Kuroda, T. Nara, and K. Oguma, *Analyst (London)* **113**, 1557 (1988).
236. U. Borgmann, *Adv. Environ. Sci. Technol. (London)* **13**, 47 (1983).
237. U. Borgmann, *Can. J. Fish. Aquat. Sci.* **38**, 999 (1981).
238. R. D. Guy and A. R. Kean, *Water Res.* **14**, 891 (1980).
239. M. Whitfield and D. R. Turner, *ACS Symp. Ser.* **557** (1979).
240. D. R. Turner and M. Whitfield, *Thalassia Jugosl.* **16**, 231 (1980).
241. G. E. Batley and T. M. Florence, *Mar. Chem.* **4**, 347 (1976).
242. B. T. Hart and S. H. R. Davies, *Aust. J. Mar. Freshwater Res.* **28**, 105 (1977).
243. D. P. H. Laxen and R. M. Harrison, *Water Res.* **15**, 1053 (1981).
244. S. K. Nielsen and W. Lund, *Mar. Chem.* **11**, 223 (1982).
245. B. T. Hart and S. H. R. Davies, *Aust. J. Mar. Freshwater Res.* **32**, 175 (1981).
246. F. L. Muller and D. R. Kester, *Mar. Chem.* **33**, 171 (1991).
247. F. L. Muller and D. R. Kester, *Environ. Sci. Technol.* **24**, 234 (1990).
248. M. Boussemart, C. Benamou, M. Richou and J. Y. Benaim, *Mar. Chem.* **28**, 27 (1989).
249. M. J. Stiff, *Water Res.* **5**, 585 (1971).
250. M. L. Simões Gonçalves, L. Sigg, and W. Stumm, *Environ. Sci. Technol.* **19**, 141 (1985).
251. W. Davison, J. Buffle, and R. de Vitre, *Pure Appl. Chem.* **60**, 1535 (1988).
252. W. Davison, *J. Electroanal. Chem.* **72**, 229 (1976).
253. W. Davison, *Limnol. Oceanogr.* **22**, 746 (1977).
254. M. P. Colombini and R. Fuoco, *Talanta* **30**, 901 (1983).
255. A. A. Kaplan, Z. S. Mikhailova, and L. F. Zaichko, *Zh. Anal. Khim.* **33**, 120 (1979).
256. G. Nembrini, Ph.D. Thesis, University of Geneva (1977).
257. F. H. Frimmel, *Vom Wasser* **53**, 243 (1979).
258. M. A. Kessick, J. Vuceta, and J. J. Morgan, *Environ. Sci. Technol.* **617**, 642 (1972).
259. J. Buffle, O. Zali, J. Zumstein, and R. de Vitre, *Sci. Total Environ.* **64**, 45 (1987).
260. W. Davison, C. Woof, and E. Rigg, *Limnol. Oceanogr.* **27**, 987 (1982).
261. B. Welz, *Atomic Absorption Spectrometry*, 2nd ed. VCH, Germany, 1985.
262. C. A. Johnson, *Anal. Chim. Acta* **238**, 273 (1990).
263. F. Ahern, J. M. Eckert, N. C. Payne, and K. L. Williams, *Anal. Chim. Acta* **175**, 147 (1985).
264. A. Kuhn, C. A. Johnson, and L. Sigg, *Adv. Chem. Ser.* **237**, 473 (1994).
265. H. P. van Leeuwen, *J. Electroanal. Chem.* **99**, 93 (1979).
266. V. M. Stackelberg, M. Pilgram, and V. Toome, *Z. Elektrochem.* **57**, 342 (1953).

267. J. Heyrovsky and J. Kuta, in *Principles of Polarography* (R. Brdička, ed.), Chapter VIII. Publ. House Czech. Acad. Sci., Prague, 1965.
268. M. G. Elenkova and T. K. Nedelcheva, *J. Electroanal. Chem.* **69**, 385 (1976).
269. A. M. Mota, J. Buffle, J. P. Kounaves, and M. L. Simões Gonçalves, *Anal. Chim. Acta* **172**, 13 (1985).
270. J. Heyrovsky and J. Kuta, in *Principles of Polarography* (R. Brdička, ed.), Chapter XIV. Publ. House Czech. Acad. Sci., Prague, 1965.
271. M. L. Simões Gonçalves and M. M. Correia dos Santos, *J. Electroanal. Chem.* **163**, 315 (1984).
272. M. M. Correia dos Santos and M. L. Simões Gonçalves, *J. Electroanal. Chem.* **208**, 137 (1986).
273. J. Heyrovsky and J. Kuta, in *Principles of Polarography* (R. Brdička, ed.), Chapter XVII. Publ. House Czech. Acad. Sci., Prague, 1965.
274. M. Eigen, *Ber. Bunsenges, Phys. Chem.* **67**, 753 (1963).
275. M. S. Shuman and I. Shain, *Anal. Chem.* **41**, 1818 (1969).
276. J. Koryta, *Collect. Czech. Chem. Commun.* **24**, 3057 (1959).
277. R. N. Adams, in *Electrochemistry at Solid Electrodes* (A. J. Bard, ed.), Chapter 4. Dekker, New York, 1969.
278. M. S. Shuman and L. C. Michael, *Int. Conf. Heavy Met. Environ.* [*Symp. Proc.*] *1st*, Toronto, *1975*, 227 (1977).
279. M. S. Shuman and L. C. Michael, *Environ. Sci. Technol.* **12**, 1069 (1978).
280. D. S. Gamble, A. W. Underdown and C. H. Langford, *Anal. Chem.* **52**, 1901 (1980).
281. A. K. Thakur, P. J. Munson, D. L. Hunston, and D. Rodbard, *Anal. Biochem.* **103**, 240 (1980).
282. E. M. Perdue and C. R. Lytle, *Environ. Sci. Technol.* **17**, 654 (1983).
283. J. C. M. de Wit, W. H. van Riemsdijk, M. M. Nederlof, D. G. Kinniburgh, and L. K. Koopal, *Anal. Chim. Acta* **232**, 189 (1990).
284. J. Buffle, in *Metal Ions in Biological Systems* (H. Siegel, ed.), Vol. 18, Chapter 6, p. 165. Dekker, New York, 1984.
285. J. Buffle, R. S. Altmann, and M. Filella, *Anal. Chim. Acta* **232**, 225 (1990).
286. J. Buffle, R. S. Altmann, M. Filella, and A. Tessier, *Geochim. Cosmochim. Acta* **54**, 1535 (1990).
287. M. M. Nederlof, W. H. van Riemsdijk, and L. K. Koopal, *J. Colloid Interface Sci.* **135**, 410 (1990).
288. M. M. Nederlof, W. H. van Riemsdijk, and L. K. Koopal, in *Trace Metals in the Environment* (J. P. Vernet, ed.), p. 365. Elsevier, Amsterdam, 1991.
289. M. M. Nederlof, W. H. van Riemsdijk, and L. K. Koopal, *Environ. Sci. Technol.* **26**, 763 (1992).
290. J. P. Pinheiro, A. M. Mota, and M. L. Simões Gonçalves, *Anal. Chim. Acta* **284**, 525 (1994).

CHAPTER

3

NONCHROMATOGRAPHIC METHODS OF ELEMENT SPECIATION BY ATOMIC SPECTROMETRY

M. DE LA GUARDIA

Department of Analytical Chemistry, University of Valencia, 46100 Burjassot, Valencia, Spain

3.1. INTRODUCTION

Methods based on atomic spectrometry (AS) are considered the most sensitive and selective for the elemental analysis of biological samples. Not surprisingly, in the last decade a large number of analytical procedures using AS have been proposed for the quantitative determination of the total content of many inorganic components of biological fluids and tissues. Such method can also provide effective means for evaluation of the role of these elements in bodily functions (1-3).

Figure 3.1 summarizes the various AS methods employed for elemental analysis of biological and environmental samples. Of these techniques, atomic fluorescence spectrometry (AFS) is presently the least commonly employed, owing to the lack of easily available instrumentation and because of insufficient knowledge by most spectroscopists as to the merits of the technique. This has discouraged manufacturers from investing in development of AFS in spite of its high analytical potential. In the last few years, absorption methods have replaced the old flame photometric techniques and recently they have been surpassed by plasma emission procedures, especially inductively coupled plasma atomic emission spectrometry (ICP–AES) and inductively coupled plasma mass spectrometry (ICP–MS). These methods are at present more developed than other plasma emission techniques such as direct current plasma (DCP) or microwave-induced plasma (MIP) spectrometries.

All the aforementioned techniques are adequate for the analysis of liquid samples (except MIP which is more convenient for gaseous samples), and they

---

Element Speciation in Bioinorganic Chemistry, edited by Sergio Caroli.
Chemical Analysis Series, Vol. 135.
ISBN 0-471-57641-7 © 1996 John Wiley & Sons, Inc.

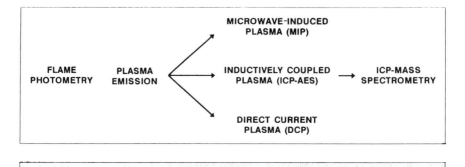

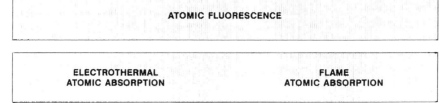

Figure 3.1. Atomic spectrometry methods currently employed for the elemental analysis of biological and environmental samples.

can also be applied to the direct analysis of solids (4–6), preferably dispersed as slurries (7, 8), especially in the case of electrothermal atomization atomic absorption spectrometry (ETA–AAS) (9).

Several appropriate digestion or solubilization methods have been developed for the analysis of the total content of inorganic species in biological samples in order to avoid losses or contaminations during the sample treatment and thus allow elements to be accurately determined.

At present, efforts to increase knowledge about the physiological activity of chemical elements have necessitated not only determination of the total content of these elements but also accurate determination of the concentration of their different chemical forms (10–12).

However, the low residence time of sample particles in the measurement zone and the high temperature of the atomization cells can cause problems in discriminating the chemical forms of an element. As a consequence, AS techniques are not convenient for the *in situ* analysis of elemental species, with only a few exceptions (13).

Generally the strategy for speciation by AS has been based on its hyphenation with separation techniques.

Figure 3.2 summarizes the separation techniques currently most often employed to preconcentrate or to isolate a specific chemical form. Chromatog-

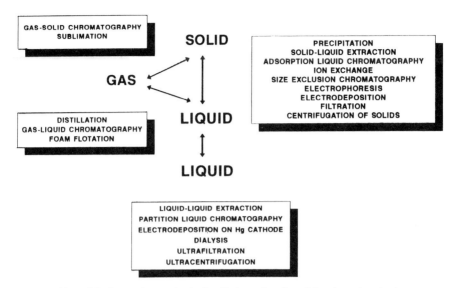

Figure 3.2. Separation methods classified as a function of the phases involved.

raphy is the most useful technique for the simultaneous identification and determination of different species.

Thus, a general consensus would seem to have been reached that AS techniques can be used as specific detectors for speciation studies by gas chromatography (GC) (14–19) or by liquid chromatography (LC) (20–25). In recent years, several studies have been carried out to develop appropriate interfaces between chromatographic instruments and flame spectrometers (25–29), electrothermal atomizers (14, 30, 31), or plasmas (32–38). These aspects are sufficiently covered in the available literature on speciation; there is also some consideration of this topic in Chapter 2 of this book, and it is discussed further in Chapters 4, 7, and 13.

Although there has been significant progress in the chromatography–AS combination, it is important to remember that chromatographic techniques represent only a minor part of the separation procedures available and that, in certain cases, some efforts must be made to apply basic chemistry in sample treatment which could provide quantitative information about specific chemical forms.

In several cases, in fact, AS measurements can be employed for speciation purposes without the necessity of chromatographic techniques. In these instances, the procedures may be based on (i) different atomization yields obtained for different chemicals; (ii) the use of selective extraction; (iii) derivatization procedures, carried out before the measurement step; (iv) selective volatilization of the different chemical forms of the elements to be determined; or (v) other less currently used separation methods.

3.2. SELECTIVE ATOMIZATION OF DIFFERENT CHEMICAL FORMS OF THE SAME ELEMENT

At the high temperatures employed in atomizers such as flames, graphite furnaces, and plasmas, the different forms of an element are usually transformed into oxides or carbides and finally into a vapor phase of atoms in the ground or excited state, independently of their different structures or oxidation states.

However, results obtained from studies on the atomization of various chemicals of a given element show some differences in the transport and atomization of different species. Chromium absorbance is not comparable for Cr(III) or Cr(VI), as the absorbance values obtained are higher for Cr(III) than for the latter, especially when measurements are carried out in reducing flames (39).

By contrast, small variations have been found in the signal obtained for different compounds of an element at the same concentration (40). This can cause problems in speciation studies carried out by coupling chromatographic techniques with AS detectors. In such cases, standardization must be accomplished for each compound individually or, alternatively, by carrying out a pretreatment of the samples to transform all the compounds to chemical forms with the same atomization behavior (41).

The aforemetioned small variations, which have been reported, e.g., in the analysis of Pb additives in gasoline, can be exploited to carry out speciation studies by direct measurement of the absorbance of samples and using different sets of standards for each of the species considered (42–46).

A general procedure for the quantitative determination of two chemical forms M and N of a given element, with different atomization yields, can be based on the measurement of the peak heights, which correspond to the samples, measured under two different sets of instrumental conditions, h_1 and h_2, and on the use of the regression parameters obtained from calibration curves for each of the compounds assayed.

Hence, the linear independent equations can be written as follows:

$$h_1 = (a_{1M} + a_{1N})/2 + b_{1M}C_M + b_{1N}C_N \quad (1)$$

$$h_2 = (a_{2M} + a_{2N})/2 + b_{2M}C_M + b_{2N}C_N \quad (2)$$

where a_{1M} and a_{2M} are the intercepts and b_{1M} and b_{2M} are the slopes, respectively, obtained for calibration plots of M measured for sets 1 and 2 of instrumental conditions, and a_{1N}, a_{2N}, b_{1N}, and b_{2N} are the corresponding parameters obtained for a series of standards of N measured under the same experimental conditions as those employed to determine h_1 and h_2.

Equations 1 and 2 provide a system with two unknowns, such that the concentrations C_M and C_N can be determined without previous separation of the two species.

As regards the determination of Pb in samples containing alkyl derivatives, it has been confirmed that alkyl compounds provide absorbance values higher than those obtained with aqueous Pb(II) in an air/C_2H_2 flame (47). They also provide higher emission signals in a direct current plasma (48). Thus, for the determination of total Pb in these samples, preliminary mineralization has been recommended in order to use aqueous solutions of Pb as standards and to compensate for the different behavior of the metal species.

In a comparison of the results obtained for various alkyl derivatives, it has been found that absorbance values for tetramethyllead (TML) are 1.4 times higher than those obtained for tetraethyllead (TEL). This fact opens the possibility of determining the concentration of TML and TEL in the same sample without the need for a previous chromatographic separation of these compounds despite their different behaviors.

When samples containing TML and TEL are directly introduced into a flame atomic spectrometer, the experimental absorbance peak height obtained can be related to the concentration of Pb as follows:

$$h_t = (a_1 + a_2)/2 + b_1 C_{TEL} + b_2 C_{TML} \tag{3}$$

where h_t is the experimentally measured peak height; a_1 and a_2 are the intercepts of the calibration graphs for TEL and TML, respectively; and b_1 and b_2 are the corresponding slopes. The four parameters of Equation 3 can be established from the calibration plots of each of the two compounds in the same run.

In the foregoing case, two unknowns are present in one equation. The similar behavior of the two Pb compounds under different instrumental conditions (such as the burner height and the gas flows) inhibits the possibility of establishing two equations of the same type from the measurement of the absorbance peak height of the samples under two different sets of instrumental conditions (see Fig. 3.3). Therefore, it is necessary to develop other strategies.

The determination of the total Pb concentration in the same sample through previous demetallation can be used to obtain another equation:

$$C_T = C_{TEL} + C_{TML} \tag{4}$$

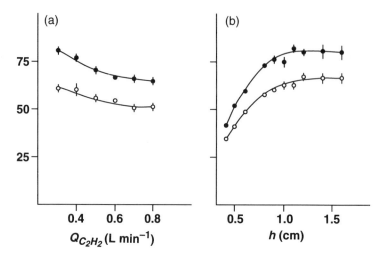

Figure 3.3. Effect of the flame parameter on the absorbance peak height of alkyl-Pb compounds: (a) effect of the acetylene flow ($Q_{C_2H_2}$); (b) effect of the burner height (h). Absorbance values are the average of five independent measurements $\pm$ their standard deviation for 80.6 mg L^{-1} Pb (TML) and 81.4 mg L^{-1} Pb (TEL), respectively (49).

where C_T is the total Pb concentration. Equations 3 and 4 provide a system that enables the concentration of both additives to be determined in the same sample (49).

3.3. SPECIATION BASED ON SELECTIVE EXTRACTIONS

The selective extraction of specific chemical forms of a given element, prior to their analysis by AS, has provided information about the concentration of different chemical forms in a sample, thus increasing the selectivity of the spectrometric measurements. A solid extraction with resins has been employed to preconcentrate Cr species from aqueous solutions. Anionic exchange resins have been also proposed to separate Cr(VI) from Cr(III) (50).

A liquid anion exchanger, Adogen 464, supported on silica gel, has been employed to quantitatively separate Cr(VI) and Cr(III) at pH = 2. Then, the column was washed with distilled water and eluted with concentrated HNO$_3$, Cr being determined by ETA–AAS (51).

Miyazaki and Barnes have carried out the differential determination of Cr(VI) and Cr(III) by ICP–AES using a polydithiocarbamate chelating resin to separate Cr(VI). Then pH of the samples was adjusted between 2.5 and 3.5, and Cr(VI) determined after digestion of the resin with HNO$_3$. Chromium(III)

Table 3.1. Liquid–Liquid Extraction Procedures Employed for Selective Determination of Cr Species

Species	Complexing Agent	Solvent	Matrix	Ref.
Cr(VI)	—	MIBK[a]	Biological material	54
Cr(VI)	Tributyl phosphate	MIBK	Blood, urine	55
Cr(VI)	Ammonium pyrrolidine dithiocarbamate (APDC)	MIBK	—	56
Cr(VI)	Diethyl dithiocarbamate (DDC)	MIBK	—	57
Cr(VI)	Aliquot 336	—	Seawater	58
Cr(VI)	Methyl tricapryl ammonium chloride	MIBK Toluene	— —	59
Cr(VI)	—	MIBK	Urine	60
Cr(III)	—	MIBK	Food	61
Cr(III)	2-Thenoyltrifluoroacetone	Xylene	Water	62

[a] MIBK: 4-methyl-2-pentanone (also called methyl isobutyl ketone, whence the acronym).

was oxidized to Cr(VI) with $KMnO_4$ in an acidic medium and then retained on the chelating resin and determined as indicated previously (52).

Isozaki et al. have developed a rapid and simple procedure for the determination of Cr(III) and Cr(VI), based on the selective extraction of Cr(III) with Chelex-100 at pH 4, followed by direct atomization of a slurry of the resin in a graphite furnace. They determined Cr(VI) by the same procedure, carrying out the extraction with Chelex-100 after reduction to Cr(III) by H_2O_2 (53).

Liquid–liquid extractions have been more frequently employed for speciation purposes than have liquid–solid procedures. As can be seen in Table 3.1, various methods have been proposed for the selective determination of Cr(III) and Cr(VI) (54–62). Other methods for the speciation of the oxidation states of Cr can be designed by coupling selective extraction procedures with AS detection (63).

For the speciation of As compounds by atomic spectrometry, a series of procedures have been developed based on the use of selective liquid–liquid extraction.

Quantitative measurements of inorganic and organic arsenicals in water and urine samples can be achieved by ETA–AAS after an HCl treatment of samples at 60–70 °C for 1 h with KI. Under the foregoing reaction conditions any inorganic arsenate is reduced to arsenite and then converted to AsI_3. On the other hand, iodide compounds are also obtained from methylated As substrates. A solvent extraction of iodide As species with $CHCl_3$ permits the separation of these from the matrix. Speciation is carried out after a selective back-extraction step. Inorganic As species (present in the organic layer as AsI_3) are hydrolyzed with water to H_3AsO_3 and then reextracted in water. The total content of As compounds is determined after a back-extraction with 0.005 M $K_2Cr_2O_7$, which forms oxidized species of methyl arsenicals less volatile than the iodides and capable of being determined by ETA–AAS. The recovery of inorganic As varies from 87 to 93% and that of the methylated arsenicals from 88.4 to 98.7% (64).

Inorganic As(III) has been selectively determined in urine samples by ETA–AAS after extraction in toluene and back-extraction with $Co(NO_3)_2$ (65).

The selective determination of inorganic As and organic compounds in foods has been carried out by Münz and Lorenzen by liquid–liquid extraction and hydride technique. Samples previously mineralized with HNO_3 are treated with ascorbic acid to reduce As(V) to As(III), followed by CH_2Cl_2 extraction from 7–9 M HCl solutions. Inorganic As is determined after a back-extraction of the CH_2Cl_2 phase with H_2O. The determination of organic As is achieved by evaporation of the remaining HCl phase, decomposition with HNO_3, and subseqnent AAS measurement by the hydride generation technique (66).

It has been also reported that the inorganic As species usually found in foods are selectively extracted with diluted H_2SO_4 from the lyophilized solid samples. Under these conditions, organically bound As is stable (67).

For the determination of total inorganic As in fish, a procedure based on its selective extraction has been proposed. Solid samples are treated with NaOH and Na_2SO_4 in a boiling water bath for 20 min. Subsequently, HCl is added and As extracted twice with APDC in MIBK. The As is back-extracted with HNO_3 and then treated with H_2SO_4. Measurements are carried out by the hydride generation technique (68).

Buchet et al. have compared several methods for the determination of various As compounds, such as As(III), As(V), methanearsonic acid, hydroxydimethylarsine oxide, and arsenilic acid in water and in urine by using ETA–AAS, after solvent extraction in toluene, as well as by the AsH_3 generation method (69).

Inorganic As(III) and As(V) are determined in biological materials after solubilization with 6 M HCl, extraction into toluene, and back-extraction with water. Inorganic As(III) is extracted into toluene from 9 M HCl solution, and inorganic As(V) is extracted under the same conditions after reduction with KI. Organic As is retained in the acidic solution. Measurements are performed after back-extraction into water by arsine generation flame AAS. Total As can be determined after digestion with HNO_3, H_2SO_4, and $HClO_4$ (70).

For the selective determination of As(III) in foods, in the presence of As(V), the liquid–liquid extraction of $AsCl_3$ in chloroform has been proposed. The analysis is carried out by hydride generation after back-extraction in water (71). (A more detailed discussion of As speciation is presented in Chapter 13 of this volume.)

Although the most popular method for the speciation of Hg is based on selective reduction to Hg(0) following the procedure proposed by Magos (72), organic Hg can be determined by solvent extraction into toluene and ETA–AAS. For this purpose, the solid residue of an acetone extract from homogenized biological materials is treated with acid bromide and $CuSO_4$, followed by extraction of Hg into toluene. For the stabilization of methyl-Hg in the furnace, dithizone is added to samples and standards and the element atomized at 950 °C (73). The aforementioned method provides an average recovery of $97 \pm 5\%$ and affords an alternative to the gas–liquid chromatographic method employed by the same research group (74).

On-line extraction procedures can help to preconcentrate the elements to be determined and also to carry out a selective separation of the different chemical forms of an element, especially when different oxidation states are considered (75, 76). Thus, a series of methods have been proposed in the last few years for the sequential determination of chemical species by FIA–AS. In this sense, in a single analytical cycle it has been possible to determine sequentially free and ethylenediaminetetraacetic acid (EDTA) complexes of Cu ions by flame AAS using an on-line chelating ion-exchange column incorporated into a monochannel manifold that retains the free but not the complexed ions (77).

The affinity of the acid form of activated alumina toward anionic Cr(VI) has been employed to provide a time-resolved emission of Cr(VI) and Cr(III) by FIA–ICP–AES. Using a single-line manifold with a microcolumn of activated alumina, Cox et al. directly determined Cr(III) without interference from Cr(VI); this last species was analyzed after injection of a 1 M ammonium solution to elute Cr(VI) from the column (78).

3.4. SPECIATION BASED ON DERIVATIZATION PROCEDURES

The introduction of a derivatization step, prior to the AS analysis of several elements, provides an enhancement of the analytical sensitivity and offers new possibilities for speciation purposes.

These procedures are based on selective reduction of the different chemical forms of a given element in order to obtain Hg(0) or covalent hydrides.

For the determination of Hg by atomic spectrometry, one of the most selective and sensitive ways is the use of cold vapor atomic absorption spectrometry (CVAAS), which is based on the low boiling point of liquid Hg and the easy reduction of mercurials to the zero oxidation state. On the other hand, for several elements, such as As, Bi, Ge, Sb, Se, Sn, Te, and—as recently reported—Cd and Pb, the prior formation of hydrides permits matrix effects to be avoided and detection power to be enhanced.

In both cases, spectrometric measurements are carried out on the reduced form of the elements. Hence, an appropriate selection of the experimental parameters for this derivatization step could provide selective procedures for different chemical forms of As, Hg, Sb, and Se.

Table 3.2 summarizes the experimental conditions proposed in the literature for the selective reduction of different chemical forms of Hg for their subsequent determination by CVAAS.

From papers published so far on Hg speciation, it can be concluded that the reduction of mercurials can be carried out selectively by using different reduction steps with $SnCl_2$ or $SnCl_2/CdCl_2$. On the other hand, pH strongly affects the reduction of mercurials by $NaBH_4$, especially in the presence of Fe(III). This opens new possibilities for Hg speciation. Finally, it would seem that a combination of the use of $SnCl_2$ and $NaBH_4$ could be very convenient when analysts are carrying out the speciation of Hg in biological samples (79–87).

In the reduction of Hg compounds by $NaBH_4$, the fact that different sensitivity values were obtained for each kind of mercurial [inorganic Hg, 0.031 absorbance units (AU) $\mu g^{-1} L^{-1}$; CH_3HgCl and CH_3CH_2HgCl, 0.026 AU $\mu g^{-1} L^{-1}$; and C_6H_5HgCl, 0.015 AU $\mu g^{-1} L^{-1}$] could indicate that a strategy combining the method of Borja et al. (49) for the direct speciation of alkyl-Pb compounds and the method developed by Oda and Ingle (87) might well lead to improved results.

Based on the Magos procedure, Coyle and Hartley have developed an automated method for the determination of total and inorganic Hg in blood and urine (86). Samples are treated with L-cysteine and trichloroacetic acid, and then the protein-free supernatant is treated with $SnCl_2$ and $CdCl_2$ for the determination of total Hg, or with $SnCl_2$ for the selective reduction of inorganic Hg.

The analysis of Sb following stibine generation has been improved so as to provide selective determination of Sb(III) and Sb(V) by means of an appropriate selection of the experimental conditions for reduction. Nakashima has selectively determined Sb(III) by AAS in the presence of a fourfold excess of Sb(V) in 0.3 M HCl and in the presence of 4 mg mL^{-1} of Zr(IV) as $ZrOCl_2$ (88). Yamamoto et al. have selectively reduced Sb(III) by $NaBH_4$ at pH 8 (using a borate buffer) (89). The same authors have proposed the selective hydride generation of Sb(III) by $NaBH_4$ at pH 2 in the presence of citric acid and the determination of total Sb in 1 M HCl (90). Analogously, Sb(III) has been reduced to SbH_3 in 1 M HCl and 0.1 M HF, the latter reagent being added to mask Sb(V) (91).

Table 3.2. Experimental Conditions Proposed for the Speciation of Mercurials by CVAAS After Selective Reduction

Species	Reduction Conditions	Matrix	Ref.
$HgCl_2$	Alkaline $SnCl_2$	Feces, kidneys, urine	79
Methoxyethyl-HgCl	L-Cysteine + alkaline Sn(II) chloride	Idem	79
C_2H_5–HgCl	L-Cysteine (heated for 1 h at 100 °C) + alkaline $SnCl_2$	Idem	79
C_6H_5–HgCl	Idem	Blood	79
Inorganic Hg	$SnCl_2$	Undigested biologicals	72
Methyl-Hg	$SnCl_2 + CdCl_2$	Idem	72
Inorganic Hg	NaCl + L-cysteine + sample + NaOH + $SnCl_2$	Blood	80
Organic Hg	Sample without inorganic Hg + $SnCl_2$ + $CdCl_2$	Idem	80
CH_3–HgCl	Steam distilled in HCl + NaCl + NaOH + $SnCl_2$	Idem	81
Inorganic Hg	Reduced in H^+ with $SnCl_2$	Idem	81
Inorganic Hg	L-Cysteine in NaOH + H^+ + $SnCl_2$	Urine	82
Inorganic Hg and C_6H_5–HgCl	$H_2SO_4 + SnCl_2$	Idem	82
Inorganic Hg	(pH 10–12) $NaBH_4$	Water	83
Total Hg	(pH 2–3) $NaBH_4$ + Fe(III)	Idem	83
C_nH_{2n+1}–HgCl	KOH + Cu(II) (heated at 100 °C for 30 min) + steam distillation in $CuSO_4$ + HCl + trapped in NH_4VO_3, H_2SO_4 and $K_2S_2O_3$	Fish	84
Inorganic Hg	$SnCl_2$	Blood, urine, hair tissues	85
Total Hg	$SnCl_2 + CdCl_2$	Idem	85
Inorganic Hg	L-cysteine + TCA[a] + $SnCl_2$	Urine, blood	85
Total Hg	L-Cysteine + TCA + $SnCl_2 + CdCl_2$	Idem	86
Inorganic Hg	KOH (heated at 90 °C for 15–30 min) + 1% NaCl + $SnCl_2 + K_2Cr_2O_7$ + HNO_3	Hair, fish urine, water	87
Organic Hg	$NaBH_4$ in KOH	Idem	87

[a] TCA: trichloroacetic acid.

In the simultaneous determination of As, Bi, Sb, Se, and Te by ICP–AES, after a previous generation of their gaseous hydride, Thompson et al. have found that in strongly acid media, provided by HCl concentrations higher than or equal to 1 M, Sb(III) and Sb(V) are both reduced. However, the emission peaks obtained for Sb(V) are lower and broader than those found for Sb(III). They have been employed for the speciation of Sb based on a combination of derivatization and calibration with a series of Sb(III) and Sb(V) standards (92).

Methyl-Sb, Sb(III), and Sb(V) species can also be differentiated by ETA–AAS after selective reduction with $NaBH_4$. Antimony(III) is reduced to SbH_3 at almost neutral pH. In the presence of KI and at pH = 1, Sb(III) and Sb(V) are reduced to SbH_3. Methyl-Sb acids, such as methylstibonic acid and dimethylstibinic acid, are reduced by $NaBH_4$ at pH = 4 to methylstibine and dimethylstibine in the absence of KI. In the procedure developed by Andreae et al. the stibine compounds are collected in a cold trap containing a chromatographic phase and separated chromatographically (93).

Further to what has been reported in Section 3.3 concerning the speciation of As, derivatization procedures turn out to be quite suited to the identification of the various chemical forms of this element. Shaikh and Tallman have developed specific methods for the determination of As in natural waters by arsine generation and ETA–AAS (94).

At pH 4–5 generation of the hydride of As with $NaBH_4$ permits the selective determination of As(III) in mixtures of As(III) and As(V) (95). On the other hand, total As can be determined by hydride generation from 5 M HCl. For the speciation of As directly in solid biological samples and to avoid the oxidation of As(III), Shaikh and Tallman propose the use of concentrated HNO_3 and H_2SO_4 (the official wet digestion procedure*) after 1:2 dilution with water (94).

Howard and Arbab-Zavar have determined As(III), As(V), monomethylarsonic, and dimethylarsinic species using a continuous flow system, by (i) selective hydride generation at pH 5 for As(III); (ii) hydride generation in 1 M HCl for total inorganic As and alkyl As compounds; (iii) condensation of arsines; and (iv) temperature-controlled evolution of arsine, methylarsine, and dimethylarsine (96).

In general, all these methods are based on hydride generation, cold trapping, and fractional volatilization of arsine and methylarsinic compounds.

It is well known that in the absence of KI and at pH = 5, $NaBH_4$ only reduces As(III). However, in strong acidic media (2 M HCl) and in the presence of KI, total As can be measured as AsH_3 (91).

---

*Prescribed by European Union Norms and Directives.

When it is necessary to determine As(III), As(V), monomethylarsonic acid (MMAA), and dimethylarsinic acid (DMAA) in the same sample, a careful selection of pH conditions, chelating agents, and redox agents must be carried out to ensure the selective reduction of these compounds by $NaBH_4$. Anderson et al. have proposed the following media for the determination of As(III), As(V), MMAA, and DMAA: (i) 5 M HCl and 0.1% m/v KI for total inorganic As determination; (ii) citric acid–0.4 M sodium citrate and pH = 6 for selective As(III) determination; (iii) 0.16 M acetic acid for the determination of As(III) and DMAA or for the selective determination of DMAA after oxidation of As(III) to As(V) with MnO_4^-; and (iv) 0.1 M mercaptoacetic acid for the determination of As(III), As(V), MMAA, and DMAA (97).

An example that highlights the possibilities offered by the traditional chemical procedures is the speciation of As(III) and As(V) by using the old reduction method proposed by Fleitman in 1851. In a paper recently published, the As(III) species are selectively reduced in a flow injection analysis (FIA) manifold by Al in a NaOH medium and the arsine is subsequently determined by AAS using a quartz cell. The same method is employed to determine the total content of As after the introduction of a KI stream, which permits the prior reduction of As(V) to As(III) (98).

Figure 3.4 schematically shows the manifold employed.

Although the power of detection of the As determination by the aforementioned procedure is not as good as that obtained with $NaBH_4$, the procedure is less expensive and avoids the use of HCl, which may accelerate the irreversible devitrification of the quartz cell.

For the determination of Se species in natural waters, Cutter has observed that when methylated Se species are present they are reduced to alkyl-Se hydrides and released from the reduced solutions at different rates, providing different responses than the H_2Se evolved from selenite (99). This fact can be

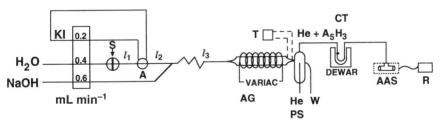

Figure. 3.4. Schematic diagram of the arsine generation manifold. *Key*: AG, electrically heated Pyrex continuous flow hydride generator filled with metallic aluminum; T, thermocouple; PS, gas–liquid phase separator; CT, liquid nitrogen cold trap; AAS, atomic absorption spectrometer; R, recorder; l_1, l_2, and l_3 denote three lengths of tubing; S, sample injection point; A, a three-way Teflon valve employed to introduce KI (98).

employed for speciation studies by selective calibration and the use of appropriate media to carry out the reduction of compounds or to provide the separation of the resulting derivatives (99).

3.5. SPECIATION BASED ON SELECTIVE VOLATILIZATION

As pointed out in the previous section, volatile compounds are very useful as derivatives to improve the sensitivity and selectivity of AS measurements.

On the other hand, as regards organometallics (which are of great interest from a toxicological point of view), these compounds are in general liquids or low-boiling solids, so that their direct determination can be carried out in the gas phase after heating of the sample.

In this section only speciation procedures based on the thermal volatilization of different compounds are included. Thus, on the one hand, the speciation of various covalent hydrides, formed under the same experimental conditions, is considered; on the other hand, the direct volatilization of alkyl derivatives from the bulk sample is described as a general approach to speciation procedures.

Edmonds and Francesconi have suggested that in the determination of As species by hydride generation and AAS the fractional volatilization of arsines, after separation in a cold trap, can provide the selective estimation of methylated As substrates (100, 101).

Arsenate, arsenite, MMAA, and DMAA, prereduced to arsine and to the corresponding methylarsines at pH 1–2 by $NaBH_4$, can be retained in a liquid-nitrogen-cooled U-tube trap and then may be transported out of the trap by a carrier gas. The speciation of the different compounds is accomplished by warming the U-tube trap [AsH_3 has a boiling point of $-55\,°C$; CH_3AsH_2, of $2\,°C$; $(CH_3)_2AsH$, of $35.6\,°C$; $(CH_3)_3As$, of $70\,°C$; and $C_6H_5AsH_2$, of $148\,°C$]. Analysis is carried out by atomic emission in a direct current discharge cell (102).

For the determination of Sn(IV) and organo-Sn compounds Hodge et al. have developed a flame AAS method based on the prior reaction with $NaBH_4$ in $2 \times 10^{-2}\,M$ acetic acid to produce volatile hydrides that are subsequently trapped in a glass-wool-packed column at the temperature of liquid nitrogen (103). After this step, the SnH_4 and organo-Sn hydrides are selectively evolved by increasing the temperature of the hydride trap. A series of compounds, such as Sn(IV) and the halides of methyl-, dimethyl-, trimethyl-, diethyl-, triethyl-, n-butyl-, di-n-butyl-, tri-n-butyl-, and phenyl-Sn can be separated by increasing temperature from $-52\,°C$, the boiling temperature of SnH_4, to $150\,°C$, for phenyl-Sn trihydride, and $250\,°C$ for dibutyl-Sn dihydride. The limits of detection obtained in a hydrogen/air flame (at a wavelength of 286.3 nm) vary from 0.4 ng for Sn(IV) to 2 ng for tri-*n*-butyl-Sn chloride. A similar approach has been employed by Braman and Tompkins (104). At pH = 6.5 (2 M tris(hydroxymethyl)aminomethane hydrochloride buffer solution) inorganic Sn and methyl-Sn compounds are reduced to the corresponding volatile hydride SnH_4 [boiling point (bp) $-52\,°C$], CH_3SnH_3 (bp 0 °C), $(CH_3)_2SnH_2$ (bp 35 °C), and $(CH_3)_3SnH$ (bp 59 °C) by reaction with $NaBH_4$. Subsequently, they are cryogenically trapped in a U-tube and separated upon warming. The limit of detection obtained is approximately 0.01 ng Sn when a hydrogen/air flame emission detector is used. Measurements are done at 809.5 nm, but the various Sn compound provide different analytical sensitivities, which open up new possibilities for speciation based on the diverse evolution, transport, and atomization processes of the Sn derivatives. Thus, the sensitivity is 6.6 emission units per nanogram of Sn (EU ng^{-1}) for methyltin; for dimethyl-Sn it is 18 EU ng^{-1}; for trimethyl-Sn it is 8.5 EU ng^{-1} (98).

All the aforementioned procedures are based on the selective volatilization of hydrides unselectively generated by a derivatization step. The low boiling points of a large number of organometallics offer great possibilities for

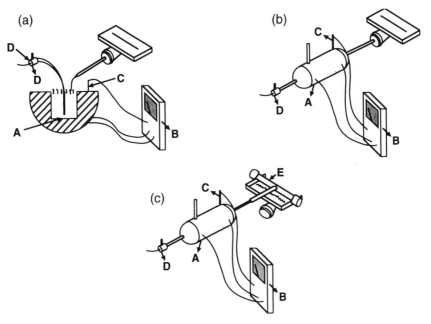

Figure 3.5. Volatilization systems employed for the direct introduction of samples in vapor form by using (a) a vertical reactor directly coupled to a nebulizer; (b) a horizontal reactor coupled to a nebulizer; or (c) the same reactor coupled to a heated quartz tube. *Key*: A; reactor; B, temperature control; C, thermocouple; D, needle valve; E, quartz cell (105).

carrying out speciation studies without a previous derivatization step. On this basis, new systems for the direct introduction of samples as a vapor phase have been proposed.

Figure 3.5 shows three different volatilization systems described in the literature for the direct introduction of samples as a vapor, by means of a nebulizer, or in a heated quartz tube (105).

In the aforementioned systems the exact control of the volatilization temperature helps in discriminating compounds that have different boiling points, which is very useful in the analysis of alkyl compounds (106, 107).

Results summarized in Table 3.3 for the direct analysis of tetramethyltin (TMT) and tetraethyltin (TET) show that this approach offers both the possibility of determining the total content of Sn (analogous sensitivity values were obtained for TMT and TET at 136 and 200°C using the quartz cell) and also the differentiation between these two chemical forms of Sn (especially when samples are introduced through the nebulization chamber or when low reactor temperatures are employed preliminarily to introduction of samples in the quartz cell).

The foregoing procedure has been applied to total Pb analysis in samples containing TML and TEL (using system b in Fig. 3.5, at 150°C and with a carrier nitrogen flow of 6 L min^{-1}) and for the speciation of TML and TEL (using system a in Fig. 3.5, at 180°C and with a carrier nitrogen flow of 3.5 L min^{-1}) (107).

3.6. MISCELLANEOUS METHODS

Other speciation methods involving analysis by AS are based on simple gel filtration or ultrafiltration procedures, coprecipitation, and certain electrochemical separation methods such as electrodeposition and electrophoresis (108–112).

Filtration methods are very appropriate for the determination of metallic elements bound to high-molecular-weight molecules such as proteins.

In the determination of Cu and Zn in human blood the identification of the free ionic species and those that are protein bound can be carried out after gel filtration of the samples by ETA–AAS, as suggested by Gardiner et al. (108).

In recent years Pt compounds have been extensively used for the treatment of certain tumors. Since only the so-called free (i.e., bound to low-molecular-weight organic moieties) Pt has cancerostatic activity, in the analysis of biological fluids of patients treated with Pt salts it is necessary to quantify these species separately from those bound to macromolecules. Ionic exchange and ultrafiltration techniques have been used in combination with traditional AS techniques for the determination of the different Pt fractions in plasma and serum (112–115).

Table 3.3. Analytical Parameters Obtained by Using a Horizontal Reactor and Introducing the Sample Through a Nebulizer and in a Quartz Cell

Compound[a]	Injected Volume (μL)	Flow of Carrier Gas (L min^{-1})	Sensitivity (μg^{-1})	Coefficient of Variation	Detection Limit (ng)
TMT	20	6	0.064	5.7	70
TET	20	6	0.039	3.3	192
TMT	20	4.6	0.052	2.5	96
TET	20	4.6	0.033	3.7	154
TMT	50	4.6	0052	1.5	86
TET	50	4.6	0.017	2.7	266
TMT	50	3.5	0.033	2.6	138
TET	50	3.5	0.011	4.5	421

Compound[b]	Reactor Temperature (°C)	Sensitivity (μg^{-1})	Coefficient of Variation	Detection Limit (ng)
TMT	75	0.42	16.7	15
TET	75	0.21	8.7	28
TMT	110	0.75	4.3	9
TET	110	0.44	4.0	15
TMT	116	0.46	1.6	13
TET	116	0.39	2.2	15
TMT	136	0.41	4.8	15
TET	136	0.39	1.9	15
TMT	160	0.19	9.5	16
TET	160	0.32	2.3	10
TMT*	200	0.19	4.5	21
TET*	200	0.20	4.5	20

Source: Mauri et al. (105).

[a] Values were obtained by introducing the samples through the nebulization chamber and by working with a reactor temperature of 230 °C, an acetylene flow of 0.7 L min^{-1}, and a burner height of 15 nm.

[b] Values were obtained using a quartz cell and an acetylene flow of 0.7 L min^{-1}—except those marked by an asterisk, which were obtained with an acetylene flow of 0.2 L min^{-1} (105).

Coprecipitation of Cr(III) with Fe(III) oxide has been proposed to determine selectively by AAS this oxidation state of Cr in the presence of Cr(VI) (110).

Electrodeposition of Cr with Hg onto pyrolytic graphite-coated tubular furnaces permits the speciation of Cr(III) and Cr(VI) compounds because, at pH 4.7 and using a deposition potential of -1.8 V, both oxidation states of Cr are reduced to metallic Cr, whereas, at the same pH and at -0.3 V, only Cr(VI) is selectively reduced (111).

On the other hand, Pinta et al. have analyzed traces of Cu, Fe, and Pb by ETA–AAS and Zn by flame AAS, after a preliminary electrophoretic separation of the various seric proteins of crustacean decapods, in order to quantify the metal aliquot taken up by the proteins (112). Analyses have been carried out on electrophoretogram portions dissolved in HNO_3, and results obtained for the different elements considered have been compared with the electrophoretic migration pattern of seric proteins.

3.7. CONCLUDING REMARKS

All the information given so far shows that the speciation of chemical elements in biological systems can be obtained through the hyphenation of any number of separation techniques with AS techniques, and not only through the combination of chromatography and atomic spectrometry.

The development of automated procedures of analysis based on FIA techniques offers new possibilities for the on-line treatment of samples. As will be extensively discussed in Chapter 4, we can expect that there will soon be substantial improvements in simple and low-cost procedures for elemental speciation based on on-line separation and AS determination.

In this sense, some separation procedures that have always been very tedious, e.g., those based on the selective distillation of chemical species, might well be expedited by automated speciation analyses of metallic elements (68, 116, 117).

On-line thermal treatment of samples and gas–liquid separators provide an easy way to achieve the selective analysis of compounds that boil at different temperatures, e.g., TEL and TML. On the other hand, the method proposed by Lunde for the separation and analysis of organic-bound and inorganic As species in marine organisms by neutron activation and X-ray fluorescence analysis (based on the extraction of As as $AsCl_3$ and subsequent removal by distillation at 100 °C) can be adapted for the analysis of As species by flame AAS and ETA–AAS after on-line microwave-assisted distillation of $AsCl_3$ (117).

ACKNOWLEDGMENTS

The author acknowledges the constant help of Professor Amparo Salvador of the University of Valencia and also the careful revision of the text carried out by Professor Nick Serpone of Concordia University, Montreal, Canada, Professor Edmondo Pramauro of Turin University, Italy, and Professor Dimiter Tsalev of Sofia University, Bulgaria.

REFERENCES

1. D. L. Tsalev, *Atomic Absorption Spectrometry in Occupational and Environmental Health Practice*. CRC Press, Boca Raton, FL, 1984.
2. J. C. Van Loon, *Selected Methods of Trace Metal Analysis*. Wiley, New York, 1985.
3. G. W. Ewing, ed., *Environmental Analysis*, Academic Press, New York, 1977.
4. B. V. L'vov, *Talanta* **23**, 109 (1976).
5. F. J. Langmyhr and G. Wibetoe, *Prog. Anal. At. Spectrosc.* **8**, 193 (1985).
6. Z. A. de Benzo, M. Velosa, C. Ceccarelli, M. de la Guardia, and A. Salvador, *J. Anal. Chem.* **339**, 235 (1991).
7. R. Martínez Avila, V. Carbonell, M. de la Guardia, and A. Salvador, *J. Assoc. Off. Anal. Chem.* **73**, 389 (1990).
8. L. Ebdon, M. E. Foulkes, and S. Hill, *Microchem. J.* **40**, 30 (1989).
9. C. Bendicho and M. T. C. de Loos-Vollebregat, *J. Anal. At. Spectrom.* **6**, 353 (1991).
10. O. Hutzinger, ed., *The Handbook of Environmental Chemistry*. Springer-Verlag, Berlin, 1980.
11. F. E. Brinekman and M. Bernhart, eds., *The Importance of Chemical Speciation in Environmental Processes*. Springer-Verlag, Berlin, 1985.
12. G. E. Batley, *Trace Element Speciation: Analytical Methods and Problems*. CRC Press, Boca Raton, FL, 1989.
13. S. Arpadjan and V. Krivan, *Anal. Chem.* **58**, 2611 (1986).
14. G. Schwedt and H. A. Russel, *Fresenius' Z. Anal. Chem.* **264**, 301 (1973).
15. D. A. Segar, *Anal. Lett.* **7**, 89 (1984).
16. F. J. Fernández, *At. Absorpt. Newsl.* **16**, 33 (1977).
17. J. C. Van Loon, *Anal. Chem.* **51**, 1139 (1979).
18. L. Ebdon, S. J. Hill, and R. W. Ward, *Analyst (London)* **111**, 1113 (1986).
19. Y. K. Chan and P. T. S. Wong, in *Trace Element Speciation: Analytical Methods and Problems* (G. E. Batley, ed.), Chapter 7, pp. 219–244. CRC Press, Boca Raton, FL, 1989.
20. J. C. Van Loon, *Can. J. Spectrosc.* **26**, 22A (1981).
21. K. Fuwa, H. Haraguchi, M. Morita, and J. C. Van Loon, *Bunko Kenkyu* **31**, 289 (1982).

22. L. Ebdon, S. Hill, and R. W. Ward, *Analyst (London)* **112**, 1 (1987).
23. Y. K. Chau, *Sci. Total Environ.* **49**, 305 (1986).
24. K. J. Irgolic, *Sci. Total Environ.* **64**, 61 (1987).
25. L. Ebdon, S. Hill, A. P. Walton, and R. W. Ward, *Analyst (London)* **113**, 1159 (1988).
26. W. A. Aue and H. H. Hill, *Anal. Chem.* **45**, 729 (1973).
27. H. Kawaguchi, T. Sakamoto, and A. Mizuike, *Talanta* **20**, 321 (1973).
28. D. R. Jones, IV and S. E. Managhan, *Anal. Chem.* **48**, 1887 (1976).
29. G. E. Parris, W. R. Blair, and F. E. Brinckman, *Anal. Chem.* **49**, 378 (1977).
30. L. Ebdon, R. W. Ward, and D. A. Leathard, *Analyst (London)* **107**, 129 (1982).
31. F. E. Brinckman, W. R. Blair, K. L. Jewett, and W. P. Iverson, *J. Chromatogr. Sci.* **15**, 493 (1977).
32. W. De Jonghe, D. Chakraborti, and F. Adams, *Anal. Chim. Acta* **115**, 89 (1980).
33. C. I. M. Beenakker, *Spectrochim. Acta* **31B**, 173 (1977).
34. D. L. Windsor and M. B. Denton, *J. Chromatogr. Sci.* **17**, 492 (1979).
35. C. H. Gast, J. C. Kraak, H. Hoppe, and F. J. M. J. Maessen, *J. Chromatogr.* **185**, 549 (1979).
36. I. S. Krull and S. Jordan, *Am. Lab.* **12**, 21 (1980).
37. D. W. Hansler and L. T. Taylor, *Anal. Chem.* **53**, 1223 (1981).
38. D. O. Duebelbeis, S. Kapila, D. E. Yates, and S. E. Manahan, *J. Chromatogr.* **351**, 465 (1986).
39. D. L. Tsalev, *Atomic Absorption Spectrometry in Occupational and Environmental Health Practice*, Vol. 2, p. 53. CRC Press, Boca Raton, FL, 1984.
40. H. Haraguchi, Y. Watanabe, Y. Nojiri, T. Hasegawa, and K. Fuwa, *Recent Adv. Anal. Spectrosc., Proc. Int. Conf. At. Spectrosc., 9th*, Tokyo, *1981*, p. 94 (1982).
41. J. C. Van Loon, *Selected Methods of Trace Metal Analysis*, p. 303. Wiley, New York, 1985.
42. American National Standard, ASTM D 1949–64.
43. K. Campbell and J. M. Palmer, *J. Inst. Petrol.* **58**, 193 (1972).
44. M. Kashiki, S. Yamazoe, and S. Oshima, *Anal. Chim. Acta* **53**, 95 (1971).
45. E. Lindemanis, *Analytical Method 113/71*. E. I. du Pont de Nemours and Co.
46. R. J. Lukasiewicz, P. H. Berens, and B. E. Buell, *Anal. Chem.* **47**, 1045 (1975).
47. V. Berenguer, J. L. Guiñón, and M. de la Guardia, *Fresenius' Z. Anal. Chem.* **294**, 416 (1979).
48. M. de la Guardia, A. Salvador, and V. Berenguer, *Analusis* **9**, 1 (1981).
49. R. Borja, M. de la Guardia, A. Salvador, J. L. Burguera, and M. Burguera, *J. Anal. Chem.* **338**, 9 (1990).
50. J. F. Pankow and G. E. Janaur, *Anal. Chim. Acta* **69**, 97 (1974).
51. P. Battistoni, S. Bompadre, G. Fava, and G. Gobbi, *Talanta* **30**, 15 (1983).
52. A. Miyazaki and R. Barnes, *Anal. Chem.* **53**, 364 (1981).

53. A. Isozaki, K. Kumagai, and S. Utsumi, *Anal. Chim. Acta* **153**, 15 (1983).
54. F. J. Feldman, E. C. Knoblock, and W. C. Purdy, *Anal. Chim. Acta* **38**, 489 (1967).
55. G. Devoto, *Boll. Soc. Ital. Biol. Sper.* **44**, 1251 (1968).
56. T. R. Gilbert and A. M. Clay, *Anal. Chim. Acta* **67**, 289 (1973).
57. K. Hiro, T. Owa, M. Takaoka, T. Tanaka, and A. Kawahara, *Bunseki Kagaku* **25**, 122 (1976).
58. G. J. de Jong and U. A. Th. Brinkman, *Anal. Chim. Acta* **98**, 243 (1978).
59. S. S. Chao and E. E. Pickett, *Anal. Chem.* **52**, 335 (1980).
60. C. Minoia, M. Colli, and L. Pozzoli, *At. Spectrosc.* **2**, 163 (1981).
61. F. J. Jackson, J. I. Read, and B. E. Lucus, *Analyst (London)* **105**, 359 (1980).
62. R. V. Whiteley, Jr. and R. M. Merrill, *Fresenius' Z. Anal. Chem.* **331**, 7 (1982).
63. V. M. Ras and M. N. Sastri, *Talanta* **27**, 771 (1980).
64. A. W. Fitchett, E. H. Daughtrey, Jr. and P. Mushak, *Anal. Chim. Acta* **79**, 93 (1975).
65. R. R. Lauwerys, J. P. Buchet, and H. Roels, *Arch. Toxicol.* **41**, 239 (1979).
66. H. Münz and W. Lorenzen, *Fresenius' Z. Anal. Chem.* **319**, 385 (1984).
67. H. A. M. G. Vaessen and A. Van Ooik, *Z. Lebensm. Unters. Forsch.* **189**, 232 (1989).
68. P. J. Brooke and W. H. Evans, *Analyst (London)* **106**, 514 (1981).
69. J. P. Buchet, R. Lauwerys, and H. Roels, *Int. Arch. Occup. Environ. Health* **46**, 11 (1980).
70. A. Yasui, C. Tsutsumi, and S. Toda, *Agric. Biol. Chem.* **42**, 2139 (1978).
71. W. Holak and J. J. Specchio, *At. Spectrom.* **12**, 105 (1991).
72. L. Magos, *Analyst (London)* **96**, 847 (1971).
73. G. T. C. Shum, H. C. Freeman, and J. F. Uthe, *Anal. Chem.* **51**, 414 (1979).
74. J. F. Uthe, J. Solomon, and B. Grift, *J. Assoc. Off. Anal. Chem.* **55**, 583 (1972).
75. V. Carbonell, A. Salvador, and M. de la Guardia, *J. Anal. Chem.* **342**, 529 (1992).
76. G. E. Pacey and B. P. Bubins, *Int. Lab.*, 26 (1984).
77. E. B. Milosavljevic, J. Růžička, and E. H. Hansen, *Anal. Chim. Acta* **169**, 321 (1985).
78. A. G. Cox, J. G. Cook, and C. W. McLeod, *Analyst (London)* **110**, 331 (1985).
79. J. C. Gage and J. M. Warren, *Ann. Occup. Hyg.* **13**, 115 (1970).
80. U. Ebbestad, N. Gundersen, and T. Torgrimsen, *At Absorpt. Newsl.* **14**, 142 (1975).
81. K. Mitani, *Eisei Kagaku* **22**, 65 (1976).
82. A. Campe, N. Velghe, and A. Claeys, *At. Absorpt. Newsl.* **17**, 100 (1978).
83. H. Mizunuma, H. Morita, H. Sakarai, and S. Shimomura, *Bunseki Kagaku* **28**, 695 (1979).
84. D. L. Collett, D. E. Fleming, and G. A. Taylor, *Analyst (London)* **105**, 897 (1980).

85. J. P. Farant, D. Brissette, L. Moncion, L. Bigras, and A. Chatrand, *J. Anal. Toxicol.* **5**, 47 (1981).
86. P. Coyle and T. Hartley, *Anal. Chem.* **53**, 354 (1981).
87. C. E. Oda and J. D. Ingle, Jr., *Anal. Chem.* **53**, 2305 (1981).
88. S. Nakashima, *Analyst (London)* **105**, 732 (1980).
89. M. Yamamoto, K. Urata, and Y. Yamamoto, *Anal. Lett.* **14**, 21 (1981).
90. M. Yamamoto, K. Urata, K. Murashige, and Y. Yamamoto, *Spectrochim. Acta* **36B**, 671 (1981).
91. K. Tsujii, *Anal. Lett.* **14**, 181 (1981).
92. M. Thompson, B. Pahlavanpour, S. J. Walton, and G. F. Kirkbright, *Analyst (London)* **103**, 568 (1978).
93. M. O. Andreae, J. F. Asmodé, P. Foster, and L. Van't dack, *Anal. Chem.* **53**, 1766 (1981).
94. A. U. Shaikh and D. E. Tallman, *Anal. Chim. Acta* **98**, 251 (1978).
95. J. Aggett and A. C. Aspell, *Analyst (London)* **101**, 341 (1976).
96. A. G. Howard and M. H. Arbab-Zavar, *Analyst (London)* **106**, 213 (1981).
97. R. K. Anderson, M. Thompson, and E. Culbard, *Analyst (London)* **106**, 1143 (1986).
98. M. Burguera, J. L. Burguera, M. R. Brunetto, M. de la Guardia, and A. Salvador, *Anal. Chim. Acta* **261**, 105 (1991).
99. G. Cutter, *Anal. Chim. Acta* **98**, 59 (1978).
100. J. S. Edmonds and K. A. Francesconi, *Nature (London)* **265**, 436 (1977).
101. J. S. Edmonds and K. A. Francesconi, *Anal. Lett.* **48**, 2019 (1976).
102. R. S. Braman, D. L. Johnson, C. C. Foreback, J. M. Ammons, and J. L. Bricker, *Anal. Chem.* **49**, 621 (1977).
103. V. F. Hodge, S. L. Saidel, and E. D. Goldberg, *Anal. Chem.* **51**, 1256 (1979).
104. R. S. Braman and M. A. Tompkins, *Anal. Chem.* **51**, 12 (1979).
105. A. R. Mauri, C. Mongay, and M. de la Guardia, *Microchem. J.* **42**, 176 (1990).
106. M. de la Guardia, A. R. Mauri, and C. Mongay, *Analusis* **18**, 440 (1990).
107. A. R. Mauri, M. de la Guardia, and C. Mongay, *J. Anal. At. Spectrom.* **4**, 539 (1989).
108. P. E. Gardiner, J. M. Ottaway, G. S. Tell, and R. R. Burns, *Anal. Chim. Acta* **124**, 231 (1981).
109. D. A. Hull, N. Muhammad, J. G. Lanesa, S. D. Reich, T. T. Finkelstein, and S. Fandrich, *J. Pharm. Sci.* **70**, 500 (1981).
110. R. Fukai and D. Vas, *J. Oceanogr. Soc. Jpn.* **23**, 298 (1967).
111. G. E. Batley and J. P. Matousek, *Anal. Chem.* **52**, 1570 (1980).
112. M. Pinta, D. Baron, C. Riandey, and W. Ghidalia, *Spectrochim. Acta* **33B**, 489 (1978).

113. S. J. Bannister, Y. Chang, L. A. Sternson, and A. J. Repta, *Clin. Chem. (Winston-Salem, N. C.)* **24**, 877 (1978).
114. C. Dominici, A. Alimonti, S. Caroli, F. Petrucci, and M. A. Castello, *Clin. Chim. Acta* **152**, 207 (1986).
115. S. Caroli, A. Alimonti, F. Petrucci, F. La Torre, C. Dominici, and M. A. Castello, *Ann. Ist. Super. Sanità* **25**, 487 (1989).
116. American National Standard, ASTM D 1949–79.
117. G. Lunde, *J. Sci. Food Agric.* **24**, 1021 (1973).

CHAPTER

4

DEVELOPMENT OF NEW METHODS OF SPECIATION ANALYSIS

I. T. URASA

*Department of Chemistry, Hampton University,
Hampton, Virginia 23668*

4.1. INTRODUCTION

The toxicity, bioavailability, and distribution of an element in solution can only be determined if all its various forms can be identified and quantified (1–9). Since an element can exist in a sample in many different forms, including particulate, dissolved, complexed, and ionic forms, as well as in different oxidation states, it is imperative that the analytical technique used is able to provide accurate experimental data on all of these forms. It is standard practice, moreover, to employ more than one analytical procedure, often requiring two or more analytical techniques, in order to achieve this objective. Ideally, *in situ* measurements would give the most accurate results because these would avoid errors that can arise during sample collection, processing, and analysis. However, in the absence of suitable methods with *in situ* measurement capability, a number of alternative approaches have been employed by analytical and environmental chemists. These range from the conversion of all forms of the element into one form, to physical separation followed by quantification of the separated species.

In this chapter a review of the various approaches for element speciation will be given to show their merits. This will be followed by a detailed discussion of a method used in the author's laboratory that employs liquid chromatography (LC) in combination with atomic emission spectrometry (AES). Finally, specific applications of element speciation protocols in a variety of analytical problem will be discussed.

Element Speciation in Bioinorganic Chemistry, edited by Sergio Caroli.
Chemical Analysis Series, Vol. 135.
ISBN 0-471-57641-7 © 1996 John Wiley & Sons, Inc.

4.2. FACTORS AFFECTING ELEMENT SPECIATION

4.2.1. Sample Treatment and Processing

Assuming that the sample being analyzed is in liquid form, two general analytical measurement approaches can be used: (i) the sample is introduced directly into the measurement system without any processing, or (ii) the sample is filtered and/or acidified before introduction into the measurement system. Sample filtration or acidification depends on several factors including the nature of the information sought for, sample characteristics, and the analytical technique employed.

Sample filtration is used as a means of separating dissolved species from undissolved species and to eliminate solid material that might adversely affect the quality of the analytical data. Typically, membrane filters with a 0.45 μm pore size are used for environmental water samples (10). Element speciation on the basis of particle size distribution has been studied using filters of nominal pore sizes ranging from 0.01 to 100 μm. Variations were observed that depended on the element measured, the type of filter used, and sample characteristics (11–17).

From a study conducted in the author's laboratory it was shown that for unfiltered samples, where average particle size is larger than 0.2 μm, these particles serve as adsorption sites for elements such as Al, Fe, and Mn, which have a tendency to adsorb onto solid surfaces (18). As these particles agglomerate and settle, they are removed from solution and can prevent the measurement of metal ions adsorbed on them. Such agglomeration and settling effects can be minimized if the sample is shaken prior to its introduction into the measurement system and provided that the measurement system is not adversely affected by the presence of particulate matter in the sample.

The effect of sample filtration on the concentration of Fe measured in natural water samples was studied and reported by Urasa and O'Reilly (18). As shown in Fig. 4.1, the Fe concentration was found to decrease exponentially with time when an unfiltered sample was used, whereas for filtered samples the measured Fe concentration remained constant. The decrease in the measured Fe was attributed to several processes, including agglomeration and subsequent gravitational settling of the metal-containing particles, adsorption on the walls of the container, and loss that may occur during transport of the sample into the measurement system. It was found in this study that a 0.45 μm pore filter can remove substantial amounts, sometimes more than 90% of the measurable Fe, which is regarded as beng associated with the "solids" in the solution.

Whether a settling effect is observed for a given sample depends on the concentration of the particles in it. This is demonstrated in Fig. 4.2, showing

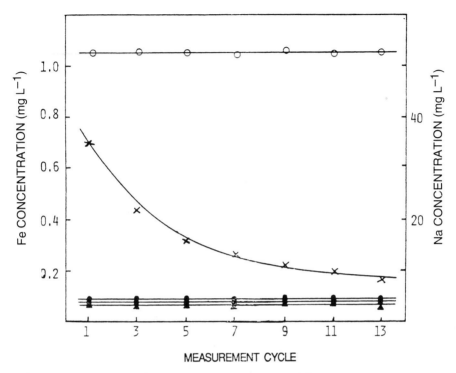

Figure 4.1. Concentration of Fe found in a natural water sample in measurements made at 70 s intervals. Iron concentrations: unfiltered sample (×); filtered with 0.80 μm (●), 0.45 μm (■), and 0.20 μm (▲). Sodium concentration found in filtered and unfiltered sample is also shown for comparison (○).

a natural water sample for which particle distribution differs considerably from the sample shown in Fig. 4.1. The fractionation of the Fe content of this sample with respect to the particle size distribution indicated that the settling effect was less pronounced as particle size distribution shifted toward smaller particle size (Fig. 4.2). When similar settling effect studies were performed in the case of other elements such as P, no loss was observed, demonstrating the variety of patterns that may exist among different elements.

4.2.2. Sample Acidification

Sample acidification is done with the aim of dissolving large particles and minimizing analyte loss that can occur by adsorption onto the container walls. Sample acidification has also been used as a way to achieve selective leaching of elements from sediments and soils (19). Several investigators have conduc-

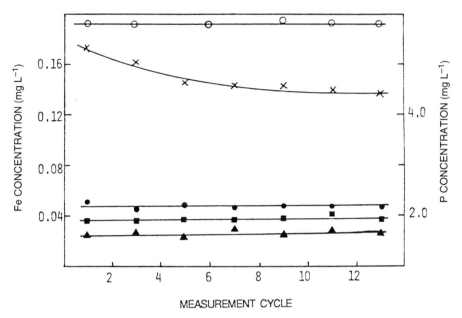

Figure 4.2. Concentrations of Fe and P found in a natural water sample in measurements taken at 70 s intervals. Iron concentrations: unfiltered sample (×); filtered with 0.80 μm (●), 0.45 μm (■), and 0.20 μm (▲). Phosphorus concentration in unfiltered sample (○).

ted studies to determine how specifically solution pH can affect the determination of trace metals in sediments (20–30). Those studies have demonstrated that this influence depends mainly on the properties of the element of interest, sample characteristics, the type of acid used, and the measurement procedure employed.

Although sample acidification is desirable, and indeed necessary in some cases for the reasons just given, it is undesirable when the objective of the analysis is to determine the various forms of an analyte present in the sample. Acidification affects not only particle size distribution but also the form (oxidation state and the amount present as free ions) in which the element is in solution. A study of the relationship between particle size distribution and acidification using Fe as the measured analyte has shown that as acid concentration increases, the fraction of analyte in the filtrate increases, as shown in Fig. 4.3 (18). This occurs as a result of the dissolution of the large particles upon which Fe is adsorbed, thereby making the metal species labile.

Acid effects can vary depending on the type of acid used. The two most commonly employed acids, namely, HCl and HNO_3, have opposing effects in some cases. Whereas HNO_3 has a tendency to oxidize certain metal species, e.g., Fe(II) to Fe(III), HCl has an opposite effect on Fe(III). Also, as will be

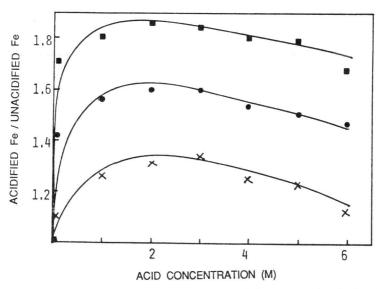

Figure 4.3. Effect of sample acidification on the measured Fe concentration in filtered and unfiltered natural water samples: filtered, acidified with HCl (■); filtered, acidified with HNO_3 (●); unfiltered, acidified with HCl (×).

shown below, HCl can add other complicating effects in view of the release of Cl^- ions, which can take part in complex formation with certain metals.

4.3. ANALYTICAL ASPECTS

4.3.1. Measurement Methods

The fundamental requirement in element speciation is the need to quantitatively determine each of the forms of a given element independently and without interference from the other forms. In this regard, an ideal element speciation method is defined as one that can provide the desired information without altering the original sample in any way.

In the absence of such a method, element speciation has relied on a combination of analytical techniques and methodologies, including wet chemical, chromatographic, electrochemical, and spectroscopic procedures. In many instances, physical-chemical approaches have been employed, whereby all forms of the element of interest are converted into one species, which is then quantified. The fractions of the forms present in the sample originally are then determined by mathematical relationships. One such approach employed a flow-injection procedure requiring a two-step process during which Cr(VI) was converted to Cr(III), followed by the formation of a colored complex (4). This complex was then measured spectrophotometrically. The Cr(VI) species was determined by the difference between spectroscopic data of the original sample and those obtained after the analyte conversion. Subramanian combined complex formation, using ammonium pyrrolidinedithiocarbamate (APDC) with electrothermal atomization atomic absorption spectrometry (ETA–AAS) to selectively determine Cr(III) and Cr(VI) (21). Selective complexation of the two Cr species was achieved by careful optimization of the ligand concentration, pH, and extraction time. Selective extractions of this nature have been employed by others using a dithiocarbamate–methyl isobutyl ketone (DTC–MIBK) system (22).

The formation of metal complexes with organic ligands followed by controlled volatilization of the complexes is another procedure that has been employed to determine different species of a given metal (23–25). Electrochemically active species can be selectively electrodeposited on an appropriate surface where they are then quantified using a suited measurement method. This approach has been employed in combination with ETA–AAS to determine Cr species in solution (26).

The conversion of element species from one form to another prior to quantification can have serious drawbacks, including incomplete reaction, introduction of contaminants, interference from other elements present, and generally a complex and tedious sample treatment procedure. An approach that circumvents these weaknesses is one that physically separates the individual species present, followed by direct quantification. With the development of high-performance liquid chromatography (HPCL) and other chromatographic methods, a variety of separation modes have been employed to distinguish among element species, followed by calorimetric, spectroscopic, or electrochemical detection of the separated species (27–29). Where metal species are involved, chromatographic separation can take place either by direct separation of metal ions using ion-exchange columns or by adsorption (reversed- or normal-phase) LC if the metal species are complexed with organic ligands. While numerous methodologies have been developed employing HPCL where complexed metals in clinical, environmental, and agricultural samples are separated, the main drawback of this approach stems from the incompatibility of commonly used mobile-phase system with the flame and plasma sources commonly used in atomic spectrometry.

A relatively new and fast-growing chromatographic approach that seems to circumvent the HPLC limitations is ion chromatography (IC). The fundamental separation scheme involved here is ion exchange. However, unlike classical ion-exchange processes where the separation occurs under gravity, IC processes are carried out under high pressures similar to those used in HPLC.

Another major difference between the two analytical techniques is that, whereas in HPLC some organic modifiers are added as required in the mobile phase, in IC the mobile phase consists of inorganic buffer systems and dilute acidic and basic solutions. In this way, no serious limitations are experienced from the standpoint of the detector systems, as most spectroscopic and electrochemical detectors are fairly compatible with aqueous solutions. In addition to mobile phases being simple and easy to work with, IC shows other advantages over HPLC as an analytical technique; in fact, the chromatographic peaks obtained are generally well separated and simple; hence data analysis is simplified as well. The method can be used to separate organic and inorganic ions, complex as well simple ions, and neutral species; it is generally simple, requiring relatively little training.

These features of IC have made the technique particularly suitable for metal speciation in aqueous systems. IC columns are now available that can separate metal ions in the forms of simple hydrated cations and anionic and cationic metal ion complexes (30–32). Whether they are used to separate alkali and alkali earth elements or transition metals, IC columns usually contain sulfonated polystyrene polymeric materials of low capacity or silica materials with sulfonated functional groups bonded to their surface. Monovalent and divalent cations are eluted with dilute mineral acids or sometimes dilute solutions of ethylenediaminetetraacetic acid (EDTA). On the other hand, the separation of transition metals has relied on the use of resins with chelating agents that form complexes with the transition metal ions, followed by elution with an appropriate mobile phase containing suitable chelating ligands (33–39). Iminodiacetic acid resin (Chelex-100 or Dowex A-1) has been used extensively as a chelating agent for the separation of transition metals in environmental or biological samples that may contain high concentrations of alkali and alkali earth metals (36, 37).

Likewise, column performance evaluation for the separation of inorganic ions has been carried out, showing a distinction between IC columns and other ion-exchange columns (40–42). The type of detector used in IC separations is influenced by the type of column used and also by the species being separated. Conductivity detection is perhaps the most commonly used method of monitoring chromatographic effluents, especially when eluents of low ionic strength are employed. Other detection methods include potentiometric measurements, atomic spectrometry (AAS or AES) and photometric measurements after postcolumn derivatization of the separated species (43–47).

The use of conductivity detection is limited not only by the low ionic strength requirement of the detector but also by its insensitivity to analytes derived from weak acidic and weak basic species that have low dissociation characteristics (48). For such analytes, other detection mechanisms such as colorimetry, amperometry, coulometry, and potentiometry have proved useful (49–54). One approach used to detect transition metal species after separation on an IC column is to colorimetrically monitor these species after postcolumn derivatization with a suitable ligand (55–61). The ligand most commonly used for this purpose is 4-(2-pyridylazo)resorcinol (PAR). This ligand forms water-soluble complexes with a large number of metals. For metals such as Fe, different complexes are formed with Fe(II) and Fe(III), making this approach suitable for the quantification of different species of the same metal present in a given sample.

As was pointed out earlier, however, the conversion of an analyte from one form to another has some disadvantages. In this particular case, an additional drawback is associated with the difference in reaction stoichiometry between

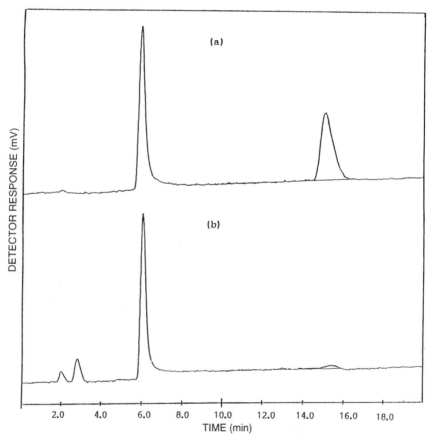

Figure 4.4. Ion chromatogram of $1.0\,\text{mg}\,L^{-1}$ Fe(II)/$1.0\,\text{mg}\,L^{-1}$ Fe(III) in (a) 0.1 M HCl and (b) 1.0 M HCl. Column: HPIC-CS5 (Dionex Corp.). Mobile phase: 6.0 mM pyridine-2,6-dicarboxylic acid + 50 mM acetic acid + 50 mM sodium acetate.

the ligand used and the various metal species present. Figure 4.4 shows the chromatograms obtained for equal amounts of Fe(II)–PAR and Fe(III)–PAR complexes. The chromatographic peak areas are not equal, resulting from the different behaviors of the moieties toward the PAR ligand. This would require different calibration curves for each of the metal species present, making the analysis tedious.

4.3.2. Element-Selective Detectors

In order to preserve analyte integrity and at the same time make the analysis simple, speciation protocols have been developed that have the ability to physically separate the analyte species of interest and then quantify them independently. This is made possible by interfacing chromatographic systems with detectors that only respond to the species of a targeted element present in the column effluent. In this way, all the chemical entities containing the element of interest are measured with equal efficiency. Such detectors are referred to as element-selective detectors (ESD).

A number of publications have appeared in the literature reporting on the use of ESDs for LC (62–75). Included in this group of detectors is flame AAS (FAAS) (62–66). While this type of detector has proved to be useful especially for easily atomizable elements, it possesses low detection power for many elements. ETA–AAS is another approach that has been employed as an ESD (67, 68). Even though high detection power can be achieved with this type of detector, the protocol required in ETA–AAS work, i.e., the separated evaporation, ashing, and atomization steps, prevent direct aspiration of the chromatographic effluent into the detector system. Plasma emission methods, mainly direct current plasma (DCP) and inductively coupled plasma (ICP) have also been employed and have turned out to be more successful than ESDs (69–75). Plasma atomic emission has several advantages over the other ESDs: both metals and nonmetals can be analyzed; high detection power can be achieved due to the high excitation energy of plasmas; and generally the type of interference problems encountered with flame methods are not experienced with plasma excitation sources.

DCP emission has been used extensively in the author's laboratory as a stand-alone emission source and as an ESD for both IC and HPLC (76–80). DCP atomic emission as an element-selective detection method was first reported by Uden et al., who used it in the detection of transition metal complexes separated with HPLC (74). This detection system has an inherent advantage in that all forms of a given element are equally excited when introduced into the excitation zone. The analytical signal obtained is only based on the atom population of the targeted element. An additional advantage is that even with different sample matrices, which would result from changing the chromatographic mobile phase, the analytical signals obtained are not affected.

The capability of DCP–AES as an ESD has been reported in several publications showing the exploitation of this approach in several environmental and clinical analytical problems. Figure 4.5 shows chromatographic data obtained when DCP is interfaced with IC to simultaneously determine As(III) and As(V) in the presence of high concentrations of other common anions. Generally, the determination of As(III) using IC equipped with detectors such as conductivity devices is very difficult due to the low ionization constant of the $HAsO_2$ species. With DCP–AES, however, response is based on the atomic emission of the element, and therefore its form is inconsequential. In this way, the only need is for the As species to be physically separated. Figure 4.5(b) shows the inability of the conductivity detector to measure the

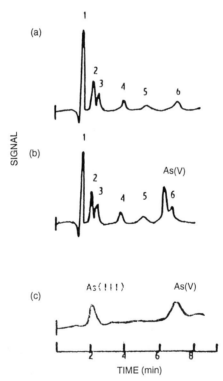

Figure 4.5. IC separation of As(III) and As(V) in the presence of other anions: (a) $1.0\,mg\,L^{-1}$ each of F^- (1), Cl^- (2), NO_2^- (3), PO_4^{3-} (4), NO_3^- (5), and SO_4^{2-} (6), with conductivity detection; (b) same as (a) plus $10\,mg\,L^{-1}$ each of As(III) and As(V) with conductivity detection; (c) same as (b) with DCP–AES detection. Mobile phase: $3.0\,mM\ NaHCO_3/2.4\,mM\ Na_2CO_3$. Column: HPIC-AS4 (Dionex Corp.).

As(III) species, while only the As(V) form that has a higher ionization constant is detected. The superiority of DCP–AES over the conductivity detector with respect to the speciation of the As species is demonstrated in Fig. 4.5(c). The weakly ionized As(III) species, which cannot be detected by measurement of the conductivity, is quantified by the DCP detector by measurement of the atomic emission of the element. It should be also noted in Fig. 4.5 that the presence of Cl^-, F^-, NO_2^-, NO_3^-, PO_4^{3-}, and SO_4^{2-} is inconsequential to the detection of the As species.

An inherent advantage of the DCP detection approach is demonstrated in Fig. 4.6 for the speciation of Se(IV) and Se(VI). Because the response of the

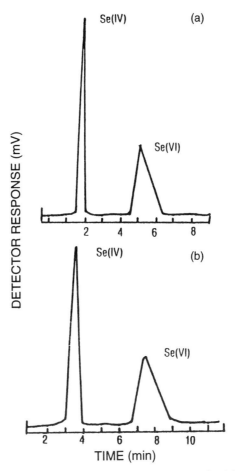

Figure 4.6. IC peaks of Se(IV) as Se(VI) obtained with (a) conductivity detection and (b) DCP–AES detection. Column: HPIC-AS4. Mobile phase: 3.0 mM $NaHCO_3$/2.4 mM Na_2CO_3.

DCP–AES detector takes place on the basis of the atom population of the analyte, equal analytical signals should be expected for equal amounts of any two or more species of the element regardless of their forms. The chromatographic peaks obtained for equal amounts of Se(IV) and Se(VI) have equal peak areas, as shown in Fig. 4.6(b). For purposes of comparison, Fig. 4.6(a) shows chromatograms obtained for the same Se samples except that a conductivity detector was employed. The Se(IV) peak area in this case is only 60% of that of Se(VI). This stems from the lower dissociation constant of the Se(IV) species compared to that of Se(VI). Thus, whereas two calibration plots are needed — one for Se(IV) and another for Se(VI) — if conductivity detection is employed, only one curve is needed if DCP–AES is utilized.

4.3.3. Coupling of IC with DCP–AES

The operational requirements of DCP–AES, as far as solution uptake rate is concerned, fall well within typical LC mobile phase flow characteristics. Therefore, ordinarily no special procedures are necessary in coupling DCP–AES with IC or HPLC systems. However, because of the longer distance traveled by the column effluent as it makes its way into the plasma excitation area, and also owing to the aerosol production mechanism, a certain peak broadening occurs. This effect can be minimized by keeping the sample tube to its minimum practical length and by minimizing aerosol chamber volume.

Another factor that should be monitored when DCP–AES is interfaced with a chromatographic system is baseline drift and noise level. Both baseline drift and noise can be reduced to a minimum by careful plasma optimization and by choosing a suitable spectral line of the element of interest. The plasma can be kept stable and reproducible over a reasonably long period of time, as depicted in Fig. 4.7.

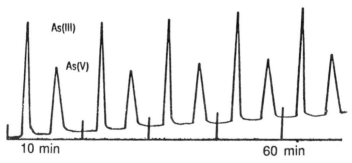

Figure 4.7. IC peaks of As(III) and As(V) obtained at 10 min intervals with DCP–AES detection. Column and mobile phase are similar to those of Fig. 4.6.

4.3.4. Detection Power

Analytical signals obtained with DCP–AES coupled to a chromatographic system are considerably smaller than those obtained when the sample is directly aspirated (76). Thus, when DCP–AES is employed as an ESD, low detection power can be experienced due to chromatographic processes (peak broadening, dilution of the analyte in the mobile phase, and incomplete recovery of the injected analyte) and also due to the interface system used, i.e., low nebulizer efficiency, loss of analyte in the aerosol chamber, dilution of the analyte by the plasma gases, and inefficient introduction of the analyte into the excitation zone.

Low detection power can be improved by using low mobile phase flow rates, by using large sample introduction loops, or by preconcentrating the analyte prior to introduction onto the analytical column. Generally, a reduced mobile phase flow rate is not a viable option because most nebulizers used with DCP to generate an aerosol will not operate efficiently at solution uptake rates below $1.0 \, mL \, min^{-1}$. However, the use of large sample loops and sample preconcentration have proved to be quite efficient means of improving chromatographic measurement sensitivity (77–80).

Large sample loops ensure that a large sample amount is placed on the column. In this way, detection power can be improved by a factor of 10 by increasing the sample loop capacity from $100 \, \mu L$ to $1.0 \, mL$ and still maintaining a linear relationship between the analytical signal (peak height) and loop size (78). The linear relationship implies that even with large loops, peak broadening and column overloading do not occur to any significant degree.

Two approaches have been employed to preconcentrate the sample prior to its injection onto the analytical column. One method resorts to a preconcentration column connected off-line with the analytical column. After the sample is concentrated on this generally shorter column (made of the same packing material as that used for the analytical column), it is connected on-line with the analytical column and the chromatographic process is then resumed. This method is not only cumbersome and long but is also not suitable for poorly retained analytes. Such analytes are lost during the preconcentration process. The other approach, which has been extensively employed in the author's laboratory, involves an on-line preconcentration procedure requiring multiple injections of the dilute analyte solution and using water as the mobile phase. In this way, the analyte builds up in the analytical column; then, if this is followed by injection of a relatively high concentration of an appropriate eluting ion, the accumulated analyte will elute as a sharp peak at a concentration higher than in the original sample. This approach has been found to improve detection power by 1–2 orders of magnitude in most cases and can be applied to the determination of a larger number of metal ions (78).

4.3.5. Influence of the Mobile Phase on Detector Performance

Chromatographic peak shape, retention time, and peak symmetry are influenced greatly by mobile phase characteristics. These parameters are altered if mobile phase strength and composition are changed. Generally, high mobile phase concentration leads to a reduction in retention time of the analyte with concomitant sharp peaks. For universal detectors such as conductivity detectors, however, an overly concentrated mobile phase may not be suitable since this would bring about an enhanced background signal, which in turn would lead to reduced measurement sensitivity. This limitation is not experienced when DCP–AES is employed, owing to its element selective nature (79). Figure 4.8 illustrates this point for Fe(II) and Fe(III) species separated on an IC column using two different concentrations of the mobile phase.

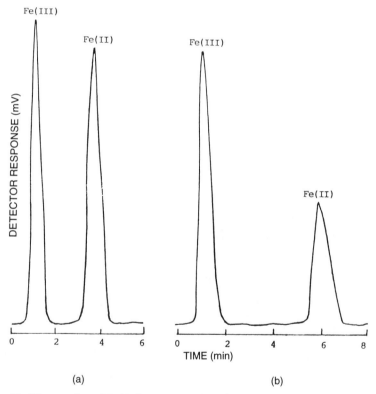

Figure 4.8. IC separation of Fe(II) from Fe(III) obtained with DCP–AES detection. Sample: 5 mg L^{-1} Fe(II) + 5 mg L^{-1} Fe(III). (a) Mobile phase: 10 mM oxalic acid + 7.5 mM trilithium citrate. (b) Mobile phase: 5.0 mM oxalic acid + 3.75 mM trilithium citrate.

4.3.6. Influence of the Chemistry of the Analyte and Sample Matrix

It has been demonstrated that by using a chromatographic system in combination with an element-selective detector, many of the detector-related limitations of element speciation are eliminated. However, if different species of an analyte cannot be chromatographically separated or are sensitive to sample conditions, then a protocol must be developed that defines the conditions used and how these conditions influence the data obtained.

Sample processing and the analytical measurement procedure employed can strongly influence the type of speciation data obtained. As already discussed, sample acidification, filtration, heating, and extraction are common pretreatment procedures, especially when environmental samples are involved. Acidification is necessary for sample preservation and digestion.

Unfortunately, this may cause changes in oxidation state and/or conversion of all the forms of the analyte into one form, this being obviously undesirable in speciation work. The degree to which such changes occur in turn depends on the nature of the analyte and the extent of the procedure applied. It has been observed that when Fe samples are acidified with HCl, the ratio Fe(II)/Fe(III) increases as a result of Fe(III) undergoing reduction (81–83). While such conversions may be desirable for total Fe determination, they introduce errors in the quantification of the individual Fe(II) and Fe(III) species.

Many elements, especially transition metals, are capable of undergoing various chemical reactions. Iron is particularly prone to hydrolysis, oxidation [Fe(II) to Fe(III)], and formation of complex ions with organic and inorganic ligands. In actual analyses, these processes can affect the sample and the analytical standards differently, depending on the respective concentrations and the solution conditions.

Acidification of Fe standards is essential in order to prevent hydrolysis, which occurs readily in aqueous solution, as depicted in Fig. 4.9. Over a period of less than 24 h, more than 70% of Fe(III) is removed from solution by the process of hydrolysis. Only about 10% of the Fe(II) species undergoes such transformation.

Such hydrolysis can be stopped or minimized. The question is, if HCl is used for this purpose, what concentration is optimal, given that HCl also can have a reducing effect on Fe(III)? Studies have been conducted in the author's laboratory to investigate this aspect. While HCl concentration of 0.1 M proved to be sufficient for the suppression of hydrolysis of low concentrations of Fe(II) and Fe(III) species, higher acid concentrations caused other very dramatic changes on Fe. For Fe(II), increase in acid concentration results in the formation of some Fe(II) moiety, which (as shown in Fig. 4.10) is poorly retained on the ion-exchange column. This moiety is believed to be an Fe(II)

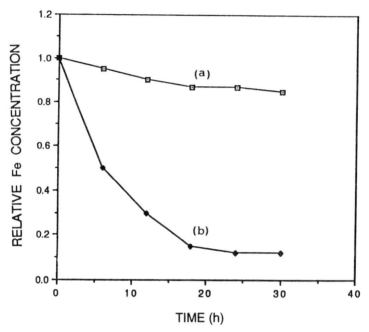

Figure 4.9. Hydrolysis of Fe in deionized water: (a) Fe(II); (b) Fe(III).

chlorocomplex, as has been reported by others in the literature (83–85). The transformation of Fe(II) into the chlorocomplex depends on the Fe(II)/HCl molar ratio. The Fe(II) mass balance was verified by using the DCP–AES detector.

Although Fe(III) undergoes reduction in HCl, the rate of this conversion is not as pronounced as it is for the Fe(II) species. Less than 10% of Fe(III) was found to undergo reduction in 1.0 M HCl. However, the speciation of Fe in 1.0 M HCl by IC would present a more serious practical problem, since the transformed Fe(II) chlorocomplex elutes close to the Fe(III) species, resulting in overlapping peaks as depicted in Figs. 4.11 and 4.12. Table 4.1 shows the concentration dependence of the transformation of the Fe(II) species. These data suggest that in a typical analysis, sample acidification can produce a mixture of Fe(II) species, one fraction consisting of the chlorocomplex species and the other consisting of untransformed Fe(II). This can further complicate the analysis.

Whereas the Fe species just discussed can be measured with equal efficiency by using the element-selective detection (DCP–AES) approach, the acid concentration dependence of the Fe(II) species would require different calibra-

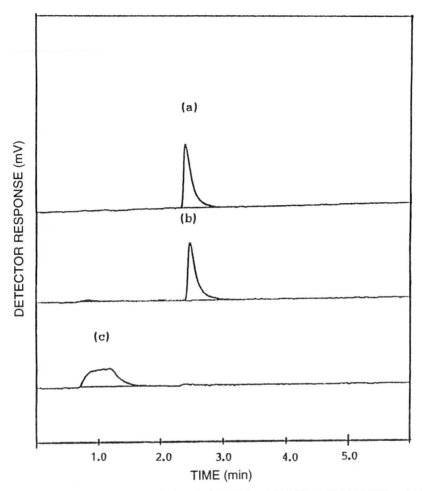

Figure 4.10. Ion chromatograms of 1.0 mg L^{-1} Fe(II) in (a) 0.1 M HCl, (b) 0.5 M HCl, and (c) 1.0 M HCl obtained with HPIC-C2 column (Dionex Corp.). Mobile phase: 10 mM oxalic acid + 7.5 mM trilithium citrate.

tion curves. Figure 4.13 shows the analytical curves prepared for Fe(II) and Fe(III) in 1.0 M HCl, comparing them with the analytical curves prepared in 0.1 M HCl. These curves show that the determination of Fe(II) in 1.0 M HCl would be done under conditions of reduced detection power, whereas that of Fe(III) would be inflated due to the peak of the Fe(III) originally present in the sample overlapping with the peaks of the Fe(II) chlorocomplexes formed in 1.0 M HCl. In 0.1 M HCl, however, both Fe species would be determined

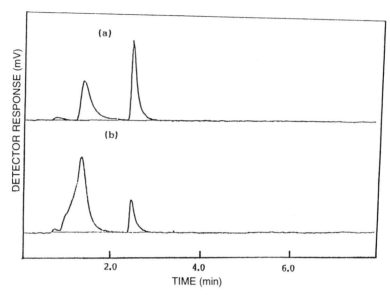

Figure 4.11. Ion chromatograms of 1.0 mg L^{-1} Fe(II)/1.0 mg L^{-1} Fe(III) in (a) 0.1 M HCl, and (b) 1.0 M HCl. Chromatographic conditions are similar to those of Figure 4.10.

with equal sensitivity with a fairly linear analytical curve, as depicted in Fig. 4.13.

The foregoing discussion regarding Fe speciation leads to the conclusion that in order to show the effects of sample conditions and their influence on the chemistry of the speciated elements, the results should be qualified as "conditional." Thus, if the main influential parameter is sample acidity, quantification of the analysis data should specify the acid type and concentration at which the speciation data were obtained. The final report should then be presented as such, e.g., "speciation of Fe in 0.1 M or 1.0 M HCl."

Other transition metals have been found to behave in a similar manner when in acidic media. Such is the case for Cr, which can exist as CrO_4^{2-} or $Cr_2O_7^{2-}$ depending on the solution pH. Vanadium undergoes a transformation from V(IV) (VO^{2+}) to V(V) (VO_3^-) when placed in acid. This transformation has been monitored by separating the V species at various HCl concentrations using the IC–DCP–AES method. Because a cation separator column was used in this study, the appearance of an unretained V species suggests the formation of anionic species in 1.0 M HCl, which is postulated to be $VOCl_4^{2-}$. Such transformation studies are very important because they provide information that may be used in the determination of the measurability and even the bioavailability of the chemical species of interest.

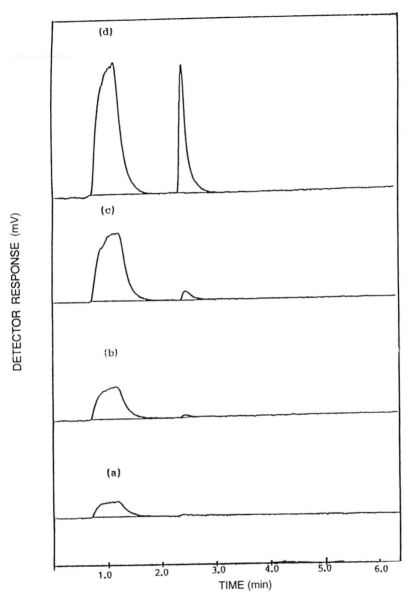

Figure 4.12. Ion chromatograms of varying concentrations of Fe(II) in 1.0 M HCl: (a) 1.0 mg L^{-1} Fe(II); (b) 2.0 mg L^{-1} Fe(II); (c) 5.0 mg L^{-1} Fe(II); (d) 10.0 mg L^{-1} Fe(II). Chromatographic conditions are similar to those of Fig. 4.11.

Table 4.1. Influence of HCl Concentration on the Fraction of Fe(II) Transformed

Fe(II)/HCl Molar Ratio	Fraction of Fe(II) Transformed (%)
1.78×10^{-5}	100.0
3.56×10^{-5}	97.6
8.90×10^{-5}	95.5
1.78×10^{-4}	76.4

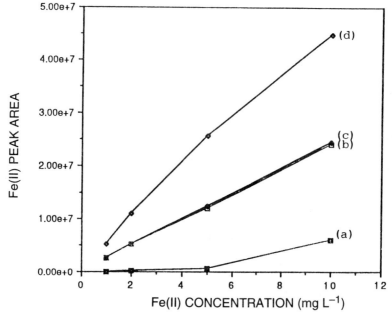

Figure. 4.13. Analytical curves for Fe(II) and Fe(III) in HCl: (a) Fe(II) in 1.0 M HCl, peak at 2.4 min; (b) Fe(II) in 0.1 M HCl, peak at 2.4 min; (c) Fe(III) in 0.1 M HCl, peak at 1.3 min; (d) Fe(II) + Fe(III) in 1.0 M HCl, peak at 1–3 min. Chromatographic conditions are similar to those of Fig. 4.10.

4.4. APPLICATIONS

4.4.1. Chromium Speciation

Chromium speciation is ecologically consequential in view of the wide industrial applications of this metal, which can ultimately have a significant impact on the environment. Of the two common inorganic forms of Cr, i.e., Cr(III) and

Cr(VI), the latter is believed to be toxic. Therefore, the analysis of environmental samples for the total content of the element only gives a partial picture of its potential impact. Many different analytical protocols have been set up in an effort to quantify the two Cr species. These attempts have included the conversion of Cr(VI) to Cr(III), followed by the formation of a colored complex, which is then measured spectrophotometrically (20); a combination of selective complex formation with AAS (21, 25); electrodeposition of electrochemically separated Cr(III) and Cr(VI) on a graphite tube, followed by AAS measurements (26); and a variety of chromatographic methods (86–91).

A procedure for separating Cr(III) and Cr(VI) has been developed in the author's laboratory that utilizes either anion- or cation-exchange columns coupled with IC–DCP–AES to operate and quantify the Cr species directly without species conversion. The procedure includes a protocol for on-column sample preconcentration, followed by rapid column regeneration with 1.0 M HCl when a cation column is used or 0.5 M NaOH when an anion column is used.

Both clinical and environmental samples have been analyzed in this way. The clinical sample consisted of human serum material obtained from the (U.S.) National Institute of Standards and Technology (NIST) as the standard reference material SRM 909. Environmental samples included natural water also obtained from the NIST as SRM 1643(b) and wastewater samples obtained from a local industrial waste stream.

Table 4.2 presents the data obtained for SRM 909 and SRM 1643(b). Both Cr(III) and Cr(VI) were found in SRM 909 in approximately equal concentra-

Table 4.2. Analytical Results of the Determination of Cr Species in Standard Reference Materials

Standard Reference Materials, NIST	Certified Cr Concentration, NIST ($mg\,L^{-1}$)	Experimental Results[a]		
		Cr(III) ($mg\,L^{-1}$)	Cr(IV) ($mg\,L^{-1}$)	Total Cr (Determined)[b] ($mg\,L^{-1}$)
SRM 909 (human serum)	0.090 ± 0.006	0.06 ± 0.003	0.05 ± 0.002	0.10 ± 0.01
SRM 1643(b) (natural water)	0.019 ± 0.001	0.02 ± 0.005	ND[c]	0.020 ± 0.005

[a] The values after the $\pm$ signs are standard deviations of three replicate measurements.
[b] Total shown in this column was determined by direct aspiration of the sample into the DCP. Analysis was made in the integration mode.
[c] ND: not detected.

tions. For SRM 1643(b), the element was mainly in the Cr(III) form. These sets of data were obtained with IC–DCP–AES, whereby DCP–AES served as an element-selective method of detection for the Cr eluted from the IC column. Direct determination of the Cr was also done by aspirating the sample directly into the DCP, disconnected from the IC column. The data obtained here pertain to the total Cr content and are in close agreement with the sum of the individually determined species.

For the industrial wastewater samples, the analyses were done in three ways. In the first mode, the unfiltered sample was aspirated directly into the DCP system. The Cr concentration obtained in this way was for total Cr present, including ionic, particulate, and other unfilterable forms. In the second mode, the sample was first filtered with a 0.2 μm pore membrane filter before being aspirated into the DCP systems. This provided information on the unfilterable fraction of the total Cr present. In the third mode, the sample was filtered and then injected into the cation separator column. This gave information on the Cr(III) and Cr(VI) present. The results obtained are summarized in Table 4.3. These data clearly point to the ability of the DCP system to provide information on the total metal content and also speciation data for that metal.

Table 4.3. Analytical Results of the Determination of Cr Species in Industrial Waste Stream Samples

Sample	Analysis Mode[a]		
	DCP Direct Injection, Unfiltered[b] (mg L^{-1})	DCP Direct Injection, Filtered[c] (mg L^{-1})	IC–DCP–AES Column Injection, Filtered[d] (mg L^{-1})
I	111.6 ± 7.0	107.7 ± 4.3	Cr(VI): 99.4 ± 5.0 Cr(III): 2.6 ± 0.1
II	450.0 ± 9.0	404.5 ± 8.0	Cr(VI): 367.8 ± 11.0 Cr(III): 38.7 ± 2.0
III	115.5 ± 3.0	103.2 ± 3.0	Cr(VI): 93.6 ± 5.0 Cr(III): 2.6 ± 0.1

[a] The values after the ± signs are standard deviations of the mean of three replicate measurements.
[b] Sample was aspirated directly into the DCP.
[c] Sample was aspirated into the DCP after filtration with a 0.2 μm pore filter.
[d] Filtered sample was injected into the cation column; the effluent was aspirated directly into the DCP.

4.4.2. Monitoring of the Element Transformation Rate

The ability of an element-selective detector to respond only to a targeted element makes it suitable for monitoring the transformations that an element can undergo as a result of chemical or biological changes. Figure 4.14 is a representation of the oxidation of Fe(II) in dilute HNO_3. This process was monitored by injecting a small amount of the Fe(II) acidic solution into the cation chromatographic column connected to the DCP–AES every several minutes, followed by monitoring the chromatographic components eluted by measuring the atomic emission signal at a fixed Fe wavelength. The ability of the detector to measure different forms of a given element with equal efficiency allowed the disappearance of the Fe(II) species and the appearance of the Fe(III) species to be monitored. Such measurements can be employed to

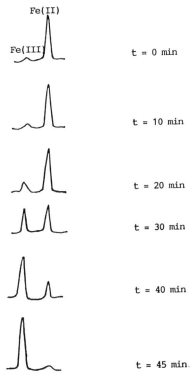

Figure 4.14. Oxidation of Fe(II) in 0.5 M HNO_3 monitored by IC–DCP–AES. Sample: 1.0 mg L^{-1} Fe(II). Column: HPIC-CS5 (Dionex Corp.). Mobile phase: 10 mM oxalic acid + 7.5 mM trilithium citrate.

provide kinetic data for chemically as well as biologically induced transformation processes.

4.4.3. Application to Clinical and Related Measurements

The use of IC–DCP–AES to determine trace elements in biological and clinical samples has been well demonstrated, as was discussed for Cr above.

In the search for a cure for cancer and related diseases, compounds containing heavy metals such as Au and Pt have been found to have unique activities that are being developed for the treatment of these diseases (92–96). A number of Pt compounds have been studied for this purpose, including cisplatin, carboplatin, iproplatin, and tetraplatin. Clinical use of these materials has been hampered by our as yet poor understanding of their chemistry. Some of them, e.g., cisplatin, have very low solubility in aqueous solutions. Others, such as tetraplatin, decompose when placed in solution.

The IC–DCP–AES method was used to study the solution chemistry of $trans$-($\pm$)-1,2-diaminocyclohexane-Pt(IV) tetrachloride (tetraplatin). This is one of the family of Pt compounds undergoing extensive evaluation for their chemotherapeutic activity. One essential measurement of new materials of this kind is the determination of their purity. However, in view of the limited thermal and solution stability of tetraplatin, accurate purity determinations have been difficult to conduct using absolute methods such as differential scanning calorimetry (DSC).

Tetraplatin undergoes a degradation process in solution leading to the formation of several fragments, some or all of which may contain Pt. These fractions were separated on an HPLC column interfaced with the DCP–AES–ESD to monitor Pt via AES measurements. Stock solutions were prepared by dissolving enough of the solid in deionized water to make a $1-5\,\mathrm{mg\,mL^{-1}}$ solution. For those studies requiring freshly prepared samples, the solution was used immediately after preparation. In those cases where maximum degradation was desired, the solution was allowed to age for 3–5 days before using it.

When tetraplatin solutions were stored for a few hours, they formed several moieties all of which contained Pt. Similar data were obtained when three types of column were used, namely, C-18, PRP-1, and IC columns, as depicted in Figs. 4.15, 4.16, and 4.17. The Pt selectivity of the DCP–AES detector allowed the purity of the tetraplatin material to be determined. Also, since all the moieties containing Pt were detected with equal efficiency, the approach can be used to determine the response factors of the chromatographic effluents if the chromatographic run is done by using other detectors, such as an ultraviolet (UV) detector.

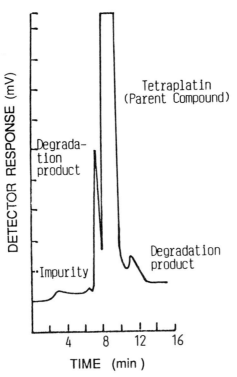

Figure 4.15. Chromatogram of aged 2.0 mg L^{-1} tetraplatin solution obtained with a C-18 column and DCP detector.

In this particular case, the DCP–AES was connected in tandem with a UV detector. In this way, it was possible to distinguish the peaks that contained Pt from those that did not contain the metal. Further, the DCP data can be used to determine the Pt content of the degradation products corresponding to the peaks detected by the UV detector. As an example, Table 4.4 shows the peak areas of the chromatograms obtained with the C-18 column used in combination with both DCP–AES and UV detectors. A tabulation of the Pt contained in each peak as measured by the DCP–AES and a corresponding response factor for the UV detector are also given. The potential of the ESD nature of DCP–AES and its applications in providing information for other detector systems are well demonstrated in this example.

Several other investigators have used ICP–AES in combination with HPLC to separate and quantify Pt-based moieties in biological fluids and tissues by monitoring the Pt content of these moieties (97–99). This capability has allowed the performance of critical clinical studies to be assessed, includ-

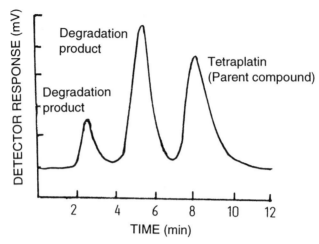

Figure 4.16. Chromatogram of aged 2.0 mg L^{-1} tetraplatin solution obtained with a PRP-1 column and DCP detector.

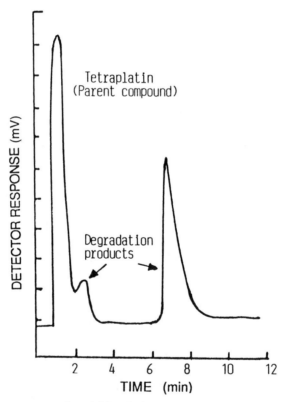

Figure 4.17. Chromatogram of aged 2.0 mg L^{-1} tetraplatin solution obtained with CS-1 ion-exchange column and DCP detector.

Table 4.4. Chromatographic Peak Areas and Response Factors of Major Decomposition Products of Tetraplatin (TP)[a]

Peak	UV Response[b]	DCP–AES Response[b]	Response Factor[c]
1	106.00	33.10	0.35
2[d]	524.00	462.00	1.00
3	31.40	209.00	0.75

[a] See also Fig. 4.17.
[b] Arbitrary units.
[c] Response factor = $\dfrac{\text{DCP–AES Response}}{\text{UV response}} \times \dfrac{\text{TP UV response}}{\text{TP DCP–AES response}}$.
[d] Tetraplatin.

ing investigations of the cytotoxicity and pharmacokinetics of Pt-containing cancer drugs (100, 101).

4.4.4. Solution Chemistry of Phosphorus

Phosphorus is one of the elements whose chemical activity can have significant consequences for biological and environmental systems. IC–DCP–AES is applicable to the study and measurement of P compounds, both organic and inorganic. Figure 4.18 shows the ion chromatographic pattern of a mixture of AMP, ADP, and ATP obtained by monitoring the atomic emission of P with the IC-coupled DCP–AES system. As with the metallic elements already discussed, all the P species are measured with equal efficiency. Analytical curves prepared for the various P species were all identical, with a regression coefficient of 0.9999. Figure 4.19 displays the analytical data obtained when a solution of Tide detergent was injected into the IC column. Several P species were identified in the chromatogram, including orthophosphate, pyrophosphate, and some polyphosphates represented by the rather poorly defined peak.

4.4.5. Determination of Trace Heavy Metals in Water

A number of publications have appeared in the literature in the past 10 years reporting on the use of hyphenated techniques to determine trace heavy metals in water. In most cases, a separation technique is coupled to a spectroscopic technique, the former serving to separate the metal species, while the latter is used as a detector. The first scheme involves the injection of water samples into the separation system, followed by a direct introduction of the effluent into the detector. Separation systems in this scheme may be HPLC, IC, or FIA (flow

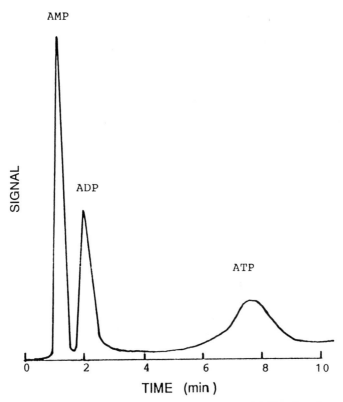

Figure 4.18. Ion chromatographic separation of AMP, ADP, and ATP (adenosine mono-, di-, and triphosphate) obtained with DCP–AES detection of P. Column: HPIC-AS7 (Dionex Corp.). Mobile phase: 0.5 M HNO_3.

injection analysis). Detectors, on the other hand, may include FAAS, DCP–AES, and ICP–AES; in those cases where the metal species can be measured spectrophotometrically, e.g., Cr(VI), UV absorption has also been employed. A summary of hyphenated techniques employed for the determination of Cr species has been published by Sperling et al. (102).

The second scheme used in trace metal determination with hyphenated techniques requires precolumn or postcolumn derivatization of the metal species. The primary requirement in precolumn derivatization is that stable complexes are formed between the metal species and the derivatizing agent. Commonly used precolumn derivatization agents include dithiocarbamic acids, 8-quinolinol, and β-diketones (103–105). These compounds have been used extensively in the determination of multivalent metal ions such as Al(III), Fe(III), Mn(VII), and Co(III) (106). In all of these cases, metal complexes are

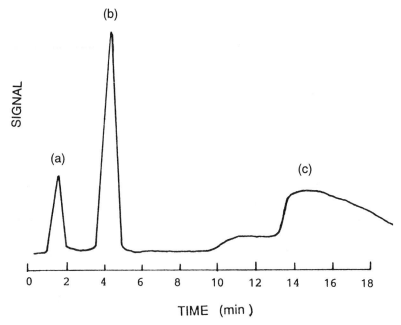

Figure 4.19. Ion chromatogram of a solution of Tide detergent obtained with DCP–AES detection of P. Peaks: (a) orthophosphate; (b) pyrophosphate; (c) polyphosphates. Column: HPIC-AS7. Mobile phase: 0.5 M HNO_3.

formed outside the chromatographic column. They are then separated, typically using a reversed-phase HPLC method.

In postcolumn derivatization, metal ions separated on a chromatographic column, usually an ion-exchange column, are converted into a form more amenable to detection. While a number of complexing ligands can serve as suitable agents in this conversion, the one that appears to be most commonly used is PAR (107). This ligand forms colored complexes with several metals. Typically, detection is done by measuring absorbance at 520 nm. A number of publications have appeared in the literature reporting on the use of IC coupled to UV–Vis detectors employing this approach (55–61).

4.4.6. Determination of the Binding Capacity of Natural Waters

Organic and inorganic ligands in water can interact with metal ions to form chelates. This process can be beneficial in that it renders the metal ion unavailable, thereby reducing its toxicity, should any exist. On the other hand, complexing ligands in water can also bring about a dissolution of particulate- and sediment-bound metals, a process that can lead to increased metal pollution in natural water. Ligands involved in these processes may be either naturally occurring or materials introduced in water by industrial discharges. In either case, the dissolved metal content of natural water can be highly influenced by these materials (108).

The role of naturally occurring ligands on the binding capacity of natural water systems has been studied by several investigators (109–112). Two general approaches have been taken in measuring the binding capacity of natural waters. In one approach, the organic and inorganic forms of metal in water are separated and the fraction of the metal in the organic form is quantified. In the other approach, the stability constants of complexes formed between metal ions and naturally occurring ligands are determined by quantifying the equilibrium metal concentration after interaction with a known amount of ligand. Ion-selective electrodes have also been used to monitor the equilibrium concentrations of metal ions in the presence of complexing agents (113). Good stability data have been reported with this method, even though it is limited to those metals for which reliable ion-selective electrodes are available.

A great deal of work remains to be done in this area, especially due to current environmental pollution concerns.

4.5. CONCLUSIONS

Table 4.5 lists elements that have been studied using the IC–DCP–AES approach, including the species measured and the attainable detection limits.

The information given here points to the value of using element-selective detectors in speciation work and the superior capability of the DCP–AES method in this regard. By using a chromatographic system in combination

Table 4.5. Speciation of Selected Elements Using IC–DCP–AES[a]

	Measurable Concentration (mg L^{-1})[b]	
Element	Without Preconcentration	With Preconcentration
As(II), As(V)	1.0	0.05
Cr(III), Cr(VI)	0.1	0.01
Fe(II), Fe(III)	0.1	0.005
Mn(II), Mn(VII)	0.05	0.005
Pt(IV)	10.0	1.0
V(IV), V(V)	0.1	0.05

[a] In all cases, a 1.0 mL loop was used for sample injections.
[b] Measurements were made at optimum chromatographic and spectroscopic conditions.

with DCP–AES not only are all the chromatographic effluents containing a targeted element measured with equal efficiency, but also different types of chromatographic columns, representing different separation processes, can be employed, including IC, C-18, and polymeric columns.

The data obtained can provide information on species present, their rate of formation, transformations occurring as a result of chemical or biological changes, purity determination, and calibration of other detection systems.

An area to which the IC–DCP–AES approach should be particularly suited is the study of the pollution effects of trace heavy metals in aquatic systems, especially the marine environment. Extensive study is needed in this regard, focusing on the influence of naturally occurring ligands such as humic acids. To the best knowledge of the author, in fact, no research has been reported in which the matrix effects of humic acids and similar compounds on plasma AES have been studied.

REFERENCES

1. G. G. Leppard, ed., *Trace Element Speciation in Surface Water and Its Ecological Implications*. Plenum, New York, 1983.
2. J. C. Van Loon, ed., *Chemical Analysis of Inorganic Constituents of Water*. CRC Press, Boca Raton, FL, 1982.
3. D. K. D. Lee, ed., *Metallic Contaminants and Human Health*. Academic Press, New York, 1972.
4. P. D. Goulden, *Environmental Pollution Analysis*. Heyden, Philadelphia, 1978.
5. W. Stumm and P. A. Brauner, in *Chemical Oceanography* (J. P. Riley and G. Skirrow, eds.), 2nd ed., pp. 173–238. Academic Press, New York, 1974.
6. R. Andrew, K. E. Biesinger, and G. E. Glass, *Water Res.* **11**, 309 (1977).
7. W. Sunder and R. R. L. Guillard, *J. Mar. Res.* **34**, 511 (1976).
8. T. M. Florence and G. E. Batley, *Talanta* **24**, 151 (1977).
9. Y. K. Chan and L.-S. Chan, *Water Res.* **8**, 383 (1974).
10. D. P. Laxen and M. I. Chandler, *Anal. Chem.* **54**, 1350 (1983).
11. W. Stumm and H. Bilinski, in *Advances in Water Pollution Research* (S. H. Jenkins, ed.), pp. 39–52. Pergamon, Oxford, 1973.
12. I. P. H. Laxen and R. M. Harrison, *Sci. Total Environ.* **19**, 59 (1981).
13. A. J. Bale and A. W. Morris, *Estuarine Coastal Shelf Sci.* **13**, 1 (1981).
14. E. R. Sholkovitz, W. A. Boyle, and N. B. Price, *Earth Planet Sci. Lett.* **49**, 130 (1978).
15. G. Figueres, J. M. Martin, and M. Meybeck, *Neth. J. Sea Res.* **12**, 329 (1978).
16. I. Ayoyama, I. Sakai, and Y. Inove, *Mizu Shori Gijutsu* **17**, 913 (1976).
17. K. Y. Chen, C. S. Young, T. K. Jan, and N. Rohatgi, *J. Water Pollut. Control Fed.* **46**, 2663 (1974).

18. I. T. Urasa and A. M. O'Reilly, *Talanta* **33**, 593 (1986).
19. J. H. Trefry and S. Metz, *Anal. Chem.* **56**, 754 (1984).
20. B. P. Bubnis, M. R. Straka, and G. E. Pacey, *Talanta* **30**, 841 (1983).
21. K. S. Subramanian, *Anal. Chem.* **60**, 11 (1988).
22. T. Tande, J. E. Patterson, and T. Torgrimson, *Chromatographia* **13**, 607 (1983).
23. S. Arpadjan and V. Krivan, *Anal. Chem.* **58**, 2611 (1986).
24. W. R. Wolf, *J. Chromatogr.* **134**, 159 (1977).
25. R. J. Lloyd, R. M. Barnes, P. C. Uden, and W. G. Elliot, *Anal. Chem.* **56**, 2025 (1978).
26. G. E. Battey and P. J. Matousek, *Anal. Chem.* **52**, 1570 (1980).
27. J. C. Van Loon, *Anal. Chem.* **51**, 1139A (1979).
28. W. S. Gardner and P. F. Landrum, *Anal. Chem.* **54**, 1196 (1982).
29. B. D. Karcher and I. S. Krull, *J. Chromatogr. Sci.* **25**, 472 (1987).
30. G. Schmuckler, *J. Liq. Chromatogr.* **10**, 1887 (1987).
31. P. Kolla, J. Koehler, and G. Schomburg, *Chromatographia* **23**, 465 (1987).
32. H. Shintani, *J. Chromatogr.* **341**, 53 (1985).
33. F. A. Cotton and G. Wilkinson, *Basic Inorganic Chemistry*. Wiley, New York, 1976.
34. J. Inczédy, *Analytical Applications of Complex Equilibria*, p. 348. Wiley, New York, 1976.
35. G. Schmuckler, *Talanta* **12**, 281 (1965).
36. R. E. Sturgeon, S. S. Berman, A. Desauniers, and D. S. Russel, *Talanta* **27**, 85 (1980).
37. A. J. Paulson, *Anal. Chem.* **58**, 183 (1986).
38. H. L. Loewenschuss and G. Schmulker, *Talanta* **11**, 483 (1970).
39. R. F. Hirsch, E. Gancher, and F. R. Russo, *Talanta* **17**, 483 (1970).
40. P. R. Haddad, P. E. Jackson, and A. L. Heckenberg, *J. Chromatogr.* **346**, 139 (1985).
41. D. T. Gjerde, J. S. Fritz, and G. Schmuckler, *J. Chromatogr.* **186**, 509 (1979).
42. D. T. Gjerde, J. S. Fritz, and G. Schmuckler, *J. Chromatogr.* **187**, 35 (1980).
43. S. Reiffenstuhl and G. Bonn, *Fresenius' Z. Anal. Chem.* **332**, 130 (1988).
44. B. V. Kondratjonok and G. Schwedt, *Fresenius' Z. Anal. Chem.* **332**, 333 (1988).
45. W. F. Lien, B. K. Boerner, and J. G. Tarter, *J. Liq. Chromatogr.* **10**, 3213 (1987).
46. G. J. Sevenich and J. S. Fritz, *J. Chromatogr.* **371**, 361 (1986).
47. P. R. Haddad and P. W. Alexander, *J. Chromatogr.* **324**, 319 (1985).
48. H. Small, T. S. Stevens, and W. C. Bauman, *Anal. Chem.* **47**, 1801 (1975).
49. R. D. Rocklin and E. L. Johnson, *Anal. Chem.* **55**, 4 (1983).
50. J. S. Fritz and J. N. Story, *Anal. Chem.* **46**, 825 (1974).
51. S. Elchuck and R. M. Cassidy, *Anal. Chem.* **51**, 1434 (1979).

52. K. Suzuki, H. Aruga, and T. Shirai, *Anal. Chem.* **55**, 2011 (1983).
53. R. J. Williams, *Anal. Chem.* **51**, 836 (1983).
54. J. E. Girard, *Anal. Chem.* **51**, 836 (1979).
55. M. D. Palmieri and J. S. Fritz, *Anal. Chem.* **59**, 2226 (1987).
56. M. D. Palmieri and J. S. Fritz, *Anal. Chem.* **60**, 2244 (1988).
57. J. W. O'Laughlin, *Anal. Chem.* **54**, 178 (1982).
58. T. Midonikawa, E. Tanoue, and Y. Sugimura, *Anal. Chem.* **62**, 1337 (1990).
59. K. Kawazu and J. S. Fritz, *J. Chromatogr. Sci.* **77**, 397 (1973).
60. J. S. Fritz and J. N. Story, *Anal. Chem.* **46**, 825 (1974).
61. G. J. Sevenich and J. S. Fritz, *Anal. Chem.* **55**, 12 (1983).
62. K. L. Jewett and F. E. Brickman, in *Liquid Chromatography Detectors* (T. M. Vickey, ed.), pp. 205–241. Dekker, New York, 1983.
63. D. R. Jones, IV, and S. E. Manahan, *Anal. Chem.* **48**, 502 (1976).
64. D. R. Jones, IV, H. C. Tung, and S. E. Manahan, *Anal. Chem.* **48**, 7 (1976).
65. D. R. Jones, IV, H. C. Tung, and S. E. Manahan, *Anal. Chem.* **48**, 1897 (1976).
66. D. J. Read, *Anal. Chem.* **47**, 186 (1975).
67. G. R. Ricci, L. S. Sheppard, G. Colovos, and N. E. Nester, *Anal. Chem.* **53**, 611 (1981).
68. A. A. Grabinski, *Anal. Chem.* **53**, 966 (1981).
69. K. Yashida and H. Baraguchi, *Anal. Chem.* **56**, 258 (1984).
70. D. W. Housler and L. T. Taylor, *Anal. Chem.* **53**, 1223 (1981).
71. P. C. Uden, B. D. Quimby, R. M. Barnes, and W. G. Elliot, *Anal. Chim. Acta* **101**, 99 (1978).
72. C. H. Gast, J. C. Kraak, H. Hoppe, and F. J. M. Maesen, *J. Chromatogr.* **185**, 549 (1979).
73. P. C. Uden and I. E. Bigley, *J. Chromatogr.* **94**, 29 (1977).
74. P. C. Uden, I. E. Bigley, and F. H. Walters, *J. Chromatogr.* **100**, 555 (1978).
75. G. J. Sevenich and J. S. Fritz, *Anal. Chem.* **55**, 12 (1983).
76. M. Yamamoto, H. Yamamoto, and Y. Yamamoto, *Anal. Chem.* **56**, 832 (1984).
77. I. T. Urasa and F. Ferede, *Anal. Chem.* **59**, 1563 (1987).
78. I. T. Urasa and S. H. Nam, *J. Chromatogr. Sci.* **27**, 30 (1989).
79. I. T. Urasa, V. D. Lewis, and S. H. Nam, *J. Chromatogr. Sci.* **27**, 468 (1989).
80. I. T. Urasa, V. D. Lewis, J. De Zwaan, and S. E. Northcott, *Anal. Lett.* **22**, 597 (1989).
81. C. O. Moses, A. T. Herlihy, J. S. Herman, and A. L. Mills, *Talanta* **35**, 15 (1988).
82. J. E. McMahon, *Limnol. Oceanogr.* **12**, 437 (1967).
83. W. Davidson and E. Rigg, *Analyst (London)* **101**, 634 (1976).
84. C. L. Stanley and R. F. Krich, *J. Chem. Phys.* **34**, 1450 (1961).
85. A. Glasner and P. Aviner, *Talanta* **11**, 761 (1964).

86. J. C. Van Loon, B. Radziuk, N. Kahn, I. Lichwa, F. J. Fernandez, and J. D. Karker, *At. Absorpt. Newsl.* **16**, 79 (1977).
87. D. Naranjit, Y. Thomassen, and J. C. Van Loon, *Anal. Chem.* **57**, 21 (1985).
88. S. Hirata, Y. Umezaki, and M. Ikeda, *Anal. Chem.* **58**, 2602 (1986).
89. S. D. Hartenstein, J. Růžička, and G. D. Christian, *Anal. Chem.* **57**, 21 (1985).
90. I. S. Krull, D. Bushee, R. N. Savage, R.G. Schleicher, and S. B. Smith, *Anal. Lett.* **15**(A3), 267 (1982).
91. I. S. Krull, K. W. Panaro, and L. L. Gershman, *J. Chromatogr. Sci.* **21**, 460 (1983).
92. R. J. Puddephatt, *The Chemistry of Gold*. Elsevier, New York, 1978.
93. B. Rosenberg, L. Van Camp, J. E. Trosko, and V. H. Mansour, *Nature (London)* **222**, 385 (1969).
94. C. J. F. Barnard, M. J. Cleare, and P. C. Haynes, *Chem. Br.* **22**, 1001 (1986).
95. S. J. Lippard, *Pure Appl. Chem.* **59**, 731 (1987).
96. G. Eastband, Jr., *Drugs Future* **12**, 139 (1987).
97. C. Dominici, A. Alimonti, S. Caroli, F. Petrucci, and M. A. Castello, *Clin. Chim. Acta* **152**, 207 (1986).
98. F. Alimonti, F. Petrucci, C. Dominici, and S. Caroli, *J. Trace Elem. Electrolytes Health Dis.* **1**, 79 (1987).
99. S. Caroli, A. Alimonti, M. A. Castello, C. Dominici, F. Petrucci, and A. Pettirossi, *Proc. Ital.-Hung Symp. Spectrochem., Biomed. Res. Specrochem., 3rd*, Ispra, *1987*, p. 239 (1987).
100. A. Alimonti, C. Dominici, F. Petrucci, F. La Torre and S. Caroli, *Acta Chim. Hung.* **128**, 527 (1991).
101. S. Caroli, A. Alimonti, F. Petrucci, F. La Torre, C. Dominici, and M. A. Castello, *Ann. Ist. Super. Sanita* **25**, 487 (1989).
102. M. Sperling, S. Xu, and B. Welz, *Anal. Chem.* **64**, 3101 (1992).
103. G. Schwedt, *Chromatographia* **11**, 145 (1978).
104. A. Berthod, M. Kolosky, J. L. Rocca, and O. Vittori, *Analusis* **7**, 395 (1979).
105. R. C. Gurira and P. W. Carr, *J. Chromatogr.* **20**, 461 (1982).
106. E. B. Sandell and H. Onishi, *Photometric Determination of Traces of Metals*, 4th ed., Part 1. Wiley, New York, 1986.
107. M. Tanaba and M. Tanaka, *Anal. Lett.* **13**(A6), 427 (1980).
108. H. Blutstein and R. F. Shaw, *Environ. Sci. Technol.* **15**, 1100 (1981).
109. R. F. Montoura, in *Marine Organic Chemistry* (E. K. Duursma and R. Dawson, eds.), Chapter 7. Elsevier, Amsterdam, 1981.
110. R. A. Saar and J. H. Weber, *Environ. Sci. Technol.* **16**, 510A (1982).
111. I. T. Neubecker and H. E. Allen, *Water Res.* **17**, 1 (1983).
112. J. Buffle, *Complexation Reactions in Aquatic Systems: An Analytical Approach*, Ellis Horwood Ser. Anal. Chem. Ellis Horwood, Chichester, UK 1988.
113. M. Takashi, T. Eiichiro, and Y. Sugimura, *Anal. Chem.* **62**, 1737 (1990).

CHAPTER

5

NEUTRON ACTIVATION ANALYSIS AND RADIOTRACER METHODS IN BIOINORGANIC CHEMISTRY

H. A. DAS AND J. R. W. WOITTIEZ

Netherlands Energy Research Foundation, ECN,*
1755 ZG Petten, The Netherlands

5.1. INTRODUCTION

Neutron activation analysis (NAA) and radiotracer techniques possess considerable potential for use in speciation analysis of biological materials. Due to the rather secluded nature of this branch of analytical chemistry, it seems useful to cast a few retrospective glances at the milestones of its development over the years (1).

5.1.1. Historic Background

Most analytical methods rely on the chemical properties of atoms or molecules, the energy involved being that of the chemical bond, 4 eV, or part of it. This is reflected not only by a sensitivity to changes in valency and bonding but also by the accompanying matrix and interference effects. The first partial exception to this limitation is X-ray fluorescence spectrometry (XRFS), where energies up to 100 keV are encountered. Although this makes the method useless for chemical speciation, it does not eliminate matrix effects by adsorption or interferences by peak overlapping. The advent of the nuclear reactor in the 1950s and the inherent possibility of neutron activation with energy jumps of ca. 1 MeV opened up an exciting vista. The new method was heralded as a substantial advance in analytical chemistry because of three factors:

---

* ECN: Energieonderzock Centrum Nederland.

Element Speciation in Bioinorganic Chemistry, edited by Sergio Caroli.
Chemical Analysis Series, Vol. 135.
ISBN 0-471-57641-7 © 1996 John Wiley & Sons, Inc.

(i) its unique specificity combined with superior sensitivity; (ii) the innate protection against blanks; and (iii) the possibility of performing an analysis in a purely instrumental way.

From the first applications in geochemistry, the method branched out into virtually every domain of science, particularly that of the life sciences and, later, of environmental protection. The annual number of publications rose exponentially as nuclear facilities became available in more and more countries and afforded higher neutron densities. Simultaneously, the application of radiotracers spread rapidly, particularly in biological and medical sciences. Small amounts of radioactivity, which could be handled with only a modest set of precautions, became an essential tool.

By the time NAA reached its first stage of maturity, in the late 1970s, some of its inherent drawbacks had already become apparent. Apart from its restriction to a few nuclear centers and the long turnover time, the most obvious limitation was its insensitivity to the chemical status of the (trace) elements under consideration, once hailed as its main advantage. Thus, NAA was caught between the competition from the rapidly improving optical methods, on the one hand, and the new demand for chemical speciation, on the other. NAA seemed to be reduced to some routine instrumental work and occasional reference analyses in about 50–70 institutes all over the world. At the same time, the use of radiotracers in transport phenomena studies was much less affected, as they could be utilized in many laboratories at a reasonable cost.

The process of adaptation to this new situation can be traced back to the early 1970s. It was already realized in those years that the unrivaled specificity of NAA and its usually high sensitivity should confer enough of an advantage to enable separation to be performed prior to irradiation, sacrificing the inherent protection against blanks (2, 3). In this way, NAA could be linked up with the growing demand for speciation.

If we now attempt to draw up a balance sheet after some 20 years of development, it appears that speciation by NAA has contributed substantially to at least three fields: (i) study of biological materials, water, and "new" solids developed for the electronics industry, most of which was achieved by NAA; (ii) application of charged particles and high-energy photons to inorganic solids (almost entirely restricted to them (4); and (iii) analysis of thin preparations by proton-induced X-ray emission (5). The use of radio-labeled compounds in metabolic studies has become another well-established branch of radioanalytical chemistry, as it allows the study of the fate of a particular component within a living organism to be carried out (6). The main budgetary difficulty lies, however, in the necessity of keeping up facilities for NAA in nuclear centers, rather than in maintaining radiotracer laboratories.

5.1.2. Definitions and Limitations

The term *speciation* has at least three different meanings, as it may refer to a spatial, physical, or chemical distribution of species. Layer-by-layer ablation analysis of solids and the determination of concentration variations in a biopsy or obduction sample are examples of the first interpretation. The second definition is applied to the analysis of separated particulate matter and colloidal fractions from environmental water samples. Chemical speciation, in turn, is employed in metabolic studies and in the analysis of physically homogeneous materials, such as serum and filtered water.

Here we shall place emphasis on physical and chemical speciation in serum by gel filtration chromatography and NAA and related applications of radiotracers and radiolabeled compounds, as performed in the radiochemical laboratory of ECN. On the other hand, the vast field of radioisotope applications in metabolic studies is left aside in this chapter.

5.1.3. General Aspects

Data. Elemental analysis of human body fluids and tissues mainly concerns blood, plasma, serum, and urine. In clinical laboratories Ca, K, Na, and P (as inorganic phosphate) and to some extent Cl and Mg are determined routinely in human serum. Much less common is the determination of minor and trace elements. There is no generally accepted definition of the term *trace element*, as the elemental content may differ by orders of magnitude over the various body fluids and tissues. It is, however, more or less accepted that the term be used for elemental concentrations below 100 ppm (parts per million), i.e., below 0.01% w/w (7, 8). This convention excludes C, Cl, H, K, N, Na, P, and S from being classified as trace elements in human serum (9, 10). Mertz defines an element to be at the trace level if its concentration does not exceed the Fe concentration in the same body fluid or tissue (11).

In human serum, application of this criterion leads to the exclusion of Br, Ca, Mg, and Si from being classified as trace elements, since their concentrations exceed the Fe concentration of ca. 1 μg g^{-1}. Table 5.1 summarizes some recent compilations of data on trace elements in various types of biological material and the literature on postirradiation separation schemes.

A much-applied classification of elements, according to their importance to living organisms, is into *essential* and *nonessential* categories. Underwood considers C, Ca, Cl, H, K, Mg, N, Na, O, P, and S to be essential major elements and As, Co, Cr, Cu, Fe, I, Mn, Mo, Ni, Se, Si, Sn, V, and Zn to be essential trace elements (12).

For those elements considered to be essential, it is thought that their concentration is held constant in living organisms by a homeostatic mechan-

Table 5.1. Some Recent Compilations of Data on Elemental Concentrations in Biological Materials

J. R. W. Woittiez	List of published mean values for elemental contents in human serum as obtained by neutron activation analysis. *ECN* [*Rep.*] **ECN-147**, pp. 215–241 (1984). (*Available upon request.*)
Y. Muramatsu and R. M. Parr	Summary of available biological reference materials. *IAEA* [*Rep.*], Vienna **IAEA/RL/128** (1985).
V. Iyengar and J. R. W. Woittiez	Trace elements in human clinical specimens: evaluation of literature data to identify reference values. *Clin. Chem.* **34**, 474 (1988).
C. Minoia and E. Sabbioni	Trace element reference values in tissues from inhabitants of the European Community: a study of 46 elements in urine, blood and serum of Italian subjects. *Sci. Total Environ.* **95**, 89 (1990).
S. Caroli, A. Alimonti, E. Coni F. Petrucci, O. Senofonte, and N. Violante	The assessment of reference values for elements in human biological tissues and fluids: a systematic review. *Crit. Rev. Anal. Chem.* **24** (1994), 363–398.

Some Recent Reports on Postirradiation Schemes in NAA Studies of Biological Materials:

R. G. Greenberg and R. Zeisler	A radiochemical procedure for ultratrace determination of chromium in biological materials. *J. Radioanal. Nucl. Chem.* **1244**, 5 (1988).
R. Zeisler, R. R. Greenberg, and S. F. Stone	Radiochemical and instrumental neutron activation analysis procedures for the determination of low level trace elements in human livers. *J. Radioanal. Nucl. Chem.* **124**, 47 (1988).
R. Cornelis	Radiochemical methods, especially neutron activation analysis, in *Quantitative Trace Analysis of Biological Materials* (H. A. McKenzie and L. E. Smythe, eds.), Chap. 14. Elsevier, Amsterdam, 1988.
N. Lavi, M. Mantel, and Z. B. Alfassi	Determination of selenium in biological materials by neutron activation analysis. *Analyst* (*London*) **113**, 1855 (1988).
J. R. W. Woittiez	Use of reference materials for quality control of elemental analysis by neutron activation with radiochemical separation. *Fresenius' Z. Anal. Chem.* **338**, 57 (1990).
N. Lavi and Z. R. Alfassi	Determination of trace amounts of Cd, Co, Cr, Fe, Mo, Ni, Se, Ti, V and Zn in blood and milk by neutron activation analysis. *Analusis* **115**, 817 (1990).

INTRODUCTION

ism. This is reflected by some narrow-ranged elemental concentrations in human blood, e.g., Cl, K, and Na. For trace elements in human organs, including blood, normal distribution occurs for essential elements and lognormal distribution for nonessential elements (12). However, the apparent shape of a distribution curve of an element in human blood, is dependent on the accuracy of the analytical technique, especially sample preparation.

Reliability. The interaction between the average concentration and the observed scatter is illustrated by a plot of the range, r, in the analytical data against the median value, Me (Fig. 5.1), as taken from the compilations of Iyengar et al. (9) and Versieck et al. (13).

The obvious question is whether the observed deviations in the mean of mean values are to be explained from the reported standard deviations (SDs) in the individual means. If this explanation, expressed as the ratio of external to internal variance, holds, then the reported data for a given element are consistent. If X_i is defined as the reported mean value and W_i as the statistical

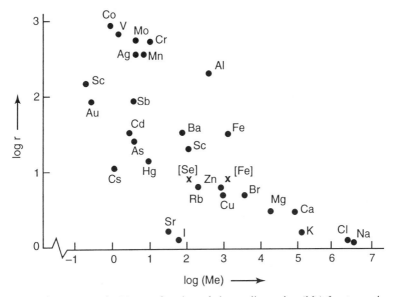

Figure 5.1. The range ratio (r) as a function of the median value (Me) for trace element determination in human serum with NAA. Median values are in nanograms per milliliter (1). The elements within brackets and marked by a cross are measured by nonradiochemical determinations.

weight, then the average of all experimental X-values, $\bar{\bar{X}}_w$, is

$$\bar{\bar{X}}_w = \frac{\sum_1^{N'} W_i \bar{X}_i}{\sum_1^{N'} W_i} = \sum_1^{N'} \left[\frac{W_i}{\sum_1^{N'} W_i}\right] \bar{X}_i \tag{1}$$

with the relevant weighted SD expressed as:

$$s_w(\bar{X}_i) = \left[\frac{\frac{1}{N'-1}\sum_1^{N'} W_i(\bar{X}_i - \bar{\bar{X}}_w)}{\frac{1}{N'}\sum_1^{N'} W_i}\right]^{\frac{1}{2}} \tag{2}$$

The weighting factors used are given by

$$W_i = \frac{s_0^2}{s_i^2} \tag{3}$$

with N' being the number of well-documented means.

To obtain an impression about the consistency of reported values, the ratio of the external variance to the internal variance in the mean of means is calculated:

$$S_{ext} = s_{ext}^2(\bar{\bar{X}}_w) = \frac{s_w^2(\bar{X}_i)}{N'} \tag{4}$$

Application of the law of the propagation of errors to Equation 1 by using Equation 3 leads to

$$S_{int} = s_{int}^2(\bar{\bar{X}}_w) = \sum_1^{N'} \frac{W_i^2}{\sum_1^{N'} W_i} s_i^2 = \sum_1^{N'} \frac{W_i s_0^2}{\left(\sum_1^{N'} W_i\right)^2} = \frac{s_0^2}{\sum_1^{N'} W_i} = \frac{1}{\sum_1^{N'} \left(\frac{1}{s_i^2}\right)} \tag{5}$$

The ratio of two variances follows the F-distribution of Fisher–Snedecor. Consistency is tested by comparison with a critical F-value, determined by a confidence level of $1 - \alpha$, and the degrees of freedom $N' - 1$ and ∞, respectively (12). Table 5.2 shows the results for Ca, Cu, Fe, I, Na, Se, and Zn.

At $\alpha = 0.05$, the hypothesis that results are not consistent must be rejected for I. For the other elements results are not consistent. Besides, differences between Me, $\bar{\bar{X}}$, and $\bar{\bar{X}}_w$ occur, indicating that it may not be justified to

Table 5.2. Calculation of the Median Mean and Unweighted and Weighted Mean of Means, with Corresponding Standard Deviations and Experimental F-Values of Reported Mean Values of Essential Elements[a,b]

Element	Me	N	$\bar{x} \pm s(\bar{x}_i)$	$\bar{\bar{x}}_w \pm s_w(\bar{x}_i)$	N'	$\dfrac{S_{ext}(\bar{\bar{x}})}{S_{int}(\bar{\bar{x}})}$
Ca	97	9	99 ± 17	111 ± 13	5	3.4
Cu	1.05	24	1.21 ± 0.45	1.25 ± 0.13	18	3.8
Fe	1.41	24	1.79 ± 1.02	1.67 ± 0.54	17	3.6
I (ng)	58	7	58 ± 7	59 ± 8	3	0.2
Na	3.255	12	3.145 ± 0.330	2.937 ± 0.149	5	3.0
Se	104	42	124 ± 54	116 ± 22	28	3.1
Zn	0.99	34	1.04 ± 0.52	0.84 ± 0.13	25	3.5

[a] Concentration in micrograms per milliliter, given as the median mean (Me).
[b] Here N = the number of measurements.
For the other terms, see Equations 1–5.

calculate an arithmetic mean, as results are probably not symmetrically distributed.

A Way to Discriminate Between Essential and Nonessential Elements. Liebscher and Smith have proposed a method of discriminating between essential and nonessential elements based on their respective normal and log-normal distributions (14). The method is based on the calculation of the arithmetic mean ± SD, the geometric mean $x \pm$ SD, and the median value within the borderlines of the first ($n/4$) and third ($3n/4$) quartile. If an element is symmetrically (or normally) distributed, the median mean, the arithmetic mean, and the geometric mean should all be close in value and $n/4$ and $3n/4$ should be predicted from the arithmetic mean (±0.68 SDs). If an element is nonessential, the median mean and the geometric mean should be close whereas the arithmetic mean should deviate; also $n/4$ and $3n/4$ should be predicted from the geometric mean ($x \pm 0.68$ SD). This has proved to work well for at least obvious cases.

The remainder of Liebscher and Smith's proposal, which suggests the testing of a distribution form resulting from a known "status," seems valid for (trace) elements in human serum. Both the demand for closeness of the mean of means and the median mean and the requirement that $n/4$ and $3n/4$ be predictable are fulfilled. Woittiez has found this to be a useful tool when controlling for serum data in the literature (1).

5.2. OUTLINE OF NEUTRON ACTIVATION ANALYSIS

5.2.1. Principles

NAA is based on the formation of unstable (radioactive) nuclei by neutron capture. The reaction rate varies by orders of magnitude over the range of stable isotopes. Moreover, it depends on the neutron energy and thus on the neutron spectrum, which covers thermal (energy $<0.4\,\text{keV}$) and epithermal (energy up to $100\,\text{keV}$) neutrons (1). The decay of the radionuclei to stable isotopes often comprises the emission of one or more characteristic γ-rays. The number of radioactive nuclei formed, N^*, is proportional to the original number of the corresponding stable reactive nuclei, N. As the decay rate is given by

$$\frac{dN^*}{dt} = -\lambda N^* \tag{6}$$

it follows that the intensity of the photon emission is proportional to N^* and thus to N. The factor λ is called the decay constant. Gamma-ray spectrometry results in a net number, A, of recorded emissions, or "counts." Thus, A is proportional to N, and where N is proportional to the element concentration c_0, it follows that

$$A = kc_0 \tag{7}$$

By standardizing the irradiation and counting conditions, k may be kept constant over a series of experiments in which samples and standards are compared. The constant k can be written as the product of four factors, k_1 to k_4, which represent the influences of isotopic composition, reactivity and irradiation conditions, decay, and γ-ray spectrometry. The factor k_1 has the dimensions $(\mu\text{g}\,\text{g}^{-1})^{-1}$, whereas the other three are dimensionless. Their definitions are $k_1 = N/c_0$; $k_2 = (N^*)_0/N$; $k_3 = (N^*)_t/(N^*)_0$; and $k_4 = A/(N^*)_t$. Here, $(N^*)_0$ is the number of radioactive nuclei formed during the irradiation; $(N^*)_t$ is the remaining number of nuclei after a decay time t_2.

Isotopic Composition. The number of stable, reactive nuclei, N, is related to the elemental concentration c_0 by

$$N = \frac{c_0 G N_{Av} a}{M} \tag{8}$$

where G is the weight in grams; N_{Av} is Avogadro's number; a is the isotopic abundance; and M is the atomic weight. Then,

$$k_1 = \frac{G N_{Av} a}{M} \tag{9}$$

Reactivity and Irradiation Conditions. These factors are covered by

$$k_2 = \left(\frac{1-\exp(-\lambda t_1)}{\lambda}\right)\int_0^\infty \sigma_E \phi_E \, dE = \left(\frac{1-\exp(-\lambda t_1)}{\lambda}\right)\phi_{th}\left[\sigma_{th} + I\frac{\phi_E}{\phi_{th}}\right] \quad (10)$$

Here, t_1 is the irradiation time (s); ϕ_{th} and σ_{th} are thermal flux (cm^{-2} s^{-1}) and thermal cross section (cm^2), respectively; the product $I\phi_E$ represents the epithermal activation probability; the epithermal neutron flux is ϕ_E (with I being the resonance integral). The influence of the thermal neutron flux and the neutron spectrum (the term in square brackets) should be distinguished. As a rule, the latter may be considered constant, but the flux may vary appreciably for each irradiated aliquot. This implies the need for individual flux monitoring for the (relative) assay of ϕ_{th} per aliquot.

Decay. It follows from the experimental decay law that

$$k_3 = \exp(-\lambda t_2) \quad (11)$$

where t_2 is the decay time (s).

Gamma-Ray Spectrometry. The expression for k_4 consists of three factors. The first concerns ideal measurement of the total count result $A = \xi a_\gamma t_m \lambda (N^*)_t$, where ξ is the detector efficiency (γ-ray energy dependent), a_γ is the γ-ray abundance, and t_m is the measuring time. The second is for decay during measurements, $(1-\exp(t_m))/t_m$. The third term concerns electronic imperfection, giving rise to terms for residual dead-time losses, f_τ, and pileup, f_p. It follows that k_4 can be written thus:

$$k_4 = \frac{A}{(N^*)_t} = \xi a_\gamma (1-\exp(-\lambda t_m)) f_\tau f_p \quad (12)$$

The combination of k_1 to k_4 gives the expression for k:

$$k = \frac{GN_{Av}a}{M}\frac{(1-\exp(-\lambda t_m))}{\lambda}\phi_{th}$$
$$\times \left[\sigma_{th} + I\frac{\phi_E}{\phi_{th}}\right](\exp(-\lambda t_2))\xi a_\gamma (1-\exp(-\lambda t_m)) f_\tau f_p \quad (13)$$

Activation analysis is normally practiced as a relative technique. Comparing the result obtained for a sample, A_x, to that for a standard, A_s, we obtain

$$\frac{A_x}{A_s} = \frac{k_x(c_0)_x}{k_s(c_0)_s} \tag{14}$$

If irradiation and counting conditions are equal, insertion of Equation 13 into Equation 14 results in

$$\frac{A_x}{A_s} = \frac{G_x(\phi_{th})_x \cdot \exp(-\lambda(t_2)_x)\,(f_\tau)_x\,(f_p)_x\,(c_0)_x}{G_s(\phi_{th})_s\,\exp(-\lambda(t_2)_s)\,(f_\tau)_s\,(f_p)_s\,(c_0)_s} \tag{15}$$

$$(c_0)_x = \frac{A_x}{A_s}\frac{G_s(\phi_{th})_s\,\exp(-\lambda(t_2)_s)\,(f_\tau)_s\,(f_p)_s}{G_x(\phi_{th})_x\,\exp(-\lambda(t_2)_x)\,(f_\tau)_x\,(f_p)_x}(c_0)_s \tag{16}$$

The influence of the factors from Equation 16 on the precision and accuracy of the analysis are briefly considered next.

5.2.2. Apparatus

Figure 5.2 shows the γ-spectrometry layout as used in NAA of biological materials. Figure 5.3 gives a γ-spectrum obtained with it, divided over 2000–8000 channels. The various parts of the measurement procedure are discussed in practical terms by Crouthamel et al. (15) and Krugers (16). The only aspects that need to be mentioned here refer to the corrections for pileup and dead-time losses and the inherent statistical error in the peak area.

The pileup losses are compensated for by observing the net peak area of the electronic pulser signal, situated at some convenient energy in the high part of the γ-spectrum. If the lifetime preset counting period is t_p and the pulser frequency 50 Hz, one should collect $50\,t_p$ counts. The lower number of counts that is actually observed reflects the fractional pileup loss, which is equal for all peaks.

Residual dead-time losses occur when the counting time/half-life ratio is higher than ca. 0.1. Fractional losses are usually from 1 to 10%. Correction is based on manual or automatic readings of the total dead-time meter (17, 18).

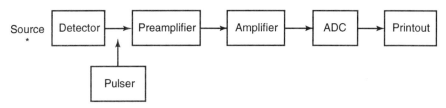

Figure 5.2. Layout of γ-spectrometry. (ADC: analog–digital converter.)

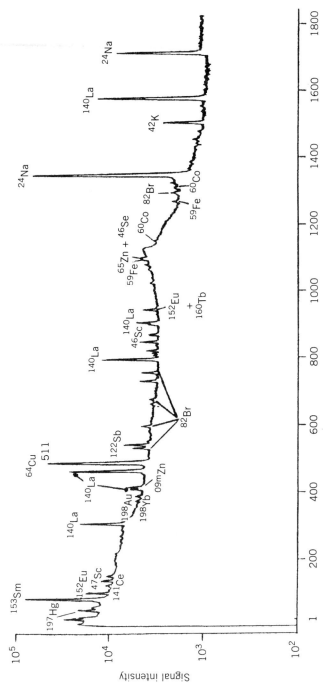

Figure 5.3. The γ-ray spectrum of a trace element concentrate from the SRM orchard leaves. Sample: orchard leaves (80.1 mg). Carbon concentrate: pH 4.5–5 (50 mg). Reagents: ammonium pyrrolidinedithiocarbamate (APDC), 30 mg; oxine, 5 mg; cupferron, 20 mg; peroxyacetyl nitrate (PAN), 5 mg; ascorbic acid, 600 mg. Counting time: 4000 s. Ge/Li well-type 4000 channel analyzer. Cooling: 48 h. Irradiation: 12 h at 5×10^{12} cm^{-2} s^{-1}.

The statistical error due to the linear (or sometimes nonlinear) interpolation of the compton background is a function of the background/peak area ratio and the number of channels in the peak. Details are given in Faanhof et al. (19).

5.2.3. Irradiation Facilities for NAA

5.2.3.1. Nuclear Reactors

Most nuclear reactors that are available for NAA are of the poolside type. Their size and power may vary widely. The difference in flux ranges over 3 orders of magnitude. The smallest commercially available reactor offers a maximal flux of ca. $10^{11}\,cm^{-2}\,s^{-1}$. At the other end of the scale the high-flux reactors reach more than $10^{14}\,cm^{-2}\,s^{-1}$. Next to the flux, the governing factors in using a reactor for activation analysis are as follows:

a. Capacity. This is the total volume of the irradiation facilities.

b. Neutron spectrum. The fast neutrons/thermal neutrons ratio determines the relative importance of any reactions other than (n, γ) reactions. A purely thermal flux is an unattainable ideal.

c. Gamma heating. The neutron flux is associated with a γ dose rate that causes heating and chemical transformation of the aliquot. The pressure buildup by evaporation and decomposition of water sets a limit to the irradiation time and the material of the capsules. A high γ/n ratio indicates that quartz ought to be used instead of polyethylene or nylon capsules.

d. Duty cycle. The predictable and regular availability of the reactor is of decisive importance for its practical significance.

The possibility of an epithermal facility is another criterion for application of the reactor involved. Table 5.3 gives a survey of the facilities available in the high-flux reactor (HFR) at Petten. For routine (I)NAA it is mandatory to have a good modular system of polyethylene capsules and corresponding shuttles. Capsules are of high-pressure polyethylene and are prepared in a specialized workshop to prevent contamination. For occasional long irradiations in a high flux it is necessary to use quartz capsules or ampoules, stored in Al cans. Irradiations are carried out in two ways: (i) batchwise in cans or capsules placed in fixed positions near the reactor core or in the slow pneumatic rabbit system (PRS-1), or (ii) one by one in the fast rabbit system (FASY).

A complete radioanalytical laboratory for biological material needs both modes. Batchwise irradiations are usually applied for the production of radioisotopes with half-life $(T_{1/2}) >$ ca. 5 h and in small research reactors with a soft irradiation regime for even shorter lived isotopes. Rabbit systems are

Table 5.3. Survey of Irradiation Facilities in the HFR Petten

Facility	Thermal Neutron Flux (in cm^{-2} s^{-1})	Fast Neutron Flux (in cm^{-2} s^{-1})	Time of Irradiation	Container	Remarks
Pneumatic rabbit system 1 (PRS-1)	5×10^{13}	5.9×10^{12}	Up to 60 min	Polyethylene capsules in polyethylene shuttle (34 mm i.d. × 120 mm)	Time of transport to laboratory: 2 min
Fast pneumatic rabbit system (FASY)	3.5×10^{13}	3.4×10^{12}	Up to 15 min	Polyethylene shuttle (10 mm i.d. × 19 mm)	Time of transport to counting position: 200 ms
Poolside rotating facility (PROF)	2–3×10^{12}	4–8×10^{11}	Up to 24 h	Polyethylene bottle (67 mm i.d. × 115 mm)	Lead shielding and rotation at 60 rpm
Poolside isotope facility (PIF)	2.2–7×10^{13}	0.3–0.7×10^{13}	Units of 24 h	Quartz capsules (3 mm i.d. and 8 mm i.d.) in Al cans (23 mm i.d. × 72 mm)	Total capacity: 30 cans

mostly of the pneumatic type, with return times on the order of seconds. Batch irradiations constitute the bulk of the activation capacity.

The quality of a radioanalytical institute is primarily governed by that of the irradiation facilities. To judge these, four factors must be considered: (i) spatial capacity; (ii) the (average) neutron flux; (iii) the neutron spectrum and its variation; and (iv) γ heating.

For one-by-one irradiations in fast rabbit systems, two other criteria are added: (i) the reproducibility of the return at the counting position and (ii) the contamination due to scraping from the shuttle tube.

It appears that up to 40 aliquots can be handled as one series. Thus, a batch facility should have enough space to contain about 60 aliquots, provided that multielement standards are used (see Section 5.3.4). The PROF device used at the poolside of the HFR 45 MW reactor at ECN consists of a 400 mL polyethylene flask in a dry Al pipe that is shielded by 2 cm Pb and turned at ca. 1 rpm during irradiation. The obvious drawback of such a large facility is the variation of the neutron flux. The alternative, used in medium-size research reactors, is the "Lazy Susan" facility, which rotates around the core. The vertical flux gradient, however, remains. For a Triga Mark III reactor it is ca. 0.4–0.5% mm^{-1}. Such a reactor has a very reproducible flux over long periods. The flux gradient may be eliminated by calibrating the positions within the capsules. A more general solution that is mandatory for high-flux reactors is the use of separate flux monitors. As a flux monitor is based on a single nuclear reaction, it will work correctly only if the neutron spectrum does not change with position or time. A flux monitor should thus be (i) reactive to thermal, epithermal, and—if possible—fast neutrons, giving rise to different radionuclides; (ii) of low self-absorption; (iii) of reasonable activity at the end of the irradiation; (iv) of reproducible mass; (v) easy to handle; and (vi) cheap.

A convenient choice is pure Fe as small rings of 10–30 mg that may be clamped around the capsules or inserted in the snap caps. The reactions involved are ^{58}Fe(n, γ)^{59}Fe, $T_{1/2} = 45$ days, and ^{54}Fe(n, p) ^{54}Mn, $T_{1/2} = 313$ days. For the 25 mg rings, used at ECN, the standard deviation in the mass is $\leqslant 0.3\%$. Iron rings are not suitable for rabbit irradiations. This so-called monostandard procedure has been dealt with by many authors; for a survey of these studies, see De Corte (20). Several monitors are commercially available.

5.2.4. Practical Aspects of NAA of Biological Materials

5.2.4.1. NAA with Short-Lived Radionuclides at ECN

Irradiations are done in the PRS-1 facilities mentioned earlier. Standard irradiation time is 2 min, in a thermal neutron flux of 5.9×10^{13} cm^{-2} s^{-1} and a fast neutron flux of 0.53×10^{13} cm^{-2} s^{-1}. The samples, mostly liquids sealed

Table 5.4. Radionuclides, with Half-lives and γ-Ray Energies, Formed by Irradiation in the PRS Device

Nuclide	Half-life (min)	γ-Ray Energy (keV)
^{28}Al	2.246	1779
^{52}V	3.75	1434
^{66}Cu	5.1	1039
^{27}Mg	9.46	844; 1014
^{80}Br	17.6	616
^{128}I	25.0	443
^{38}Cl	37.18	642
^{85}Sr	67.7	232
^{56}Mn	154.8	847; 1811
^{42}K	742	1525
^{24}Na	902	1369

in 80 × 15 mm polyethylene capsules, are processed one by one. The decay and counting times are 4 and 10 min, respectively, in a ca. 1.6 μg (50 μL)$^{-1}$ enriched ^{84}Sr(NO$_3$)$_2$ solution. The norm to which flux is corrected depends on the sample and counting volumes; pipetting errors in these volumes are also corrected for by use of the internal flux monitor. If sample weights are replaced by sample volumes, V_x and V_s, and the count results for Sr, $(S_{Sr})_x$ and $(A_{Sr})_s$, represent relative flux differences, Equation 16 becomes

$$(c_0)_x = \frac{A_x}{A_s} \frac{V_s}{V_x} \frac{(A_{Sr})_s}{(A_{Sr})_x} \frac{(f_{\tau,Sr})_s}{(f_{\tau,Sr})_x} \frac{(f_\tau)_s}{(f_\tau)_x} \frac{(f_p)_s}{(f_p)_x} (c_0)_s \qquad (17)$$

Table 5.4 gives the half-lives and γ-ray energies of the radionuclides of interest, formed by irradiation in the PRS-1 facilities. By purely instrumental NAA the presence of milligram amounts of NaCl in serum gives rise to dominating ^{24}Na and ^{38}Cl activities, masking all the other γ lines. Therefore, Na and Cl have to be separated from the matrix, or a group of elements (or a single element) has to be isolated. Were this done after irradiation, one would be obliged to act quickly, risking high dose rates (of ca. 1 R h^{-1}), and having to do radiochemistry, including mineralization, in a few minutes. For that reason it is preferable to remove Na and Cl prior to irradiation, as we chose to do.

5.2.4.2. NAA with Long-Lived Radionuclides at ECN

All irradiations are performed in the PROF facility. The PROF, though 10 times lower in flux, was preferred over the PIF, with its harder spectrum and higher γ heating, as the former device produces less contamination and irradiation damage. In the PIF, samples must be irradiated in relatively large (96 × 3 mm) sealed quartz capsules (minimum-weight capsule/sample ratio of ca. 10), resulting in flux differences over the sample and increased chances of contamination. The samples, heavily damaged, must then be recovered after irradiation by breaking the capsules. In the PROF, on the other hand, ca. 250 mg samples are irradiated in 8 × 11 mm polyethylene capsules (minimum weight capsule/aliquot ratio ca. 1.1), which can stand a 12 h irradiation when closed only with a snap cap. Thus, if necessary, samples can be unpacked easily.

Standard irradiation time is 12 h, in a flux of 4×10^{12} cm^{-2} s^{-1}; 22 samples can be irradiated simultaneously. The decay times are 1–4 weeks, with counting times from 10^3 to 10^4 s. Flux corrections are made by external monitoring, through ca. 24 mg Fe flux rings, packed with each sample. The norm, to which the count rate of the monitor is corrected, is constant. Equation 16 now becomes

$$(c_0)_x = \frac{A_x}{A_s} \frac{G_s}{G_x} \frac{(A_{Fe})_s}{(A_{Fe})_x} \frac{\exp(-\lambda(t_2)_s)}{\exp(-\lambda(t_2)_x)} \frac{(f_p)_s}{(f_p)_x} (c_0)_s \tag{18}$$

Correction factors due to residual dead-time losses can be neglected here. Relative flux differences are given by the count results for the Fe flux rings, $(A_{Fe})_s$ and $(A_{Fe})_x$. Table 5.5 lists the radionuclides of interest formed by PROF irradiation, including their half-lives and the γ energies used for analysis. The spectrum is dominated by ^{24}Na activity for a decay period of up to 1 week.

Table 5.5. Radionuclides, with Half-lives and γ-Ray Energies, Formed by Irradiation in Poolside Rotating Facilities

Nuclide	Half-life	γ-Ray Energy (keV)
^{24}Na	15.03 h	1369
^{82}Br	35.34 h	776; 554
^{32}P	14.3 days	No γ
^{86}Rb	18.7 days	1077
^{59}Fe	44.6 days	1099; 1292
^{124}Sb	60.3 days	603; 1691
^{75}Se	120 days	265; 136
^{65}Zn	244 days	1115
^{134}Cs	2.06 years	605; 796
^{60}Co	5.272 years	1332; 1173

Decay from 2 up to 4 weeks allows for the detection or determination of the aforementioned elements. The analysis may be speeded up by group separation or removal of ^{24}Na and ^{32}P. This will generally also improve detection limits. However, for the accurate determination in serum of elements at the nanogram per gram level that have radionuclides with half-lives of from 0.5 to 5 days, separation of the individual elements and γ-spectrometry in well-type NaI detectors is necessary, with the disadvantage of losing multielement measurement by Ge(Li) detectors. The standard irradiation and measurement program is based on irradiation of serum samples for 12 h in a neutron flux of $5 \times 10^{12}\,\mathrm{cm^{-2}\,s^{-1}}$. Samples are counted three times: (i) for 1000 s after about 1 week decay in a 45 cm^3 coaxial Ge(Li) detector; (ii) for 10,000 s after about 3 weeks decay in a 89 cm^3 coaxial well-type Ge(Li) detector; (iii) for 10,000 s after about 5 weeks decay in the well-type detector in an Al tube to reduce the bremsstrahlung spectrum of ^{32}P.

The γ-ray spectra are analyzed by a computer program that determines the total peak area above a linearly interpolated background. The γ-spectrum of serum after a week decay period features ^{24}Na and ^{82}Br peaks. Both elements can be determined precisely, with a typical error in the concentration of 1–2%. Especially the Na peak can be used to detect possible systematic errors due to lyophilization.

After a decay period of 3 weeks, γ-ray energies of ^{86}Rb, ^{65}Zn, and ^{59}Fe in serum can also be measured. The precision for an Rb determination is characterized by a total analytical error of 3–7%; for Zn, this error is 1–3%. The Fe level is at the limit of determination, resulting in errors of up to 20%. If samples are counted in the Al tubes after 4–5 weeks decay, the 136 keV and 264 keV γ-ray energies of ^{75}Se also become measurable. Typical errors in the determination are 10% for 50 ng g^{-1} to 6% for 120 ng g^{-1}. The γ-ray energies of Co, Cr, Cs, and Sb can also be found in the spectra: the Cr peak originates as a blank from the polyethylene vessel and cannot be used for analysis; Cs and Sb, occurring in human serum at a nanogram per milliliter level, and Co, which is present below that level, are difficult to measure.

5.2.4.3. Standards

In the case of long-lived radionuclides, standards are aqueous solutions of one or more elements, in different concentrations, pipetted on very pure activated carbon and dried. Samples and standards are irradiated simultaneously in similar vessels so as to minimize differences in geometry. For elemental analysis via short-lived radionuclides in most liquid samples, standard solutions of the corresponding elements are pipetted to a volume similar to that of the matrix of concern. Small deviations in the total volume to be irradiated and counted are corrected by use of the Sr-flux monitor.

The final results of analysis are calculated from the counting results with errors for sample and standard, after correcting for electronic losses and adjusting to the chosen flux value. The counting results from the samples are divided by the (average) specific count rate for the standard, with the same flux value and measuring conditions.

5.2.4.4. Sampling and Sample Preparation

Blood samples are taken from a vein in the arm, with the donor in a recumbent state, the blood being obtained with a Teflon intravenous catheter. The first 5 mL are discarded; then 10 mL blood are collected in a 15 mL Teflon centrifuge tube, which is closed by a polyethylene snap cap. The sample is transported to the clean room, where within 1 h after collection it is centrifuged (2×15 min, 3200 rpm). The supernatant liquid (about 5 mL) is pipetted in a weighed Teflon beaker, weighed, frozen, lyophilized for ca. 65 h, and weighed again. After being powdered with a Teflon stirring rod, an amount of 150–200 mg lyophilized serum is weighed in a polyethylene vial and irradiated.

Transfer of the lyophilized homogenized dry serum from the Teflon beaker to the irradiation vessel is done by passage through a polyethylene funnel [upper i.d. = 34 mm; lower i.d. = 3 mm; length = 60 mm].

Polyethylene irradiation vessels are washed with distilled water, then dried and weighed. After the lyophilized serum is added to them, they are weighed, closed with a snap cap, washed again, and stored in a polyethylene vessel with a screw cap washed with distilled water.

Except for Co and Cr, none of the elements detected gave rise to significant contributions originating from the polyethylene irradiation vessel. Table 5.6 presents blank values or lower detection limits for blank values.

Table 5.6. Blank Values or Lower Detection Limits for Blank Values in Polyethylene Irradiation Vessels[a]

Element	Blank Value of Polyethylene	Number of Positive Results
Br (μg)	0.022 ± 0.006	3
Na (μg)	3.4 ± 3.0	11
Rb (ng)	<10	0
Zn (ng)	13 ± 2	2
Se (ng)	<4	0
Fe (μg)	0.2 ± 0.05	2
Co (ng)	<1	11
Cs (ng)	<0.4	0
Sb (ng)	1.0 ± 0.6	3

[a]The total number of blanks measured is 11.

5.2.4.5. The Error Budget

5.2.4.5.1. Statistical Errors. As regards weighing, all is done on the same balance, with a constant deviation of 0.03 mg. For a 400 mg dried serum sample weighed in a polyethylene container, the error is estimated to be 0.02%. Wet serum is processed by measuring the weight difference of a serum-filled and empty Teflon beaker. For a 5 g sample (ca. 4.9 mL), the relative standard deviation is less then 0.01%.

As for pipetting, those analyses of serum that feature short-lived radionuclides are based on volumes of 0.5–2.0 mL, each volume always being pipetted with the same micropipette. The micropipettes are calibrated for demineralized water and serum. Pipettes used for flux monitoring, 50 μL, and standards, 50 or 100 μL, are calibrated for demineralized water. In general, weighing errors can be ignored but pipetting is critical. If the sample size has to be based on volume, as in gel-filtration procedures, one should use a large (i.e., 500 μL) pipette. The total relative variance in the final counting results from various sources, but it is mainly determined by the uncertainty in the evaluated net peak areas of the photopeak of interest, the pulser peak, and the flux monitor photopeak. Table 5.7 surveys this issue.

Table 5.7. Error Budget for INAA Based on γ-Ray Spectrometry

Quantity	Statistical Error	Systematic Bias
Concentration of the standard	≤0.01	—
Weight of standard	<0.1	—
Weight of sample	<0.1	Depends on lyophilization procedure; usually ≤1
Geometry	<0.5	None if volumes are equal
Irradiation time	<0.1	—
Decay time	<0.1	—
Counting time	<0.1	—
Peak area of sample	1–10	Usually ≤5
Peak area of standard	~1	—
Peak area of iron flux monitor	<0.5	—
Peak area of pulser peak	<0.5	—
	3–10	≤5

5.2.4.5.2. Systematic Errors. Systematic errors can originate from many sources. The overall accuracy is controlled by running standards, both single and multielement, with each batch and comparing their specific count rates to well-defined values based on numerous determinations under fixed geometry, which are stored in a library. Besides, accuracy checks have to be done by analyzing aliquots from a standard reference material (SRM). The main sources are pipetting, counting geometry, and variations in aliquot mass due to lyophilization and subsequent uncontrolled uptake of moisture. For the SRM-1577 bovine liver, issued by the National Institute of Standards and Technology (NIST) (U.S.), the relative moisture uptake amount to ca. 14% over ca. 3h after lyophilization (6). Corrections for pileup and residual dead-time losses are capable of eliminating these effects almost completely. Table 5.7 summarizes some of these observations.

5.3. SEPARATION OF PROTEINS AND ITS CONTROL BY RADIOTRACERS AND NEUTRON ACTIVATION

5.3.1. General Remarks

When one sets out to investigate the association of human serum proteins and trace elements, several approaches are possible. They will vary from separation of all proteins from the serum matrix (usually by precipitation) to isolation and purification of one protein as practiced in immunochemistry. Three main conditions have to be fulfilled for the detection of protein-bound trace elements: (i) elemental contamination should be avoided as far as possible, which excludes, e.g., protein collection on gel slabs, as in electrophoresis, the use of precipitating agents, or metal-containing equipment; (ii) serum samples ought to be large enough to enable the detection of elements in protein fractions, which have low concentrations in whole serum (for fractionation this results in volumes of at least 1 mL; and (iii) a separation technique should be used that is based on mild separating conditions. As element–protein associations might be weak, ion-exchange techniques and affinity chromatography and techniques using electrical fields (isoelectric-focusing) should be excluded.

In light of these three conditions, gel filtration chromatography is the separation technique of choice. A disadvantage is its poor resolving power and thus its considerable sample dilution. Gel filtration is applied in two ways, namely, as a desalting technique to determine the total protein-bound element content and as a protein-fractionating technique to determine the elemental content in different serum protein fractions. For desalting Biogel is used with an upper fractionation limit of 1800 Da, whereas the protein separation is done on LKB Ultrogel AcA-34 with a fractionation range of 2×10^4 to 3.5×10^5 Da (21, 22).

5.3.2. Gel Filtration Chromatography

The technique is described in detail by Determann (23), whereas the high-performance size-exclusion chromatography aspects are considered elsewhere (21). The theoretical aspects are dealt with by Shibukava (24) and Groh and Halasz (25). Zeisler has described the application of gels in electrophoresis (26). The important parameter is the availability constant K_{av}, defined as $K_{av} = (V_r - V_0)/(V_t - V_0)$, with V_r representing the retention volume of the solute, V_t the total (geometric) volume of the column, and V_0 the dead-volume (27).

It has been found that for a group of molecules with similar shapes (e.g., globular proteins) there is a range of molecular weights where a linear relation between K_{av} and the logarithm of the molecular weight exists (23). Deviation in linearity occurs near total exclusion and total permeation. A disadvantage of the use of K_{av} is the necessity for the determination of V_0 for each experiment. For this purpose the synthetic polymer Blue Dextran is frequently used, thus introducing contamination into the column. As an alternative, the retention volumes of the peaks in a human serum protein spectrum have been related to the position of the salt peak, $K_s = V_r/V_{salt}$. The salt peak always shows up in a protein spectrum containing inorganic salts and small molecules. The setup is shown in Fig. 5.4, and a typical chromatogram in Fig. 5.5.

The buffer used was a 0.1 M Tris–acetic acid buffer with 0.5 M ammonium acetate at pH 7.0 (27). The use of Na and Cl in the buffer had to be avoided, as these elements give rise to interfering γ-radiation after NAA. To lower blank values, the buffer was changed to 0.02 M Tris–acetic acid with 0.1 M ammonium acetate at pH = 7.4.

5.3.3. Gel–Element Interaction of Gel Filtration Materials

Many studies have tested the validity of the assumption that the gel material is inert, under various experimental conditions. George showed the absorption of Sr on Sephadex G-25 and G-50 when a milk sample containing 0.6 mg Sr mL^{-1} was eluted with phosphate saline buffer (28). Evans and Fritze found Cu contamination in a protein spectrum originating from a Biogel P-150 column (29). Sabbioni and Marafante reported recoveries of 85% of human protein-bound Fe and 95% of rat protein-bound V after gel filtration in 10 mM Tris–HCl buffer at pH 8 on Sephadex G-150 (30). Johnson and Evans showed Cu and Zn binding by various Sephadex, Biogel, and LKB materials, with different contents under different conditions (31). Norheim and Steinnes,

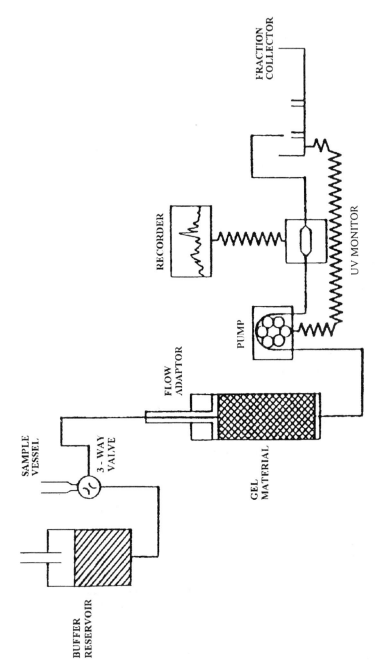

Figure 5.4. Gel filtration chromatography setup.

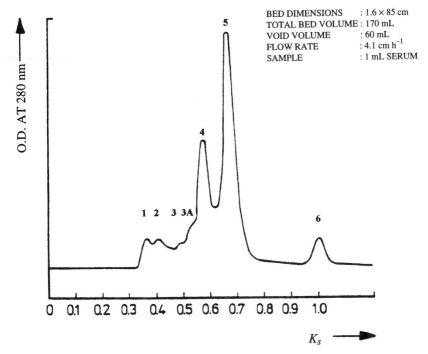

Figure 5.5. Protein spectrum after elution of human serum on Ultrogel AcA-34.

however, reported satisfactory recovery of protein-bound As, Cd, Hg, Se, and Zn from liver after gel filtration on Sephadex G-75 in a 0.01 N Tris, 0.05 N NaCl buffer with 100 mg L^{-1} NaN$_3$ (32). Gardiner and Ottaway found ⩽200% recoveries for Cu, Fe, and Zn after fractionation of human serum proteins on Sephadex G-100 in 0.055 M Tris–HCl buffer at pH 7.4 (33). These authors state the reason to be both the ion-exchange capacity of the beads and the elemental content of the beads themselves. Both elemental sources are exploited by passing proteins. Firouzbakht et al. find $85 \pm 6\%$ of iodoamino acid hormonal I compounds in urine after elution in H$_2$O and NH$_4$OH from Biogel P-2 (100–200 mesh) (34). Goldschmidt et al. mention quantitative recovery of tri- and tetraiodothyronine from human serum (35). They fractionated on Sephadex G-25 with 0.01 M NaOH as an eluent. Kamel et al. report 93% recoveries of gold in proteins after gel filtration of human serum on Sephadex G-150 with 0.1 M Tris–HCl buffer (pH 8.1) (36). Schmelzer showed severe contamination problems for Zn in human serum proteins after gel filtration in 0.1 M NH$_4$Ac on Sephadex G-50 (37).

It can thus be concluded that gel filtration of protein-bound elements may lead to erroneous results due to both contamination and losses of elements.

5.3.4. Radiotracer Experiments

Aliquots of 1–2 mL serum are spiked with 50 μL carrier-free solution of a radionuclide in a known chemical form. Quantities of 0.5 mL are then used for desalting on Biogel P-2 after various incubation times. Similar experiments are performed with the saline solution and the buffer used. Particularly the interaction of Zn with the gel matrix is found to be considerable. Yet, the recovery of this and other elements is satisfactory.

Identical experiments for fractionation on Ultrogel AcA-34 gave the results reported in Table 5.8.

The elements I, Mn, Na, and Rb are recovered quantitatively from both serum and buffer. Each of these elements is recovered in symmetrical peaks in the salt fraction. The time lapse between carrier addition to serum and sample application to the column is too short to permit any protein-bound Mn to be found. Also, selenate and selenite give activity in the salt fraction only, with recoveries ranging from 84 to 93%. Antimony has recoveries in serum and buffer of 95%. A small amount of Sb shows up in protein fractions after serum fractionation. Iron in buffer is completely retained. If eluted in serum it is partly recovered in protein fractions. As the major amount is eluted in the void volume, it is likely that the Fe is displaced from the column, rather than bound by proteins. Zinc shows a very broad interaction peak, with K_s values ranging up to 1.23. If eluted in serum, part of the Zn (15%) is recovered in the albumin

Table 5.8. Influence of the Sample Matrix on Recovery of Elements

Matrix Element	Buffer Total Recovery (%)	Serum Total Recovery (%)
Na	99 ± 1^a	99 ± 1^a
Zn	87 ± 3^b	87 ± 6^b
Fe	0^a	92 ± 4^a
Sb	96 ± 1^a	94 ± 1^a
Se (IV)	90 ± 3^b	91 ± 2^b
Se(VI)	84 ± 1^b	93 ± 1^b
Co	79 ± 4^b	69 ± 3^b
Rb	97 ± 5^a	99 ± 7^a
I	98 ± 1^b	101 ± 3^b
Mn	100 ± 4^c	100 ± 1^c

[a] $V_t = 165$ mL; $V_0 = 58$ mL; flow rate = 3.9 cm h^{-1}.
[b] $V_t = 113$ mL; $V_0 = 44$ mL; flow rate = 3.8 cm h^{-1}.
[c] $V_t = 414$ mL; $V_0 = 158$ mL; flow rate = 2.6 cm h^{-1}.

fraction, while the salt peak of Zn is broad. Some influence of the contact time is noticeable for Se.

5.3.5. Use of NAA to Detect Protein-Bound Elements

By using the procedure described in Section 5.4, the influence of contamination on the desalting procedure can be investigated. Table 5.9, taken from Waddell and Hill (38), gives the results for seven elements including Na. It follows from the table that Cu, Mn, and V will partially show up in protein fractions. To detect similar gel–element interaction for *in vivo* bound elements, a 1 mL serum sample was desalted. A total volume of 15.7 mL effluent, divided into seven fractions, was collected. The sum of the elemental contents in the fractions was compared to the elemental content deduced from analysis of whole serum, both determined by NAA of long-lived radionuclides. Results are given in Table 5.10.

For the protein-bound elements, the recovery of Zn is good. The difference in Se values is not significant. The recovery for Na, Br, and Rb is good. Bromine is protein bound for about 1.5%. The low recovery of Fe indicates absorption by the gel material.

In conclusion, desalting and fractionation are both subject to contamination and yield variations due to varying experimental conditions. A rigorously standardized procedure, checked by standard additions and radiotracer experiments, is mandatory for reliable work. The conditions chosen at ECN are summarized in Table 5.11.

Table 5.9. Affinity of Human Serum Proteins Toward Inorganic Elemental Compounds Added *in Vitro*[a]

Sample	Element						
	I	Br	Mn	Cu	V	Cl	Na
Unspiked serum protein (t_0)	17.3 ± 1.1	31.3 ± 1.9	0.3 ± 0.1	440 ± 15	0.25 ± 0.03	1.2	0.9
Added (A)	90.9	90.0	90.9	364	18	7.5	—
Incubation time 0.1 h (t_1)	18.4 ± 1.2	31.6 ± 1.9	2.7 ± 0.2	688 ± 19	1.5 ± 0.1	0.4	0.3
Incubation time 3 h (t_2)	18.1 ± 1.3	29.9 ± 2.6	18.4 ± 0.5	759 ± 20	1.1 ± 0.1	0.5	0.2
Incubation time 6 h (t_3)	15.7 ± 1.0	30.0 ± 2.1	14.7 ± 0.4	702 ± 20	1.0 ± 0.1	0.5	0.3

[a] Results are expressed in nanograms per 0.5 mL serum, except for Na and Cl (micrograms per 0.5 mL). Errors given are internal standard deviations.

Table 5.10. Comparison of the Total Recovery of Elements after Desalting with the Whole-Serum Elemental Content[a]

Fraction Number	Element					
	Na	Br	Rb	Zn	Se	Fe
1	0.6 ± 0.04	—	—	—	—	—
2	1.2 ± 0.1	0.082 ± 0.005	—	1.05 ± 0.03	0.112 ± 0.014	1.1 ± 0.2
3	3027 ± 135	—	0.237 ± 0.016	0.01 ± 0.01	—	—
4	236 ± 6	4.1 ± 0.1	—	—	—	—
5	2.9 ± 0.05	2.5 ± 0.05	—	—	—	—
6	1.7 ± 0.1	—	—	—	—	—
7	1.7 ± 0.1	—	—	—	—	—
Total	3271 ± 135	6.68 ± 0.12	0.237 ± 0.016	1.06 ± 0.03	0.112 ± 0.014	1.1 ± 0.2
Whole serum	3200 ± 42	6.62 ± 0.18	0.233 ± 0.019	1.03 ± 0.03	0.131 ± 0.009	2.8 ± 0.3

[a] The second fraction represents the protein fraction. All values are in micrograms per milliliter. The values for whole serum are calculated by multiplying the original data in micrograms per gram by the serum specific density, 1.026.

Table 5.11. Column Parameters for Human Serum Protein Fractionation and Desalting

Column	K 16-100	LKB 2137-026	C-10
Diameter (cm)	1.6	2.6	1.0
Length, L (cm)	94.4	22.7	10.4
V_t (mL)	190	121	8.2
Flow rate (mL h^{-1})	8.3	16.4	11.2
Linear flow rate (cm h^{-1})	4.1	3.1	14.3
Contact time (h)	23.0	7.4	0.73
Purpose	Fractionation	Fractionation	Desalting
Fraction size (mL)	6.6	6.6	2.2
ΔK_s/fraction	0.035	0.054	0.27

5.4. AN EXPERIMENT INVOLVING NAA OF DESALTED HUMAN SERUM

5.4.1. Purpose

After desalting or fractionation, Cl and Na are present only in small amounts in the protein fractions obtained. This allows for instrumental NAA of these fractions. The limiting conditions, apart from the limits of determination, for a reliable elemental analysis are buffer blank values and sample contamination. On the other hand, it follows from the investigations on gel filtration chromatography that for some elements contamination of the protein fractions may be considerable. A practical test of precision and accuracy is thus mandatory prior to any routine application. This section summarizes our experience at ECN with desalting procedures and subsequent analyses.

5.4.2. Samples

Six fasting donors were sampled, resulting in a total of 60 mL serum. Serum is stored frozen in Teflon beakers at $-25\,°C$. All sample handling, including gel filtration, is done in the clean laboratory. Samples are applied to columns or sample vessels by pipetting with 0.55 and 1.0 mL micropipettes, always using the same pipettes and pipette tips. After use, the tips are rinsed with double-distilled water and stored in a closed plastic box. Before use, they are cleaned again with double-distilled water and washed with buffer.

5.4.3. The Procedure Using NAA

5.4.3.1. NAA with Short-Lived Radionuclides

To 8 mL polyethylene tubes, cleaned with 0.1 M suprapure HNO_3 and washed with double-distilled water, 50 µL of a $^{84}Sr(NO_3)_2$ flux monitor solution containing 1.6 µg ^{84}Sr $(50 \mu L)^{-1}$ is added. In these tubes, placed in a shielded fraction collector, the effluent is collected. Then, after collection is completed, the tubes are closed with polyethylene stoppers and their content is homogenized. In case of fractionation, 5 mL aliquots are pipetted into similarly cleaned polyethylene irradiation tubes. In case of desalting, the total collected effluent is used. The tubes are sealed and irradiated during 2 min in a flux of $5 \times 10^{13} cm^{-2} s^{-1}$. After a decay time of 4 min, 2 mL of the desalted sample or 4 mL of the fractionated sample are counted on a 45 cm^3 coaxial Ge(Li) detector connected to a 4096 channel Canberra 1800 analyzer. In order to avoid geometry differences, standards and blanks are measured in ^{84}Sr-spiked effluent fractions, eluted in the prevoid volume, i.e., the volume between the sample application and the first protein peak. Thus, the Sr-flux monitor serves as an internal standard for the entire procedure, as small differences in fraction volume and counted volume are monitored as well.

5.4.3.2. NAA with Long-Lived Radionuclides

Effluent fractions are collected in 8 mL polyethylene tubes, cleaned with 0.1 M suprapure HNO_3, and washed with double-distilled water. Here, as a whole-procedure monitor, 50 µL amounts of $RuCl_3$, containing 1 µg Ru $(50 \mu L)^{-1}$, are added to the tube before fraction collection. After collection is completed, the tubes are closed with polyethylene stoppers, their content is homogenized, and 5 mL are pipetted in previously cleaned polyethylene irradiation vessels. For desalting, the total collected protein-containing effluent is used. Samples are frozen, lyophilized for 16 h at 60 µm Hg pressure, and sealed. Standard solutions, pipetted in Ru-spiked prevoid volume fractions and blanks are processed the same way. Irradiation is done in a thermal neutron flux of $4 \times 10^{12} cm^{-2} s^{-1}$ for 12 h. After decay times of 1 and 2 weeks, samples are counted for 2000 and 7000 s, respectively, on a 45 cm^3 Ge(Li) detector connected to a 4096 channel Canberra 1800 analyzer. After 3 weeks, samples are counted for 10,000 s on an 89 cm^3 coaxial well-type Ge(Li) detector, connected to a Northern Scientific 4096 channel analyzer.

5.4.4. Results

Data were obtained by using short-lived radionuclides for both desalting and fractionation. The use of long-lived radionuclides was limited to desalting, as preliminary experiments for fractionation showed low recoveries for Fe and Zn, and amounts of Se per fraction were too small to be determined accurately.

5.4.4.1. Desalting on Bio P-2

5.4.4.1.1. Short-Lived Radionuclides. The γ-ray spectrum shows the presence of the radionuclides ^{28}Al, ^{80}Br, ^{38}Cl, ^{66}Cu, ^{128}I, ^{56}Mn, ^{24}Na, ^{85}Sr, and ^{52}V. From these, ^{28}Al originates from buffer blank values, but some ^{28}Al is also formed from the ^{32}P (n, α) ^{28}Al reaction. The usual sample size is 0.5 mL.

Table 5.12 shows the blank values for the aforementioned elements in ng mL^{-1} buffer, based on seven measurements. The number of results—other than "not detected"—is given. Also given for each element are the detection limits in desalted serum samples as calculated from Currie's equation, $L_{AD} = 4.65\sigma(B)/A_{sp}$, for a buffer and protein fraction (39). Here, $\sigma(B)$ and A_{sp} are respectively standard deviation in the background under the photopeak of interest and the specific count rate of the radionuclides under test.

Limits of detection for an effluent fraction in the case of fractionation are given as well. The fourth column in Table 5.12 shows the ratio of the elemental content in whole serum to the buffer blank value in the whole desalted fraction. It is clear from the table that Mn and V are likely to remain undetermined, especially if only a fraction of the elemental amount in serum is occurring in a protein-bound form.

In turn, the precision of the elemental analysis procedure is estimated by comparing the internal and external standard deviations (Section 5.1.3) of

Table 5.12. Blank Values and Detection Limits for Elements Featuring Short-Lived Radionuclides

Element	Blank Value (ng mL^{-1} Buffer)	Number of Results (Out of Seven)	Serum/Blank Ratio	Detection Limit According to Currie (ng mL^{-1} Effluent)		
				Desalting		Fractionation Blank
				Blank	Protein	
Na	40 ± 15	7	3×10^4	18	53	6
Cl	53 ± 24	7	3×10^4	13	54	4.5
Cu	1.1	1	4×10^2	12	20	4.2
Br	6.4 ± 0.1	5	4.6×10^2	1.5	3.2	0.5
I	0.3 ± 0.1	1	0.7×10^2	0.9	1.1	0.3
Mn	0.3 ± 0.1	7	0.9	0.3	0.4	0.09
V	0.02 ± 0.01	2	0.8	0.06	0.2	0.02

irradiated reference solutions of the elements Br, Cl, Cu, I, Mn, and V. Four independent reference solutions are irradiated (the various factors contributing to the internal standard deviations are discussed in Section 5.2.4.5). Furthermore, to monitor additional errors due to the separation step, five serum samples from one pool were desalted and analyzed, and the internal and external standard deviations were estimated. Here, it should be noted that the uncertainty due to the pipetting of 0.5 mL serum is included in the sample size term for the internal standard deviation. The agreement between external (observed) and internal (predicted) standard deviations was found to be satisfactory.

As regards accuracy, this was checked by analyzing the protein fraction of samples of different size, 0.5 and 1.0 mL from the same serum pool. This exercise demonstrated the sensitivity of Mn and V for blanks and contamination, already apparent from Table 5.12.

Data for serum from six donors are given in Table 5.13. Results for pooled serum are given for comparison. The decontamination factor for Na is ca.

Table 5.13. Results of Element Analysis of Desalted Serum from Six Apparently Healthy Donors[a]

Element	Sample	$\dot{x}_w$	$\bar{x}_{uw}$	N	Standard Deviation of $\bar{x}_w$	F	Sample Standard Deviation	Range
I	Donors	37.8	37.0	6	2.2	6.2	5.3	32.8–
	Pool	34.2	34.4	5	0.6	0.4	1.4	44.8
Br	Donors	60.3	59.9	6	2.5	1.8	6.0	51.4–
	Pool	55.2	55.8	5	3.0	2.5	6.7	67.8
Cu	Donors	1.03	1.03	6	0.04	7.2	0.09	0.88–
	Pool	1.13	1.13	5	0.04	4.6	0.08	1.12
Mn	Donors	1.6	1.6	5[b]	0.1	3.5	0.2	1.4–
	Pool	2.1	2.1	5	0.3	10.9	0.7	1.8 (+5.0)
Na	Donors	2.0	1.9	6	0.2	22	0.4	1.6–
	Pool	2.0	1.8	5	0.2	37	0.5	2.6
Cl	Donors	2.8	2.8	6	0.2	20	0.5	2.2–
	Pool	3.4	3.1	5	0.6	134	1.5	3.2
V	Donors	0.7	0.7	6	0.1	11.0	0.3	0.3–
	Pool	0.7	0.7	5	0.1	1.7	0.1	1.1

[a] Values are given in nanograms (I, Br, Mn, V) or micrograms (Cu, Na, Cl) per milliliter of serum.
[b] One value, 5.0 ng mL^{-1}, has not been included.

Table 5.14. Buffer Blank Values, Ratios of Element Concentrations for Buffer and Serum, and Limits of Detection

Element	Buffer Blank Value (ng mL^{-1})	Number of Positive Results	Serum/Buffer Concentration Ratio	Limit of Detection (ng mL^{-1} Effluent) Desalting		Fractionation Blank
				Blank	Protein	
Na	220 ± 30	4	6.5×10^3	0.07	0.1	0.02
Br	13 ± 2	4	2.3×10^2	0.06	0.1	0.02
Rb	ND^a	—	—	16.5	16.4	5.6
Zn	9 ± 5	1	50	6.1	13.8	2.1
Se	ND^a	—	—	4.4	7.5	1.5
Fe	70 ± 30	1	13	92	90	31.1
Sb	0.3 ± 0.1	4	9.4	0.3	0.3	0.1

^{a}ND: not detected.

1.6×10^3, and that for Cl is ca. $12. \times 10^3$. The values for V are much too high compared to the ca. 40 pg mL^{-1} determined by Cornelis (7). This is also the case for Mn, where concentrations of ca. 0.6 ng mL^{-1} have been reported (7, 9, 40). This is in line with the warnings obtained from the radiotracer experiments discussed in Section 5.3.

5.4.4.1.2. Long-Lived Radionuclides. The γ-ray spectra of 1.0 mL desalted serum after a 12 h irradiation period and three different decay and counting times feature ^{82}Br, ^{59}Fe, ^{24}Na, ^{86}Rb, ^{75}Se, ^{69m}Zn, and ^{65}Zn. Occasionally, γ-rays of ^{124}Sb and ^{60}Co are detected; the elements occur at the nanogram per milliter level and originate from the buffer blank. Cesium-134 is not observed.

As for the blank values of the buffer solution, these are shown in Table 5.14.

Here, samples were counted in 8 mL polyethylene irradiation tubes, cleaned before and after irradiation in a 1:1 suprapure HNO$_3$ and double-distilled water mixture. To evaluate precision, the internal and external standard deviations were compared for reference solutions. Agreement was again satisfactory for the six individuals sampled. Moreover, standard addition experiments, as mentioned in Section 5.3, were applied to assess the accuracy of measurements.

Results obtained so far are summarized in Table 5.15. It is clear from the comparison between the experimental F-values for donors and pooled samples that the analytical procedure is not powerful enough to detect differences between individuals for protein-bound Fe, Se, or Zn. It is thus preferable to use average results rather than single results when comparing elemental concentrations in whole serum and the fraction of a protein-bound

Table 5.15. Element Concentrations in the Protein Fraction of Desalted Serum of Six Individuals and Six Samples from One Pool[a]

Element	Sample	$\bar{x}_w$	$\bar{x}_{uw}$	N	Standard Deviation in $\bar{x}_w$	F	Sample Standard Deviation	Range
Na	Donors	2.14	2.10	6	0.26	49	0.65	1.29–
	Pool	1.23	1.23	6	0.14	72	0.34	3.05
Br	Donors	0.078	0.076	6	0.008	34.2	0.021	0.041–
	Pool	0.083	0.084	6	0.003	6.1	0.008	0.100
Rb	Donors	ND[b]	<0.04	6	—	—	—	—
	Pool	ND[b]	<0.04	6	—	—	—	—
Zn	Donors	1.19	1.17	6	0.07	26	0.18	0.91–
	Pool	1.19	1.16	6	0.07	23	0.16	1.41
Se	Donors	0.106	0.106	6	0.007	1.62	0.019	0.070–
	Pool	0.103	0.103	6	0.006	1.12	0.014	0.123
Fe	Donors	1.4	1.4	6	0.1	0.24	3.0	1.10–
	Pool	1.4	1.2	6	0.1	0.27	0.2	1.46

[a]Results, except for F and N, are given in micrograms per milliliter of serum.
[b]ND: not detected.

element. Rubidium is not found to be protein bound. It is completely eluted in the salt fraction.

Table 5.16 gives the averaged results for protein-bound element contents and the whole-serum concentrations for the six donors sampled. It should be noted that the whole-serum analysis was performed on different samples from the same donors.

In conclusion, desalting is effective, leading to Cl and Na decontamination factors of ca. 1.5×10^3. Desalting allows for the instrumental determination of Br, Cu, Fe, I, Mn, Se, and Zn in the protein fraction. For these elements, except I and Se, the desalting step introduces a measurable uncertainty in the analytical procedure, suggesting variable recoveries. Thus, to obtain sufficient precision, expressed as consistency in internal and external standard deviations, the recovery per element per sample should be determined and used to correct the final analytical result.

Experiments to measure the accuracy of the analysis of protein-bound elements show a bias for Mn, pointing to enrichment during chromatography. Sodium and chlorine are recovered in low quantities, this being evidence of effective desalting. Vanadium recoveries are also low, which probably means that only buffer blanks are measured.

The uncertainty in the analytical results for different donors, as deduced from experimental F-values, is small enough only for I and probably for Br to

Table 5.16. Comparison Between Whole-Serum and Protein-Bound Element Contents[a]

Element	Whole-Serum Content		Protein-Bound Content	
	Mean	Range	Mean	Range
Na	3200 ± 139	3119–3240	2.1 ± 0.7	1.3–3.1
Br	6.47 ± 1.40	5.14–2.50	0.076 ± 0.021	0.041–0.100
Rb	0.27 ± 0.10	0.18–0.44	ND[b]	—
Zn	0.88 ± 0.11	0.70–1.01	1.17 ± 0.18	0.91–1.41
Se	0.106 ± 0.019	0.077–0.128	0.106 ± 0.019	0.070–0.123
Fe	2.7 ± 0.6	1.8–3.4	1.4 ± 0.3	1.10–1.46

[a] Values are in microgram per milliliter. Original data for whole serum (in microgram per gram) have been multiplied by the serum specific density, 1.026.
[b] ND: not detected.

enable differences to be found between individuals. For the other elements, the sample standard deviations for samples of one pool and of different donors are approximately the same.

In human serum, Cu and Se are mainly protein bound (99% and 100%, respectively). The same is true for Zn. However, from recoveries exceeding 100% it cannot be excluded that inorganic Zn from the column is appearing as protein-bound Zn. Also Mn levels in protein fractions are high compared to values for whole serum in the literature, suggesting that there is total protein binding of exogenous elements.

5.5. ELEMENTS IN HUMAN SERUM PROTEINS OR OTHER PROTEINS AND RELATED RADIOTRACER EXPERIMENTS

5.5.1. General Remarks

The living organism possesses a mechanism for absorption transport, storage, and excretion of essential elements that functions so as to keep their concentration in body tissues and fluids optimal. In blood, this is expressed by a small concentration range in which elements are present, mainly for elements occurring at the intracellular level as well as for those with the highest concentration in plasma. For some elements in both groups, it is known that plasma proteins play an important role in metabolism of the elements. Two examples are transferrin and ceruloplasmin, which are transport proteins for Fe and Cu, respectively.

The improvement of the detection power of analytical techniques for trace elements has also made possible study of the interactions between trace element concentrations and activities of element-containing proteins. For example, this is demonstrated by the work of Kasperek et al. who measured the glutathione peroxidase activity and the Se content in blood of children suffering from maple-syrup–urine disease, or phenylketonuria (41).

Furthermore, several elements have been acknowledged to be essential in metalloproteins, e.g., Zn in carboxypeptidase. The metalloproteins perform specific biochemical functions. Detailed discussions of these functions have been given by Bowen (10) and, more recently, by Underwood (12). Surveys of the state of knowledge of this subject have been presented by Cornelis (7) and Iyengar et al. (9).

A critical review of work on single trace elements, updated to 1983, has been given elsewhere (1). Here, we shall only mention some conspicuous radiotracer applications. The outstanding specificity of radioactivity measurements makes radiotracers the ideal tools for speciation analysis in all its aspects.

5.5.2. Distribution of Trace Elements

The relative distribution of an element in blood, serum, and serum proteins can be derived from the following parameters:

a. The total blood volume elemental content. The data are obtained by multiplying whole-blood element concentrations with the total blood volume. Most concentrations are taken from the compilation made by Iyengar et al. (9).

b. The amount of element in serum of 1 mL blood, related to the element content per milliliter of blood. The value is calculated from whole blood and serum element concentrations, using a hematocrit value of 0.44. Values for serum concentrations are mainly from Iyengar et al. (9).

c. The amount of element bound to serum or plasma proteins, related to the serum concentration. The sources of these data are cited in Iyengar et al. (9).

Such distributions are given in Table 5.17. Present primarily in serum are Cl, Cu, I, and V, whereas Cs, Co, Fe, K, Mn, Rb, and Zn are roughly equally divided, with a slight preference for the red cells. The results for amounts of protein-bound elements reflect the sometimes controversial data in the literature. Furthermore, the elements are not equally well documented. A detailed discussion for each element is given in Section 5.3.

The information is scarce about the function of element association with serum proteins, other than for element-specific reasons, e.g., in transferrin. According to van den Hamer and Houtman, albumin is the usual carrier of metabolically active trace elements (39). Thus, compared to metalloproteins, the elements are loosely bound to albumin and easily transferred to binding

Table 5.17. Distribution of Elements in Blood and Blood Fractions

Element	Reference	Total Blood Volume Amount[a]	Whole Serum to Whole Blood Ratio[b]	Protein Bound Amount (%)
Na (g)	(9)	10.5	92	None
Cl (g)	(9)	15.4	71	None
K (g)	(9)	8.6	7	None
Mg (g)	(9)	0.2	44	<10
Zn (g)	(9)	0.04	9	Complete
Cu (mg)	(9)	5.4	66	Complete
Fe (g)	(9)	2.4	0.1	Complete
I (mg)	(9)	0.3	65	ca. 90
Se (mg)	(9)	0.9	40	>80
Rb (g)	(9)	0.01	4	None
Br (mg)	(9)	25	46	ca. 2
Mn (mg)	(43)	0.05	4	Complete
V (μg)	(7)	0.3[c]	70	Not known
Sb (mg)	(42)	0.024	36	Not known
Cs (mg)	(43)	0.015[c]	17	Not known
Co (μg)	(43)	2.1	32	Complete

[a] Whole blood concentration is multiplied by the total blood volume. Total blood volume is calculated for a 70 kg male, with 76 mL blood per kilogram of body weight, thus resulting in 5320 mL (7, 41).
[b] Given as a percentage. The ratio of the element amount in serum from 1 mL blood and the amount per milliliter of blood is calculated assuming a hematocrit value of 0.44 (41).
[c] Recalculated from the element content in erythrocytes and serum or plasma.

sites in the various tissues. This holds for Cu and Zn. As for the other protein-bound elements that occur in serum, apart from Fe and I, the situation is much less clear, mostly due to insufficient documentation and to specific difficulties in trace element analytical techniques.

5.5.3. Some Applications of Radiotracers

Next, some examples are listed that are illustrative of the issues discussed in preceding sections.

- *Sodium* (^{22}Na, $T_{1/2} = 2.6$ years). Desalting efficiency can be checked with this (carrier-free) radioisotope, as discussed in Section 5.3.

- *Vanadium* (^{48}V; $T_{1/2} = 16.0$ days). Various compounds have been labeled and used in metabolic studies, e.g., transferrin, xanthate, and etioporphyrin (6). Most of the intravenously injected ^{48}V in blood of rats is found in serum, probably transferrin bound (12). Cornelis et al. determined about 70% of human blood V to occur in serum, indicating 40 pg mL^{-1} amount (42). Sabbioni reported 90% of whole-blood V to be in plasma, resulting in approximately 5 ng mL^{-1} (43); after gel filtration on Sephadex G-150, this author states that most of the V is recovered in the transferrin-containing fraction.
- *Chromium* (^{51}Cr, $T_{1/2} = 27.7$ days). Chromium-51-labeled red blood cells have been used to estimate the red cell mass (44).
- *Manganese* (^{54}Mn, $T_{1/2} = 312$ days). Cotzias and Miller as well as Krivan and Geiger showed from dialysis experiments with serum, and without an $MnSO_4$ spike, that Mn is extensively bound in a nondialyzable form (45, 46).
- *Iron* (^{59}Fe, $T_{1/2} = 48$ days). Serum Fe is bound more than 99% to transferrin, a glycoprotein (molecular weight: ca. 76,000 Da) with two binding sites for Fe (12). A second Fe-containing protein in serum is ferritin, a non-heme Fe storage compound (molecular weight: ca. 460,000 Da). The ferritin-bound Fe represents less than 1% of the total serum Fe. Yet, ferritin levels in serum correlate well with Fe stores in the human body. Chesters and Will added ^{59}Fe *in vitro* to porcine plasma and recovered Fe in the transferrin peak and, to a much smaller extent, in a high-molecular-weight fraction, probably ferritin (47). Sabbioni isolated transferrin from rat plasma by gel filtration on Sephadex G-255 with subsequent ion-exchange on DEA-Sephadex A-0 and analytical disk electrophoresis (43). The radioactive Fe, previously administered by intraperitoneal injection, was recovered only in the transferrin fraction.
- *Cobalt* (^{60}Co, $T_{1/2} = 5.2$ years). Most tracer applications have to do with vitamin B12, which in plasma is bound to transcobalamin (48).
- *Zinc* (^{65}Zn, $T_{1/2} = 244$ days). By incubating serum, single protein solutions, and amino acid solutions with ^{65}Zn, Prasad and Oberleas showed, on the basis of the percentage of ultrafilterable ^{65}Zn, that amino acids bind small amounts of serum Zn and that the addition of physiological amounts of amino acids decreases the amount of *in vitro* Zn binding of dialyzed single protein solutions, except for ceruloplasmin and α_2-macroglobulin (49). The authors suggest the presence of two binding sites for Zn on predialyzed albumin, ceruloplasmin, and immunoglobulin G (IgG).
- *Arsenic* (^{73}As, $T_{1/2} = 80$ days; As, $T_{1/2} = 17.8$ days). Betaine and choline have been labeled with ^{73}As by Goetz and Novin (50), and dimethylarsinic

acid has been labeled with ^{74}As by Vahter et al. (51). In both cases the authors were engaged in metabolic studies.

- *Selenium* (^{75}Se, $T_{1/2} = 120$ days). Many investigations have been carried out with this radionuclide. Selenium is thought to take the place of S in S-containing amino acids like cysteine and methionine, and in proteins, possibly in the form of S—Se—S bonds (12). Sandholm found intravenously applied ^{75}SeO$_3^{2-}$ in mice to be rapidly taken up by the erythrocytes (52). Subsequently, albumin transported Se to the liver, where the maximum content occurred after 15 min. Liver Se was then excreted in the blood, mainly in plasma. Burk and Consolazio injected ^{75}SeO$_3^{2-}$ intraperitoneally into rats (53). After 10 min Se occurred as non-protein-bound forms in plasma, after which it was concentrated by the liver. Liver-released Se in plasma appeared to be bound to a protein of molecular weight ca. 80,000 Da after some 2–3 h. Also, it was shown to be bound to a macroglobulin after 24 h. The element was loosely bound in both protein fractions.

- *Bromine* (^{82}Br, $T_{1/2} = 35.3$ h). Apart from its use in desalting experiments, this radionuclide may serve as a tracer for extracellular water due to its slow reaction with proteins (54).

- *Iodine* (^{131}I, $T_{1/2} = 8.0$ days). The use of this radionuclide has been abundant. In the domain of protein research, investigations on I-labeled antigens should be mentioned (55).

- *Caesium* (^{137}Cs, $T_{1/2} = 30.2$ years). By *in vivo* radiotracer experiments with ^{137}Cs, Yamagata and Iwashima established that 2% of blood Cs is found in plasma (56). NAA of natural blood Cs pointed to the same distribution value between red cells and plasma.

5.6. CONCLUSIONS

Radioanalysis is making two important contributions to bioinorganic chemistry, namely, the determination of trace elements by activation analysis and the observation of transport phenomena by radiotracers. The first of these is predominantly based on NAA. It relies on well-coordinated cooperation between the two disciplines of clinical biochemistry and activation analysis at a nuclear center (57–59). Radiotracers have a wider field of application in that they can be used in any laboratory by adopting basic precautions for the handling of small quantities of radioactivity. Moreover, they can be used in the living organism as well as to control chemical yields and contamination in the analytical laboratory.

REFERENCES

1. J. R. W. Woittiez, *Elemental analysis of human serum and serum protein fractions by thermal neutron activation*. Thesis, Amsterdam (1983); *ECN* [*Rep.*] **ECN-147** (1984).
2. M. Gallorini, M. Bonardi, and E. Sabbioni, *Proc. Int. Comp. Nucl. Methods Environ. Energy Res. 5th*, p. 141 (1984).
3. J. J. M. de Goeij, *IAFA–TECDOC, Qual. Assur. Biomed. Neutron Act. Anal.*, Vienna, p. 121 (1984).
4. T. Sato and T. Kato, *Anal. Sci.* **4**, 307 (1988).
5. *Health Relat. Monit. Trace Elemt. PIXE*, Vienna, IAEA-TECDOC-330 (1985).
6. E. Sabbioni, R. Pietra, F. Mousty, and F. Colombo, *Biol. Trace Elem. Res.* **12**, 199 (1987).
7. R. Cornelis, *Neutron activation analysis of trace elements in human tissues*. Thesis, Ghent University, Belgium (1980).
8. R. F. M. Herber and A. E. Wibowo, *Paramedica* **1**, 23 (1979).
9. G. V. Iyengar, W. E. Kollman, and H. J. M. Bowen, *The Elemental Composition of Human Tissues and Body Fluids*. Verlag Chemie, Weinheim, 1978.
10. H. J. M. Bowen, *Trace Elements in Biochemistry*. Academic Press, London, 1966; *J. Radioanal. Chem.* **19**, 215 (1974).
11. W. Mertz, R. Brätter, and P. Schramel, in *Trace Element Analytical Chemistry in Medicine and Biology* (P. Brätter and P. Schramel, eds.), p. 727. de Gruyter, Berlin, 1980.
12. E. J. Underwood, *Trace Elements in Human and Animal Nutrition*, 4th ed. Academic Press, New York, 1977.
13. J. Versieck, F. Barbier, A. Speecke, and J. Hoste, *Acta Endocrinol. (Copenhagen)* **76**, 783 (1974).
14. K. Liebscher and H. Smith, *Arch. Environ. Health* **17**, 881 (1968).
15. C. E. Crouthamel, F. Adams, and R. Dams, *Applied Gamma-Ray Spectrometry*. Pergamon, Oxford, 1970.
16. J. Krugers, ed., *Instrumentation in Applied Nuclear Chemistry*. Plenum, New York 1973.
17. J. B. Luten, J. B. Gouman, H. A. Das, F. de Witte, and J. Wÿkstva, *Radiochem. Radioanal. Lett.* **21**, 61 (1975).
18. J. R. W. Woittiez, *J. Radioanal. Chem.* **57**, 191 (1979).
19. A. Faanhof, H. A. Das, and H. A. van der Sloot, *J. Radioanal. Chem.* **54**, 289, 603 (1989).
20. F. De Corte, *The k_0 standardization method: a move to the optimization of NAA*. Thesis, Ghent University, Belgium (1987).
21. Bio Rad Laboratories, Catalogue H, p. 33 (1982).
22. L. K. B. Ultrogel, Instruction Manual 1-2204-EOI (1976).

23. M. Determann, *Gel Chromatography*, 2nd ed. Springer-Verlag, Berlin, 1969.
24. M. Shibukava, *Anal. Chem.* **53**, 1620 (1981).
25. R. Groh and I. Halasz, *Anal. Chem.* **53**, 1325 (1981).
26. R. Zeisler, *J. Radional. Nucl. Chem.* **112**, 95 (1987).
27. *Gel Filtration: Theory and Practice.* Pharmacia Fine Chemicals, 1980–1982.
28. W. H. S. George, *Nature (London)* **195**, 155 (1962).
29. D. J. R. Evans and K. Fritze, *Anal. Chim. Acta* **44**, 71 (1969).
30. E. Sabbioni and E. Marafante, *J. Toxicol. Environ. Health* **8**, 419 (1981).
31. P. E. Johnson and G. W. Evans, *J. Chromatogr.* **188**, 405 (1980).
32. G. Norheim and E. Steinnes, *Anal. Chem.* **47**, 1688 (1975).
33. P. E. Gardiner and J. M. Ottaway, *Anal. Chim. Acta* **124**, 281 (1981).
34. M. L. Firouzbakht, S. Kayhan Garmestani, E. P. Rack, and A. J. Blotcky, *Anal. Chem.* **53**, 1746 (1981).
35. H. M. J. Goldschmidt, C. J. A. Van den Hamer, J. J. M. De Goeij, J. P. W. Houtman, and C. Zegers, *Int. J. Appl. Radiat. Isot.* **30**, 496 (1979).
36. H. Kamel, D. H. Brown, J. M. Ottaway, and W. E. Smith, *Analyst (London)* **102**, 645 (1977).
37. W. Schmelzer, *Anwendung der isoelektrischen Fokussierung und der Neutronen Aktivierungsanalyse zur Untersuchung proteingebundener Spurenelemente*. Thesis, Technical University, Berlin (1976).
38. W. J. Waddell and C. Hill, *J. Lab. Clin. Med.* **48**, 311 (1956).
39. W. J. A. van den Hamer, J. P. W. Houtman, in *Trace Element Analytical Chemistry in Medicine and Biology* (P. Brätter and P. Schramel, eds.), p. 233. de Gruyter, Berlin, 1980.
40. E. Deutsch and G. Geyer, *Laboratoriumsdiagnostik, Normalwerte und Interpretation.* Steinkopf, Berlin, 1969.
41. K. Kasperek, E. Land, and L. E. Feinendegen, in *Trace Element Analytical Chemistry in Medicine and Biology* (P. Brätter and P. Schramel, eds.), p. 75. de Gruyter, Berlin, 1980.
42. R. Cornelis, J. Versieck, L. Mees, J. Hoste, and F. Barbier, *J. Radioanal. Chem.* **55**, 35 (1980).
43. E. Sabbioni, in *Nuclear Activation Techniques in the Life Sciences*, p. 179, IAEA, Vienna, 1979.
44. H. Doornenbal, in *Body Composition in Animals and Man* (R. T. Reid, ed.), p. 218. Nat. Acad. Sci., Washington, DC, 1968.
45. G. C. Cotzias, S. T. Miller, and J. Edwards, *J. Lab. Clin. Med.* **67**, 836 (1966).
46. V. Krivan and H. Geiger, *Fresenius' Z. Anal. Chem.* **305**, 339 (1981).
47. J. K. Chesters and M. Will, *Br. J. Nutr.* **46**, 111 (1981).
48. G. Pethes, *Tech. Rep. Ser. I.A.E.A.* **197**, 3 (1980).
49. A. S. Prasad and D. Oberleas, *J. Lab. Clin. Med.* **76**, 416 (1970).

50. L. Goetz and H. Novin, *Int. J. Appl. Radiat. Isot.* **34**, 1509 (1983).
51. M. Vahter, E. Marafante, and L. Dencker, *Arch. Environ. Contam. Toxicol.* **13**, 259 (1984).
52. M. Sandholm, *Acta Pharmacol. Toxicol.* **33**, 1 (1973).
53. R. F. Burk and C. F. Consolazio, *Fed. Proc., Fed. Am. Soc. Exp. Biol.* **31**, 692 (1972).
54. G. M. Ward, R. M. Argenzio, and J. E. Johnson, *Isot. Stud. Physiol. Domest. Anim., Proc. Symp., 1972*, p. 73 (1972).
55. W. M. Hunter and F. G. Greenwood, *Nature (London)* **194**, 495 (1962).
56. N. Yamagata and K. Iwashima, *Nature (London)* **211**, 528 (1966).
57. L. A. Currie, *Anal. Chem.* **40**, 586 (1968).
58. G. C. Cotzias, P. S. Papavasiliou, E. R. Hughes, L. Tang, and D. C. Borg, *J. Clin. Invest.* **47**, 992 (1968).
59. E. Damsgaard, *Risoe Rep.* **271**, 17 (1973).

CHAPTER

6

QUALITY CONTROL OF RESULTS OF SPECIATION ANALYSIS

PH. QUEVAUVILLER, E. A. MAIER, and B. GRIEPINK

European Commission, Standards, Measurements and Testing Programme.*
B-1049 Brussels, Belgium

6.1. INTRODUCTION

The number of determinations of chemical species in environmental matrices carried out in routine and research laboratories has increased considerably in the last few years. However, the quality of results has often been neglected in environmental, food, and biomedical analyses. Good reproducibility of an analysis is not sufficient. It helps, of course, to make results comparable over a limited area or a limited period ("trend" monitoring), but for full comparability of results over time and location, and thus for a sound and general interpretation of the findings, accuracy is a must; this has been widely demonstrated in the literature (1–3). Too many scientists have stated that good reproducibility in time was sufficient to show trends and demonstrate effects of actions taken by authorities to improve the quality of the environment or food. Such statements overlook modeling applications, theory development, and the like, and ignore improvements in equipment and methodology.

To not only achieve good reproducibility but also good accuracy, various measures are necessary. It is clear that good quality control (QC) of speciation analysis has not yet been achieved.

Typical cases can illustrate the lack of accuracy that occurs in the determination of inorganic and organic traces in environmental matrices (3, 4). These examples are by no means selected; they occur quite commonly in many fields of analysis, including the determination of species in environmental matrices.

---

* Formerly BCR: Bureau Communautaire de Référence.

Element Speciation in Bioinorganic Chemistry, edited by Sergio Caroli.
Chemical Analysis Series, Vol. 135.
ISBN 0-471-57641-7 © 1996 John Wiley & Sons, Inc.

When results differ to such an extent, they are not trustworthy. Poor performance by analytical laboratories creates economic losses: extra analyses, destruction of food and goods, court actions, etc. Moreover, in the past too often wrong although highly reproducible results have lead to misinterpretation of environmental processes.

In recent years, the determination of chemical species of elements (in particular As, Hg, and Sn species) has become of increasing concern due to their high toxic impacts (5). Some of these compounds (e.g., methyl-Hg and tributyl-Sn) are now included in the blacklist of compounds to be monitored in the marine environment according to European Union (EU) Directive (an amendment of Directive 76/464/EEC). Consequently, a wide variety of analytical techniques have been developed recently and are described in the literature, e.g., for As, Hg, Pb, Se, and Sn speciation (6–13).

Intercomparisons are valuable tools for validating methods or the way they are applied in a laboratory as well as for the verification of results. The results of recent intercomparisons have shown that the speciation analyses performed are far from being accurate. In order to improve this situation a good QC system should be introduced in each laboratory. The main guidelines of QC are addressed here, together with examples of sources of discrepancies detected in the determination of some chemical species.

6.2. THE NEED FOR DEVELOPMENT OF NEW ANALYTICAL TECHNIQUES IN SPECIATION ANALYSIS

The term *speciation* comprises a wide variety of compounds or chemical forms of elements. Here the term *chemical species* will refer to a specific form (monoatomic or molecular) or configuration in which an element can occur, or to a distinct group of atoms consistently present in different matrices (14). The determination of "extractable trace metals" will also be dealt with, as described below, although here strictly speaking the word speciation cannot be applied. These group determinations, i.e., those using operationally defined procedures (single and sequential extractions), were often used in the 1980s as a practical (compromise) way to identify forms as "labile" (e.g., for Al) or "bioavailable" [e.g., EDTA (ethylenediaminetetraacetic acid)-extractable trace metal contents]; in some cases such an approach can yield sufficient information to enable planners to arrive at a sound environmental policy. In most cases the determinations have to be more selective however, and speciation analyses will include forms of elements with different oxidation states (e.g., As, Cr, Se) or individual organometallic species (such as Hg and Sn species).

The need for the determination of individual chemical species occurs especially where these species are known to be very toxic to humans and biota, such as As(III), Cr(VI), Se(VI), methyl-Hg, or tributyl-Sn. Consequently, the increasing concern for an improved control of environmental contamination levels has led to the development of new analytical techniques for the determination of a wide variety of compounds, which is also reflected by a considerable increase in the number of determinations.

These techniques involve many analytical steps, such as extraction, derivatization, separation, and detection. These steps should be performed in such a way that decay of the unstable species does not occur. However, the lack of suitable reference materials for speciation analysis hampers the control of the quality of measurements, which in turn creates important economic losses due to the poor comparability of data produced.

Owing to the need for QC in speciation analysis, international organizations, namely, the Standards, Measurements and Testing (SM&T) Programme (formerly BCR) of the European Commission, the National Institute of Standards and Technology [NIST (United States)], the National Institute for Environmental Studies [NIES (Japan)], and the National Research Council of Canada (NRCC) have developed some certified reference materials (CRMs), which will be referred to in the following sections. The demand is far from being met however, and many other attempts are being made to improve the quality of measurements.

Thus, the general concepts of quality assurance (QA) will be dealt with here, particularly as regards strategies applied, e.g., at the SM&T Programme for improvement of the quality of analyses of a wide variety of elements and compounds, including speciation analyses.

6.3. OVERVIEW OF QUALITY ASSURANCE PRINCIPLES

6.3.1. General Remarks

Two basic parameters should be considered when one is discussing analytical results: *accuracy* (absence of systematic errors), and *uncertainty* (coefficient of variation or confidence interval) caused by random errors and random variations in the procedure. In this context, accuracy is of primary importance. However, if the uncertainty of a result is too high, it cannot be used for any conclusion concerning the quality of the environment or of food, nor can it be used for the diagnosis of a patient's illness. An unacceptably high degree of uncertainty renders the result useless.

In the performance of an analysis, all basic principles of calibration, of elimination of sources of contamination and losses, and of correction for interferences should be followed (2).

A prerequisite for a good result is a correct calibration. Although careful calibration is considered to be obvious, experience has shown that, with the introduction of modern automated equipment, it has become a part of the analytical chain to which insufficient attention is being paid. This observation is particularly meaningful for speciation analysis, in which the necessary pure compounds are often not available.

Indeed, it may be worthwhile to stress yet again the oft-stated admonition, that compounds of well-known stoichiometry should be used—those of which any possible water content is known. Careful consideration should be given to the choice of the method of calibration, using pure calibrant solutions, matrix-matched solutions, or standard additions. Moreover, internal standards should be used when appropriate. Some examples of internal standards and necessary calibrants for speciation analysis are given in Table 6.1.

Very often yields of reactions (e.g., derivatization) must be determined. This can be considered as being a part of the calibration process. In such a case the calibrant should be taken through the whole determination procedure. However, in order to be better able to investigate sources of error, it would be more advantageous to calibrate the final determination step with the compound measured in reality (i.e., the derivative) and to run the recovery experiment with the analyte. Although more time and labor consuming, such procedures should be performed at least at regular intervals to detect possible errors and are therefore considered part of the laboratory's QA system.

Table 6.1. Examples of Internal Standards and Calibrants Necessary for Speciation Analyses

Calibrants Available as Internal Standards	Calibrants Needed
CH_3HgCl	$C_2H_5HgR^a$
C_2H_5HgCl	—
Phenyl-HgCl	—
$(CH_3)_2Hg$	—
As_2O_3/As_2O_5, $(CH_3)_2As$	$(CH_3)_3As=O$
Arsenobetaine	AsH_3
Arsenocholine	$As(CH_3)_xH_y$
Bu_3SnCl	Bu_3SnH
Bu_3SnR^b	Bu_3Sn–morin complex
Monobutyl-, dibutyl-, cycloexyl-, phenyl-Sn	Phenylbutyl-Sn

[a] $R = -CH_3, -C_2H_5$, phenyl.
[b] $R = $ penthyl, ethyl, methyl.

6.3.2. Statistical Control

When a laboratory works at a constant level of high quality, fluctuations in the results become random and can be predicted statistically (15). This implies first of all that limits of determination and detection should be constant and well known, while rules for rounding off final results should be based on the performance of the method in the laboratory. Furthermore, if such a situation exists in which there is an absence of systematic fluctuations, normal statistics (e.g., regression analysis, t- and F-tests, or analysis of variance) can be applied to evaluate results wherever necessary (16). Whenever a laboratory is in statistical control, results are not necessarily accurate; they are, however, reproducible. The ways to verify accuracy will be described in the following paragraphs.

Control charts should be used as soon as the method is under control in the laboratory using reference materials of good quality (i.e., stable, homogeneous, and relevant with respect to matrix and interferences).

A CRM can be used to assess accuracy. It must be emphasized that reproducible or accurate results, obtained respectively with reference materials (RMs) or CRMs, are not always sufficient in QA. It is necessary that the composition of the control materials be close to the composition of the unknown sample. This similarity should involve matrix composition, possibly interfering major and minor substances, and the like.

A control chart provides a graphic way of interpreting the method's output in time, so that the reproducibility of the results and the method's precision over a period of time and with different operators can be evaluated.

To do such an evaluation, one or several materials of good homogeneity and stability should be analyzed with each batch of unknown materials. Some 5–10% (depending on the frequency of situations being out of control) of all analytical runs should be used for this purpose (16). These checks can be done with a Shewhart control chart in which the results of calibrant analyses (X) or the difference between duplicate values (R) are plotted over time.

The X-chart additionally presents the lines corresponding to a risk of 5 or 1% that the results are not contained in the whole population of results obtained over a period of time. These lines are for "warning" and "action," respectively. The results of a method are considered to be out of control if (i) the upper or lower control limit is exceeded ("action"); (ii) the same "alarm line" is exceeded twice in succession; and (iii) 11 successive measurements are on the same side of the line.

Figure 6.1 shows an example of a control chart for tributyl-Sn (TBT) determination in oyster and sediment samples; in this example, the upper warning limit was set at 2 SDs (standard deviations) and the upper control limit at 3 SDs of the mean of the first 10 analyses (17).

200 QC OF SPECIATION ANALYSIS

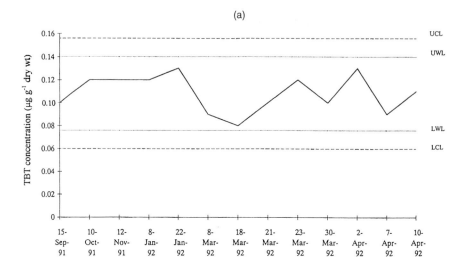

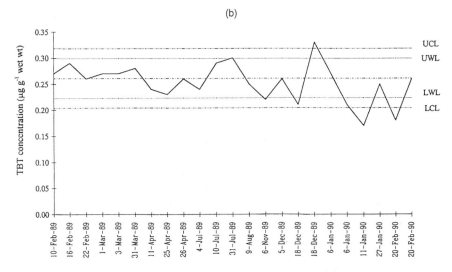

Figure 6.1. Quality control chart for TBT in (a) sediment and (b) oyster (*Crassostrea gigas*) samples. In this example, the upper and lower warning limits (UML and LWL) were set at 2 SDs, whereas the upper and lower control limits (UCL and LCL) were set at 3 SDs of the mean of the first 10 analyses. These control charts were provided by M. Waldock; courtesy of the MAFF Fisheries Laboratory (17).

Cusum charts are also used to detect drifts in methods and trends by plotting the sum of the differences between the results obtained and a reference value in relation with time (16).

6.3.3. Comparison with Results of Other Methods

The use of a Shewhart chart enables the ascertainment of whether or not a method is still in control; it is not, however, able to detect a systematic error that is present from the moment of introduction of the method in a laboratory, as has already been pointed out. Results should be verified by other methods.

All methods have their own particular sources of error. For instance, for some techniques, errors may occur due to incomplete derivatization, a step which is not necessary for other techniques, such as high-performance liquid chromatography (HPLC); the latter technique, however, may display errors such as incomplete separation, which is not encountered, or to a lesser extent, in the aforementioned techniques involving derivatization.

An independent method should be used to verify the results of routine analysis. If the results of both methods are in good agreement, it can be concluded that the results of the routine analysis are unlikely to be affected by a contribution of a systematic nature (e.g., insufficient extraction). This conclusion is stronger when the two methods differ widely, such as derivatization–gas chromatography (GC), GC–atomic absorption spectrometry (AAS) and HPLC–inductively coupled plasma mass spectrometry (ICP–MS).

If the methods have similarities, such as an extraction step, a comparison of the results would most likely lead to conclusions concerning the accuracy of the method of final determination but not as regards the analytical result as a whole.

6.3.4. Use of Certified Reference Materials

Results can only be accurate and comparable worldwide if they are traceable. By definition, traceability of a measurement is achieved by an unbroken chain of calibrations connecting the measurement process to the fundamental units. In the vast majority of chemical analyses, the chain is broken because in the treatment the sample is physically destroyed by dissolutions, calcinations, etc. To approach the full traceability it is necessary to demonstrate that no loss or contamination has occurred in the course of the sample treatment.

The only possibility for any laboratory to ensure traceability in a simple manner is to verify the analytical procedure by means of a so-called matrix RM certified in a reliable manner.

The laboratory that measures such a CRM by its own procedure and finds a value in disagreement with the certified value is thus warned that its measurement includes an error, of which the source must be identified. CRMs having well-known properties should be used to (i) verify the accuracy of results obtained in a laboratory; (ii) monitor the performance of the method (e.g., cusum control charts); (iii) calibrate equipment that requires a calibrant similar to the matrix; (iv) demonstrate equivalence between methods; and (v) detect errors in the application of standardized methods [of the International Organization for Standardization (ISO), the American Society for Testing and Materials (ASTM), etc.].

The conclusion on the accuracy obtained on the unknown sample should always be a conservative one: if the laboratory finds wrong results on a CRM, it is by no means certain of a good performance on the unknown; if, however, the laboratory finds a value in agreement with the certified value (according to ISO Guide 33), it should realize that, owing to discrepancies in composition between the CRM and unknown, there is a risk that the result on the unknown may be wrong (18). The use of as many as possible relevant CRMs is therefore necessary for a good QA.

6.3.5. Intercomparisons

When all necessary measures have been taken in the laboratory to achieve accurate results, the laboratory should demonstrate its capabilities in intercomparisons, which are also useful to detect systematic errors. In general, besides the sampling error, three sources of error can be detected in all analyses: (i) sample pretreatment (extraction, separation, cleanup, preconcentration, etc.); (ii) final measurements (calibration errors, spectral interferences, peak overlap, baseline and background corrections, etc.); and (iii) the laboratory itself (training and educational level of workers, care applied to the work, awareness of pitfalls, management, clean bench facilities, etc.).

When laboratories participate in an intercomparison, different sample pretreatment methods and techniques of separation and final determination are compared and discussed, as well as the performance of these laboratories. If the results of such an intercomparison are in good and statistical agreement, the collaboratively obtained value is likely to be the best approximation of the truth.

Before conducting an intercomparison the aims should be clearly defined. An intercomparison can be held: (i) to detect the pitfalls of a commonly applied method and to ascertain its performance in practice, or to evaluate the performance of a newly developed method; (ii) to measure the quality of a laboratory or a part of a laboratory (such as audits for accredited laboratories); (iii) to improve the quality of a laboratory in collaborative work with mutual learning processes; and (iv) to certify a reference material.

In an ideal situation, where the results of all laboratories would be under control and accurate, only intercomparisons of types ii and iv would be held. Under current conditions, however, types i and iii still play an important role.

6.4. POTENTIAL SOURCES OF ERROR IN SPECIATION ANALYSIS

The basic techniques used in speciation analysis were developed in the early 1980s. These include sample pretreatment steps, i.e., extraction either with organic solvents (e.g., toluene or dichloromethane) or various types of acids (e.g., acetic or hydrochloric acid), derivatization procedures (e.g., hydride generation or ethylation), separation steps (GC or HPLC), followed by a detection of the compound or the element by a wide variety of methods, such as AAS, mass spectrometry (MS), flame photometric detection (FPD), flame ionization detection (FID), inductively coupled plasma atomic emission spectrometry (ICP–AES), ICP–MS, and electron capture detection (ECD). Each step of the analytical procedure includes specific sources of error, some of which will be described below; a summary of the main sources of error observed is given in Table 6.2.

Examples have been taken from the summary of roundtable discussions at a workshop on speciation held in Arcachon, France, in 1990 (19).

Table 6.2. Summary of Steps in an Analytical Procedure for the Determination of Forms of Elements and the Main Sources of Error Observed

Steps	Methods	Sources of Error
Storage	Wet storage	Instability of compounds (volatization, degradation)
Drying procedure	Freeze-drying	Instability of compounds (volatization, degradation)
Pretreatment	Extraction	Incomplete extraction, change of original speciation, losses in cleanup
Derivatization	Hydride generation, ethylation, cold vapor	Inhibition, incomplete transformation, decomposition
Separation	GC, HPLC, cold trapping	Decomposition of the species, adsorption on column
Detection	Element-specific (AAS, ICP–AES, MS) or -nonspecific (FID, ECD) procedures	Interferences, atomization, ionization problems

Most examples given here will deal with the analytical methods for the determination of As, Hg, and Sn species, as it is with these species that most work on accuracy and identification of pitfalls and sources of error has been done so far.

6.4.1. Extraction

The extraction should be done in such a way that the analyte is separated from the interfering matrix without loss or contamination, without change of the species, and with the minimum of interferences.

A considerable variety of acid extraction procedures have been utilized for sediment and biota analysis, particularly when AAS is used as the final determination step; these have involved acids such as HCl or CH_3COOH or an $HCl-CH_3COOH$ mixture for organo-Sn compounds (20–22). NaOH has also been used (23).

Other procedures, e.g., for techniques involving GC or HPLC, are based on extraction with an organic solvent, such as dichlomethane, chloroform, toluene, and hexane (24–26).

A good assessment of QA implies that the extraction recoveries are verified; this can be done by spiking a sample of similar composition as the sample analyzed with a known content of the analyte of concern. After equilibration and extraction the analyte can be determined. The major drawback is that the spike is not always bound in the same way as the naturally occurring compounds.

Alternatively, and only if the extraction procedure does not change the matrix composition and appearance, the recovery experiment may be carried out on the previously extracted real sample by spiking, equilibrating, and extracting. The recovery assessment can often be overestimated, however, and this risk should be faced. CRMs may again be a tool to ascertain accuracy; but they are only useful in cases where they contain incurred, and not spiked, species.

Recovery experiments are, however, necessary. Systematic studies using, e.g., incurred labeled organometallic compounds are desirable to assess the recoveries with a high degree of precision. Such studies would be costly and have not been performed as yet.

The extraction recovery may vary from the one chemical species of the same metal to another, e.g., mono-, di-, and tributyl-Sn display different behaviors (27). Consequently, recovery should be assessed independently for each compound as well as for compounds present in a mixture.

It is also necessary to find a compromise between a good recovery (sufficiently strong attack) and the preservation of the speciation. In fact, the use of overly concentrated HCl for the extraction of biological material has been shown to alter the methyl-Hg content, as demonstrated by the use of radiolabeled methyl-Hg (28).

The effects of extraction are recognized to be different according to the types of species. They are less important for stable or inert compounds, such as TBT, more important for stable or not inert species [e.g., Cr(VI)], and of paramount importance for not stable or not inert species (e.g., Al species).

Enzymatic matrix attack was found promising for biological matrices. Trypsin has been used for the determination of arsenobetaine in fish tissues (29).

Cleanup of extracts by using ion-exchange chromatography or other chromatographic techniques has been proposed for biological materials; however, this might lead to losses such as those observed for TBT (27). Cleanup could become simplified or even be redundant (reducing the risk of decay) if a more specific method of final determination could be found.

For the development of good extraction methods, materials with an incurred analyte (i.e., bound to the matrix in the same way as the unknown), preferably labeled (radiolabeling would enable the recovery to be verified), would be necessary. Such materials are not available, the extraction method used should be validated by other independent methods.

Supercritical extractions could well be the methods of choice. Such methods are currently being developed, e.g., for butyl-Sn compounds, and seem to offer good possibilities for extracting the species without alteration; however, they still need to be validated (30).

Even if the recoveries of spiked compounds as equilibrated in a certain matrix are total, there is no evidence that the incurred compound will be extracted with the same efficiency. Therefore, a systematic collaborative study is necessary of various extraction methods as applied in experienced laboratories in order to validate these methods.

We should stress here that extraction methods used for speciation suffer in principle from the same sources of error and have the same pitfalls as the methods used for the determination of traces of organic compounds. The reader is referred to such literature (31, 32).

6.4.2. Derivatization

Derivatization procedures can be employed to separate trace elements from their matrices and interferences, and to concentrate the analyte species. Some derivatization methods such as hydride generation (HG) are used to generate volatile species that are more easily separated from each other by chromatography. Reactions employed nowadays are mostly centered around the addition of simple groups, involving derivatization alkylation with Grignard reagents (e.g., pentylation, ethylation, or butylation).

The HG procedure is carried out in acid media, generally using $NaBH_4$ as the reducing and hydride transfer agent to yield metal and metalloid hydrides. The main advantage of this reaction is that metal–carbon cleavage does not occur and the speciation is therefore maintained. A wide range of element species can be determined using this approach, such as As, Ge, Sb, Se, and Sn. When a proper pH in the reaction vessel is selected, the range of compounds that yield hydrides is restricted. In the case of As, for instance, reduction at pH 1 yields arsine from both arsenite and arsenate, whereas at pH 5 only arsenite is reduced. However, this derivatization has major drawbacks in instances when there are high contents of organic compounds in the matrix, as shown for As and butyl-Sn compounds (33, 34).

$NaBH_4$ is not able to convert As in arsenobetaine or arsenocholine into arsine or any other volatile compound. The binding in these molecules is simply too strong. Therefore, in one sense HG is a good tool for separating As species, which can be very useful in, e.g., fish analysis, where only the content of toxic (i.e., inorganic) As is of importance. However, for the determination of all As-containing species (i.e., distribution of As), HG cannot be applied without pretreatment, e.g., by ultraviolet (UV) irradiation (35, 36).

Grignard reactions, such as pentylation, are widely used for the determination of alkyl-Pb and -Sn species; the reaction yields products that can be separated relatively easily by GC. Water destroys the reagent, and therefore the species of interest has to be removed from water-based matrices. This may be achieved, e.g., by extraction of a diethyldithiocarbamate complex into an organic phase prior to derivatization, as in the case of alkyl-Pb species determination (37). This back-extraction increases the risks of contamination or losses. Moreover, the yield of derivatization should again be verified, as it is presently hampered by the lack of suited calibrants.

The use of $NaBH_4$ overcomes the problem of hydrolytic instability of the Grignard reagents, allowing ethylation to be carried out in an aqueous medium. This reaction has been employed for the analysis of alkyl-Pb, -Sn, and -Hg species (38, 39). Here again, the lack of suitable calibrants hampers a verification of the yield of derivatization.

When working with derivatization procedures, we should realize that the reactions are far from being well controlled. An analyst does not know whether in the particular case at hand the binding of the analyte with matrix components is sufficiently accessible to the applied reagent, whether the reaction of matrix components (e.g., oils) with the reagent does not prevent the latter from reacting with the analyte, and so on. Despite the large number of publications, the reaction mechanisms of derivatization are not well understood, which certainly also holds for HG procedures.

A validation or at least a careful study of the procedure as it is used in the laboratory for a given matrix with its particular interferences is necessary.

Again, spiking experiments could be used here, but, as stated earlier, they can give only limited information.

In general, the risk of producing wrong results increases with the number of steps in a determination and with their complexity. Therefore, if derivatization can be avoided it is worth considering such a possibility.

6.4.3. Separation

Separation is performed where the determination of different species cannot be performed with sufficient selectivity, e.g., an Sn-selective detector (AAS) cannot distinguish between different alkyl or aromatic Sn species, such as TBT, triphenyl-Sn (TPhT) and dibutyl-Sn (DBT).

The separation of chemical species of elements can only be performed by techniques that do not destroy the chemical forms. Except for cold trapping (CT) methods (for water analysis), the separation is performed after extraction and suitable cleanup of the extract.

The separation of chemical species of elements is usually performed by three basic methods: chromatography [e.g., anion-exchange, ion-pairing reversed-phase liquid chromatography (LC) or capillary GC], capillary zone electrophoresis (CZE), or CT.

GC has become a powerful tool in the determination of traces of organic compounds. Unfortunately, such developments are not sufficiently applied in speciation analysis. Whereas for most environmental applications packed columns have been abandoned for the determination of traces of organic compounds because of poor separation and time-consuming procedures, they are still widely used for speciation. For example, most laboratories still apply such columns for the determination of methyl-Hg. Even a tedious and badly understood "conditioning" of the column (of which the effect or mechanisms are not clear) using $HgCl_2$ is accepted. Needless to say that statistical control is hardly achieved in such cases.

GC techniques that allow a better understanding of the separation process to be obtained with a better reproducibility of column performances ought soon to be introduced for speciation. The use of capillary columns as well as of internal standards, a general practice when working with organic traces, urgently needs to be introduced on a large scale in speciation work. Petersen and Drabaek have recently shown the feasibility and positive outcome of such work for the determination of methyl-Hg (40).

As in organic analysis, precautions should be taken to preserve the compound integrity in the column. For example, in GC separation of Sn species a heat-induced decay may occur leading to the deposition of Sn oxides in the capillary column, which in turn causes peak tailing.

Thermostable and volatile compounds may be separated by gas–liquid chromatography (GLC). Several stationary phases are available on fused silica capillary columns. The separation power relies on the polarity of the compounds and of the stationary phase. This separation method often requires a derivatization step.

For LC (e.g., HPLC) systems there is no need for derivatization prior to separation. Unfortunately, the availability of stationary phases in HPLC is less important (e.g., ion exchangers, ion pairing) than in GC; consequently, separation problems may still exist for some species, such as arsenobetaine and As(III). However, LC is better suited for element-specific detections, such as ICP–AES or ICP–MS, AAS, X-ray fluorescence spectrometry (XRFS), neutron activation analysis (NAA), or electrochemical detectors.

CT has been used successfully for the determination of, e.g., alkyl-Sn and some Se, Pb, and As compounds (23, 41). The technique presents advantages both to concentrate the species and to sequentially separate them according to their specific volatility.

One drawback of this method is that only volatile forms of elements (hydrides, ethylated or methylated forms, etc.) may be separated; other molecules of low volatility, e.g., TPhT or arsenobetaine, cannot be separated. In addition, CT requires a derivatization step. Both steps are difficult to validate, and it is still unclear which physical and chemical parameters may hamper, for a given matrix, the formation and separation of volatile forms. Moreover, the separation based on evaporation temperatures may not be sufficient to distinguish two compounds of similar volatility, e.g., $BuMeSnH_2$ and MPhT (42). Although the technique is not always applicable, its simplicity and the fact that it can operate on-line with derivatization steps makes it a recommended method for a variety of compounds.

Other powerful techniques that have proven to be suited for a wide range of applications in organic analysis should be considered for speciation. The capabilities of CZE have been demonstrated for nonvolatile, stable polar or chargeable compounds. Recently, this technique was successfully tried for As species such as arsenobetaine and arsenocholine, which could hardly be separated in another way (43). However, the retention time is not always well under control; after fractionation, this may cause errors in identification and quantification.

The transfer of technology currently in use for organic compounds to the field of determination of species will allow for a better understanding of the procedures and should lead to easier and more rugged methods in speciation.

QC procedures for such methods, taking into account the various sources of error, have been successfully developed for organic compounds (31, 32). These QC procedures might well also be applied to a hyphenated method to set a high-quality standard.

6.4.4. Detection

The detectors used for speciation analysis are either element specific (e.g., AAS) or nonspecific (e.g., FID, FPD, or ECD). In general, the determinand should enter alone into the detector to avoid interferences; multicomponent detection (e.g., ICP–AES, ICP–MS, or NAA) always has larger uncertainties. In speciation analysis, the choice of the detector strongly depends on the chemical forms to be determined and on the mode of separation used.

Electrothermal AAS (ETA–AAS), although a sensitive technique, is generally not recommended for speciation analysis, as the method cannot be applied in a continuous (on-line) mode. The necessary manipulations, caused by the off-line character of the method, considerably increase the risks of errors.

Whenever applied, the precautions for the measurement are the same as for inorganic analysis; the choice of the matrix modifier, the temperature program, etc., should follow the same rules as those for the determination of the element content.

Flame AAS and quartz furnace AAS are often used as element-sensitive detectors. When various parameters are chosen and set properly, these techniques can be performed on-line; provided that a proper separation (e.g., GC or CT) is achieved, it can be used to determine the various species containing the same element. ICP–AES or ICP–MS can be used on-line after HPLC separation, using a proper interface. MS can be specific in certain cases and even would allow for on-line QA in the isotope dilution mode.

Voltammetry can be used in some cases as a species-sensitive detector, provided that the species to be determined are sufficiently electroactive and that electrode reactions proceed with a sufficiently high rate. This technique has been successfully used, e.g., for the determination of TBT in water and sediment (44).

Classical detectors after LC or GC separation may be applied for the determination of some chemical forms of elements, e.g., FPD or FID detection for TBT.

6.4.5. Prelimary Calibrants and Internal Standards

General aspects of calibration can be applied to speciation analysis. Balances should be frequently calibrated as well as all volumetric glassware. The number or laboratories that do not use volumetric glassware in calibration but rather use gravimetric dilution of reagents is rapidly increasing, as the advantages in terms of accuracy are well understood.

In general and over many years the SM&T experience is that in about 25–30% of all cases erroneous results were attributed to calibration errors.

This has also been reflected in the SM&T intercomparison on extractable trace elements in soil and sediment, where many laboratories were found to be affected by errors when analyzing a common calibrant solution. In fact, most discrepancies that occurred could be attributed to a biased final determination (45).

This illustrates the importance of calibration. All efforts made to obtain a good sample and perform the extraction under the proper conditions are spoiled if the calibration is wrong.

First of all calibrants should be of good purity and stoichiometry should be verified. Compounds containing water of crystallization should be stored under such conditions that stoichiometry is maintained. When stock calibrant solution is prepared preferably two stocks should be made independently, one serving to verify the other; when the dilution is carried out (preferably gravimetrically) it is recommended that again two independent solutions be made for verification purposes. The alternative would be to verify the new calibrant solution using the previous one.

Calibrant solutions should be made prior to use, even if the solutions are acidified. The laboratory should carefully consider the calibration mode chosen: (i) standard additions; (ii) calibration curve; or (iii) bracketing standards. There is no calibration mode that can be recommended in all cases. All suffer from typical sources of error, e.g.: (i) linearity of the calibration curve (extrapolation difficulties, chemical form of calibrant added); (ii) especially in the nonlinear part of a curve, points have to be measured quite often, for sometimes even minor changes of the calibrant matrix (e.g., relatively weak complexing agents) can change the calibration considerably, which leads to the recommendation to reproduce the matrix in some detail in matrix-matched solutions; (iii) although bracketing standards in either mode i or mode ii are excellent ways to correct for fluctuations in the measurement system, for many routine laboratories they can be very tedious.

The best recommendation we can make is probably to validate the methods for each type of matrix and for the extraction agent applied. Once this is done, the chosen calibration method is more trustworthy.

Matrix effects as revealed by loss of signal in ECD, FPD, or AAS detection, or by inhibition of the derivatization (e.g., HG) due to interferences of aromatic compounds, may strongly affect the calibration. If pure solutions of the determinand are used for calibration, the determination may be wrong by some orders of magnitude. Standard addition techniques are therefore the only way to control the validity of the detection, but only if the addition has been performed with the proper identical form of the compound to be determined. The analyst must keep in mind that, by using standard addition procedures, levels of concentration may be reached that are no longer within the linear range of the detector response (e.g., for ECD or AAS). It is therefore of primary importance to evaluate this linear range before starting the analysis.

6.4.6. Further Efforts of Improve the State of the Art and Method Validation

As already described, the analytical methods used for the determination of chemical forms of elements are based on subsequent steps that may vary from one procedure to another and may be tested independently by step-by-step exercises (46).

Each step of the analytical procedure should be examined separately to verify the laboratory's performance and detect the sources of error that may arise. This can be done in a collaborative way once the laboratories have taken all the steps necessary to evaluate the performance of their methods in their own laboratory. Then, the laboratory may participate in intercomparisons (see Section 6.3.5) to verify the accuracy of their methods: in these exercises, different samples previously characterized for their homogeneity and stability, are prepared and sent to the participating laboratories. Typical matrices are as follows: (i) solutions containing one or several pure compound to be measured to evaluate the performance of the final detection; (ii) similar solutions as in item i, including potential interfering compounds to evaluate the performance of the final detection and of the separation techniques; (iii) cleaned extracts to fully test the performance of the separation and the final detection using real samples; (iv) raw extracts to verify the cleanup procedure; (v) real matrices homogeneously enriched and equilibrated with the analyte(s) to be determined to test the total analytical procedure and determine recovery; and (vi) real samples.

These several steps in a collaborative exercise enable the sources of error to be identified and consequently help the laboratories to remove them. The results of these exercises are discussed in technical meetings with all the participating laboratories. Such intercomparisons may also enable the participants to remove any bias linked to a certain method.

Systematic errors such as contamination, losses, incomplete extraction, decay of compounds, and calibration errors can be identified by comparing different (pre)treatment procedures (e.g., solvent and acid extraction), separation techniques (GC or LC) and final detections (e.g., AAS, ICP–AES, or ICP–MS).

The first step to be undertaken when organizing interlaboratory comparisons is to verify that the solutions and/or samples used are stable over time, i.e., than no adsorption or contamination occurs and that the speciation has been preserved. The experience at the SM&T Programme has shown that solutions of Cr(III)/Cr(VI), Se(IV)/Se(VI), methyl-Hg, butyl-Sn, and trimethyl-Pb species are stable at ambient temperature in the dark for periods of at least 6 months when precautions particular to the compounds are taken (47–50). The stability of TBT has also been verified in sediment (49). In the case of As, the stability of some species is more difficult to achieve. As for as pure solutions of As(III), As(V), monomethylarsonic acid (MMAA), dimethylarsinic acid (DMAA), arsenobetaine, and arsenocholine are concerned, no instability could be detected for samples stored at 20 °C in the dark. However, mixtures of different As species were shown to display degradation problems for some of the compounds, particularly upon light exposure [e.g., degradation of methylated As compounds; oxidation of As(III) to As(V)]. These difficulties demonstrate the need to monitor the stability of the materials distributed to laboratories, which has to be carried out by laboratories experienced in that particular field of analysis.

Figure 6.2(a–d) depicts step-by-step approaches for the development and validation of methods. These approaches have particular relevance for speciation analyses. The following subsections describe results obtained in collab-

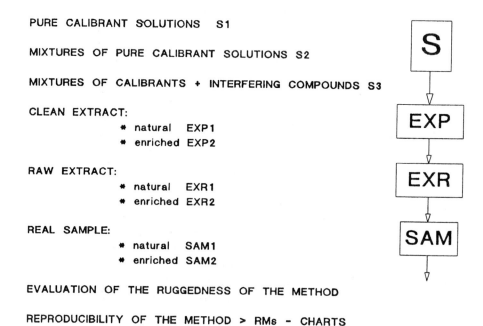

Figure 6.2. [pp. 212–214] Step-by-step approach (a–d) for the development and validation of analytical methods, an intralaboratory procedure (46). (ID–MS: isotope dilution mass spectrometry; chrom. = chromatographic.)

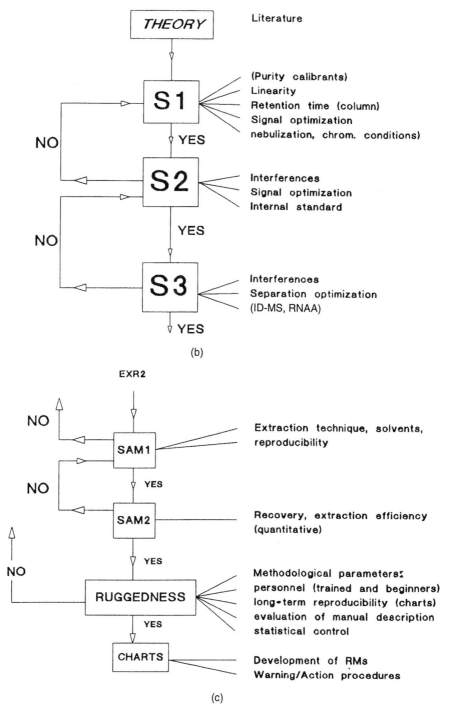

Figure 6.2. (Continued)

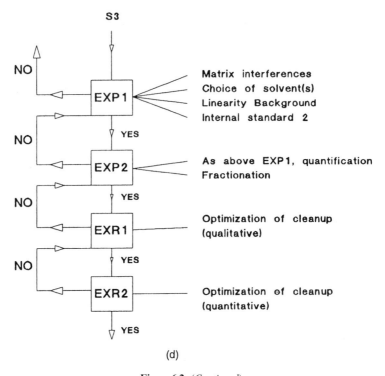

Figure 6.2. (*Continued*)

orative step-by-step projects organized by the SM&T Programme to improve the analyses of As, Hg, and Sn species.

6.4.6.1. Determination of As(III), As(V), and Organo-As Compounds

Organo-As compounds are used in medicine, cosmetic products, and agriculture. The most toxic form of As is As(III). Arsenic is also present in the environment as methylated forms, biologically or chemically mediated (in soil, sediment, or biological tissues). Other organic forms, such as arsenobetaine and arsenocholine, are present in animal tissues and are determined by environmental and food analysts, although these As species are not toxic.

In order to improve the quality of As speciation analyses, the SM&T Programme has organized a series of intercomparisons with solutions containing pure analytes of As(III), As(V), MMAA, DMAA, arsenobetaine, and arsenocholine after a thorough investigation of the best way to stabilize the species (43).

The analytical techniques involved were mainly based on derivatization steps (HG), FI and ion-exchange methods, chromatographic separation (GC and HPLC) and a variety of final detection methods (AAS, ICP–AES, ICP–MS, XRF, and instrumental NAA).

Four intercomparisons highlighted the analytical difficulties of As-speciation analyses, as shown by the wide spread of data. As outlined before, the main sources of error were assumed to be linked to the derivatization step (mostly HG). These exercises demonstrated that considerable improvements are necessary to achieve good analytical QC.

6.4.6.2. *Determination of Butyl- and Phenyl-Sn Compounds*

TBT is known to be highly toxic to marine organisms even at very low concentrations. This compound originates from the use of antifouling paints on ships and is a source of major mortality of shellfish population. Several European laboratories monitor the TBT levels in water to verify whether its concentration is in accordance with current regulations (e.g., France, the United Kingdom; and now the Netherlands).

Intercomparisons have been organized within the SM&T Programme, first dealing with the analysis of solutions containing pure analytes (TBT and mixtures of mono-, dibutyl-, and triphenyl-Sn); the stability of TBT in the samples had previously been demonstrated. A second exercise on a TBT-enriched sediment was also successfully concluded (49).

Finally, a third exercise dealt with the determination of TBT in a harbor sediment. The techniques tested were mainly based on extraction steps (organic solvent or acid), derivatization steps (e.g., HG, alkylation with a Grignard reagent), separation (GC or HPLC) followed by various detection methods [e.g., spectrofluorimetry, quartz furnace AAS, ETA–AAS, FID or FPD, differential pulse polarography (DPP), MS].

No systematic errors could be detected for the determination of TBT in simple synthetic solutions (first intercomparison). However, it should be noted that possible interferences by, e.g., aromatic compounds in AAS detection were not evaluated in this exercise.

The determination of TBT in a spiked sediment (second intercomparison) did not reveal any systematic errors. As the sediment contained a high level of organic matter, it was concluded that detection, separation, and derivatization procedures were satisfactory. However, the extraction of spiked TBT was thought to be easier than a "naturally" incurred compound; moreover, the content of TBT (ca. $3\,\mu g\,g^{-1}$ as TBTAc) was not considered to be representative of natural contamination levels, which are usually 10–50 times lower. Therefore, a third exercise was performed, using a naturally contaminated harbor sediment; in this intercomparison, a wide spread of data obtained with techniques involving derivatization (HG and ethylation) was demonstrated. A better agreement was found for techniques based on organic solvent extraction, capillary GC or HPLC, and FPD, MS, or ICP–MS detection. It is clear that the derivatization procedures were not under control for the determination of TBT in such a complex matrix (containing high amounts of organic matter) at such low levels (ca. 20 ng g^{-1} as TBT).

The pitfalls suspected to be due to derivatization were confirmed by an extensive study in which the inhibition of HG yield was shown to be up to 60% in some cases (51). Further work was considered to be necessary to improve the quality of butyl-Sn determination in sediment, which might be done by using, e.g., coastal sediment and/or biological samples (mussels or oysters) with butyl-Sn contents representative of the levels found in the environment.

6.4.6.3. Determination of Methyl-Hg

Methyl-Hg can be directly released into the environment (e.g., from a Cl_2-producing industry), or it can originate from the biomethylation of inorganic Hg in animal tissues. This compound accumulates in the food chain and leads to highly toxic effects on biota and humans. Therefore, methyl-Hg is determined in many laboratories all over the world.

A project to improve the quality of methyl-Hg determination in Europe has continued within the SM&T Programme since 1987. The intercomparisons dealt first with solutions of pure analyte (CH_3HgCl) and later with cleaned fish extracts, raw fish extracts, and methyl-Hg-enriched raw extracts (48).

Methods used for the determination of methyl-Hg were based on solvent extraction, in some cases derivatization (e.g., ethylation), chromatographic separation (GC with packed or capillary columns), combined with final detection methods such as ECD, cold vapor (CV) AAS, and NAA with radiochemical separation (RNAA). Other techniques such as coupled HPLC–AAS or HPLC–AES also used.

The analysis of MeHgCl solutions did not allow any systematic error to be revealed at the detection step. As for the determination of TBT, the likelihood of interferences due to organic compounds was not taken into account in this exercise.

The results of the second round-robin exercise on extracts revealed a wide spread of results and the presence of uncontrolled sources of error. At this stage, the separation with packed columns and the pretreatment of the column with $HgCl_2$ were suspected to be the cause of this dispersion.

It was therefore decided to run a third round-robin exercise with other fish extracts in order to remove the remaining bias. In this exercise, systematic errors due to the use of a packed column could be demonstrated, as shown in a Youden diagram plotting the results of spiked extract vs. raw extract analysis

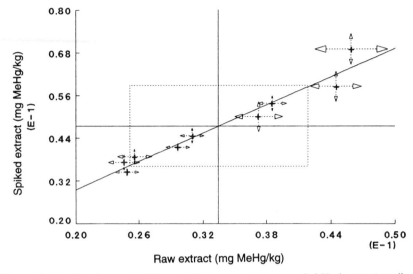

Figure 6.3. Youden plot, methyl-Hg in spiked raw extract vs. methyl-Hg in raw (unspiked) extract: horizontal and vertical continuous lines are the means of the laboratory means, dotted lines being the SD of these means. The length of the arrows of each set of data is equal to the SD of five replicate determinations performed by the individual laboratories (48).

(Fig. 6.3). It was apparent that capillary columns were more reliable, and their use was therefore recommended (40).

Mussel and tuna samples were also analyzed to evaluate the long-term reproducibility of methods; results proved that the state of the art was good enough to envisage a certification campaign.

6.5. ERRORS IN TRACE ELEMENT SPECIATION IN SEDIMENTS AND SOILS

The determination of extractable trace element contents in soil is currently performed in many laboratories to assess the bioavailable metal fraction (and related potential phytotoxic effects) and the importance of leaching of polluted soil or sediment after landfill application (e.g., for controlling the contamination of groundwaters). This assessment of metal mobility and bioavailability is performed on the basis of single and sequential extraction schemes that have been developed for soil and sediment analysis (52–54).

Preliminary SM&T Programme studies have compared different extraction schemes (45). The outcome was such that a larger project to compare results of different methods using defined extraction protocols seemed to be

Table 6.3. Sequential Extraction Procedures Suggested by (A) Tessier et al. (52) with Modifications by (B) Förstner and Wittman (55) and (C) Meguellati et al. (54)

Method	1	2	3	4	5
A	$MgCl_2$ 1 mol L^{-1} pH 7 "exchangeable"	NaOAc 1 mol L^{-1} pH 5 "carbonate"	$NH_2OH \cdot HCl$ 0.04 mol L^{-1} 25% HOAc "Fe/Mn oxide"	H_2O_2 8.8 mol L^{-1} + HNO_3/NH_4OAc "organic matter & sulfide"	$HF/HClO_4$ "Residual silicate phase"
B	NaOAc 1 mol L^{-1} pH 5 "exchangeable + carbonate"	$NH_2OH \cdot HCl$ 0.1 mol L^{-1} "easily reducible"	NH_4Ox/HOx 0.1 mol L^{-1} pH 3 in dark "moderately reducible"	H_2O_2 8.8 mol L^{-1} pH 7 "organic matter & sulfide"	HNO_3 "residual silicate phase"
C	$BaCl_2$ pH 7 "exchangeable"	H_2O_2 8.8 mol L^{-1} + HNO_3 "organic matter"	NaOAc pH 5 1 mol L^{-1} "carbonate/ silicate"	$NH_2OH \cdot HCl$ 0.1 mol L^{-1} 25% NOAc "Fe/Mn oxide"	Ashing followed by HF/HCl "Residual silicate phase"

feasible. An illustration of the "equivalence" of the different schemes as tested on reference sediment and soil samples is given in Table 6.3. On the basis of these results, it was clear that whereas similar conclusions on, e.g., the importance of trace metal mobility could be drawn by using the different schemes, the data obtained could not be accurately compared. This study highlighted the need for the adoption of common extraction schemes to allow a worldwide comparison of data to be carried out. Two schemes (single and sequential extractions) were chosen to be used uniformally by 40 participants in an intercomparison exercise dealing with a polluted soil sample and a sediment sample. The extraction procedures are the following: soil sample (single extraction), 0.05 mol L^{-1} EDTA as an ammonium salt solution, 0.43 mol L^{-1} acetic acid, 1 mol L^{-1} ammonium acetate (pH 7); sediment sample (sequential extractions), 0.11 mol L^{-1} aqueous CH_3COOH, 0.1 mol L^{-1} aqueous $NH_2OH \cdot HCl$ (pH 2), 1 mol L^{-1} aqueous CH_3COONH_4 (pH 5), 8.8 mol L^{-1} aqueous H_2O_2.

The stability of the total metal content had been previously documented in soil and sediment samples (55). However, very little was known on the stability of extractable forms of metals. Therefore, the stability of these forms was preliminarily verified (45).

A first intercomparison enabled evaluation of the performance of laboratories both for the extraction and the final detection methods (AAS and ICP–AES). Results were promising with regards to soil analysis, where a single extraction step is sufficient; however, improvements were considered to be necessary for sediment analysis, where a sequential procedure is followed. As mentioned previously, the discrepancies observed were mainly due to calibration errors, this again underlying the importance of a careful calibration of the final method of determination. Since participants also received a trace metal solution to verify the performance of the final determination, it could be shown that 40% of all errors were made in the calibration step and not in the extraction procedures studied.

The experience gained in this project will most probably lead to a set of methods that can be uniformly applied. Proper bodies [e.g., ISO and CEN (Comité Européen de Normalisation)] may use the results of these tests to normalize some methods of extraction with the ensuing better comparability of data and better interpretation. This item has been extensively discussed in a workshop held in Sitges, Spain, in 1992 (56).

6.6. CONCLUSIONS

It is now internationally recognized that the determination of total trace element contents is not sufficient for assessing environmental contamination levels and understanding the biogeochemical pathways of trace elements. Consequently, the number of determinations of chemical species carried out in routine and research laboratories is constantly increasing. The techniques used for these determinations are in most cases based on several analytical steps that considerably increase the risk of error.

The examples given illustrate the lack of accuracy that occurs in the determination of some chemical species. It is clear that good QC of speciation analyses has not yet been achieved. The need to improve the quality of these analyses should encourage laboratories to participate in intercomparisons, which are valuable tools for validating methods and disseminating expertise. Owing to the great difficulties posed by speciation analyses, efforts should be continued to make results comparable worldwide. Another way to demonstrate accuracy of results is to use CRMs; however, very few CRMs are presently available in the field of speciation analysis. To the authors' knowledge the only materials certified up to 1992 for chemical species are a harbor sediment certified for mono-, di-, and tributyl-Sn, fish tissue certified for its content of tributyl-Sn, and lobster materials certified for trace metals and methyl-Hg (57–60). The SM&T Programme is also working on the production of materials certified for butyl-Sn species (coastal sediment) and methyl-Hg (tuna fish).

CRMs certified for other chemical species are urgently needed. Increased collaborative work and use of suitable CRMs may well be the only means to ensure, in the future, a high quality of speciation analyses.

REFERENCES

1. B. Griepink, *Fresenius' Z. Anal. Chem.* **338**, 486 (1990).
2. E. A. Maier, *Trends Anal. Chem.* **10**, 340 (1991).
3. B. Griepink and M. Stoeppler, in *Hazardous Metals in the Environment* (M. Stoeppler, ed.). Elsevier, Amsterdam, 1992.
4. D. E. Wells, J. De Boer, L. G. M. Th. Tuinstra, L. Reutergardh, and B. Griepink, *Fresenius' Z. Anal. Chem.* **332**, 591 (1988).
5. P. J. Craig, ed., *Organometallic Compounds in the Environment*, p. 368. Longman, 1986.
6. O. F. X. Donard and R. Pinel, in *Environmental Analysis Using Chromatography Interfaced with Atomic Spectroscopy* (R. Harrison and S. Rapsomanikis, eds.), p. 189. Ellis Horwood, Chichester, UK, 1988.
7. M. Leroy, O. F. X. Donard, M. Astruc, and Ph. Quevauviller, *Pure Appl. Chem.* in press (1995).
8. S. C. Apte, A. G. Howard, and A. T. Campbell, in *Environmental Analysis Using Chromatography Interfaced with Atomic Spectroscopy* (R. Harrison and S. Rapsomanikis, eds.), p. 189. Ellis Horwood, Chichester, UK, 1988.
9. M. Morita, *Pure Appl. Chem.* in press (1995).
10. I. Drabaek and Å. Iverfeldt, in *Hazardous Metals in the Environment* (M. Stoeppler, ed.). Elsevier, Amsterdam, 1992.
11. M. Radojevic, in *Environmental Analysis Using Chromatography Interfaced with Atomic Spectroscopy* (R. Harrison and S. Rapsomanikis, eds.), p. 189. Ellis Horwood, Chichester, UK, 1988.
12. W. Frech and D. C. Baxter, *Pure Appl. Chem.* in press (1995).
13. R. Muñoz, C. Camara, O. F. X. Donard, and Ph. Quevauviller *Anal. Chim. Acta* **286**, 371 (1994).
14. S. Hetland, I. Martinsen, B. Radzuk, and Y. Thomassen, *Anal. Sci.* **7**, 1029 (1991).
15. J. K. Taylor, *NBS Spec. Publ. (U.S.)* **260**, 100 (1985).
16. T. H. Hartley, *Computerized Quality Control: Programs for the Analytical Laboratory*, 2nd ed., p. 99. Ellis Horwood, Chichester, UK, 1990.
17. M. Waldock, Quality control charts. MAFF Fisheries Laboratory, personal communication.
18. *Certification of Reference Materials: General and Statistical Principles*, ISO-Guide 33 (1985).
19. Ph. Quevauviller, O. F. X. Donard, E. A. Maier, and B. Griepink, *Mikrochim. Acta* **109**, 169 (1992).

20. O. F. X. Donard, in *Fifth Colloquium Atomspektrometrische Spurenanalytik* (B. Welz, ed.), p. 395. Perkin-Elmer, Überlingen, 1990.
21. M. Astruc, R. Pinel, and A. Astruc, *Mikrochim. Acta* **109**, 73 (1992).
22. R. Ritsema, *Mikrochim. Acta* **109**, 61 (1992).
23. O. F. X. Donard, *Thesis*, University of Bordeaux, **136** (1987).
24. R. Rubio, A. Padró, J. Albertí, and G. Rauret, *Mikrochim. Acta* **109**, 39 (1992).
25. E. H. Larsen, *Mikrochim. Acta* **109**, 47 (1992).
26. I. Tolosa, J. Dachs, and J. M. Bayona, *Mikrochim. Acta* **109**, 87 (1992).
27. A. Astruc, R. Lavigne, V. Desauziers, R. Pinel, and M. Astruc, *Appl. Organomet. Chem.* **3**, 267 (1989).
28. E. Bailey and A. G. F. Brooks, *Mikrochim. Acta* **109**, 121 (1992).
29. L. Ebdon, personal communication.
30. J. M. Bayona, J. Dachs, R. Alzaga, and P. Quevauviller, *Anal. Chim. Acta* **286**, 319 (1994).
31. D. E. Wells, J. De Boer, L. G. M. Th. Tuinstra, L. Reutergardh, and B. Griepink, Improvements in the Determination of Chlorobiphenyls Leading to the Certification of Seven CBs in Two Fish Oils and in Sewage Sludge, *EUR Report*, 12496. CEC, Brussels, 1989.
32. T. A. Rymen, B. Griepink, and S. Fachetti, *Chemosphere* **20**, 1291 (1990).
33. A. G. Howard and S. D. W. Comber, *Mikrochim. Acta* **109**, 27 (1992).
34. Ph. Quevauviller, F. Martin, C. Belin, and O. F. X. Donard, *Appl. Organomet. Chem.* **7**, 149 (1993).
35. G. Rauret, R. Rubio, A. Padró, and J. Albertí, *Proc. Int. Symp. Microchem. Tech., 12th*, Córdoba, *1992* (1992).
36. N. Violante, F. Petrucci, F. La Torre, and S. Caroli, *Spectroscopy* **7**, 36 (1992).
37. W. M. R. Dirkx, R. J. A. Van Cleuvenbergen, and F. C. Adams, *Mikrochim. Acta* **109**, 133 (1992).
38. S. Rapsomanikis, O. F. X. Donard, and J. H. Weber, *Anal. Chem.* **58**, 35 (1986).
39. J. R. Ashby and P. J. Craig, *Sci. Total Environ.* **78**, 219 (1989).
40. J. H. Petersen and I. Drabaek, *Mikrochim. Acta* **109**, 125 (1992).
41. R. Harrison and S. Rapsomanikis, eds., *Environmental Analysis Using Chromatography Interfaced with Atomic Spectroscospy*. p. 189. Ellis Horwood, Chichester, UK, 1988.
42. Ph. Quevauviller, R. Ritsema, R. Morabito, W. M. R. Dirkx, S. Chiavarini, J. M. Bayona, and O. F. X. Donard, *Appl. Organomet. Chem.* **8**, 541 (1994).
43. E. A. Maier, B. Griepink, M. Leroy, Ph. Morin, M. Amran, S. Favier, R. Heimburger, and H. Muntau, *Proc. Int. Conf. Euroanal., 7th*, Vienna, *1990* (1990).
44. M. Ochsenkühn-Petropoulou, G. Poulea, and G. Parikassis, *Mikrochim. Acta* **109**, 93 (1992).
45. A. Ure, Ph. Quevauviller, H. Muntau, and B. Griepink, Improvements in the

Determination of Extractable Contents of Trace Metals in Soil and Sediment Prior to Certification, *EUR Report.* CEC, Brussels, 14763 (1993).
46. E. A. Maier, Ph. Quevauviller, and B. Griepink, *Anal. Chim. Acta,* **283**, 590 (1993).
47. S. Dyg, R. Cornelis, and B. Griepink, *Proc. Int. Conf. Met. Speciation Environ., NATO Workshop Speciation,* Izmir, Turkey (1989).
48. Ph. Quevauviller, I. Drabaek, H. Muntau, and B. Griepink, *Appl. Organomet. Chem.* **7**, 413 (1993).
49. Ph. Quevauviller, B. Griepink, E. A. Maier, H. Meinema, and H. Muntau, *Proc. Int. Conf. Euroanal. 7th,* Vienna, *1990* (1990).
50. R. J. A. Van Cleuvenbergen, W. M. R. Dirkx, Ph. Quevauviller, and F. C. Adams, *Int. J. Environ. Anal. Chem.* **47**, 21 (1992).
51. F. Martin, C. M. Tseng, Ph. Quevauviller, C. Belin, and O. F. X. Donard, *Anal. Chim. Acta* **286**, 343 (1994).
52. A. Tessier, P. G. X. Campbell, and M. Bisson, *Anal. Chem.* **51**, 844 (1979).
53. U. Förstner, in *Chemical Methods for Assessing Bioavailable Metals in Sludges* (R. Leschber, R. A. Davis, and P. L'Hermitte, eds.), p. 1. Elsevier, Amsterdam, 1979.
54. M. Meguellati, D. Robbe, P. Marchandise, and M. Astruc, *Proc. Int. Conf. Heavy Met. Environ.,* Heidelberg, p. 1090 (1983).
55. U. Förstner and C. T. W. Wittman, eds., *Metal Pollution in the Aquatic Environment,* p. 473. Springer-Verlag, Berlin and New York, 1983.
56. Ph. Quevauviller, G. Rauret, and B. Griepink, *Int. J. Environ. Anal. Chem.* **50**, 231 (1993).
57. National Research Council Canada, *Marine Sediment Reference Materials for Trace Metals and Other Constituents,* PACS-1. NRCC, Ottawa, Canada, 1990.
58. National Institute for Environmental Studies, *NIES Certified Material,* No. 11, p. 305., NIES, Yatabemachi, Tsukuba, Ibaraki, Japan, 1990.
59. S. S. Berman and R. E. Sturgeon, *Fresenius' Z. Anal. Chem.* **332**, 546 (1988).
60. National Research Council Canada, *Lobster Hepatopancreas Marine Reference Material for Trace Metals and Other Elements,* TORT-1. NRCC, Ottawa, Canada, 1987.

CHAPTER

7

ALUMINUM AND SILICON SPECIATION IN BIOLOGICAL MATERIALS OF CLINICAL RELEVANCE

A. SANZ-MEDEL and B. FAIRMAN

Department of Physics and Analytical Chemistry, University of Oviedo, Oviedo, Spain

KATARZYNA WRÓBEL

Institute of Chemistry, Białystok, Poland

7.1. INTRODUCTION

Althogh Al and Si are two of the three most abundant elements in the lithosphere, during human evolution both have been mostly excluded from biochemical processes. Exposure to forms of Al and Si different from those normally found in the environment (where they are usually found together as aluminosilicates) is increasing due to modern technologies based on these elements (1). Although there have been many efforts to establish possible essential functions of Al and Si, interestingly enough in humans there does not seem to be a specific active pathway for their uptake and retention (2). Conversely, effective ways exist in healthy individuals to excrete these elements when they are taken up by the body via the alimentary tract or through the bloodstream, as summarized in Table 7.1 (2–5).

The risk of Al accumulation in the body has been associated with several factors. These include treating patients with Al-containing antacids, long-term industrial exposure, total parenteral nutrition hemodialysis treatment for renal failure patients, etc. (1, 3). The accumulation of Al in uremic patients may result in dialysis-related diseases, including dialysis encephalopathy syndrome (DES), non-Fe-related microcytic anemia, and Al-induced bone diseases (1, 3, 6). The main sources of Al that can affect uremic patients are Al-containing drugs and dialysate fluids (7). The high bioavailability of "free Al" in the physiological environment has led to the assumption that it may compete with

---

Element Speciation in Bioinorganic Chemistry, edited by Sergio Caroli.
Chemical Analysis Series, Vol. 135.
ISBN 0-471-57641-7 © 1996 John Wiley & Sons, Inc.

Table 7.1. Circulation of Al and Si in the Healthy Subject and Their Content in Some Body Tissues

	Approximate Amount of the Element	
	Al	Si
Dietary intake	3–5 mg/day	500 mg/day
Gastrointestinal uptake (% of intake)	0.5–1.0%	1–4%
Excretion of feces (% of intake)	>90%	>90%
Urinary excretion	20–50 μg/day	9 mg/day
Total body content	30–40 mg	Several grams
Serum content	2–6 μg L^{-1}	100–500 μg L^{-1}
Lung content	16–24 μg g^{-1}	47–68 μg g^{-1}
Bulk brain content	1.5 μg g^{-1}	12–27 μg g^{-1}
Bone content	1–6 μg g^{-1}	
Components increasing gastrointestinal uptake	Citrate, malate	
Components decreasing gastrointestinal uptake	Tannin, silicic acid, fluoride	Mg, Ca, and Al ions

Source: Data from Berman (2), D'Haese (3), Brätter and Schramel (4), and Combarn et al. (5).

other metals, causing changes in the activity of some enzymes as well as affecting bone mineralization processes (3, 6, 8–11). Aluminum has been shown to be a cytoskeletal toxin and to interact with DNA, probably by cross-linking the protein chains (12–15). It should also be mentioned that the metal is suspected of involvement in a number of neurological disorders, e.g., Alzheimer's disease (AD), and some liver diseases (1, 16).

In the case of Si, its biological role is even more unknown that that for Al. Silicon is thought to enter the organism via the alimentary tract, most probably in the form of monosilicic acid (17). Generally, Si is considered to be an element of relatively low toxicity (see Table 7.1). It can however, cause the lung disease silicosis by the inhalation of siliceous dust during long-term occupational exposure. An active role of Si in both bone mineralization processes and connective tissue metabolism has been demonstrated (18, 19). An essential role in nucleic acid function has also been suggested (17). There is still no strong evidence, however, for the essentiality of Si in the human body. Recently, Birchall has suggested that the element may have no direct biological function but may reduce the bioavailability of Al (18).

A significant increase of Si concentration in body fluids and a decline in the amount excreted in urine have been reported for uremic patients (20–23). Enhanced Si levels in serum and cerebrospinal fluid of patients with neurological disorders have been also detected (21, 24). All these findings, strikingly similar to those for Al, suggest the possible interaction between both elements in the human body. Speciation studies of Al and Si together in aqueous solutions have shown the formation of hydroxyaluminosilicates from about pH 4.0 upward (18). It is of physiological importance that even in the presence of citrate (at pH 7.4), stable soluble hydroxyaluminosilicates have been detected. On the other hand, Si has been found co-localized with Al as aluminosilicates in the senile plaque cores of chronic AD brains (25, 26). These reports have formed a new challenge in the biochemistry of these elements. The question is whether Si affects the toxicity of Al, and if so, how. To shed further light on all these aspects, there is a strong need for reliable total analyses of both elements and particularly for knowledge of their *speciation and fate* within the human body.

In this context the state of the art of analytical techniques for the determination of the total amount and different species of Al and Si in body tissues will be outlined in the following sections. The analysis of total levels in tissues will be dealt with first, and some conclusions will be drawn from past efforts and today's practices that might be useful to those attempting to avoid new errors. Furthermore, a critical look will be taken at the Al speciation techniques currently being utilized for serum samples (unfortunately there are no such data available in the case of Si), with special regard to deferoxamine (DFO), now widely used in the treatment of Al overload. Finally, the discussion will focus on the proposed links between Al, Si, and AD, the analytical speciation methods being employed for this type of investigation, and the possible role that Si may have in the toxicity of Al.

7.2. TOTAL ALUMINUM AND SILICON ANALYSES IN BIOLOGICAL MATERIALS

Before speciation analysis of trace metals can be addressed, it is obvious that total elemental analysis should be adequately taken into account. The determination of Al and Si in biological samples has proved to be a difficult task, mainly because of their comparatively low concentration in biological samples and high content in the environment. Therefore, special care in the control of contamination has proved mandatory during sampling, storage pretreatment, and measurement procedures (3, 27). Precautions to prevent contamination problems have been listed as follows: (i) use only double-distilled or deionized water; (ii) use laboratory plastic ware after decontamination, preferably with

10% HNO_3 and deionized water; (iii) work under clean-room conditions; (iv) minimize manipulation of the sample; (v) minimize the number of reagents added during pretreatment procedures; and (vi) check for contamination at each stage of the procedure.

Application of quality control (QC) measures in the laboratory is one of the best ways to ensure the quality of analytical results (28). This includes, as a mandatory tool in the quest for the necessary analytical accuracy, the availability of reference materials (RMs) containing physiological levels of the element, as has been strongly recommended for Al analysis in biological materials (28–30). At present, there are some biomedical RMs certified for Al, although to our knowledge no RMs are available for Si in clinical samples.

Adherence to these contamination-avoidance practices may well help to prevent gross contamination errors like those responsible for the amazingly wide range of "normal" human serum Al values that have been reported since 1972 (as shown in Table 7.2).

Table 7.2. Published "Normal" Levels of Human Serum Al According to Various Authors

Authors	Analytical Technique	Normal Level Reported ($\mu g\ L^{-1}$)	References[a]
Berlyne	NAA	1460	1
Clarkson	NAA	72	2
Kaehny	ETA–AAS	6	3
Gorsky	ETA–AAS	28	4
Andersen	ETA–AAS	140	5
Versieck and Cornelis	ETA–AAS	2	6
Leung and Henderson	ETA–AAS	6.5	7
Bettinelli et al.	ETA–AAS	17.3	8
D'Haese et al.	ETA–AAS	2	9

[a]References:
1. G. M. Berlyne, *Lancet* **2**, 494 (1970).
2. E. M. Clarkson, *Clin. Sci.* **43**, 519 (1972).
3. W. D. Kaehny, *N. Engl. J. Med.* **296**, 1389 (1977).
4. J. E. Gorsky, *Clin. Chem. (Winston-Salem, N.C.)* **24**, 1485 (1978).
5. J. R. Anderson, *N. Engl. J. Med.* **301**, 728 (1979).
6. J. Versieck and R. Cornelis, *Anal. Chim. Acta* **116**, 217 (1980).
7. F. Y. Leung and A. R. Henderson, *Clin. Chem. (Winston-Salem, N.C.)* **28**, 2139 (1982).
8. M. Bettinellis, V. Baroni, F. Fontana, and P. Poisetti, *Analyst (London)* **110**, 19 (1985).
9. P. C. D'Haese, F. L. Van de Vyver, F. A. De Wolff, and M. E. De Broe, *Clin. Chem. (Winston-Salem, N.C.)* **31**, 24 (1985).

7.2.1. Total Aluminum Determination

Although Al intoxication may enhance the element concentration in different regions of the body, the Al levels to be analyzed in healthy humans, e.g., in serum, plasma, urine, and tissues, are very low (at the microgram per liter level). In today's clinical practice, normal and pathological Al concentrations should be clearly identifiable and distinguishable. This task necessitates especially sensitive performance from the analytical techniques selected for the determination of this element at such low levels in complex biological matrices.

Gravimetric, titrimetric, fluorimetric, photometric, flame-absorption spectrometric, and chromatographic methods have been used for the quantification of Al, but they are not well suited for ultratrace determinations in biological samples because either they lack the required sensitivity or the measurements are strongly dependent upon sample composition or both (31). The biological matrix together with the high oxidation potential of Al (+1.66 V) make it impossible to directly use voltamperometric techniques. Neutron activation analysis (NAA) offers excellent sensitivity for Al (with a reported absolute detection limit of about 0.02 μg) and in many instances allows the direct analysis of small, microgram-size samples to be carried out (2, 32, 33). The drawbacks of this method for Al, however, include the short half-life of ^{28}Al (2.27 min), the conversion of P and Si present in the sample to ^{28}Al, and interferences from Na and Cl ions (3). Spectrofluorimetry has proved to be a very sensitive method for Al determination, but its application to real samples is limited due to interference problems that necessitate prior separation of the element (34).

The vast majority of Al determinations in biological samples have been carried out by using electrothermal atomization atomic absorption spectrometry (ETA–AAS) and inductively coupled plasma atomic emission spectrometry (ICP–AES) (27). The advantages of ICP–AES include its capability for multielemental analysis and the lack of chemical interferences; although its detection power for Al is comparatively weak, combination of ICP–AES with electrothermal volatilization and/or previous analyte separation and preconcentration, preferably in an on-line system, can mitigate such limitations (27, 35–38).

The high analytical sensitivity (16 pg/0.0044 A), together with relatively short analysis time, small sample size, and relatively simple pretreatment procedures, has made ETA–AAS the most commonly used analytical technique for Al determination in biological materials (1). The main difficulties originally experienced in using this technique, e.g., matrix interferences, formation of carbonaceous residues in the cuvette during organic decomposition, and poor precision of measurements, have been widely overcome with the use of modern ETA–AAS instrumentation and matrix modifiers and the application of various pretreatment procedures (2, 28, 39–50).

Sanz-Medel et al. have critically compared ICP–AES and ETA–AAS for the determination of Al in serum (27). They concluded that, where the element levels in serum, water, or dialysate fluids are above 30 μg L^{-1}, ICP–AES should be the method of choice, but for lower Al concentrations they recommended ETA–AAS because this technique possesses a superior detection power (0.5–1.0 μg L^{-1} of Al). The consensus that seems to have evolved among analysts favoring ETA–AAS for Al determinations in normal serum is that only a simple sample dilution (1:1) with ultrapure water is necessary(27, 28, 39, 51, 52). Some representative results of most reliable Al determinations in clinical samples by various techniques, but mainly ETA–AAS, are presented in Table 7.3 (51–62). The normal Al level in serum is still difficult to estimate as the element concentrations in the body are highly dependent on the geographic distribution (i.e., water, air, and diet supplementation). The reference value for serum Al as given by H.T. Delves [see Massey and Taylor (1)] and Versieck (28) is < 10 μg L^{-1}, whereas D'Haese (3) lowers that value to 2 μg L^{-1}.

Table 7.3. Representative Results of Al Determinations in Clinical Samples

Sample Type	Source[a]	n	Analytical Technique	Mean Al Content (μg L^{-1} ± SD)	References
Serum	Control	10	NAA	25	53
			ETA–AAS	2.1	54
			ETA–AAS	2.1 ± 2.2	46
		19	ETA–AAS	9.8	55
			ETA–AAS	1.6 ± 1.3	56
		10	ETA–AAS	2.0 ± 0.4	51
		40	ETA–AAS	17.3 ± 6.1	41
		15	Fluorimetry	6.8 ± 2.8	57
		2	ETA–AAS	21.5	58
			ETA–AAS	5.0 ± 1.0	21
	Dialyzed	68	ETA–AAS	29 ± 16	41
		100	ETA–AAS	62 ± 10.9	21
		35	Fluorimetry	48.3 ± 44.2	57
	AD	20	Fluorimetry	25.8 ± 7.8	57
Blood	Control	10	ETA–AAS	12.1 ± 1.5	51
		5	ETA–AAS	3.6 ± 10.7	58
Urine	Control	3	ETA–AAS	8.3 ± 1.4	42

(Continued)

Table 7.3. (*Continued*)

Sample Type	Source[a]	n	Analytical Technique	Mean Al Content ($\mu g\,L^{-1} \pm SD$)	References
Bulk brain	Control		ETA–AAS	1.9 ± 0.7	59
	DES	5	ETA–AAS	10 ± 27	59
	DES	6	ETA–AAS	21.0 ± 18.5	60
	DES	2	ETA–AAS	45.6 ± 2.1	42
	Dialyzed	2	ETA–AAS	24.8 ± 3.3	42
Gray matter	Control	6	ETA–AAS	2.1 ± 1.0	49
	Dialyzed	1	ETA–AAS	22.9 ± 3.8	49
	Dialyzed	12	ETA–AAS	5.1	61
White matter	Control	6	ETA–AAS	1.7 ± 0.5	49
	Dialyzed	1	ETA–AAS	5.8 ± 1.1	49
	Dialyzed	12	ETA–AAS	3.5	61
Bone	Control	10	ETA–AAS	<1.0	62
			ETA–AAS	2.4 ± 1.2	2
			ETA–AAS	9.3 ± 1.5	2
	Renal failure	15	ETA–AAS	2.7 ± 2.0	2
	Dialyzed	27	ETA–AAS	35 ± 29	62
		29	ETA–AAS	67 ± 46	2
		18	ETA–AAS	12.9 ± 1.0	2

[a]DES: dialysis encephalopathy syndrome; AD: Alzheimer's disease.

7.2.2. Total Silicon Determination

As a result of the complexity of biological matrices and the chemistry of Si, there are still few analytical techniques suitable for its determination in clinical samples. The first report of this type of analysis was published in 1955 by King, who used the colorimetric reaction of Si with molybdic acid (63). Nowadays, spectroscopic techniques are preferred for the measurement of the element in biological samples. These include ETA–AAS, ICP–AES, direct current plasma atomic emission spectrometry (DCP–AES), and X-ray spectrometry (XRS) (24, 36, 64–77). The Si contents normally found in different biological fluids by using various techniques are summarized in Table 7.4. From this table it can be seen that, as for Al, the most commonly used analytical technique for Si determination in these samples has been ETA–AAS. This is in spite of the element's tendency to form refractory carbides, which can result in a decrease of the analytical signal (78). This drawback can be circumvented by the pretreatment of the inner graphite surface of the atomization tube and/or platform with elements that form carbides thermodynamically more stable than those of Si, as well as the use of a suitable matrix modifier (79–81). Another problem is the strong influence of matrix components on the analytical signal (64, 76). In 1990, Gitelman and Alderman

Table 7.4. A Review of Si Contents as Obtained by Different Analytical Techniques in Biological Samples

Sample Type	Source	n	Analytical Technique	Mean Al Content (mg L^{-1} ± SD)	References
Serum	Control		Flame–AAS	0.88	73
			Spark source MS	0.43	73
			Flame photometry	0.22 ± 0.10	74
		33	ETA–AAS	0.11	75
			N_2O–C_2H_2 flame–AAS	0.60 ± 0.13	75
		21	ETA–AAS	0.310 ± 0.082	76
			ETA–AAS	0.20 ± 0.06	21
			ETA–AAS	0.275 ± 0.022	20
	Renal failure		ETA–AAS	0.480 ± 0.085	20
		36	ETA–AAS	0.52	22
	Dialyzed		ETA–AAS	0.839 ± 0.057	20
		100	ETA–AAS	0.620 ± 0.238	21
Plasma	Control		ETA–AAS	0.265 ± 0.017	23
Urine	Control	20	ETA–AAS	25.9 ± 27.5	76
			ETA–AAS	21.4 ± 3.9	23
		15	ETA–AAS	11.0 ± 7.5	65
Saliva	Control	6	ETA–AAS	2	64
Cow serum	Control	20	ETA–AAS	1.63	77
	Mastic	21	ETA–AAS	1.02	77

described an analytical procedure for Si in human plasma and urine by Zeeman ETA-AAS (65). They used Mo-coated tubes and platforms with multicomponent matrix modification [including EDTA (ethylenediaminetetraacetic acid), phosphates, and Ca salts]. The detection limits for Si, as expected by ETA-AAS in standard aqueous solutions, are in the range of 1.2–2.7 μg L^{-1} (65, 81, 82).

It has been pointed out that ICP-AES constitutes better analytical approach than ETA-AAS for the determination of Si because of its higher tolerance of matrix interferences and suitability for refractory elements (66). For ICP-AES, however, sample pretreatment is usually required to avoid conventional nebulization problems (66, 69). Fortunately, in body fluids the Si concentration is relatively high (Table 7.4), enabling Tanaka and Hayashi to successfully use simple dilution of the samples to diminish matrix-induced problems for ICP-AES determination (67). Matusiewicz and Barnes have introduced microvolumes of biological fluids to the ICP torch via electrothermal vaporization, obtaining a detection power of 500 μg L^{-1} for 5 μL samples (36). One clear advantage of ICP-AES is derived from its multielemental capability, which facilitates the simultaneous determination of Al, Si, and other key elements in order to investigate their interactions within the human body.

It is also important to realize that the total elemental concentration is only part of the picture and that speciation studies, together with a knowledge of the ultrastructural distribution of the element, are of great importance. Apart from speciation, the knowledge of such ultrastructural localization in target organs is essential if the toxicity of Al and the influence of Si on such toxicity are going to be finally elucidated. The main analytical techniques, together with some key references, that are used for ultrastructural studies of Al in tissues are listed in Table 7.5 (1, 24, 83–93).

Table 7.5. Analytical Techniques Used in Ultrastructural Distribution Studies of Al and Si in Tissues

Techniques	References
Atomic emission spectrometry with lasers	24, 83
Energy dispersive X-ray microanalysis	1, 72, 84–87
Laser microprobe mass spectrometry	86, 88
Secondary ion mass spectroscopy	1
Scanning electron microscopy coupled with X-ray spectrometry	89
Histochemical techniques	86, 90
Immunocytochemical procedures	84, 91–93
Magnetic resonance spectroscopy	85

7.3. ALUMINUM SPECIATION TECHNIQUES

The monitoring of total Al in certain body tissues has long been recognized by physicians to be a useful diagnostic and prognostic indicator of various diseases that have been associated with Al toxicity. However, the usefulness of total-element determinations is of limited value, as it is the bioavailability of Al that governs its toxicity, this last being in turn dictated by its physicochemical form and transportation mechanisms within body fluids and tissues. Unfortunately, Al speciation analysis in biological samples is beset with difficulties. One major difficulty is that normal tissue levels of Al are very low. As stated earlier in this chapter, the concentration of the element in blood serum is reported to be less than 10 μg L^{-1} for healthy individuals. This requires special analytical ability, even more sophisticated when one is trying to analyze small fractions of an element at these concentration levels (as is the case for speciation analyses). Problems associated with Al speciation analysis will be reviewed here, as well as the techniques and methods currently being used for the determination of the element species present in body fluids, particularly serum.

7.3.1. Analytical Techniques for Speciation of Aluminum

Several basically different analytical techniques have been employed so far in studies of the speciation of Al in serum which have produced rather conflicting results.

The speciation techniques used for Al in serum normally include a separation technique and an on-line or off-line sensitive detection system. According to their basic separation mechanisms, these have been classified as nonchromatographic and chromatographic techniques (as shown in Table 7.6) (94–108).

7.3.1.1. Nonchromatographic Techniques: Determination of Ultrafiltrable Serum Aluminum

The ultrafiltration, ultramicrofiltration, and dialysis methods all mimic the dialysis procedure by separating the serum into two fractions. Therefore, these have been labeled as nonchromatographic techniques (Table 7.6). Separation is accomplished by passing aliquots of serum samples through a membrane sieve (normally having a nominal cutoff of 10,000–30,000 Da), either under gas pressure or by centrifugal force. The Al fraction that passes through the membrane is called the *ultrafiltrable fraction* and is deemed to be the fraction that would be removed from the blood plasma during hemodialysis.

Table 7.6. Aluminum Speciation Techniques for Serum Samples and Literature Citations

Speciation Method	References
Nonchromatographic group:	
Ultrafiltration	94–99
Ultramicrofiltration	94, 95, 100
Dialysis	101
Chromatographic group:	
Gel chromatography	95, 97, 100, 102–106
HPLC	16, 95, 107, 108
Immunoaffinity chromatography	97
Gel electrophoresis	105

Here we come to the first controversy about Al speciation in serum – a controversy concerning the exact fraction of total element that is ultrafiltrable in normal serum, in uremic patient serum, or in patients with some forms of brain degenerative disease. Keirsse et al. undertook a survey of the performance of the ultrafiltration Al speciation methods employed (102). They reported that results cited in the literature ranged from 2.5 to 85% of the total for the serum Al ultrafiltration fraction, seemingly depending on the type of analytical method used. They also concluded that these divergent results could be ascribed to two reasons: first, methods and procedures that had been used might not have been sensitive or selective enough for the determination of Al at the ultratrace levels found in biological fluids; secondly, even more important, because of the ubiquity of the element in the environment, contamination might well have been a major cause of error.

This theme has been taken up and critically studied by Pérez Parajón et al. (94). They compared and evaluated two different ultrafiltration methodologies currently used for the fractionation of Al in serum. One system was a conventional Model 8050 stirred cell (Amicon Corp., Danvers, Massachusetts 01923), pressurized with nitrogen. The other was an Amicon micropartition ultramicrofiltration system (MPS-1). In this system the ultrafiltration is achieved by centrifugation; the determination of the Al content of the serum fractions and the total serum was by ETA–AAS, following the procedure previously described by Sanz–Medel et al. (27).

As described for total-element analysis, Pérez Parajón et al. also took extreme precautions to avoid contamination by extraneous Al in all of the main stages of the procedure (27, 94). Using the conventional ultrafiltration method (nominal cutoff: 20,000 Da), they found that the Al ultrafiltrable fraction ranged from 20.8 to 47.9% [mean, 35.0; standard deviation (SD),

±11.3] of the total Al concentration in normal pooled serum, whereas the nonultrafiltrable fraction amounted to 91.7–99.2% (mean, 96.0; SD, ±2.7). Similar results were obtained using Amicon YM10 membranes (nominal cutoff: 10,000 Da). These results compare favorably with those of other workers who used the same system (97, 99). However, they also demonstrate the poor precision of the method, as well as the fact that the sum of the Al fractions obtained added up to more than 100% of the element total content (97, 100). This seemed to indicate that despite the rigorous precautions taken, the ultrafiltrable fraction had become seriously contaminated by external Al at some stage in the procedure (probably from the ultrafiltration membrane).

By using the ultramicrofiltration technique it was found that the Al ultrafiltrable fraction of normal serum ranged from 5.8 to 12.3% (mean, 8.3; SD, ±2.1) and of uremic serum pools from 8.5 to 16.3% (mean, 13.3; SD, ±2.5). Moreover, the sum of the ultrafiltrable and nonultrafiltrable Al fractions was now nearly identical to that of the total serum content.

Pérez Parajón et al. pointed out that earlier workers possibly had underestimated the Al contamination problem, as Al in the nonultrafiltrable fraction was not always determined to check for complete Al balance (91, 97, 99, 102, 109). The mean value for normal serum ultrafiltrable Al of 8.3%, as obtained by ultramicrofiltration, is very close to the values reported by Leung et al. (100), using a similar method, by Elliot et al. (98) and Khalil-Manesh et al. (96), using conventional ultrafiltration, by Bertholf et al. (106), using gel chromatography, or by Graf et al. (101), using dialysis *in vivo*. Some of these authors also reported that for uremic patients the observed ultrafiltrable fraction is slightly higher (96). From these studies it appears that ultramicrofiltration holds several advantages when compared to conventional ultrafiltration: it is simpler, quicker, and more efficient in minimizing the Al contamination risks, which are the most probable cause for the inconsistency of the analytical data available at the present time for the metal in biological matrices.

7.3.1.2. Chromatographic Techniques: Aluminum–Protein Speciation

Having established that only 10% or less of Al in normal serum is ultrafiltrable, that hemodialysis fails to remove significant quantities of serum Al, and that the element passes into the plasma against the concentration gradient, analysts thought it reasonable to assume that the majority of total Al in serum should be bound to a high-molecular-weight (HMW) protein (97, 101, 107, 110, 111). The analytical methods used for this type of Al separation/speciation are classified in the "chromatographic" group, as defined in Table 7.6. It is here that a new controversy about Al in serum arose. Early workers in this field, using the gel filtration method, claimed to have found that a sizable fraction of the nonultrafiltrable element in the HMW fraction was bound to the protein albumin and also to transferrin (112, 113). More recently, other workers, either using similar gel supports or other types of supports, have classified the nonultrafiltrable element as "HMW protein bound" and have even identified the Al-binding protein in serum to be transferrin (97, 100, 102, 106). This identification required the use of other complementary separation or detection techniques. These include the individual identification of proteins and Al after postcolumn fractionation of the eluent, immunoaffinity chromatography, gel electrophoresis, and ^{67}Ga tracer studies (97, 100, 105, 113).

Keirsse et al. have expressed the view that gel filtration methodologies are acceptable for Al speciation in biological fluids because the overall mass balance is acceptable while no evidence of a redistribution of the metal species on the gel supports has been observed (102). However, Bertholf et al. have claimed that the application of gel filtration chromatography to the study of Al binding in serum is limited because the method cannot evaluate specific binding constituents, especially in the low-molecular-weight (LMW) region (106). Other criticisms of this technique range from the lack of general resolution to the fact that Al from the buffer can be absorbed by the column packing and by serum transferrin (97, 102).

These drawbacks have prompted researchers to look for alternative techniques offering higher resolution in the separation of serum proteins, as is the case with high-performance liquid chromatography (HPLC). Although the range of applications of HPLC technique has expanded dramatically over the last 10 years, they have only been applied sparingly to the separation of metal-carrying proteins in serum (16, 95, 107, 108, 114). HPLC appears to offer important advantages over the traditional size-exclusion chromatography methods (e.g., gel filtration) for analytical protein fractionation. One such method has been developed by Blanco González et al. (95). The separation of the serum proteins immunoglobin G, albumin, and transferrin was carried out using a TSK DEAE-3SW ion-exchange column with a sodium acetate gradient (0–0.5 mol L^{-1}) in Tris–HCl buffer, pH = 7.4, plus 0.1 mol L^{-1} sodium bicarbonate. The sodium bicarbonate was added after reports that bicarbonate was required for the binding of Al to transferrin (113, 115). The protein detection and determination was accomplished by ultraviolet (UV) absorption spectroscopy at 280 nm. The type of chromatogram and protein separation achieved by this research team are given in Fig. 7.1 (95). Atomic spectrometric methods applied to the corresponding eluent fractions showed that Al is clearly associated with the transferrin fraction but not with albumin. This is shown by Fig. 7.2, where an examination of the retention time of the Al peak from Fig. 7.2 is clearly seen to coincide with the appearance time of the transferrin peak as shown in Fig. 7.1. Hewitt and Day, using a fast protein column (Mono Q 5/5) and a slightly different buffer also showed that the

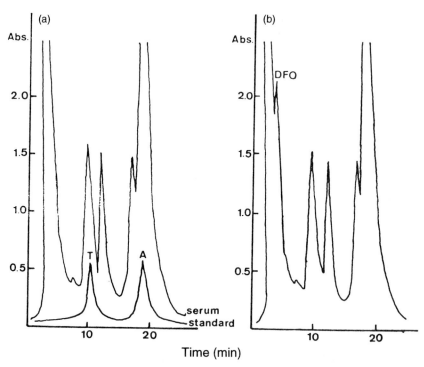

Figure 7.1. Ion-exchange HPLC separation of serum proteins (a) without and (b) with added DFO: A = albumin; T = transferrin; and DFO = desferrioxamine (200 mg L^{-1}). Reprinted by permission of García Alonso et al. (107).

separation of transferrin and albumin could easily be achieved in such columns (108). These authors and Sanz-Medel's group used off-line ETA–AAS detection for the determination of Al in fractions of the eluent (95, 108). The result obtained, confirmed in both cases, is that transferrin is the sole protein carrier of Al in serum. García Alonso et al. took the HPLC technique one stage further by developing an on-line flow injection analysis (FIA) detection system for Al (34). This is based upon micellar enhanced fluorimetric detection of the metal with 8-hydroxyquinoline–sulfonic acid complex in a micellar medium of CTAB. This procedure avoids the need for the cumbersome and time-consuming (when compared to FIA) off-line ETA–AAS technique, but is less selective.

Finally, it should be mentioned here that a few authors have indicated the presence also of an LMW Al-bound species of around 8000 Da. This species has been identified after ultrafiltration separation and after gel chromatogra-

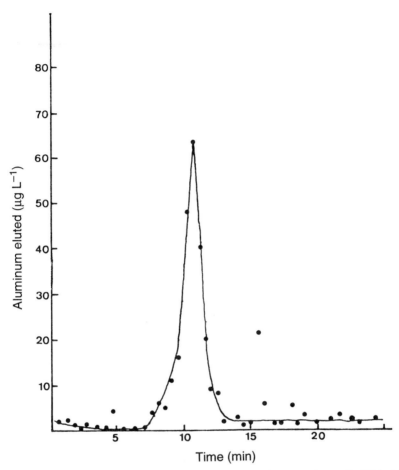

Figure 7.2. Elution profile of Al from spiked serum (1000 μg L^{-1} of Al) as detected by ETA–AAS. Reprinted by permission of García Alonso et al. (107).

phy (96, 105, 112). However, it should be stressed that no evidence for such species has been reported by workers using ultramicrofiltration or HPLC speciation techniques (94, 95, 108, 112).

It seems clear today that gel chromatography experiments do not allow analysts to make an unequivocal identification of the Al binding with HMW protein, mainly owing to the poor resolution between albumin and transferrin achieved using this separation technique.

Confirmatory data that transferrin is the only HMW serum Al-binding protein have been provided by thermodynamic calculations. It has been shown by Martin et al. that, at the usual transferrin concentration in plasma, only 30% of transferrin's two metal-ion binding sites are occupied by Fe^{3+} and, as transferrin is the strongest Al^{3+} binder in blood plasma, Al–transferrin should be the dominant species in blood (116). Martin also stated that because approximately 0.1 mmol L^{-1} of citrate is present in plasma, this oxygen-donor ligand becomes the main LMW plasma binder of metal ions such as Al (117). In his excellent theoretical study, Martin showed, via thermodynamic calculations, that with the citrate/transferrin molar ratio present in blood plasma, citrate (although the strongest LMW Al binder in serum) would release both Al^{3+} and Fe^{3+} to transferrin. He also calculated that albumin was too weak a binder to compete for Al with ligands such as hydroxide, phosphate, or citrate at physiological concentrations. These conclusions have been confirmed by a more recent study by Harris and Sheldon (118). They evaluated which of the aforementioned Al species would be present in serum at a typical 5 μmol L^{-1} total Al^{3+} concentration by using the computer program ECCLES. They calculated that 94.2% of the total Al in plasma would be bound to the bicarbonate–transferrin complex, with the next-best metal binder being citrate with 4.9% of the total. These authors also postulated that because of competition from hydroxide, which would also complex Al, one could expect dissociation of the Al–transferrin complex during separation processes that reduce the sample's bicarbonate concentration. This might account for some fractionation studies showing that up to 20% of the element in serum was bound to LMW molecules.

Although the different computer simulations and equilibrium calculations available sometimes disagree on the binding coefficients for the various Al complexes (due in part to different experimental conditions and procedures), mostly the same conclusions seem to be reached as to which Al species are dominant in blood serum, namely, Al–transferrin for the HMW fraction and Al–citrate for the LMW ultrafiltrable fraction.

7.3.1.3. Low-Molecular-Weight Aluminum Species: Aluminum Citrate

In recent years, attention has been given to the speciation of the LMW Al forms in plasma, particularly with regard to the citrate complex. This interest arose when it was found that in uremic patients the ultrafiltrable Al fraction in serum is raised from about 8 to 15% of the total (94, 95). On the other hand, the Al–citrate complex has aroused great concern because of reports that in the presence of the citrate ligand (L) the metal absorption in the gastrointestinal tract increases substantially, resulting in raised blood Al levels (119). Martin highlighted the disparities in earlier work in describing the citrate binding of Al^{3+} in blood serum (120). This author pointed out the special biological significance of the neutral AlL⁰ complex (at pH = 3.0, the mole fraction of this complex is nearly 0.6), which could provide an effective pathway for the entry of Al into the circulatory system via the upper region of the gastrointestinal tract. Therefore, there is a need for the speciation of the LMW–Al complexes in order to determine which complexes are present in the ultrafiltrable fraction and their relative physiological importance.

Unfortunately, at the time of writing, there has been no published method for the speciation of these LMW–Al complexes in serum. Aluminum fluoride, Al oxalate, and Al citrate species have been separated from the $Al(H_2O)_6^{3+}$ complex by ion chromatography in water, but with little resolution between the species (121). Gel chromatography has been used to separate Al citrate from the DFO complex, but this technique also lacked resolution with LMW species (122). Recently, two other approaches have been tried. Datta et al. applied HPLC with off-line ETA–AAS detection of the Al fractions (123). They tried a combination of nine different stationary phase supports with numerous mobile phases. Best results were obtained with α- and β-cyclodextrin stationary phases with a water/methanol (1:1 v/v), 0.1 mol L⁻¹ triethylamine and glacial acetic acid (pH = 4.0) mobile phase. However, the retention characteristics and recovery of Al (50–65%) were not considered acceptable for reliable speciation (or determination) of Al citrate in biological samples. Venturini and Berthon, by using a kinetic protocol to account for the formation of $Al(OH)_3$ species, have recalculated the stability constants of a series of LMW ligands (124–126). They have used these new values for computer simulation of the dominant species with ligands such as citrate and phosphate (127). According to their computer model, citrate does favor gastrointestinal absorption of Al, but, on the other hand, it induces Al urinary excretion. They pointed out that this type of result only highlights the necessity for further speciation studies to explain how a ligand such as citrate could have such an antagonistic role with regard to Al intoxication in humans (127).

7.3.1.4. *Aluminum-Complexing Drugs: Speciation of Aluminum–Deferoxamine*

Because Al accumulation can lead to a variety of neurological and physiological disorders, clinicians have been looking for methods to aid the removal of the metal from tissues and body fluids. The effective direct removal of Al from the body by hemodialysis has been shown to be unlikely, as more than 90% of the metal in serum is protein bound (107). Early attempts to remove Al by using chelating reagents such as penicillamine were apparently not successful (100). However, Al can be mobilized and eliminated by dialysis when the

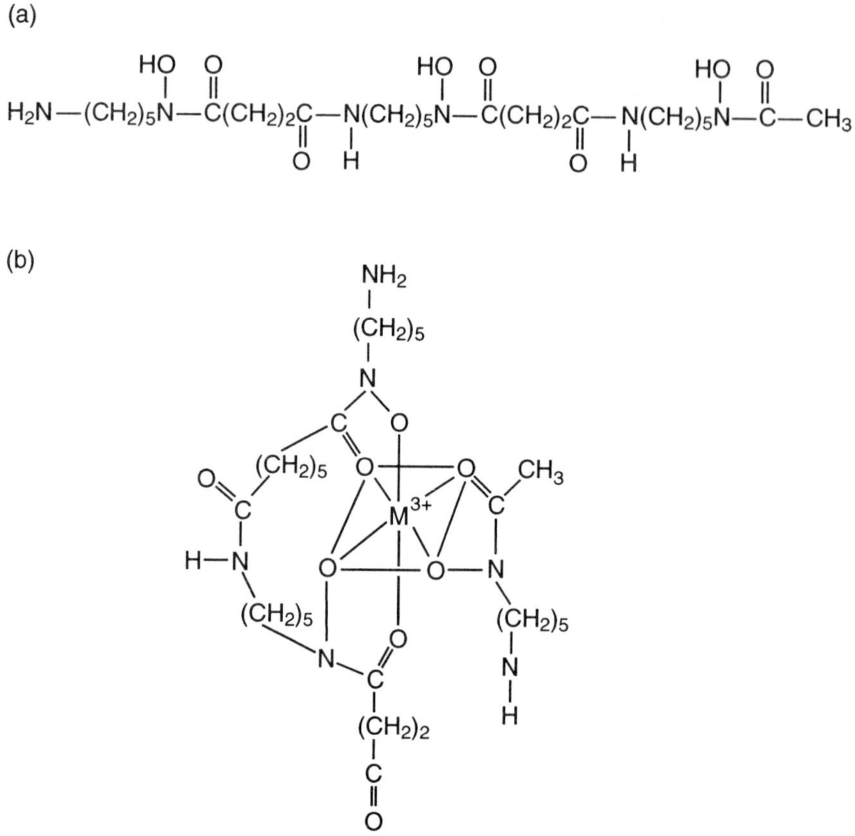

(a)

$$H_2N-(CH_2)_5\underset{\underset{H}{|}}{N}-\underset{\underset{O}{||}}{C}(CH_2)_2\underset{\underset{O}{||}}{C}-N(CH_2)_5\underset{\underset{H}{|}}{N}-\underset{\underset{O}{||}}{C}(CH_2)_2\underset{\underset{O}{||}}{C}-N(CH_2)_5\underset{\underset{H}{|}}{N}-\underset{\underset{O}{||}}{C}-CH_3$$

with OH on each N (hydroxamate).

(b)

Figure 7.3. (a) Molecular formula of deferoxamine. (b) Diagrammatic representation of the possible structure of the M^{3+}–DFO complex.

LMW chelating agent DFO is previously administered as a drug. DFO is a fungal siderophore (the formula of which is given in Fig. 7.3) that is able to effectively bind Al. The first report of the use of the drug for the detoxification of Al was by Ackrill et al. in 1980, although DFO had been used as an Fe-binding agent, especially for thalassemic patients, for about 20 years (128–130). As in the case of speciation of Al in body fluids, some contradictory results have been reported in the literature regarding DFO treatments.

One unsettling aspect about DFO treatment is that this drug is being more and more widely given to patients undergoing regular dialysis while several important questions remain unresolved. These unanswered questions include side effects, dosage, mode of action, and DFO metabolite studies. One example of this is that the doses quoted for DFO treatment during dialysis have not yet been established on a scientific basis and so range from 0.5 g/week to 6.0 g per dialysis session (105, 131). Another example of inconsistency in this regard is the disagreement as to timing of the peak Al concentration occurrence in blood serum after DFO treatment. At present this has been reported as ranging from 3 to 48 h (105, 131). By ultrafiltration methods it has been shown that the ultrafiltrable fraction increased from 10 to 60-85% of the total after DFO addition to serum and that the total Al concentration increased as well (94, 96). However, ultrafiltration methods do not provide detailed information on the effect of such chelating therapy upon the distribution of Al in serum.

Despite its previously noted lack of resolution, gel chromatography has been used by several workers for this purpose. Bertholf et al. found that after long-term treatment with 2 g of intravenous DFO before each dialysis session, serum Al in hyperaluminemic patients reached a peak of 800 μg L^{-1} (104). The concentration eventually fell to below 100 μg L^{-1} and then the patients' neuropathic symptoms vanished. After DFO treatment, these authors identified an LMW Al-containing fragment as the Al-DFO species (104). These results have been mirrored by Leung et al. using similar techniques and procedures (122). When a Biogel P2 column was employed, the Al profiles obtained clearly showed that after DFO treatment (100 mg kg^{-1} infusion of DFO before dialysis) a shift of Al from HMW complexes to the Al-DFO complex takes place (113).

According to Sanz-Medel's group, who carried out studies of renal failure patients treated with DFO, only Al-transferrin and Al-DFO complexes were positively detected by their HPLC technique (95, 107). A typical chromatogram for serum with added DFO is given in Fig. 7.1 (in Section 7.3.1.2, above); the Al profiles, as obtained by ETA-AAS, are shown in Figs. 7.2 and 7.4, before and after DFO infusion, respectively. From these chromatograms the shift of Al from serum transferrin to the Al-DFO complex induced by DFO therapy is clearly shown. As already stated, no other major Al-binding component was observed when using this technique.

We mentioned earlier that Khalil-Mansh et al., using gel chromatography and gel electrophoresis techniques, found that a significant proportion of the serum of hyperaluminemia patients Al was associated with an 8000 Da component (96, 105). After DFO therapy, they found that this fraction increased, along with the total Al concentration in serum. They speculated about the possibility of formation of a ternary complex between DFO and the "Al-binder" protein in tissues reasoning that this complex might subsequently be released into the circulatory body fluids (105).

It is worth mentioning that parallel work exists on the speciation and identification of DFO, its Fe^{3+}/Al^{3+} complexes, and some of its metabolites. The preferred analytical technique for these studies seems to be HPLC, either

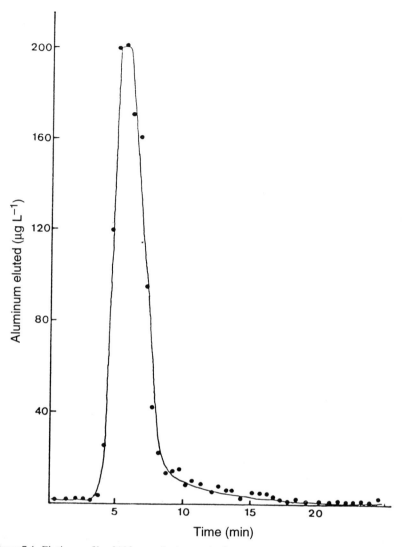

Figure 7.4. Elution profile of Al from spiked serum in the presence of DFO (1000 µg L^{-1} of Al and 2 mg L^{-1} of DFO). Reprinted by permission of Garcia Alonso et al. (107).

silica gel column or reversed-phase HPLC, based on silica-free stationary phases (132, 133). Other indirect methods for DFO speciation analysis include atomic spectrometry techniques and polarographic determinations, although lack of sensitivity seems to be a problem with these latter two approaches (134, 135).

7.4. THE ROLE OF ALUMINUM AND SILICON IN NEUROLOGICAL DISORDERS: THE QUEST FOR SPECIATION

7.4.1. Dialysis Encephalopathy and Other Diseases

The neurotoxicity of Al began to be taken seriously in the mid-1970s, when the DES was described and linked to Al intoxications (136). Water purification procedures and European Community Directives have reduced the incidence of this disorder in renal failure patients undergoing regular hemodialysis (6, 7, 112). The role of Al in this disease has been amply demonstrated, and Al overload can be treated with DFO, the Al-chelating agent producing clear, symptomatic relief of dialysis encephalopathy (128, 137–140). Other neuropathological disorders linked to Al are endemic amyotrophic lateral sclerosis (ALS), Parkinson's disease (PD), and Down syndrome (DS) (6, 14, 141). To establish or disprove the role of Al in any of these neurological disorders, data as to the metal's distribution and speciation in body tissues (particularly in the brain) would be of great importance. Generally, it is now known that LMW species of Al (especially citrate complexes) exhibit the highest bioavailability. However, as has been shown earlier in this chapter, techniques for such Al speciation analysis virtually do not exist at present.

7.4.2. Alzheimer's Disease

The major neuropathology features of AD have been confirmed as neurofibrillary tangles (NFTs) and senile plaques formed in the neurocortical and archicortical regions of the brain. The involvement of Al, as a neurotoxic agent, in the pathogenesis of AD was prompted by the identification of four target sites of accumulation in brain tissue in AD patients: (i) DNA-containing structures in cell nuclei of the central nervous system (CNS); (ii) the protein moieties of the neurofibrillary tangles; (iii) the amyloid core of the senile plaques; and (iv) cerebral ferritin (83, 86, 88–90, 142–145).

Speculation as to the possible connection between Al and Si in the pathogenesis of AD was triggered by some reports of Si and Al co-localization in senile plaque cores (25, 26). Raised concentrations of both Al and Si were also reported in NFT-bearing neurons in AD (88). Klatzo et al. have demonstrated that neurofibrillary degeneration could be induced in rabbits by injecting Al phosphate intracerebrally (146). Further investigations by high-resolution solid-state ^{27}Al nuclear magnetic resonance have confirmed the presence of aluminosilicates in the plaque cores, isolated from the cerebral cortex of chronic AD brains (85, 147). Those species were not detected in the same brain region of healthy individuals. More confirmatory data on this has been provided by energy dispersive X-ray microanalysis of Ag-stained sections and by secondary ion mass spectrometry (148). The presence of aluminosilicates in the central part of senile plaques in chronic AD brains suggests that they have a positive role in the plaque formation processes and, consequently, in the pathogenesis of Alzheimer-type disorders. Because aluminosilicates are very insoluble, this raises the question as to whether they are transported to the brain or are formed inside the CNS of patients with neurological disorders.

7.4.3. Species Favoring Aluminum and Silicon Transport to the Brain

Aluminum-loaded transferrin has been shown to exhibit almost identical binding and uptake characteristics as Fe-loaded transferrin (149). This suggests that Al might also be able to follow the Fe pathway in entering the intracellular environment (8). This has been confirmed by ^{67}Ga and ^{26}Al radiographic techniques whose mapping of the element coincides with the distribution of the transferrin receptors in the brain (6, 147, 150, 151). In addition, the localization of both ^{67}Ga and transferrin receptors was found mainly in those brain regions where a specific enhancement of Al level in chronic AD brains has been observed (147, 152).

Silicic acid is present in all body fluids at low concentrations because the small uncharged ($pK_a = 9.8$) Si(OH)$_4$ molecule is freely diffusible (6). Pullen et al., who used ^{68}Ge as a Si tracer to investigate its entry into rat brain, found that the element's passage was bidirectional with a relatively even regional distribution (152). Those reports, together with the reports on Al transport to the brain, make the *in situ* formation of aluminosilicate species inside the brain the most probable scenario. Iler has shown that the deposition of aluminosilicates is initialized by the polymerization of silicic acid (153). Subsequent adsorption of amyloid proteins on those structures may occur, leading to accumulation of fibrillary aggregates of the type observed in senile plaques (141). This accumulation mechanism is thought to be similar to that responsible for silica-induced fibrogenesis (154). Thus, it seems possible that the formation of aluminosilicate deposits in cerebral cortex may cause the accumulation of A4 amyloid fibrils, which are well-known typical constituents of senile plaques in Alzheimer-type disorders.

However, Farrar et al. found that total plasma Al levels were not elevated in AD subjects when compared to controls (155). They suggested that a difference in the chemical speciation of the metal in blood plasma could be responsible for the rise in AD brain level. Using ^{67}Ga as an Al tracer, they found that significantly less ^{67}Ga is bound to transferrin in AD samples, with a subsequent increase in levels of ^{67}Ga present as LMW species. Similar results had been reported earlier by De Wolff et al. for animal experiments, and the LMW species of Al was identified as a citrate complex (156). These results strongly

Table 7.7. Distinguishing Features of Alzheimer's Disease and Al-Induced Neuropathological Changes

Alzheimer's Disease	Al Intoxication	References
Differences in neurofibrillary tangles (NFTs):		
Occur as a pair of 100 Å filaments wound into a helix	Occur as 100 Å filaments	92, 93, 112
Differences in immunochemical characteristics of NFTs:		
Presence of senile plaques	Absence of senile plaques (macroscopic characteristics of the brain unchanged)	93, 158, 159–161
Selective neuronal vulnerability (lesions mostly localized in the cortical, subcortical, and hippocampal areas)	Widespread lesions (uniform distribution in in the cerebellum, brain stem, and spinal cord)	148
Different changes in the neurotransmitter systems:		
Normal Al levels in serum, hair, and cerebrospinal fluid	Comparatively elevated Al levels	144, 160, 162, 163

indicate that LMW species of Al, rather than its protein-bound form, are involved in the element transfer and, in consequence, accumulation into the brain in neurological disorders (119, 120).

While these observations do indeed back the hypothesis of a positive role for Al in AD, several important differences between AD and Al-induced changes make any causal relationship speculative (14, 143, 157). The main distinguishing features, as reported in the literature, are presented in Table 7.7 (92, 93, 112, 144, 148, 158–163). In brief, apart from the analytical evidence referred to in the foregoing paragraphs, there is no direct experimental proof that demonstrates the implication of Al as an etiological factor in AD.

It should be borne in mind, however, that in experimental Al toxicology the administered metal compounds, or Al toxins, must be precisely identified and able to survive in the biological fluids or in culture media in order to reach their biological target. Moreover, the slow kinetics of Al complexes, particularly hydroxides, at physiological pH (around 7.4) must always be considered because equilibration may require very long—sometimes undefined—periods of time (164).

7.5. DOES SILICIC ACID PREVENT ALUMINUM TOXICITY VIA FORMATION OF ALUMINOSILICATES?

In light of the increased serum Al and Si contents in patients with renal failure and the high affinity of silicic acid for Al (resulting in formation of aluminosilicates), recent reports have suggested a protective role for Si against Al intoxication (165). It has been shown that in model solutions of Al, silicic acid, and citrate, the formation of aluminosilicates is favored above pH = 7.0 (166). Citrate has also been shown to cause the solubilization of aluminosilicates. Birchall and Chappell have suggested that the aluminosilicates may be formed in the extracellular environment under the conditions of poor Al–transferrin binding, low citrate levels, and/or high Al concentration (8). Roberts and Williams reported an increased plasma Si content of chronic renal failure patients, which was correlated with a decreasing glomerular filtration rate (70). This again suggests that Si protects renal-failure patients from possible Al toxicity by the formation of inactive aluminosilicates and that these can exist only in the extracellular environment (8).

Carlisle and Curran investigated the effect of dietary Si and Al supplementation on the brain content of both elements in two groups of rats (167). They found that Si dietary supplementation does not change its brain regional distribution. In younger rats, the high Al and low Si diet did not cause any change in the Al brain content. On the other hand, in elderly rats this diet resulted in an increase of the metal in most brain regions examined (including those involved in AD). In the brains of rats given a high-Si and high-Al diet, there was no observed increase in Al concentration. In light of these findings, the authors suggested that dietary Si supplementation may be protective against Al accumulation in ageing brain tissue (167).

7.6. CONCLUSIONS

Nowadays there is no doubt that accumulation of Al in the human body is correlated with some known pathologies. It is also known that the bioavailability and toxicity of the element depends on its chemical form. However, many questions still remain to be answered before the mechanism of Al function in such diseases and the influence, if any, of Si on Al toxicity are firmly established.

Analytical techniques for Al and Si determination and speciation in clinical samples are needed to appropriately address the biomedical problems still awaiting solution. After the revision of methods for total elemental analysis of Al in serum, the recommended analytical technique that emerges is ETA–AAS; for Si, it is ICP–AES. For both elements, special precautions against contamination (because of their ubiquitous character and low physiological levels) should be taken.

It is clear that the speciation of Al in biological fluids has been fraught with difficulties in the past and is still in a state of development. Although earlier work seems to have been plagued with contamination problems, some sort of consensus has emerged in recent years, based on the following statements:

a. In healthy individuals less than 10% of serum Al is ultrafiltrable.

b. The HMW Al-binding protein in serum seems to be transferrin.

c. The main LMW Al binder in serum seems to be citrate.

d. DFO has a beneficial therapeutic effect on hyperaluminemic patients, linked to its ability to enhance the elimination of Al by forming an LMW Al complex that can be dialyzed.

e. Other chelating agents, e.g., the drug commercially sold as L-1, can form a complex with Al in the body fluids less toxic and more effective than that of DFO for Al detoxification treatments (168).

f. A correlation exists between Al accumulation and the occurrence of neuropathological disorders, although the role of Al in the pathogenesis of those diseases is still unclear.

g. LMW species of Al seem to be of the highest bioavailability, and hence toxicity, in humans.

h. Silicon, by interacting with aluminum in the physiological environment, may affect its toxicity and is clearly associated with Al in AD.

It is thought that LMW–Al species, because of their high bioavailability, are responsible for enhanced Al transfer and accumulation in neuropathological brains. Separation/determination methods that can distinguish between different LMW–Al toxins, e.g., Al citrate, $AlPO_4$, and "free" and hydroxy Al species, are urgently needed (169). The benefits of DFO treatment in the elimination of Al during dialysis are well established. Important questions remain unanswered concerning such treatments, however, and their solutions require more speciation work.

Regarding the hypothesis of possible Si influence on Al action in neurological diseases, further supporting evidence is needed via the development of adequate analytical techniques for simultaneous speciation of Al and Si (e.g., aluminosilicates) in clinical samples.

Finally, quality assurance strategies and production of RMs for Al species in biological tissues (especially in serum to start with) would not only help to end a lot of the controversies still plaguing this type of analysis but would also facilitate the routine determination of the more toxic Al species in clinical laboratories.

REFERENCES

1. R. C. Massey and D. Taylor, eds., *Aluminium in Food and the Environment.*, Royal Society of Chemistry, London, 1988.
2. E. Berman, in *Toxic Metals and Their Analysis* (L. C. Thomas, ed.). Heyden, London, 1980.
3. P. C. D'Haese, *Aluminium Accumulation in Patients with Chronic Renal Failure. Monitoring, Diagnosis and Therapy*. Akademisch Proefschrift, Amsterdam and Antwerp, 1988.
4. P. Brätter and P. Schramel, eds., *Trace Element Analytical Chemistry in Medicine and Biology*, Vol. 2. de Gruyter, Berlin and New York, 1983.
5. J. D. Cowburn, G. Farrar, and J. A. Blair, *Chem. Br.* **26**, 1169 (1990).
6. P. O. Ganrot, *Environ. Health Perspect.* **65**, 363 (1986).
7. *Off. J. Eur. Commun*, C/184/16 (1986).
8. J. D. Birchall and J. S. Chappell, *Lancet* **2**, 1008 (1988).
9. R. E. Viola, J. F. Morrison, and W. W. Cleland, *Biochemistry* **19**, 3131 (1980).
10. J. C. Lai and J. P. Blass, *J. Neurochem.* **42**, 438 (1984).
11. R. K. Sihag, A. Y. Jeng, and R. A. Nixon, *FEBS Lett.* **233**, 181 (1988).
12. R. L. Bertholf, *Crit. Rev. Clin. Lab. Sci.* **25**, 195 (1987).
13. U. De Boni, *Xenobiotica* **15**, 643 (1985).
14. D. R. McLachlan, W. J. Lukiw, and T. P. Kruck, *Can. J. Neurol. Sci.* **16** (Suppl. 4), 490 (1989).
15. S. J. Karlik and G. L. Eichhorn, *J. Inorg. Biochem.* **37**, 259 (1989).
16. M. R. Willis and J. Savory, *Lancet* **2**, 29 (1983).
17. J. Y. Corey, E. R. Corey, and P. P. Gaspar, eds., *Silicon Chemistry*. Ellis Horwood, Chichester, UK, 1988.
18. J. D. Birchall, *Chem. Br.* **26**, 141 (1990).
19. E. Frieden, ed., *Biochemistry of the Essential Ultratrace Elements*. Plenum, New York and London, 1984.
20. G. M. Berlyne, E. Dudek, A. J. Adler, J. E. Rubin, and M. Seidman, *Kidney Int.* **28** (Suppl. 17), S175 (1985).
21. S. Hosokawa, A. Oyamaguchi, and O. Yoshida, *Nephron* **55**, 375 (1990).
22. A. J. Adler and G. M. Berlyne, *Nephron* **44**, 23 (1986).
23. G. M. Berlyne, A. J. Adler, N. Ferran, S. Bennett, and J. Holt, *Nephron* **43**, 5 (1986).
24. C. O. Hershey, L. A. Hershey, A. Varnes, S. D. Vibhakar, P. Lavin, and W. H. Strain, *Neurology* **33**, 1350 (1983).
25. C. Masters, G. Multhaup, G. Simms, J. Pottgiesser, R. Martins, and K. Beyreuther, *Eur. Mol. Biol. Organ. J.* **4**, 2757 (1985).
26. J. M. Candy, J. A. Edwardson, J. Klinowski, A. E. Oakley, E. K. Perry, and R. H. Perry, in *Senile Dementia of Alzheimer Type* (J. Traber and W. H. Gispen, eds.). Springer-Verlag, Berlin, 1985.

27. A. Sanz-Medel, R. Rodriguez Roza, R. González Alonso, A. Noval Vallina, and J. Cannata, *J. Anal. At. Spectrom.* **2**, 177 (1987).
28. J. Versieck and R. Cornelis, *Trace Elements in Human Plasma or Serum.* CRC Press, Boca Raton, FL, 1989.
29. J. Versieck, J. Hoste, L. Vanballenberghe, A. DeKesel, and D. Van Renterghem, *J. Radioanal. Nucl. Chem.* **113**, 299 (1987).
30. J. Versieck, L. Vanballenberghe, A. DeKesel, J. Hoste, B. Wallaeys, J. Vandenhaute, N. Baeck, H. Steyaert, A. R. Byrne, and F. W. Sunderman, *Anal. Chim. Acta* **204**, 63 (1988).
31. J. Savory and M. R. Wills, *Kidney Int.* **29**(Suppl. 18), S24 (1986).
32. K. Garmestani, A. J. Blotsky, and E. P. Rack, *Anal. Chem.* **50**, 144 (1978).
33. H. Nakahara, Y. Nagame, Y. Yoshizawa, H. Oda, S. Gotoh, and Y. Murakami, *J. Radioanal. Chem.* **54**, 183 (1979).
34. J. I. García Alonso, A. Lopez García, A. Sanz-Medel, and E. Blanco González, *Anal. Chim. Acta* **225**, 339 (1989).
35. H. M. Crews, D. J. Halls, and A. J. Taylor, *J. Anal. At. Spectrom.* **5**, 75R (1990).
36. H. Matusiewicz and R. M. Barnes, *Spectrochim. Acta* **39B**, 891 (1984).
37. A. A. Brown, D. J. Halls, and A. Taylor, *J. Anal. At. Spectrom.* **2**, 43R (1987).
38. M. R. Pereiro García, A. Lopez Garcia, M. E. Diaz Garcia, and A. Sanz-Medel, *J. Anal. At. Spectrom.* **5**, 15 (1990).
39. K. S. Subramanian, *Prog. Anal. At. Spectrosc.* **9**, 237 (1986).
40. W. Slavin, *J. Anal. At. Spectrosc.* **1**, 281 (1986).
41. M. Bettinelli, U. Baroni, F. Fontana, and P. Poisetti, *Analyst (London)* **110**, 19 (1985).
42. J. R. Andersen and S. Reimert, *Analyst (London)* **111**, 657 (1986).
43. S. W. King, M. R. Wills, and J. Savory, *Anal. Chim. Acta* **128**, 221 (1981).
44. F. Y. Leung and A. R. Henderson, *Clin. Chem. (Winston-Salem, N.C.)* **28**, 2139 (1982).
45. S. Brown, R. L. Bertholf, M. R. Wills, and J. Savory, *Clin. Chem. (Winston-Salem, N.C.)* **30**, 1216 (1984).
46. F. R. Alderman and H. J. Gitelman, *Clin. Chem. (Winston-Salem, N.C.)* **26**, 258 (1980).
47. H. J. Gitelman and F. R. Alderman, *Clin. Chem. (Winston-Salem, N.C.)* **35**, 1517 (1989).
48. K. B. Pierson and M. A. Evenson, *Anal. Chem.* **58**, 1744 (1986).
49. A. A. Bouman, A. J. Plantenkamp, and F. D. Posma, *Ann. Clin. Biochem.* **23**, 97 (1986).
50. E. M. Skelly and F. T. Di Stefano, *Appl. Spectrosc.* **42**, 1302 (1988).
51. P. C. D'Haese, F. L. Van de Vyver, F. A. De Wolff, and M. E. De Broe, *Clin. Chem. (Winston-Salem, N.C.)* **31**, 24 (1985).
52. O. Guillard, K. Tiphaneau, D. Reiss, and A. Piriou, *Anal. Lett.* **17**, 1593 (1984).

53. M. K. Ward, T. G. Feest, H. A. Ellis, I. S. Parkinson, and D. N. S. Kerr, *Lancet* **2**, 841 (1978).
54. J. Versieck and R. Cornelis, *Anal. Chim. Acta* **116**, 217 (1980).
55. P. E. Gardiner and J. M. Ottaway, *Anal. Chim. Acta* **128**, 58 (1981).
56. W. Frech, A. Cedergren, C. Cederberg, and J. Vessman, *Clin. Chem. (Winston-Salem, N.C.)* **28**, 2259 (1982).
57. Y. Suzuki, T. Kamiki, and S. Imai, *Analyst (London)* **114**, 839 (1989).
58. Z. Q. Shan, S. Luan, and Z. M. Ni, *J. Anal. At. Spectrom.* **3**, 99 (1988).
59. D. R. Crapper, S. S. Krishnan, and S. Quittkat, *Brain* **99**, 67 (1976).
60. J. R. McDermott, A. I. Smith, M. K. Ward, I. S. Parkinson, and D. N. S. Kerr, *Lancet* **2**, 901 (1978).
61. I. S. Parkinson, M. K. Ward, and D. N. S. Kerr, *J. Clin. Pathol.* **34**, 1285 (1981).
62. M. E. De Broe, F. L. Van de Vyver, F. J. E. Silva, P. C. D'Haese, and A. H. Verbeuken, *Nefrologia* **6**, 42 (1986).
63. E. J. King, *Analyst (London)* **80**, 441 (1955).
64. G. P. Sighinolfi, C. Gorgoni, O. Bonori, E. Cantoni, M. Martelli, and L. Simonetti, *Mikrochim. Acta* **1**, 171 (1989).
65. H. J. Gitelman and F. R. Alderman, *J. Anal. At. Spectrom.* **5**, 687 (1990).
66. F. E. Lichte, S. Hopper, and T. W. Osborn, *Anal. Chem.* **52**, 120 (1980).
67. T. Tanaka and T. Hayashi, *Clin. Chim. Acta* **156**, 109 (1986).
68. X. Wang, A. Lásztity, M. Viczian, Y. Israel, and R. M. Barnes, *J. Anal. At. Spectrom.* **4**, 727 (1989).
69. A. Lásztity, X. Wang, M. Viczian, Y. Israel, and R. M. Barnes, *J. Anal. At. Spectrom.* **4**, 737 (1989).
70. N. B. Roberts and P. Williams, *Clin. Chem. (Winston-Salem, N.C.)* **36**, 1460 (1990).
71. M. S. Goligorsky, C. Chaimovitz, Y. Nir, R. Rappoport, R. Kol, and J. Yehuda, *Miner. Electrolyte Metab.* **11**, 301 (1985).
72. S. Galassini, *Nucl. Instrum. Methods Phys. Res. Sect. B* **43**, 556 (1989).
73. T. H. McGavach, J. G. Leslie, and K. Y. T. Kao, *Proc. Soc. Exp. Biol. Med.* **110**, 215 (1962).
74. S. Indprasit, G. V. Alexander, and H. C. Gonick, *J. Chronic Dis.* **27**, 135 (1974).
75. J. W. Dobbie and M. J. B. Smith, *Scott. Med. J.* **27**, 17 (1982).
76. J. W. Berlyne and C. Caruso, *Clin. Chim. Acta* **129**, 239 (1983).
77. J. Parantainen, E. Tenhumen, R. Kangasuiemi, S. Sankari, and F. Altroshi, *Vet. Res. Commun.* **11**, 467 (1989).
78. W. Frech and A. Cedergren, *Anal. Chim. Acta* **113**, 227 (1980).
79. G. Müller-Vogt and W. Wendl, *Anal. Chem.* **53**, 651 (1981).
80. M. Taddia, *J. Anal. At. Spectrom.* **1**, 437 (1986).
81. J. P. Parajon and A. Sanz-Medel, *J. Anal. At. Spectrom.* **9**, 111 (1994).
82. J. A. Rawa, E. L. Henn, *Anal. Chem.* **51**, 452 (1979).

83. S. Kobayashi, S. Fujiwara, S. Arimoto, H. Koide, J. Fukuda, K. Shimode, S. Yamagushi, K. Okada, and T. Tsunematsu, *Prog. Clin. Biol. Res.* **317P**, 1095 (1989).
84. R. W. Jacobs, T. Duong, R. E. Jones, G. A. Trapp, and A. B. Scheibel, *Can. J. Neurol. Sci.* **16** (Suppl. 4), 498 (1989).
85. J. A. Edwardson, J. Klinowski, A. E. Oakley, R. H. Perry, and J. M. Candy, *Ciba Found. Symp.* **121P**, 160 (1986).
86. M. E. De Broe, and J. W. Coburn, eds., *Aluminium and Renal Failure.* Kluwer Acad. Publ., Dordrecht, Boston, and London, 1990.
87. S. Kobayashi, N. Hirota, K. Saito, and M. Utsuyonna, *Acta Neuropathol.* **74**, 47 (1987).
88. D. P. Perl and A. R. Brody, *Science* **208**, 297 (1980).
89. D. P. Perl and W. W. Pendlebury, *Can. J. Neurol. Sci.* **13** (Suppl. 4), 441 (1986).
90. D. Senitz and K. Bluthner, *Zentralbl. Allg. Pathol. Pathol. Anat.* **136**, 329 (1990).
91. G. B. Van der Voet, E. Marani, S. Tio, and F. A. de Wolff, *Prog. Histochem. Cytochem.* **23** (1991).
92. D. Munoz-García, W. W. Pendlebury, J. B. Kessler, and D. P. Perl. *Acta Neuropathol.* **70**, 243 (1986).
93. D. Langui, B. H. Anderton, J. P. Brion, and J. Ulrich, *Brain Res.* **438**, 67 (1988).
94. J. Pérez Parajon, E. Blanco Gonzalez, J. Cannata, and A. Sanz-Medel, *Trace Elem. Med.* **6**, 41 (1989).
95. E. Blanco González, J. Pérez Parajón, J. I. García Alonso, and A. Sanz-Medel, *J. Anal. At. Spectrom.* **4**, 175 (1989).
96. F. Khalil-Manesh, C. Agness, and H. C. Gonick, *Nephron* **52**, 329 (1989).
97. H. Rahman, A. W. Skillen, S. M. Chammon, M. K. Ward, and D. N. S. Kerr, *Clin. Chem. (Winston-Salem, N.C.)* **31**, 1969 (1985).
98. H. L. Elliot, A. I. Macdougall, G. S. Fell, and P. H. E. Gardiner, *Lancet* **2**, 1255 (1978).
99. A. P. Lundin, C. Caruso, M. Sass, and G. M. Berlyne, *Clin. Res.* **26**, 63 (1978).
100. F. Y. Leung, A. B. Hodsman, N. Muirheud, and A. R. Henderson, *Clin. Chem. (Winston-Salem, N.C.)* **31**, 20 (1985).
101. H. Graf, H. K. Stummvoll, V. Meisinger, J. Korarik, A. Wolff, and W. F. Purggera, *Kidney Int.* **19**, 582 (1981).
102. H. Keirsse, J. Smeyers-Verbecke, D. Verbeden, and D. L. Massart, *Anal. Chim. Acta* **196**, 103 (1987).
103. L. Canavese, C. Pramotton, A. Pacilti, G. Segoloni, S. Bedino, G. Testone, S. Lamon, and S. Talaino, *Trace Elem. Med.* **3**, 93 (1986).
104. R. L. Bertholf, J. Savory, and M. R. Wills, *Trace Elem. Med.* **3**, 157 (1986).
105. F. Khalil-Manesh, C. Agness, and H. C. Gonick, *Nephron* **52**, 323 (1989).
106. R. L. Bertholf, M. R. Wills, and J. Savory, *Clin. Physiol. Biochem.* **3**, 271 (1985).

107. J. I. García Alonso, A. Lopez García, J. Pérez Parjón, E. Blanco, González, A. Sanz-Medel, and J. Cannata, *Clin. Chim. Acta* **189**, 69 (1990).
108. L. D. Hewitt and J. P. Day, *Trace Elem. Health Dis., Ext. Abstr., Nord. Symp. 2nd*, p. 47 (1987).
109. H. Rahman, S. M. Chammon, A. W. Skillen, M. K. Ward, and D. N. S. Kerr, *Proc. EDTA-ERA* **21**, 360 (1989).
110. W. Kaehny, A. Alfrey, R. Holman, and W. Shorr, *Kidney Int.* **101**, 775 (1984).
111. M. R. Pereiro García, M. E. Diaz García, and A. Sanz-Medel, *J. Anal. At. Spectrom.* **2**, 699 (1987).
112. S. W. King, J. Savory, and M. R. Wills, *Crit. Rev. Chim. Lab. Sci.* **14**, 1 (1981).
113. G. A. Trapp, *Life Sci.* **33**, 311 (1983).
114. L. Ebdon, S. Hill, and P. Jones, *Analyst (London)* **112**, 437 (1987).
115. M. Cochran, J. H. Coates, and T. Kurucser, *Life Sci.* **40**, 2337 (1987).
116. R. B. Martin, J. Savory, S. Brown, R. L. Bertholfe, and M. R. Wills, *Clin. Chem. (Winston-Salem, N.C.)* **33**, 405 (1987).
117. R. B. Martin, *Clin. Chem. (Winston-Salem, N.C.)* **32**, 1797 (1986).
118. W. R. Harris and J. Sheldon, *Inorg. Chem.* **29**, 119 (1990).
119. P. Slanina, W. Frech, L. G. Ekström, L. Loof, S. Storach, and A. Cedergren, *Clin. Chem. (Winston-Salem, N.C.)* **32**, 539 (1986).
120. R. B. Martin, *J. Inorg. Biochem.* **28**, 181 (1986).
121. P. M. Bertsch and M. A. Anderson, *Anal. Chem.* **61**, 535 (1989).
122. F. Y. Leung, A. E. Nibloch, L. Bradley, and A. R. Henderson, *Sci. Total Environ.* **71**, 49 (1988).
123. A. K. Datta, P. J. Wedlund, and R. A. Yokel, *J. Trace Elem. Electrolytes Health Dis.* **4**, 107 (1990).
124. M. Venturini and G. J. Berthon, *J. Inorg. Biochem.* **37**, 69 (1989).
125. M. Venturini and G. J. Berthon, *J. Chem. Soc., Dalton Trans.*, 1145 (1987).
126. M. Venturini, G. J. Berthon, and P. M. May, *Recl. Trav. Chim. Pay-Bas* **106**, 406 (1987).
127. D. Sandrine, M. Fillela, and G. J. Berthon, *J. Inorg. Biochem.* **38**, 241 (1990).
128. P. Ackrill, A. J. Ralston, J. P. Day, and K. C. Hodge, *Lancet* **2**, 692 (1980).
129. D. Martin, *Chem. Br.* **27**, 689 (1991).
130. R. M. Bannermann, S. T. Callender, and D. L. Williams, *Br. Med. J.* **2**, 1573 (1962).
131. B. Winterberg, H. P. Bertram, A. E. Lison, L. Spieher, E. Schalthofer, H. Raidt, and H. Zumkley, *Trace Elem. Med.* **3**, 95 (1986).
132. T. P. A. Kruck and W. Kalow, *J. Chromatogr. (Biomed. Appl.)* **341**, 123 (1985).
133. H. B. Jenny and H. H. Peter, *J. Chromatogr.* **438**, 433 (1988).
134. P. C. D'Haese, L. V. Lamberts, and M. E. De Broe, *Clin. Chem. (Winston-Salem, N.C.)* **35**, 884 (1989).
135. R. A. Romero and J. P. Day, *Trace Elem. Med.* **2**, 1 (1985).

136. A. C. Alfrey, G. R. Le Gendre, and W. D. Kaehny, *N. Engl. J. Med.* **294**, 184 (1976).
137. J. P. Masselot, J. P. Adhemar, M. C. Jaudon, D. Kleinknecht, and A. Galli, *Lancet* **2**, 1386 (1978).
138. A. Buge, M. Poisson, S. Masson, J. M. Bliebel, R. Mashaly, M. C. Jaudon, B. Lafforgue, M. Lebkiri, and P. Raymond, *Nouv. Presse Med.* **8**, 2729 (1979).
139. R. S. Arze, I. S. Parkinson, N. E. F. Cartlidge, P. Britton, and M. K. Ward, *Lancet* **2**, 1116 (1981).
140. F. J. Milne, B. Sharf, P. D. Bell, and A. M. Meyers, *Lancet* **2**, 502 (1982).
141. J. M. Candy, A. E. Oakley, D. Gauvreau, P. Chalker, H. Bishop, D. Moon, G. Staines, and J. A. Edwardson, *Interdiscip. Top. Gerontol.* **25**, (1988).
142. D. R. McLachlan, *Neurobiol. Aging* **7**, 525 (1986).
143. S. S. Krishnan, D. R. McLachlan, B. Krishnan, S. S. Fenton, and J. E. Harrison, *Sci. Total Environ.* **71**, 59 (1988).
144. D. Shore and R. J. Wyatt, *J. Nerv. Ment. Dis.* **171**, 553 (1983).
145. J. M. Candy, A. E. Oakley, J. Klinowski, T. A. Carpenter, R. H. Perry, J. R. Atack, E. K. Perry, G. Blessed, A. Fairbairn, and J. A. Edwardson, *Lancet* **2**, 354 (1986).
146. I. Klatzo, H. Wisniewski, and E. Streicher, *J. Neuropathol. Exp. Neurol.* **24**, 187 (1965).
147. J. M. Candy, J. A. Edwardson, R. Faircloth, A. B. Keith, C. M. Morris, and R. G. L. Pullen, *J. Physiol. (London)* **392**, 34P (1987).
148. V. W. Henderson and C. E. Finch, *J. Neurosurg.* **70**, 335 (1989).
149. C. M. Morris, J. M. Candy, J. A. Court, C. A. Whitford, and J. A. Edwardson, *Biochem. Soc. Trans.* **15**, 498 (1987).
150. M. Cochran, S. Neoh, and E. Stephens, *Clin. Chim. Acta* **132**, 199 (1983).
151. J. P. Day, J. Barker, L. J. A. Evans, J. Perks, P. J. Seabright, P. Ackrill, J. S. Lilley, P. V. Drumm, and G. W. A. Newton, *Lancet* **2**, 1345 (1991).
152. R. G. L. Pullen, J. M. Candy, C. M. Morris, G. Taylor, A. B. Keith, and J. A. Edwardson, *J. Neurochem.* **55**, 251 (1990).
153. R. Iler, *The Chemistry of Silica: Solubility, Polymerization, Colloid and Surface Properties and Biochemistry.* Wiley, New York, 1979.
154. J. Schmidt, C. Oliver, J. Lepe-Zuniga, I. Green, and I. Gery, *J. Clin. Invest.* **73**, 1462 (1984).
155. G. Farrar, P. Altman, S. Welch, O. Wychrij, B. Ghose, J. Lejeune, J. Corbett, V. Prasher, and J. A. Blair, *Lancet* **2**, 747 (1990).
156. F. A. De Wolff, M. F. van Ginkel, A. Brandsma, and G. B. Van der Voet. *Clin. Chem. (Winston-Salem, N.C.)* **34**, 1305 (1988).
157. H. Suzuki, M. Takeda, and Y. Nakamura, *Neurosci. Lett.* **89**, 234 (1988).
158. W. W. Pendlebury, M. F. Beal, N. W. Kowall, and P. R. Solomon, *J. Neural Transm.* 24P (Suppl.), 213 (1987).
159. W. W. Pendlebury, M. F. Beal, N. W. Kowall, and P. R. Solomon, *Neurotoxicology* **9**, 503 (1988).

160. C. L. Sholtz, M. Swash, and A. Gray, *Clin. Neuropathol.* **6**, 93 (1987).
161. J. T. Hughes, *Lancet* **2**, 490 (1989).
162. N. W. Kowall, W. W. Pendlebury, and J. B. Kessler, *Neuroscience* **29**, 329 (1989).
163. M. F. Beal, M. F. Mazurek, and D. W. Ellison, *Neuroscience* **29**, 339 (1989).
164. B. Corain, A. Tapparo, A. A. Sheikh-Osman, and G. G. Bombi, *Coord. Chem. Rev.* **112**, 19 (1992).
165. J. W. Dobbie and J. M. B. Smith, *Ciba Found. Symp.* **121P**, 194 (1986).
166. J. S. Chappell and J. D. Birchall, *Inorg. Chim. Acta* **153**, 1 (1988).
167. E. M. Carlisle and J. M. Curran, *Alzheimer Dis. Assoc. Disord.* **1**, 83 (1987).
168. J. G. Goddard and G. J. Kontoghiorghes, *Clin. Chem. (Winston-Salem, N.C.)* **36**, 5 (1990).
169. B. Corain, M. Nicolini, and P. Zatta, *Coord. Chem. Rev.* **112**, 33 (1992).

CHAPTER

8

SPECIATION OF TRACE ELEMENTS IN MILK BY HIGH-PERFORMANCE LIQUID CHROMATOGRAPHY COMBINED WITH INDUCTIVELY COUPLED PLASMA ATOMIC EMISSION SPECTROMETRY

E. CONI, A. ALIMONTI, A. BOCCA, F. LA TORRE, D. PIZZUTI, AND S. CAROLI

Istituto Superiore di Sanità, 00161 Rome, Italy

8.1. INTRODUCTION

Since essential trace elements are clearly implicated in numerous biochemical processes, there is an increasing interest in the study of their ensuing health effects (1). In this context, more and more attention is being given to the role of trace elements in the field of human etiology as adequate instrumental techniques become available that incorporate the most recent technological innovations and as knowledge quickly develops about the physiological role of trace elements.

In spite of the progress made in identifying the role played by elements in human nutrition and pathology, it is extremely difficult to achieve clinical evidence of a state of deficiency. Specific symptoms of deficiency of single elements are usually not unequivocal. In addition, there are very few laboratory tests suitable for this type of diagnosis. Presently, the number of deficiencies of trace elements that are clearly related to pathological processes in humans is quite small (e.g., Cr, Cu, Fe, I, Mn, Mo, Se, and Zn).

It is also worth recalling that while deficiencies of essential elements may impair human health, so may exposure to potentially toxic elements such as Al, Cd, Hg, and Pb (2). It is recognized that such elements, being natural components of the environment, are involved in biogeochemical cycles that cause their mobilization. Anthropic activities, in turn, often involve displacement of toxic inorganic agents that results in environmental pollution and

---

Element Speciation in Bioinorganic Chemistry, edited by Sergio Caroli.
Chemical Analysis Series, Vol. 135.
ISBN 0-471-57641-7 © 1996 John Wiley & Sons, Inc.

human exposure. This ever-increasing presence in various ecosystems implies that the levels beyond which such agents are considered to have effects on the human organism are often reached and surpassed. This is also true in the case of essential elements, the levels of which must not pass a maximum threshold. When highly concentrated they can have negative effects on health, causing pathological alterations of homeostatic mechanisms that normally are able to maintain the presence of elements to within physiological limits.

Consequently, it is desirable that the intake of trace elements be adequate and balanced and it is also necessary not to exceed the no-effect levels of toxic elements. For this purpose, some specific areas of health and scientific interest may be identified that have some priority in food chemistry. In this context, the nutritional quality of the food given in early childhood plays a very important role. Children (infants, and particularly newborns) have a high body growth rate and are therefore more sensitive than adults to potential imbalances in the intake of essential elements. This is a direct consequence of a faster metabolism, different body composition, and a greater need for energy per mass unit of body. For the same reasons, these children also exhibit a higher absorbing potential of toxic elements (3).

As regards early childhood, milk—whether human, cow, or formula milk—is no doubt the most important food since it is the only source of nutrition during the first months of a baby's life. It is therefore essential that all macronutrients (protides, glucides, and lipids) in general and all micronutrients (vitamins and elements) in particular are present in milk in adequate quantities. This is particularly true for micronutrients such as some essential elements that are not stored by the fetus during its growth inside the uterus. From this standpoint, essential elements can be divided into two groups: those such as Cu and Fe, with reserves usually sufficient to protect the baby from potential deficiencies during the first 4–6 months, and those such as Se and Zn, which require immediate reintegration right from birth so as to reach the best rate of growth (4).

This clearly shows why the determination of trace elements in milk has stimulated increasing interest and now represents an important sector of inorganic biochemistry. Despite the considerable amount of research performed in this field, there is very little nationally based, reliable, and updated information on reference concentration values in milk, and data are totally lacking at the international level regarding the various chemical forms in which trace elements may appear. Over the last few years, however, it has become increasingly obvious that the mere assessment of the total quantity of an analyte in the sample tested was no longer sufficient. Rather, researchers have been focusing on the qualitative and quantitative determination of its various forms—in brief, its speciation—according to the different biological activities, toxicity, and ecotoxicity of the forms themselves.

In general, therefore, it is highly important to know the chemical forms of a given element in the biological matrices under examination. This concept is clearer if the functional aspects linked to the presence of the elements in the organism are taken into account. For example, if Zn is linked to albumin, macroglobulins, enzymes, or low-molecular-weight ligands like citrates, the features and biological meanings that it assumes vary greatly from case to case. The determination of Fe linked to hemoglobin in blood plays a diagnostic role that is completely different from the determination of Fe linked to serum ferritin. In addition, the total concentration of Co in serum may be of interest in occupational exposure, whereas only the aliquot present in vitamin B_{12} is specifically of biological importance.

Hence, the need arises to detect chemical species that can act as potential markers of the nutritional state of the organism. So far, these species have been identified for a very limited number of elements. On the other hand, the speciation of trace elements in food matrices represents the complementary aspect of the question. The relationship between the amount of an element that, once absorbed, causes a certain effect on the organism and the quantity introduced, i.e., bioavailability, strongly depends on the chemical form of the element when ingested (5, 6). In fact, among the factors influencing bioavailability, in addition to intrinsic factors such as the animal's species, sex, age, genetic anomalies, physiological and nutritional status, and pathological status, there are also external or food-related factors, namely, chemical species and their physical, chemical, and biochemical features. The validity of this statement is apparent if one considers, e.g., the metabolic fate of Fe. Absorption at the gastrointestinal level of this element in the form of a heme compound is high and is not compromised by possible interactions with other diet components. On the contrary, if this metal appears in a nonhemetic form, absorption drops dramatically and is affected by other diet components. In this regard it should be noted that the bioavailability of many elements depends, among others, on their oxidation state. The Fe^{2+} ion is absorbed more rapidly and easily than Fe^{3+}, whereas this is not true for Se, for which the more oxidized form Se(VI) is the one best absorbed. If it is borne in mind that the organism is capable of oxidizing and reducing only a few essential elements to biologically active forms, it can easily be grasped how important it is to know in which oxidation state they exist in the system being considered, this representing a particular case of speciation.

One of the most pressing recent demands upon analytical chemistry as related to trace elements is the increasing need for highly flexible methods to verify the nature of the different chemical species under which a given element may appear in a particular system. In the following sections, we shall thus aim at verifying whether it is expedient for the purpose of speciation studies to combine inductively coupled plasma–atomic emission spectrometry (ICP–

AES) and high-performance liquid chromatography (HPLC). The analytical features of ICP–AES (multielemental capabilities, high detection power, wide dynamic range, reduced dependence on matrix composition) ideally match the excellent HPLC fractionating features. Moreover, because of the very good compatibility of the two systems in terms of sample flux, this combination appears to be an advanced research tool with high potentialities still to be fully exploited. In section 8.2, the instrumental and operational aspects are mainly considered in order to verify the performance of the combined technique and to extend its use. Then, in Section 8.3, we evaluate the results of applying the HPLC/ICP–AES coupled technique to the speciation of trace elements that are of environmental, toxicological, clinical, and nutritional interest in various kinds of milk. In particular, this method is meant to be used to detect the chemical forms under which some elements of primary importance, such as Al, Ba, Cd, Co, Cr, Cu, Fe, Mg, Mn, Ni, Pb, Se, Sr, and Zn, are present in human, cow, and formula milk and to compare the distributions obtained with parameters identifying the nutritional potential and ensuing toxicological risks.

8.2. EXPERIMENTAL WORK

8.2.1. General Aspects

So far, trace element analysis in food matrices has mainly been aimed at jointly obtaining high detection power and a high degree of specificity. Much more neglected has been the investigation of the chemical forms under which the elements are found. A significant exception has been the speciation analysis of Fe, which has been detected in the form of heme in many foods of animal origin, as reported by various studies in which the chromatographic and spectrometric methods were combined (7–9).

Gel chromatography has been particular useful for the separation of elements linked in the form of complexes with high-molecular-weight ligands. The performance of this method can be improved if radioactive isotopes are used. In this manner, the proteic complexes under which Zn and Cd are present in horse liver and kidneys and in oysters and crabs have been isolated, as well as the complexes formed by these metals with milk proteins (10–14). With use of the same technique, Cr has been found in many vegetables in the form of very stable anionic complexes (15). In addition, dialysis, ultrafiltration, and gel electrophoresis (GE) have been used for this same purpose (13, 16).

However, in the case of low-molecular-weight compounds, separation has been more conveniently achieved by using thin-layer chromatography (TLC), ion-exchange chromatography (IEC), gas chromatography (GC), or HPLC (12, 16–18). Organomercuric compounds have been separated by TLC or GC, preferably after their transformation in dithizonates (19). In order to separate an organic form from an inorganic one, extraction by solvent can sometimes be used. This applies, e.g., in the case of methylmercury, which represents the main form under which Hg is present in seafood, as the compound can be extracted in an acid medium by benzene or toluene and then determined by GC or atomic absorption spectrometry (AAS) (20–22).

In many cases, speciation studies are based on differences in chemical behavior (19, 20, 22). Sometimes, the organic fraction is determined by the difference between the total amount and the inorganic fraction of a given element. The strongly bound aliquot may be obtained by comparison of difference between the total and the "weak" (ionic and not complexed) fraction, as measured by anodic stripping voltammetry (ASV) or by differential pulse anodic stripping voltammetry (DPASV) (20, 23). In addition to the usual colorimetric tests, the oxidation state of the element can also be verified by measuring the magnetic sensitivity and by using electrochemical methods. The latter methods, such as polarography and potentiometry, together with the ASV and DPASV techniques, have made valuable contributions to trace element speciation (23).

As previously mentioned, for our experimental purposes the combined HPLC/ICP–AES technique was chosen to ascertain and quantify the element–protein associations in milk. The HPLC features, in fact, enable a direct link to be made to an ICP–AES spectrometer, thus creating a powerful instrument for speciation studies. In this study both techniques performed their usual roles as, respectively, an analytical instrument with high detection power and a separation technique of great resolution (24–26). A detailed description of the instruments used and the working conditions adopted is given in the following subsections.

8.2.2. Sampling Procedure

The following types of milk samples were analyzed: (i) human milk (colostrum) (10 samples); (ii) human milk (mature) (10 samples); (iii) formula milk (10 samples); (iv) raw cow milk (10 samples); (v) pasteurized skim cow milk (10 samples); and (vi) ultrahigh-temperature(UHT)-sterilized skim cow milk (10 samples).

Great care was given to this phase of the analytical process since any contamination, loss, or alteration phenomena in specimens would nullify the reliability of the results. To do this all samples were collected by ad hoc trained personnel who used carefully cleaned collection vessels. For transport to the laboratory, samples were then transferred to chemically decontaminated polyethylene flasks. These last were indelibly marked with an appropriate code number using a felt-tipped pen, sealed, and kept at $-20\,°C$ until determinations were performed.

In particular, a strict working protocol was established for sampling human and cow milk. As regards the former, first the nipple and surrounding aureole were cleaned with soft cottonwool and double-distilled water. The operator wore disposable talc-free gloves in order to minimize the possibility of contamination at this stage. Each time a sample was taken, about 20 g of milk were withdrawn using a glass aspirating syringe previously decontaminated by repeated rinsing with diluted solutions of HNO_3, care being taken to avoid touching (by any means) the inner walls of the glass device. In all cases detailed information on general lifestyle and dietary habits of the mothers was collected. For this purpose an inquiry form was compiled to be completed by medical personnel during the sample collection, thus identifying and immediately excluding anomalous individuals, or enabling specific factors emerging at a later stage to be traced back to their causes. Apart from some obvious modifications, the same procedure was adopted for raw cow milk.

On the other hand, as regards formula, pasteurized, and UHT-sterilized skim cow milk, samples were obtained directly from the usual commercial suppliers and stored at $+5\,°C$ until analysis.

8.2.3. Determination of the Total Element Concentrations

8.2.3.1. Sample Treatment

It is well known that elements are present in biological matrices in the form of compounds with organic substances such as carbohydrates, proteins, sterols, and enzymes, or in the form of compounds bound directly to carbon atoms. Since the mass ratio between the organic substance and the element is extremely high, in order to perform the analysis, regardless of the method employed, it is necessary in most cases to destroy the organic matrix or to extract the elements to be analyzed prior to their complexation (27).

Matrix decomposition methods can be divided into two main categories known as *wet ashing* and *dry ashing*. With wet ashing large amounts of oxidants in solution (HNO_3, H_2SO_4, etc.) and relatively low temperatures are employed. This increases the possibility of contamination by reagents but lessens the risk of losing elements present in the form of volatile compounds. Dry ashing uses gaseous oxidizing agents (mainly air and oxygen, possibily with small amounts of acids) at high temperature. This method increases the risk of losing elements as volatile compounds, but contamination from reagents is avoided (28–31).

On each series of samples, organic matrix mineralization efficiency was tested by performing dry and wet ashing. For practical reasons all the samples under examination were previously lyophilized. This enables them to be better homogenized and greatly facilitates the subsequent destruction of the organic compounds. After a systematic comparison of their respective performances, carried out as described hereafter, the dry ashing procedure was preferred. In summary, the lyophilized sample is transferred, under strictly controlled conditions, into decontaminated quartz containers and the sample wet weight is determined. The sample is then dried in an oven at 120 °C for 120 min to remove even the slightest trace of moisture and final weight is ascertained. Slow charring is performed on a hot plate after addition of the smallest amount possible of 65% HNO_3 (Suprapur Merck, Darmstadt, Germany). Calcination follows at 400 ± 10 °C in a muffle furnace internally lined with laminar quartz to minimize any possibility of releasing elements from the refractory material. At the end of the thermal treatment the sample should yield white ashes; if this is not achieved at the completion of the first cycle, a few drops of 65% HNO_3 are added and the procedure just described is repeated. Dissolution of the white ashes is then performed with 1 cm^3 of 65% HNO_3 at 40 °C. The final solution is transferred into a 25 cm^3 flask, and double-distilled water is added up to the marked volume.

8.2.3.2. *Analysis*

The first part of the study is concerned with optimization of the analytical conditions for determining trace elements by ICP–AES. To ascertain the accuracy of the mineralization process, tests were carried out on three series of certified reference materials (CRMs) produced by the European Commission Standards, Measurements and Testing Programme [(SMT) previously Community Bureau of Reference (BCR)] representing the type of matrix under consideration: (i) "skim milk powder, natural" (CRM 063); (ii) "skim milk powder, low spiked" (CRM 150); and (iii) "skim milk powder, high spiked" (CRM 151). Analytical data for all three CRMs were acceptable in the case of dry ashing, and the recovery percentages varied from 90 to 105%, depending on the element. Precision was also found to be more than satisfactory, with relative standard deviations (RSDs) always between 1 and 5% ($n = 10$). In only two cases (Cd and Ni) were the concentrations measured very close to the detection limits. In such instances precision worsened to 20–30%. The corresponding results for both accuracy and precision are reported in Table 8.1. After careful preliminary evaluation of the instrumental operative parameters for ICP–AES, the analytical conditions described in Table 8.2 were adopted.

Table 8.1. Analysis of Certified Reference Materials (CRMs)[a]

Element	CRM 063 Concentration (µg g^{-1})		CRM 150 Concentration (µg g^{-1})		CRM 151 Concentration (µg g^{-1})	
	Certified[b]	Found[c]	Certified[b]	Found[c]	Certified[b]	Found[c]
Cd	0.0029 ± 0.0012	0.003 ± 0.0006 (103%, 20%)	0.0218 ± 0.0014	0.021 ± 0.001 (96%, 5%)	0.101 ± 0.008	0.091 ± 0.005 (90%, 5%)
Co	0.0062	0.006 ± 0.001 (97%, 17%)	0.0064	0.006 ± 0.001 (94%, 17%)	0.0060	0.006 ± 0.001 (100%, 17%)
Cu	0.545 ± 0.030	0.536 ± 0.025 (98%, 5%)	2.23 ± 0.080	2.15 ± 0.10 (96%, 5%)	5.23 ± 0.080	5.09 ± 0.13 (97%, 3%)
Fe	2.06 ± 0.25	2.15 ± 0.11 (104%, 5%)	11.8 ± 0.6	12.0 ± 0.5 (102%, 4%)	50.1 ± 1.3	50.3 ± 1.5 (100%, 3%)
Mg	1120 ± 30	1098 ± 22 (98%, 2%)	—	—	—	—
Mn	0.226	0.210 ± 0.009 (93%, 4%)	0.236	0.223 ± 0.011 (94%, 5%)	0.223	0.209 ± 0.010 (94%, 5%)
Ni	0.0112	0.011 ± 0.002 (98%, 18%)	0.0615	0.064 ± 0.006 (105%, 9%)	0.056	0.058 ± 0.006 (104%, 10%)
Pb	0.1045 ± 0.003	0.109 ± 0.005 (104%, 5%)	1.000 ± 0.040	1.048 ± 0.052 (105%, 5%)	2.002 ± 0.026	2.075 ± 0.062 (104%, 3%)
Zn	42.0	41.7 ± 0.35 (99%, 1%)	49.5	50.2 ± 0.42 (101%, 1%)	50.4	48.0 ± 0.78 (95%, 2%)

[a] Results are the mean of 10 independent sample preparations.
[b] Each mean value is accompanied by its standard deviation (SD). Concentrations without SD are only qualified.
[c] Each mean value is accompanied by its SD. Values in parentheses are, in order, the recovery percentage and the relative standard deviation (RSD).

Table 8.2. Instrumentation and Working Conditions for ICP–AES

Instrumentation:
Spectrometer	JY 32 + 38 VHR (Instruments SA, Longjumeau, France)
RF (radiofrequency) generator	Durr-JY 3848; frequency, 56 MHz; nominal output, 1.6 kW
Induction coil	Five turns; OD, 32 mm; height, 30 mm
Torch	INSA. demountable
Monochromator	HR 1000 M; focal length, 1 m; Czerny–Turner mounting, equipped with a 3600 groove mm^{-1} holographic plane grating; linear dispersion in the first order, 0.27 nm mm^{-1}; spectral range, 170–450 nm
Polychromator	HR 1000 M; focal length, 0.5 m; Paschen–Runge mounting, equipped with a 3600 groove mm^{-1} holographic concave grating; linear dispersion in the first order, 0.55 nm mm^{-1}; spectral range, 170–410 nm
Computer	IBM PS/2 55SX, with Jobin–Yvon ESS vers. 3.4 software
Nebulizer	Meinhard-type, with Scott-type nebulizer chamber

Working conditions:
Spectral lines (nm)	
Monochromator	Al(I), 237.3; Cd(I), 226.5; Cr(II), 206.1; Pb(II);, 220.4
Polychromator	Ba(II), 233.5; Mg(II), 279.6; Sr(II), 407.7; Co(II), 238.9; Mn(II), 257.4; Zn(I), 213.8; Cu(I), 324.7; Ni(II), 231.6; Fe(II), 259.9; Se(II), 196.0
Argon flow rates	Plasma, 18 L min^{-1}; coating, 0.9 L min^{-1}; carrier, 0.1 L min^{-1}
Slit width	40 μm (entrance and exit for monochromator); 50 μm (entrance and exit for polychromator)

8.2.4. Determination of Chemical Species

8.2.4.1. Sample Treatment

For the speciation of elements in food matrices, the samples under consideration should not be subjected to robust pretreatment since any disturbance of the system may cause modifications and/or breaking of the weak links between the element and the organic substance. Therefore, in this study, the milk samples were analyzed as such, except for human milk, raw cow milk, and formula milk, where a previous elimination of lipids was necessary.

The samples to be defatted underwent a weak centrifugation (3000 g for 30 min at 5 °C) performed directly in the storage vessels. Lipids were removed after being solidified by cooling lightly.

8.2.4.2. Analysis

In order to optimize the proteic separation by size exclusion chromatography (SEC), a series of preliminary tests were carried out to detect the most suitable mobile phase and to evaluate the effect of pH and saline concentration on the retention time of proteins (32–34). The results obtained showed that little pH variations around the typical pH of milk (pH = 6.8 ± 1) have no effect on the chromatographic processes. On the other hand, high concentration of NaCl elution solution does cause undesirable modifications both in the conformation of proteins and in their hydrodynamic radius. For some proteins it also influences the process of chromatographic separation on SEC columns. The preferred concentration of NaCl is 0.3 M since it enables a good proteic separation to be performed and at the same time does not give rise to the aforementioned phenomena.

Comparison tests were made on different types of columns in order to optimize the chromatographic separation of proteins, since this is the essential condition for the successive transfer of analysis parameters to the HPLC/ICP–AES on-line system. The best results, in terms of separation, were obtained by coupling in series a TSK-G3000 SW$_{XL}$ column with a TSK-G4000 SW$_{XL}$ column (Tosohaas, Montgomeryville, Pennsylvania).

On the basis of these preliminary data the procedure described in Table 8.3 was chosen. This seems to represent the best compromise between the re-

Table 8.3. Instrumentation and Working Conditions for SEC–HPLC

Instrumentation:	
Pump	Biocompatible 250 (Perkin-Elmer, Norwalk, Connecticut)
Injector	U 6 K (Perkin-Elmer, Norwalk, Connecticut)
Columns	TSK Guardcolumn; TSK-G3000 SW$_{XL}$; TSK-G4000 SW$_{XL}$ (Tosohaas, Montgomeryville, Pennsylvania)
Ultraviolet/visible light (UV/Vis) detector	Model 2550, variable wavelength (Varian, Palo Alto, California)
Integrator.	Model 3396A (Hewlett-Packard, Amondale, Pennsylvania)
Working conditions:	
Buffer	0.1 M HEPES[a] + 0.3 M NaCl; pH = 6.8
Flow rate	0.7 mL min^{-1}
Sample volume	200 μL
UV detector wavelength	280.0 nm
Calibrants	Solutions of thyroglobulin, ferritin, ceruloplasmin, transferrin, myoglobin, and vitamin B$_{12}$

[a] HEPES: *N*-(2-hydroxyethyl)piperazine-*N*'-2-ethanesulfonic acid.

quirement for an adequate chromatographic separation of milk proteins and the need to leave the physical-chemical balance of the samples undisturbed.

Milk samples were preliminarily treated as described earlier in the description of the sampling procedure (Section 8.2.2), and 200 μL of each specimen was injected. The mobile phase was purified on a glass column filled with an ionic resin (Chelex-100) in order to remove the elements present in it.

Bearing in mind the scarce data available on the average concentration levels of element bound to specific milk proteins, we opted to postpone the actual speciation study to a later stage when some more information on element concentration levels might be available. For that reason, in this pilot study we deemed it more appropriate to analyze the chromatographic fractions both individually and separately so as to establish the optimum instrumental parameters to be adopted in the on-line combined technique. For this purpose, the eluent emerging from the UV/Vis detector was collected in several different fractions according to the number of chromatographic peaks obtained for each type of milk. This made it possible to perform sample preconcentration by lyophilization following calibrated dilution of the single proteic fractions. The resulting solutions were analyzed directly by means of ICP–AES.

8.3. EVALUATION OF DATA

8.3.1. Total Concentrations of Elements

Table 8.4 presents the results obtained for each element and each of the five types of milk analyzed in terms of mean, median, and 90th percentile. In general, the data do not indicate any anomalies and/or cases of contamination by toxic elements exceeding the national average. The concentrations are comparable to the values referred to in the literature, notably in a 1989 international collaborative (WHO/IAEA) report on the contents of trace elements in milk worldwide (35).

On the basis of these results, some general comments can be made. In the first place the elements under consideration can be divided into two groups. One group includes elements (Al, Cd, and Pb) that are present both in cow and human milk samples at a relatively high ratio between the minimum and maximum concentration values. This confirms their exogenous origin and consequent extreme variability of intake. The second group includes essential elements, such as Cu, Mg, and Zn, with concentration values virtualiy overlapping for similar types of samples. The reason for this is probably that the concentrations of essential trace elements in the human organism are maintained in narrow ranges thanks to homeostatic control mechanisms.

Table 8.4. Data for Element Concentration (µg mL^{-1}) in Milk Samples[a]

Element		Raw Cow Milk	Pasteurized Milk	UHT-Sterilized Milk	Human Milk	Formula Milk
Al	Mean	0.406	0.769	0.746	0.195	0.229
	Median	0.299	0.735	0.729	0.169	0.156
	90th percentile	0.937	1.104	1.222	0.780	0.852
Ba	Mean	0.498	0.457	0.481	0.026	0.427
	Median	0.375	0.382	0.403	0.013	0.395
	90th percentile	0.601	0.591	0.604	0.056	0.505
Cd	Mean	0.001	0.001	0.002	0.002	0.001
	Median	0.001	0.001	0.001	0.002	0.001
	90th percentile	0.020	0.015	0.030	0.010	0.009
Co	Mean	0.009	0.004	0.009	0.002	0.005
	Median	0.008	0.004	0.007	0.002	0.004
	90th percentile	0.019	0.010	0.015	0.009	0.015
Cr	Mean	0.018	0.032	0.029	0.027	0.030
	Median	0.016	0.028	0.027	0.016	0.029
	90th percentile	0.040	0.051	0.055	0.049	0.055
Cu	Mean	0.034	0.106	0.098	0.402	0.359
	Median	0.030	0.097	0.086	0.391	0.302
	90th percentile	0.102	0.186	0.174	0.855	0.584
Fe	Mean	0.394	0.292	0.325	0.661	2.15
	Median	0.327	0.260	0.310	0.572	1.95
	90th percentile	0.738	0.593	0.602	2.01	4.25
Mg	Mean	91.9	90.4	93.4	29.9	89.1
	Median	71.3	85.7	90.4	28.4	81.4
	90th percentile	102.5	111	127	40.5	98.9
Mn	Mean	0.016	0.021	0.020	0.010	0.115
	Median	0.014.	0.019	0.020	0.006	0.098
	90th percentile	0.031	0.031	0.035	0.059	0.157
Ni	Mean	0.035	0.044	0.043	0.037	0.045
	Median	0.029	0.041	0.040	0.030	0.044
	90th percentile	0.064	0.073	0.068	0.082	0.067
Pb	Mean	0.021	0.028	0.030	0.018	0.025
	Median	0.019	0.025	0.025	0.016	0.023
	90th percentile	0.059	0.069	0.076	0.067	0.060
Se	Mean	0.014	0.015	0.014	0.015	0.012
	Median	0.013	0.013	0.012	0.015	0.014
	90th percentile	0.031	0.040	0.036	0.027	0.029
Sr	Mean	0.236	0.279	0.280	0.105	0.268
	Median	0.258	0.267	0.278	0.099	0.247
	90th percentile	0.520	0.610	0.549	0.298	0.574
Zn	Mean	3.40	4.68	4.76	3.42	4.15
	Median	3.11	4.52	4.48	3.08	3.97
	90th percentile	7.67	7.68	8.42	9.73	6.18

[a] Results are expressed as the mean, median, and 90th percentile (10 different samples for each matrix).

The second comment regards the diversity of levels of some trace elements in the various types of samples. The phenomenon is quite evident when data on cow milk are compared to those regarding human and formula milk. It is further confirmed when raw pasteurized cow milk and UHT-sterilized cow milk are compared. This clearly reflects the influence exerted by technological production processes on the content not only of macronutrients but also of micronutrients like trace elements. This influence is strictly due to the chemical-physical treatments employed and the materials used, as well as to the environmental conditions of storage. In fact, it is possible that, as a result of the contact at high temperatures between the milk and the production equipment, a release may well occur of all the elements that normally are contained in the metal alloys of the facilities employed.

Apart from the foregoing facts, however, no other general behavior has been identified, although this is probably due to the limited number of samples analyzed. On the other hand, this part of the study was aimed only at providing general information on the total concentration of trace elements in milk. This might serve as a backup for comparison, once the different elemental species have been quantified, as relationships are delineated between the chemical form in which a certain element is present in milk and the actual bioavailability of the element itself.

Even though data available in the literature on this issue are rather abundant, they are often insufficient and contradictory. In some cases, differences in results can sum up to 2, 3, or even 4 orders of magnitude for the same element and in the same matrix. The concentration ranges of the various elements examined here are much narrower than those mentioned in the literature, since differences in the average experimental values fall within the same order of magnitude.

Moreover, the analytical method developed in this study appears to be appropriate to the goal and can therefore be safely used for similar investigations.

8.3.2. Concentration of Element Species

As regards the various chemical forms of the elements investigated, the following information was gained by examining the HPLC chromatograms shown in Figs. 8.1–8.6.

Raw Cow Milk. As a rule, six peaks can be distinguished on chromatograms of raw cow milk samples (Fig. 8.1). The first peak is related to α_s-, β-, and κ-caseins (α_s-, β-, and κ-CNs), which, in spite of their molecular weights—around 20,000 Da—elute with the column dead volume (V_0) since, as is well known, they are present in milk in a micellar form. Micelles are spherical polyspread colloidal aggregates the diameter of which varies from 20 to 600 nm, with molecular weights ranging from 10^6 to 3×10^9 Da. During the

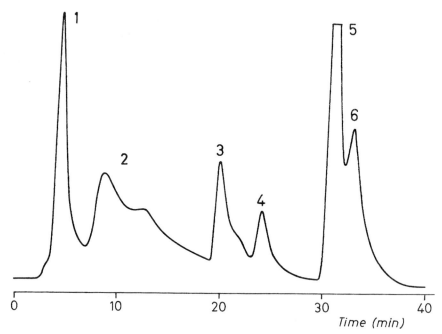

Figure 8.1. Gel filtration chromatographic pattern of raw cow milk. Peak identification: *1*, CNs; *2*, CNs, Igs, and BSA; *3*, β-LG (dimeric form) and CNs (monomeric form); *4*, α-La; *5* & *6*, nonproteic fractions.

last few years several hypotheses have been made to attempt to determine the structure and qualitative–quantitative composition of the micelles (36, 37). One of the most reliable assumptions is based on the existence of submicelles. According to this theory, the formation of micelles originates in two different phases: the monomers of α_s-, β-, and κ-CNs first aggregate to form submicelles of regular dimensions and varying composition. Then, owing to the presence of Ca^{2+} and Mg^{2+} ions, carboxylic and/or phosphate groups present on the surface of the submicelles interact, giving rise to the micelle (38, 39). So far, five types of submicelles denominated F1, F2, F3, F4, and F5 have been discovered, and it has been shown that most casein micelles present in cow's milk are formed by the F2 and F3 subunits only.

The second proteic fraction is still mainly formed by CNs, and the corresponding peak is large enough (due to a wide distribution of different molecular weights) to cover the immunoglobulins (Igs) peak completely and the bovine serum albumin (BSA) peak partially. The presence of CNs in the second fraction as well suggests that there might be a partial disaggregation of the micelles, probably caused by the chromatographic separation process itself. The third peak is essentially formed by β-lactoglobulin (β-LG) in a dimeric form, with traces of CN monomers. The fourth peak can be entirely ascribed to α-lactalbumin (α-La).

The fifth and sixth peaks represent all the substances with a relatively low molecular weight naturally present in cow milk, such as proteose/peptones, orotic acid, lactose, citrates, and mineral salts. In accord with the objectives of this study, it was not deemed necessary to further investigate the specific nature of these last two peaks, which do not stand for any proteic fraction. In addition, qualitative determinations would have been problematic due to the extremely diversified nature of such substances. The only exception made was for citrates: owing to their importance as complexants of many elements in human milk, they were identified by means of enzymatic tests.

Pasteurized Cow Milk. Chromatograms relating to samples of pasteurized cow milk (Fig. 8.2) show a qualitative situation similar to that previously mentioned for raw cow milk. On the other hand, under a quantitative profile, there are considerable differences in the dimensions of the six peaks and

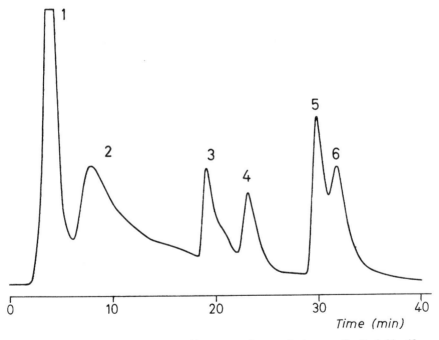

Figure 8.2. Gel filtration chromatographic pattern of pasteurized cow milk. Peak identification: *1*, CNs; *2*, CNs, Igs, and BSA; *3*, β-LG (dimeric form) and CNs (monomeric form); *4*, α-La; *5* & *6*, nonproteic fractions.

consequently in the ratios between the proteic concentrations relating to them. In particular, a decrease in whey protein levels is observed, which is due to their partial denaturation as a consequence of the heat treatment undergone by milk during the process of pasteurization. It has been fully proved and documented that pasteurization carried out even under "milder" conditions may cause irreversible thermal damage to the more thermosensible whey proteins (40). At 72 °C for 15 s Ig 1 is denatured by more than 60%, Igs 2 and 3 by about 10%, BSA by 8%, and β-LG by 3%, with a total denaturation of whey proteins of about 4%. Thermal damage in the samples under study is greater, clearly due to a particularly intense industrial process of pasteurization.

UHT-Sterilized Cow Milk. Chromatographic evidence in the case of samples of UHT-sterilized cow milk confirms what was just mentioned above and also shows an additional dramatic decrease of the whey protein levels (Fig. 8.3). The peaks associated to Igs and BSA are no longer conspicuous, and the β-LG peak decreases to a rather negligible height. This phenomenon is linked to the

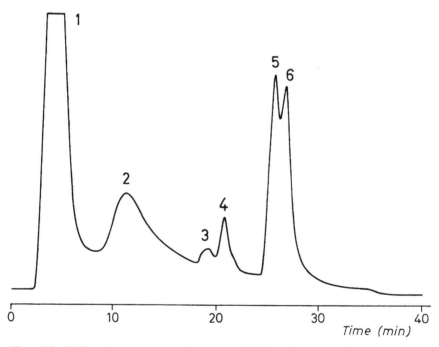

Figure 8.3. Gel filtration chromatographic pattern of UHT-sterilized cow milk. Peak identification: *1*, CNs; *2*, CNs, Igs, and BSA; *3*, β-LG (dimeric form) and CNs (monomeric form); *4*, α-La; *5* & *6*, nonproteic fractions.

intensity of UHT thermal treatment and has repeatedly been studied with the aim of verifying and quantifying the processes of proteic denaturation (41). In this context, β-LG appears to play a basic role in the various interactions with other proteins that can result when milk is heated. In recent years, many researchers have hypothesized a series of possible mechanisms, which can be summarized as follows (42–45): (i) creation of complexes between β-LG and κ-CN; (ii) aggregation among β-LG dimers; (iii) creation of complexes among β-LG and κ-CN aggregates; (iv) aggregation between β-LG and α-La; and (v) interaction among β-LG, α-La, and κ-CN aggregates.

Human Milk. Chromatograms of samples of human milk [colostrum (Fig. 8.4); mature (Fig. 8.5)] show a qualitative trend of the peaks similar to that of cow milk, except for β-LG, which, as is well known, is absent in the former. On the other hand, peak dimensions differ substantially since they represent proteic contents that differ greatly in terms of both absolute and relative amounts. In this regard, it may be appropriate to recall that mature human milk contains proteic levels that are definitely lower than those present in cow milk. In addition, single proteins have a different relative weight within

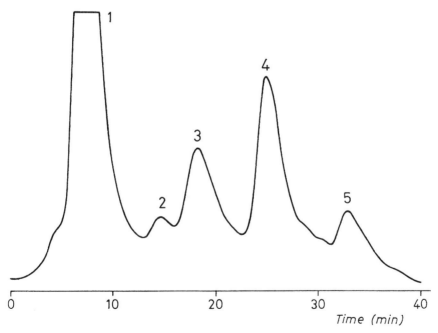

Figure 8.4. Gel filtration chromatographic pattern of human milk (colostrum). Peak identification: *1*, CNs, Igs, and lactoferrin (LF); *2*, CNs and SA; *3*, α-La; *4* & *5*, nonproteic fractions.

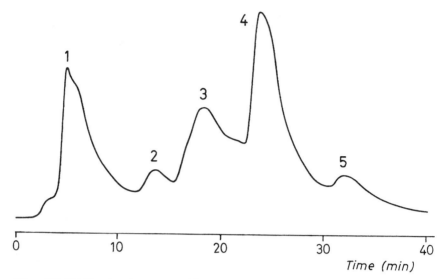

Figure 8.5. Gel filtration chromatographic pattern of human milk (mature). Peak identification: *1*, CNs, Igs, and LF; *2*, CNs and SA; *3*, α-La; *4* & *5* nonproteic fractions.

the global proteic fraction. The ratio between CNs and whey proteins is 4:1 in cow milk but decreases to 0.3:1 in human milk. The α-La in cow milk ranges between 2 and 5%. β-LG is the most representative whey protein, with a mean value of 9.6% of the global proteic content. As noted earlier, β-LG is practically absent in human milk, whereas the main whey protein is α-La (10–25% of the global proteic content). Finally, Igs and SA are present in human milk in larger quantities, accounting for, respectively, 10 and 5% of the proteic content vs. 2.2 and 0.9% in cow milk.

In addition, the two peaks referring to nonproteic fractions give evidence of larger quantities of low-molecular-weight substances in human milk. Some authors justify this phenomenon by the high concentration in human milk of free amino acids, the presence of which may well be masked by that of enzymes, vitamins, etc. in the structure of the two peaks (46). The results of the present study confirm this hypothesis, since the chromatographic and enzymatic analysis of the two last fractions showed significant concentrations of free amino acids, vitamins, and citrates, especially in mature human milk.

As can easily be seen in the chromatograms related to mature human milk (Fig. 8.5) and colostrum (Fig. 8.4), the quantitative composition of proteins is quite different. In particular, the total amount of proteins is reduced and there is a brisk fall in the percentage of higher-molecular-weight proteins (Igs, lactotransferrin, etc.). This phenomenon has been well known for a long time.

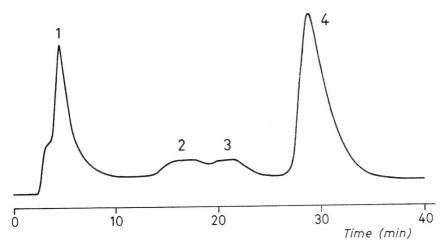

Figure 8.6. Gel filtration chromatographic pattern of formula milk. Peak identification: *1*, CNs; *2*, β-LG (dimeric form); *3*, α-La; *4*, nonproteic fraction.

In 1977, Schneegans and Lauer described in detail the variations in human milk composition during the first 70 days of nourishment (47).

Formula Milk. Chromatograms pertaining to formula milk (Fig. 8.6) differ from those already described. The casein present gives rise to a single peak with a retention volume (V_r) that corresponds to V_0, whereas whey proteins, including α-La, are present in such modest amounts that they cannot be quantified by the chromatographic technique adopted in this study. The second peak with a V_r equal to the total volume (V_t) represents a nonproteic fraction with a low molecular weight.

8.3.3. Relative Distribution of Trace Elements

Figures 8.7–8.20 show the relative distribution of trace elements in chromatographic fractions for the six types of milk investigated. From these data it can be clearly seen that there is a broad diversity of trace elements present in the samples analyzed. In order to further highlight this phenomenon, it should be considered that the maximum difference in the range of values within each series of samples is very small and that the differences noted in the various series clearly reflect a different distribution of the element–proteic components in the various types of milk.

An assessment of data obtained so far makes it possible to point out some interesting patterns of general validity. The first important finding concerns cow milk samples, for which data analysis, on the basis of analogies present in

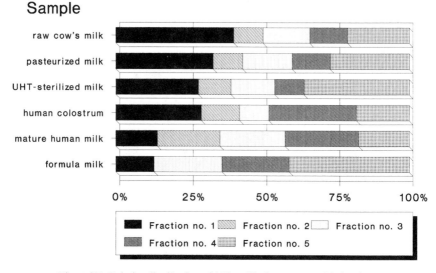

Figure 8.7. Relative distribution of Al in milk chromatographic fractions.

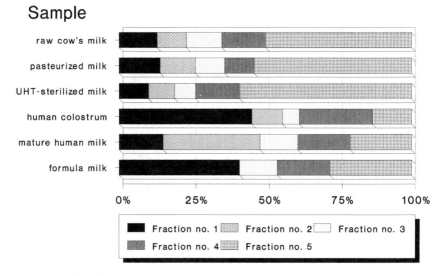

Figure 8.8. Relative distribution of Ba in milk chromatographic fractions.

EVALUATION OF DATA

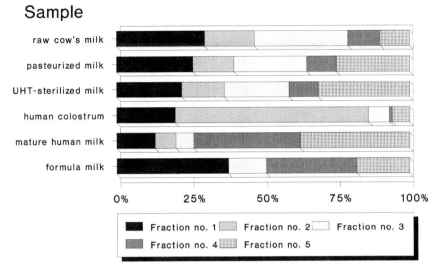

Figure 8.9. Relative distribution of Cd in milk chromatographic fractions.

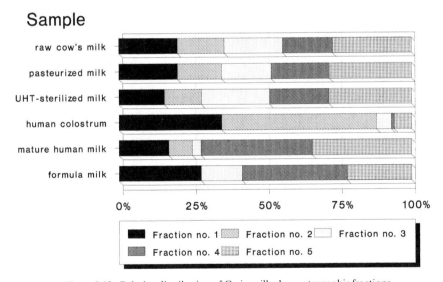

Figure 8.10. Relative distribution of Co in milk chromatographic fractions.

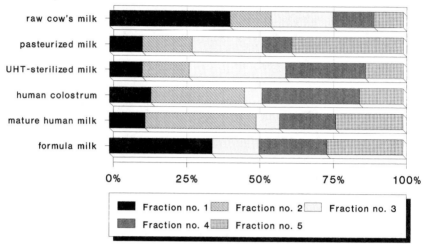

Figure 8.11. Relative distribution of Cr in milk chromatographic fractions.

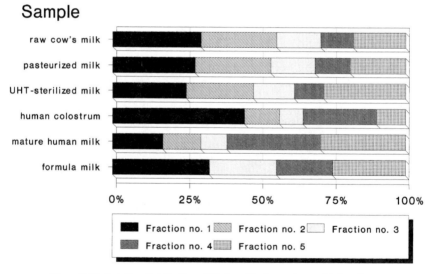

Figure 8.12. Relative distribution of Cu in milk chromatographic fractions.

EVALUATION OF DATA 277

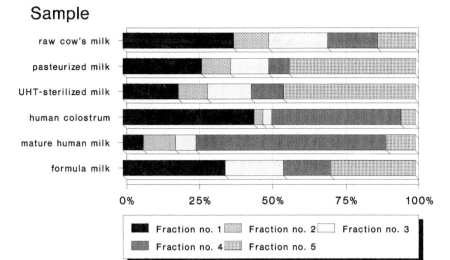

Figure 8.13. Relative distribution of Fe in milk chromatographic fractions.

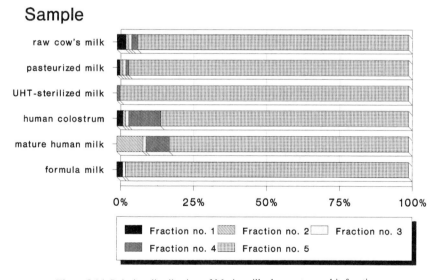

Figure 8.14 Relative distribution of Mg in milk chromatographic fractions.

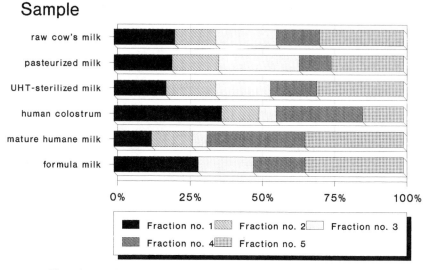

Figure 8.15. Relative distribution of Mn in milk chromatographic fractions.

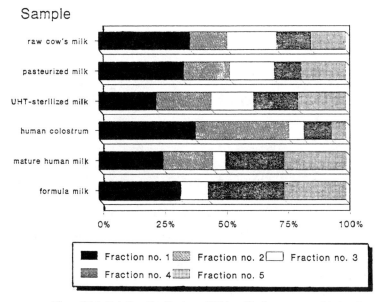

Figure 8.16. Relative distribution of Ni in milk chromatographic fractions.

EVALUATION OF DATA

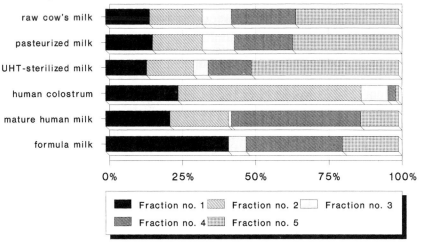

Figure 8.17. Relative distribution of Pb in milk chromatographic fractions.

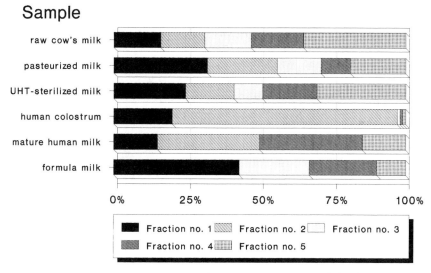

Figure 8.18. Relative distribution of Se in milk chromatographic fractions.

280 ELEMENT SPECIATION IN MILK BY HPLC/ICP–AES

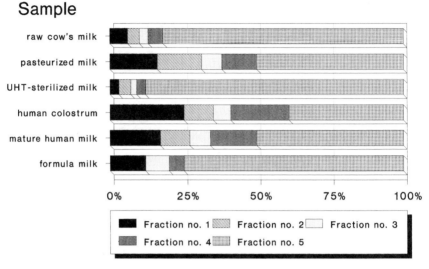

Figure 8.19. Relative distribution of Sr in milk chromatographic fractions.

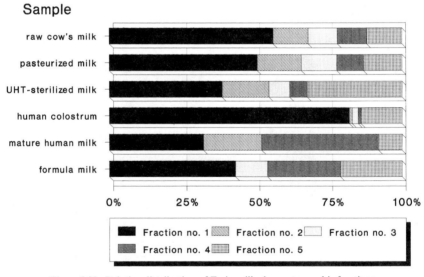

Figure 8.20. Relative distribution of Zn in milk chromatographic fractions.

the distribution, has shown the existence of two groups of elements. The first contains elements such as Al, Cd, Cr, Cu, Fe, and Ni, which are equally distributed among components with high molecular weight (CNs, Igs, and BSA) and components with low molecular weight (α-La, β-LG, proteose/peptones, orotic acid, etc.). The second group contains Ba, Ca, Mg, Mn, Pb, Li, and Sr, which bind preferably to low-molecular-weight components. The only exception is Zn, which has a peculiar distribution, as it is mostly present in components with high molecular weight.

The second consideration centers on the influence of the technological processes of pasteurized and UHT-sterilized cow milk production on element distribution. As has been widely reported, thermal treatment of milk triggers a partial denaturation of the most thermolabile proteins and, consequently, causes the weak element–protein bonds to break. The outcome of this phenomenon is that almost all the elements examined shift from high- to low-molecular-weight fractions. Following this proteic denaturation, soluble salts may be formed by free elements in the solution and the anions that are commonly present in milk. Also to be considered is the fact that the last chromatographic peak, the V_r of which refers to total permeation, reveals the highest relative increase. The more evident the phenomenon, the more dramatic the thermal treatment adopted. In addition, during its pasteurization and UHT-sterilization stages, milk is enriched in a selective way by elements prone to leaching from metallic materials. These are subsequently present in the matrix as cations.

The third experimental finding deals with the percentage distribution of elements in the proteic fractions of formula milk, which reveals a trend similar to that of UHT-sterilized cow milk. This is not surprising since food products of the same origin (cow milk) and equivalent production technologies are involved. In any case, it should be pointed out that homogeneity of distribution occurs only for some elements whereas other elements tend toward particular fracions. This apparent discrepancy is due to the integration of minerals performed in every technological procedure involved in the production of food for infants.

The fourth and last finding centers on mature milk, where the relative distribution of trace elements observed is completely different from those just mentioned. In particular, there is an equivalent distribution in the five chromatographic fractions for almost all the elements considered. This is caused by a considerable increase in the amounts of elements in those intermediate fractions that are related to substances with molecular weights ranging between 10,000 and 100,000 Da (Igs, BSA, α-La, proteose/peptones, enzymes, and citrates). It must be realized that, underlying this distribution, an important role is played by the differences in the percentage of these proteins in the composition of cow and human milk. It is also highly significant that colostrum shows an element composition different from that of mature milk because the former has a higher content of all elements in the first proteic fraction as well as of Cd, Co, Pb, and Se in the second fraction. It should be stressed, however, that beyond the aforementioned general trends, each trace element analyzed has shown some differences in behavior that doubtless will prompt further studies in this area of investigation.

8.4. CONCLUSIONS

As regards the element distribution in milk proteic fractions, the data collected in this study generally agree with those described in the international scientific literature in this field (48–57). It is advisable, however, to stress the significant shortage of information presently available, due both to the scarce number of trace elements examined (Cu, Fe, Mg, Se, and Zn) and the analytical methods generally used, the latter aspect being sometimes inadequate to give consistent results.

Our findings presented here can contribute to the clarification of some issues regarding the metabolic fate of major essential and toxic trace elements. Moreover, data collected make it possible to more credibly assess the distributions to be expected for some elements of primary importance for nutrition, health, and the environment. Assuming that in human milk, as is widely documented and acknowledged, there are specific ligands with strong affinity for elements, such as certain enzymes or citrates, that can increase the bioavailability of the elements themselves, it is obvious that formula milk differs from human milk in a highly significant way as regards their distribution typology.

Under the nutritional rubric, it should again be stressed that until recently scant attention has been given to the importance of evaluating the chemical species under which elements are present in various matrices in order to establish what fraction of the elements themselves can reach the corresponding targets in the biologically active form. This aspect has too often been underestimated and even completely neglected when determining the nutritional quality of milk. In formula milk production, while particular attention is normally given to the qualitative and quantitative conformity of proteins, lipids, carbohydrates, and vitamins to precise standards derived from the composition of human milk, for elements only the necessary total daily amount is taken into consideration. Moreover, it is of primary importance to assess the concentration levels of the various species of toxic elements in milk, both to evaluate its safety and more generally to comprehend how and to what extent polluting substances can interact with the biosphere, including their effect on humans.

In attempting to appraise current knowledge in the field of the metabolism of elements, we must state that, while exhaustive data on the functions and homeostasis of extracellular elements are available, information regarding many metal-containing enzymes and complexes of elements with proteins, especially RNA and DNA, is still very scarce. This is so particularly because their biological activity takes place inside cells, an area of investigation still in its infancy.

In the last few years it has been demonstrated that the more precisely the metabolic role of trace elements and the toxicological role of xenobiotic elements are assessed, the more the clinician is able to link these biological effects to specific pathologies still characterized by etiological uncertainties. In this regard, the further development of innovative analytical techniques will no doubt contribute substantially to the solution of long-standing problems in the biomedical disciplines.

REFERENCES

1. C. F. Mills, G. Brenner, and J. K. Chester, *Trace Elem. Man Anim., —TEMA-5, Int. Symp., 5th, 1984* (1985).
2. S. Langard and T. Norseth, *Toxicology of Metals.* Natl. Tech. Inf. Serv., Springfield, IL 1977.
3. R. K. Chandra, *Trace Elements in Nutrition of Children.* Raven Press, New York, 1988.
4. E. J. Underwood, *Trace Elements in Human and Animal Nutrition*, 4th ed. Academic Press, New York 1977.
5. B. L. O'Dell, *Fed. Proc., Fed. Am. Soc. Exp. Biol.* **42**, 1714–1715 (1983).
6. B. L. O'Dell, *Nutr. Rev.* **42**, 301–307 (1984).
7. B. R. Schricker, D. D. Miller, R. R. Rasmussen, and D. A. Van Campen, *Am. J. Clin. Nutr.* **34**, 2257–2263 (1981).
8. E. R. Morris, *Fed. Proc., Fed. Am. Soc. Exp. Biol.* **42**, 1714–1715 (1983).
9. D. D. Miller, B. R. Schricker, R. R. Rasmussen, and D. A. Van Campen, *Am. J. Clin. Nutr.*, **34**, 2248–2256 (1981).
10. D. L. Eaton and B. F. Toal, *Sci. Total Environ.* **28**, 375–384 (1993).
11. D. W. Engel, *Sci. Total Environ.* **28**, 129–140 (1983).
12. G. W. Evans and P. E. Johnson, *Pediatr. Res.* **14**, 876–880, (1980).
13. G. Harzer and H. Kauer, *Am. J. Clin. Nutr.* **35**, 981–987 (1982).
14. B. Lonnerdal, B. Hoffman, and L. S. Surley, *Am. J. Clin. Nutr.* **36**, 1170–1176 (1982).
15. G. H. Starich and C. Blincoe, *Sci. Total Environ.* **28**, 443–454 (1983).
16. B. Lonnerdal, A. G. Stanyslowski, and L. S. Hurley, *J. Inorg. Chem.* **12**, 71–78 (1980).

17. L. S. Hurley, B. B. Lonnerdal, and A. G. Stanyslowski, *Lancet* **1**, 677–678 (1979).
18. W. Holak, *Analyst (London)* **107**, 1457–1461 (1982).
19. J. O'G. Tatton, in *Mercury Contamination in Man and His Environment*, No. 201, pp. 131–134. IAEA, Vienna.
20. V. D'Arrigo, *Boll. Chim. Lab. Prov.* **34**, 245–254 (1983).
21. S. C. Hight and S. C. Capar, *J. Assoc. Off. Anal. Chem.* **66**, 1121–1128 (1983).
22. L. Magos, *Analyst (London)* **96**, 847–853 (1971).
23. T. M. Florence, *Talanta* **29**, 345–364 (1982).
24. I. S. Krull, *Trends Anal. Chem.* **3**, 76–80 (1984).
25. N. Violante, F. Petrucci, F. La Torre, and S. Caroli, *Spectroscopy* **7**, 38–41 (1992).
26. S. Caroli, F. La Torre, F. Petrucci, O. Senofonte, and N. Violante, *Spectroscopy* **8**, 46–50 (1993).
27. T. T. Gorsuch, *Destruction of Organic Matter*. Pergamon, New York, 1970.
28. W. L. Hoover, J. C. Reagor, and J. C. Garner, *J. Assoc. Off. Anal. Chem.* **52**, 708–713 (1969).
29. F. F. Farris, A. Poklis, and G. E. Griesmann, *J. Assoc. Off. Anal. Chem.* **61**, 660–663 (1978).
30. E. F. Dalton and A. J. Malonoski, *J. Assoc. Off. Anal. Chem.* **52**, 1035–1041 (1969).
31. D. D. Fetterolf and A. Syty, *J. Agric. Food Chem.* **27**, 377–380 (1979).
32. B. B. Gupta, *J. Chromatogr.* **282**, 463–475 (1983).
33. P. G. Dimenna and H. J. Segall, *J. Liq. Chromatogr.* **4**, 639–649 (1981).
34. A. T. Andrews, M. D Taylor, and A. J. Owen, *J. Chromatogr.* **348**, 177–185 (1985).
35. World Health Organization/International Atomic Energy Agency, *Minor and Trace Elements in Breast Milk*. Report of a joint WHO/IAEA study. WHO/IAEA, Geneva, 1989.
36. B. Ribadeau and B. Dumas, *Latte* **7**, 99–114 (1982).
37. O. Tomodata and O. Takayuki, *J. Dairy Res.* **56**, 453–461 (1989).
38. T. Ono, S. Odagiri, and T. Takagi, *J. Dairy Res.* **50**, 37–44 (1983).
39. T. Ono and T. Takagi, *J. Dairy Res.* **53**, 547–555 (1986).
40. H. G. Kessler and H. J. Beyer, *Int. J. Biol. Macromol.* **13**, 165–173 (1991).
41. D. G. Dalgleish, Y. Pouliot, and P. Paquin, *J. Dairy Res.* **54**, 29–37 (1987).
42. H. Singh and P. F. Fox, *J. Dairy Res.* **54**, 509–521 (1987).
43. Z. Haque, M. M. Kristjansson, and J. E. Kinsella, *J. Agric. Food Chem.* **35**, 644–649 (1987).
44. Z. Haque and J. E . Kinsella, *J. Dairy Res.* **55**, 67–80 (1988).
45. A. R. Hill, *Can. Inst. Food Sci. Technol. J.* **22**(2), 120–123 (1989).
46. P. Brätter, B. Gercken, U. Rosick, and A. Tomiak, in *Trace Element Analytical Chemistry in Medicine and Biology* (P. Brätter and P. Schramel, eds.), Vol. 5, pp. 133–143. de Gruyter, Berlin and New York, 1988.
47. E. Schneegans and H. Lauer, *Senologia* **2**, 11–22 (1977).

48. L. Dunemann and G. Schwedt, in *Trace Element Analytical Chemistry in Medicine and Biology* (P. Brätter and P. Schramel, eds.), Vol. 5, pp. 99–118. de Gruyter, Berlin and New York, 1988.
49. E. Bermann, *Appl. Spectrsoc.* **29**, 1–7 (1975).
50. P. E. Gardiner, P. Brätter, B. Gercken, and A. Tomiak, *J. Anal. At. Spectrosc.* **2**, 375–378 (1987).
51. P. Brätter, B. Gercken, A. Tomiak, and U. Rosick, in *Trace Element Analytical Chemistry in Medicine and Biology* (P. Brätter and P. Schramel, eds.), Vol. 5, pp. 119–135. de Gruyter, Berlin and New York 1988.
52. P. Blackeborough, D. N. Salter, and M. I. Gurr, *Biochemistry* **209**, 505–512 (1983).
53. M. T. Martin, F. A. Jacobs, and J. G. Brushmiller, *J. Nutr.* **114**, 869–879 (1984).
54. J. Duncan and L. S. Hurley, *Am. J. Physiol.* **235**, 556–559 (1978).
55. H. Singh, A. Flynn, and P. Fox, *Dairy Res.* **56**, 249–263 (1989).
56. P. Van Dael, G. Vlaemynck, R. Van Renterghem, and H. Deelstra, *Z. Lebensm. Unters. Forsch.* **192**, 422–426.
57. Van Dael, G. Vlaemynck, R. Van Renterghem, and H. Deelstra, in *Trace Element Analytical Chemistry in Medicine and Biology* (P. Brätter and P. Schramel, eds.), Vol. 5, pp. 136–144. de Gruyter, Berlin and New York, 1988.

CHAPTER

9

ORGANOTIN COMPOUNDS IN MARINE ORGANISMS

S. CHIAVARINI, C. CREMISINI, AND R. MORABITO

Environmental Chemistry Division, ENEA,
00060 Rome, Italy*

9.1. INTRODUCTION

In the last 20 years the importance of an accurate assessment of organo-Sn concentration levels in the environment, including many biotic systems, has increased as a consequence of the increase in the number and variety of anthropic sources.

The biological activity of organo-Sn compounds is related to the progressive introduction of organic groups at the Sn atom in any member of an $R_n SnX_{(4-n)}$ series, reaching a maximum when $n = 3$ (1, 2). Triorgano-Sn compounds are widely used in antifouling paints and in agriculture as pesticides (3, 4). In the first case, tributyltin (TBT) and, to a lesser extent, triphenyltin (TPhT) are widely employed against the most common types of fouling organisms such as mollusks and seagrass (5). In the second case, TPhT acetate (Brestan™) and TPhT hydroxide (Duter™) are used for the control of *Phytophthora infestants*, tricyclohexyl-Sn hydroxide (Plictran™) is used for the control of phytophagous mites, and 1-tricyclohexylstannyl-1,2,4-triazole (Peropal™) and hexakis(2-methyl-2-phenylpropyl)distannoxane (Vendex™) have more recently been employed as miticides (4).

TBT-based antifouling paints are used on surfaces in prolonged contact with water to inhibit the growth of fouling organisms, i.e., barnacles, tube worms, and the like. Antifouling paints are generally applied on boat hulls wherever the presence of fouling slows boat movement through the water and increases fuel consumption and maintenance, among other costs.

---

* ENEA: Ente per le Nuove tecnologie, l'Energia e l'Ambiente (National Agency for New Technology, Energy and the Environment).

Element Speciation in Bioinorganic Chemistry, edited by Sergio Caroli.
Chemical Analysis Series, Vol. 135.
ISBN 0-471-57641-7 © 1996 John Wiley & Sons, Inc.

Actually, the largest part of total organo-Sn consumption comes from nonbiocidal uses; in particular, mono- and dibutyl-Sn compounds are widely used in the stabilization of poly(vinyl chloride) (PVC) to avoid undesiderable coloration and brittleness that can occur after prolonged exposure of PVC to heat or ultraviolet (UV) light (6). However, even though biocidal uses represent only 30% of total organo-Sn world consumption, they are responsible for the largest portion of free organo-Sn compounds in the environment, due to their direct introduction into it (7).

In agricultural applications contamination can result from runoff water and crop spraying, whereas in marine antifouling paints the triorgano-Sn compounds are released directly from the paint into the water (8, 9). In water these compounds are readily degraded (by chemical, biological, or UV mechanisms) via a stepwise dealkylation process that yields the di- and monosubstituted compounds and continues down to inorganic Sn (1, 10, 11).

At the same time, however, they are easily absorbed by lipophilic phases such as the lipid fractions of organisms and sediments. In the early 1980s many authors found that TBT in antifouling paints also had deleterious effects on nontarget organisms, some of which are economically important (12, 13). It was pointed out that high TBT concentrations in water caused mortality in shellfish and microalgae, but even very low TBT concentrations (nanogram per liter levels) caused sublethal effects such as poor growth rates and low reproductive success in a wide range of marine organisms (14–16). Among these, Pacific oysters (*Crassostrea gigas*) and blue mussels (*Mytilus edulis*) were able to accumulate large amounts of TBT (15, 17, 18). The toxicity of TBT for these species and their high bioconcentration factor (BCF) resulted in the decline of some important shellfisheries, with great economic consequences (19).

The first evidence appeared in France where, from 1975 to 1982, oyster production was severely depleted by a lack of reproduction and the appearance of shell calcification anomalies in adult oysters, with consequent severe economic losses (14). As a result, France was the first country to restrict the use of TBT-based antifouling paints by legislation that prohibited the use of such paints on boats with hulls less than 25 m long (20). Similar laws were subsequently enacted by many other countries such as the United Kingdom, the United States, and Canada.

In order to evaluate the effectiveness of such legal prohibitions, several monitoring programs have been performed in Europe as well as other continents. The concentration levels of organo-Sn compounds were measured in water, sediments, and biota. In this chapter we present a summary of data concerning the toxicological effects of triorgano-Sn compounds on nontarget aquatic organisms, together with a survey of the results obtained on biological samples collected from French, British, Italian, and North American coastal environments in detail and from other countries in general. The legislative measures taken by these countries to protect the marine environment are summarized as well. Organo-Sn compounds in water and sediments are treated in the next chapter of this book (21).

9.2. TOXICITY

Many reviews containing toxicological data on organo-Sn compounds have been published (2,5,22–24). Elemental Sn and its inorganic compounds have very low toxicity for living organisms in general and humans in particular. Due to their very low solubility in lipids, they are scarcely accumulated by organisms (25, 26). Furthermore, at physiological pH, elemental Sn is not reactive and its oxides are practically insoluble (27).

On the other hand, the association of certain organic groups with the Sn atom can exert a profound influence on its chemical-physical properties, biological activity, mobility, and persistence. As already mentioned in Section 9.1, the progressive introduction of organic groups in the $R_nSnX_{(4-n)}$ series increases toxicity of the molecule, reaching a maximum for the trisubstituted compounds (28, 29). Tetrasubstituted compounds are characterized by similar symptoms, although usually with a longer delay (as compared to the trisubstituted analogue) between administration of the toxicant and the adverse effects (30). Tetraalkyl-Sn compounds probably undergo a dealkylation process *in vivo* (31, 32).

Just which one of the organic groups is linked to the Sn atom has a particular bearing on the toxicity of the organo-Sn compounds, and consequently there are distinctive differences in the sensitivity of various kinds of organisms to them, as shown in Table 9.1.

For mammals, including humans, the highest toxicity is shown by ethyl-Sn compounds: the tragic "Stalinon incident" (involving administration of an oral medication containing diethyl-Sn diiodide), which occurred in France in 1954, resulted in more than 200 intoxication cases, with 98 deaths; it was probably caused by triethyl-Sn iodide impurities (33, 34). Human exposure can also occur directly at industries where workers are producing organo-Sn chemicals or at dockyards where antifouling paints are often used; it can occur indirectly by consumption of farm produce contaminated by organo-Sn pesticides or consumption of seafood contaminated by organo-Sn-based antifouling paints. Increasing the organic chain length, especially beyond that of butyl groups, sharply decreases the toxic properties of trialkyl-Sn compounds. Tri-*n*-octyl-Sn compounds are, in fact, practically nontoxic for all organisms (1).

For marine organisms the highest toxicity is shown by compounds of TBT, TPhT, and tricyclohexyl-Sn (35). The nature of the inorganic substituent does

Table 9.1. Sensitivity Dependence of Various Types of Organisms on R (the Organic Group) in R$_3$SnX

R	Types of Organisms
Methyl	Insects
Ethyl	Mammals
n-Propyl	Gram-negative bacteria
n-Butyl	Gram-positive bacteria, fungi, mollusks, and fish
Phenyl	Fungi, mollusks, and fish
Cyclohexyl	Mites
Neophyl	Mites

not affect the toxicity of the compounds, unless it is a strongly coordinating group (1, 4, 36). The relative lipophilicity of triorgano-Sn compounds, owing to their tendency to bind with complex and simple lipids, enables them to cross biological membranes and produce toxic effects.

The main cause of the biocidal properties of such compounds seems to be the derangement of mitochondrial activity (respiration). This involves the following: mitochondrial membrane damage such as swelling and increased permeability (triorgano-Sn compounds act as surfactants); transport through membranes of anions (Cl$^-$/OH$^-$) acting as phase-transfer agents; and inhibition of the oxidative (and photosynthetic) phosphorylation of ADP to ATP (37).

Some membrane damage could be due to removal of the anionic polyphosphoinositol phospholipids from the transmembrane protein glycophorin, resulting in the breakage of bonds with cytoskeletal proteins (38). Aldridge reported that TBT concentrations higher than hundreds of micrograms per liter affected oxidative phosphorylation in mitochondria (39).

Gray et al. found that the same TBT concentration levels cause transformation of erythrocyte shapes, and fivefold higher concentrations cause loss of membrane barrier functions and hemolysis (40). The same authors reported that TBT seems to be a carrier in erythrocyte membranes and suggested that it may be responsible for the removal of anionic membrane lipids. Dialkyl-Sn compounds show the same dependence on substituent groups as trisubstituted organo-Sn compounds, although they are considerably less toxic. The main biochemical pathway affected by diorgano-Sn compounds seems to be the inhibition of α-keto acid oxidation, probably by reacting with enzymes or coenzymes possessing dithiol groups, such as reduced lipoic acid (41–45).

Other authors have attempted to identify possible binding sites for triorgano-Sn compounds on a macromolecular scale (46–50). The strongest interactions seem to be between triorgano-Sn compounds and protein residues such as histidine and cysteine, with Sn assuming a tetrahedral or cis-pentacoordinated structure (51, 52). Excellent correlation of toxic behavior, chemical-physical properties, and bioaccumulation potential has been achieved by means of molecular predictor systems employing calculation of total surface area and the like (53–57).

As regards marine organisms, triorgano-Sn compounds have proved to be highly toxic for fish, crustaceans, and mollusks. Acute toxicity tests performed on several species of fish gave values of 96 h LC_{50} (lethal concentration wherein 50% of the test organisms are killed within 96 h) ranging from 1.5 to 24 $\mu g\, L^{-1}$.

It is worth stressing here that many studies have been carried out as static bioassays, with true concentrations not determined. Under these conditions the reported value of the toxicant concentration is probably overestimated due to adsorption losses on the container walls. As at nanogram per liter levels organo-Sn adsorption losses can be up to 70%, the actual LC_{50} might well be different from the reported values. More accurate estimates can be furnished by static renewal bioassays: toxicant losses by adsorption or degradation phenomena can be minimized by a careful experimental setup. However, flow-through bioassays with analytical determination of the chemical species would seem to be a more correct approach (6).

Table 9.2 shows some of the huge quantity of data available providing acute toxicity values for aquatic organisms (58–86). These data, although of the utmost importance in the effort to determine the most sensitive species, do not tell us anything about the real danger to the aquatic environment that is caused by *chronic* (i.e., sublethal) exposure to organo-Sn compounds. Table 9.3 summarizes the most relevant results of some long-term experiments. Chronic effects are extremely difficult to assess on a quantitative basis, and the experimental conditions are much more stringent than for acute toxicity tests: the measurements of water organo-Sn compound concentration are themselves critical. Data in Table 9.3 almost invariably pertain to TBT compounds; TPhT and other trisubstituted organo-Sn compounds probably have the same effects, although they may occur at different concentrations, as the mechanism of action is quite independent of the organic group (14, 15, 17, 65, 71, 76, 77, 83–125). Much less is known about chronic effects due to di- and monosubstituted species. The factor limiting their activity is probably the reduced capability to cross biological membranes because they are less lipophilic than the aforementioned compounds.

Table 9.2. Acute Toxicity Values of Triorgano-Sn Compounds for Nontarget Organisms

Test Organisms	Compound	Concentration in Water ($\mu g\ L^{-1}$)[a]	Toxicity Value[b]	Ref.
Fish, freshwater:				
Carassius auratus (goldfish)	TBTO	75	24 h LC_{100}	58
Salmo gairdneri (rainbow trout)	TBTO	28	24 h LC_{50}	59
	TBTO	21	48 h LC_{50}	59
Salmo gairdneri (rainbow trout) fry	TCHTCl	3 nM	110 d LC_{100}	60
	TBTCl	15 nM	110 d LC_{100}	60
	TPhTCl	15 nM	110 d LC_{100}	60
	TMTCl	>75 nM	110 d LC_{100}	60
	DBTCl	>4000 nM	110 d LC_{100}	60
	DPhTCl	>4000 nM	110 d LC_{100}	60
Lebistes reticulatus (guppy)	TBTO	75	24 h LC_{100}	58
	TBTO	39	7 d LC_{50}	61
Tilapia mossambica (Mozambique mouthbrooder)	TBTO	28	24 h LC_{50}	62
Ictalurus punctatus (channel catfish)	TBTO	12	96 h LC_{50}	63
Lepomis macrochirus (bluegill)	TBTO	7.6	96 h LC_{50}	63
Fundulus heteroclitus (mummichog)	TBTO	24	96 h LC_{50}	63
Lebistes reticulatus (guppy) fry	TBTO	10–20	24 h LC_{50}	64
Salmo gairdneri (rainbow trout)	TBTCl	5	10–12 d LC_{100}	65
	TBTO	6.9	96 h LC_{50}	63
Lebistes reticulatus (guppy)	TBTCl	21	7 d LC_{50}	61
Lebistes reticulatus (guppy) fry	TBTOAc	28	7 d LC_{50}	61
Cyprinus carpio (carp)	TPhTOAc	521	24 h LC_{50}	66
	TPhTOAc	320	48 h LC_{50}	66

Table 9.2. (*Continued*)

Test Organisms	Compound	Concentration in Water ($\mu g\ L^{-1}$)[a]	Toxicity Value[b]	Ref.
Carassius auratus (goldfish)	TPhTOAc	75	24 h LC_{100}	58
Gambusia affinis (mosquitofish)	TPhTOAc	400	24 h LC_{100}	67
Anguilla anguilla (eel)	TPhTOAc	400	24 h LC_{100}	67
Carassius auratus (goldfish)	TPhTCl	250	24 h LC_{100}	68
Salmo gairdneri (rainbow trout)	TPhTOH	78	24 h LC_{50}	69

Fish, saltwater:

Test Organisms	Compound	Concentration in Water ($\mu g\ L^{-1}$)[a]	Toxicity Value[b]	Ref.
Alburnus alburnus (bleak)	TBTO	13–17	96 h LC_{50}	70
Cyprinodon variegatus (sheepshead minnow)	TBTO	0.96	21 d LC_{50}	71
	TBTO	3.2	14 d LC_{100}	71
	TBTO	18	7 d LC_{50}	71
	TBTO	1	14 d LC_{50}	71
	TBT	>31	48 h LC_{50}	72
	TBT	28.1	72 h LC_{50}	72
	TBT	25.9	96 h LC_{50}	72
Menidia menidia (Atlantic silversides)	TBT	12.7	48 h LC_{50}	72
	TBT	9.3	72 h LC_{50}	72
	TBT	8.9	96 h LC_{50}	72
Menidia beryllina (inland silversides)	TBT	7.7	48 h LC_{50}	72
	TBT	4.6	72 d LC_{50}	72
	TBT	3.0	96 h LC_{50}	72
Fundulus heteroclitus (mummichog) subadults	TBT	>32.2	48 h LC_{50}	72
	TBT	28.3	72 h LC_{50}	72
	TBT	23.8	96 h LC_{50}	72
Fundulus heteroclitus (mummichog) larvae	TBT	>32.2	48 h LC_{50}	72
	TBT	28.2	72 h LC_{50}	72
	TBT	23.4	96 h LC_{50}	72

(*Continued*)

Table 9.2. (*Continued*)

Test Organisms	Compound	Concentration in Water ($\mu g\ L^{-1}$)[a]	Toxicity Value[b]	Ref.
Brevoortia tyrannus (Atlantic menhaden) juveniles	TBT	6.8	48 h LC_{50}	72
	TBT	5.2	72 h LC_{50}	72
	TBT	4.5	96 h LC_{50}	72
Brevoortia tyrannus (Atlantic menhaden) adults	TBT	5.8	48 h LC_{50}	72
	TBT	4.7	72 h LC_{50}	72
	TBT	5.2	96 h LC_{50}	72
Cyprinodon variegatus (sheepshead minnow)	TBT	12.6–16.5	96 h LC_{50}	73
Fundulus heteroclitus (mummichog)	TBTO	23.36	96 h LC_{50}	63
Cyprinodon variegatus (sheepshead minnow)	TBTO	1.46–3.11	96 h LC_{50}	71
	TBTO	0.96	21 d LC_{50}	71
Solea solea (sole) adults	TBTO	88	48 h LC_{50}	74
	TBTO	36	96 h LC_{50}	74
Solea solea (sole) larvae	TBTO	8.5	48 h LC_{50}	74
	TBTO	2.1	96 h LC_{50}	74
Oncorhyncus tshawytscha (chinook salmon)	TBTO	1.5	96 h LC_{50}	75

Mollusks:

Test Organisms	Compound	Concentration in Water ($\mu g\ L^{-1}$)[a]	Toxicity Value[b]	Ref.
Crassostrea gigas (Pacific oyster) spat	TBTF leachate	2.0	30 d LC_{100}	14
Crassostrea gigas (Pacific oyster) adults	TBTO	290	96 h LC_{50}	74
	TBTO	1800	48 h LC_{50}	74

Table 9.2. (*Continued*)

Test Organisms	Compound	Concentration in Water ($\mu g\ L^{-1}$)[a]	Toxicity Value[b]	Ref.
Mytilus edulis (blue mussel) adults	TBTO	300	48 h LC_{50}	74
Mytilus edulis (blue mussel) larvae	TBTO	0.1	15 h LC_{50}	76
	TBTO	2.3	48 h LC_{50}	74
Crassostrea gigas (Pacific oyster) larvae	TBTO	1.6	48 h LC_{50}	74
Biomphalaria glabrata (mud snail)	TBTO	10	5 d LC_{100}	77
Crassostrea virginica (Eastern oyster)	TBT	1	96 h LC_{100}	78
Mercenaria mercenaria (clam) adults	TBT	1	96 h LC_{100}	78
Biomphalaria glabrata (mud snail)	TBTO	10–100	48 h LC_{50}	79
Ostrea edulis (European oyster) spat	TBT leachate	210	96 h LC_{50}	74
Crustaceans:				
Daphnia magna (water flea)	TBTO	1.7	48 h LC_{50}	63
Gammarus sp. (amphipod) young	TBT	12.5	48 h LC_{50}	72
	TBT	4.3	72 h LC_{50}	72
	TBT	1.3	96 h LC_{50}	72
Gammarus sp. (amphipod) adults	TBT	20.2	48 h LC_{50}	72
	TBT	10.1	72 h LC_{50}	72
	TBT	5.3	96 h LC_{50}	72
Palaemonetes sp. (grass shrimp)	TBT	>32.3	48 h LC_{50}	72
	TBT	>31	72 h LC_{50}	72
	TBT	>31	96 h LC_{50}	72

(*Continued*)

Table 9.2. (*Continued*)

Test Organisms	Compound	Concentration in Water ($\mu g\ L^{-1}$)[a]	Toxicity Value[b]	Ref.
Crangon crangon (brown shrimp)	TBTO	40	96 h LC_{50}	74
Crangon crangon (brown shrimp) larvae	TBTO	1.5	96 h LC_{50}	74
Acartia tonsa (copepod)	TBTO	1	96 h LC_{50}	80
Daphnia magna (water flea)	TBTCl	13	24 h LC_{50}	81
Acartia tonsa (copepod) subadults	TBTCl	1.1	48 h LC_{50}	82
Eurytemora affinis (copepod) subadults	TBT	1.4	48 h LC_{50}	82
	TBT	0.6	72 h LC_{50}	82
	TBT	2.5	48 h LC_{50}	82
	TBT	0.5	72 h LC_{50}	82
Acanthomysis sculpta (mysid shrimp) juveniles	TBT leachate	0.42	96 h LC_{50}	83
Hemigrapsus nudus (larvae)	TBTO	25	8 d LC_{100}	84
Homarus americanus (lobster)	TBTO	5	6 d LC_{100}	84
Nitopra spinipes (harpaticoid)	TBTO TBTF	2	96 h LC_{50}	70

Algae:

Skeletoma costatum (marine diatom)	TBTO	1–18	Algistatic within 5 d	74
	TBTO	>18	Algicidal within 5 d	74
	TBTO	0.1	No growth	85
	TBTO	0.33	72 h LC_{50}	86
	TBTO	5	48 h LC_{100}	85
Paulova lutheri	TBTO	5	48 h LC_{100}	85

Table 9.2. (*Continued*)

Test Organisms	Compound	Concentration in Water ($\mu g\ L^{-1}$)[a]	Toxicity Value[b]	Ref.
Thalassiosira pseudonana (marine diatom)	TBTO	1.03	72 h EC_{50}	86
Dunaliella tertiolecta	TBTO	5	48 h LC_{100}	85

Source: Derived from UNEP/FAO/IAEA, *MAP Technical Reports Series*, No. 33, UNEP, Athens, 1989, with additional data.

[a]Concentrations in other units (mM) are indicated in this column.

[b]Here "h" and "d" stand for hours and days, respectively. LC_{50} and LC_{100} are respectively the lethal concentrations in which 50% and 100% of the test organisms die within the time periods indicated. EC_{50} is the effect concentration, i.e., the concentration of added Sn that reduced the algal growth by 50%.

9.3. METABOLISM AND ACCUMULATION

It is well known that the capability of a chemical compound to be bioaccumulated is related to its lipophilicity. Microcosm experiments have shown that bioaccumulation is usually well represented by the octanol/water partition coefficient (K_{ow}), in which octanol simulates the lipidic phase. As a decrease in a chemical's solubility in water usually increases its fat solubility, so bioaccumulation too has been shown to be inversely related to the substance's solubility in water. Furthermore, a compound must be resistant to an organism's metabolic and excretory processes for a sufficiently long period so that it can be accumulated at high concentrations (126).

Thus, the bioaccumulation process will depend on the lipophilicity of the substance and on its resistance to metabolic and excretory processes (127). Metabolism of hydrophobic xenobiotics usually involves reactions leading to more water-soluble compounds, which can be rapidly eliminated. The principal enzymatic system involved is the cytochrome P_{450}–dependent monooxygenase, which is able to hydroxylate, e.g., TBT to α-, β-, γ-, and δ-hydroxy derivatives (128). These derivatives can be conjugated to sugars (and then excreted) or may spontaneously degrade to dibutyl-Sn (α- and β-derivatives). Many of the previously reported sublethal effects, including imposex and reproductive anomalies, are probably due to interference (inhibition or

Table 7.3. Sublethal Effects of Organo-Sn Compounds on Aquatic Organisms

Compound	Aquatic Organisms	Observed Effects	Ref.
TBTO	*Pecten maximus*	*Mortality*: mortality rate of juvenile scallops > 20% after 15 weeks; no excessive mortality rate for adult scallops *Adverse effects*: no marked changes in growth; only slight differences in shell height and total tissue weight	87
TBTO	*Crassostrea gigas*	*Mortality*: no excessive mortality rate *Adverse effects*: remarkable growth reduction, with all growth ceasing after 10 weeks; about 60% reduction of total tissue weight	88
TBTO	*Crassostrea gigas* (spat)	At 10–200 ng L^{-1} (as TBT)— *Oxygen consumption*: significant reduction and correlation with TBT concentration levels down to 50 ng L^{-1} *Feeding rate*: same as above *Compensation for hypoxia*: same as above even at 10 ng L^{-1} *Growth rate*: same as above down to 20 ng L^{-1} *Shell thickening*: same as above down to 10 ng L^{-1}	
TBTO	*Gammarus oceanicus*	*Oxygen consumption*: no significant effects *Larval survival*: observed reduction starting from 300 ng L^{-1} concentration levels	89
TBTO	*Biomphalaria glabrata*	At 100 ng L^{-1} (as TBT)— *Mortality*: about 60% *Adverse effects*: 80% reduction in eggs laying At 1–10 ng L^{-1}— *Adverse effects*: inhibition of maturation, growth, and fecundity	77

TBTO	*Rhithropenopeus harrisii*	At 500–5000 ng L^{-1}— *Hormesis* (apparent increase in the growth rate) *Metamorphosis*: affected after exposure to 10 μg L^{-1} TBTO and 20 μg L^{-1} TBTS	90
TBTO	*Campanularia flexuosa*	*Hormesis* (at low concentration levels)	91
TBT	*Nucella lapillus*	Induction of imposex^a at about 20 ng L^{-1} Reproductive failure when tissue concentration of TBT > 0.1 ng g^{-1}	92
TBT	*Salmo salar*	Growth test in antifouling-treated nets in aquaculture— *Mortality*: a few cases *Adverse effects*: feeding rate drastically reduced after 4 days; extensive hyperplasia at multiple points	93
TBT	*Salmo salar*	Growth test in antifouling-treated nets in aquaculture, at 10–1000 ng L^{-1}— *Mortality*: a few cases *Adverse effects*: feeding rate drastically reduced after 4 days; extensive hyperplasia at multiple points	94
TBTCl	*Salmo gairdneri* (rainbow trout, fry)	NOEL^b < 200 ng L^{-1}— *Mortality*: after 10 days, with extensive kidney damage (at 5000 ng L^{-1}) *Adverse effects*: increase of relative liver weight due to hyperplasia; hypertrophy of hepatocytes (at 200–2000 ng L^{-1}); decrease in growth rate; diminished storage of liver glycogen	65
TBT	*Oncorhynchus tshawytscha*	At 1500 ng L^{-1}— *Adverse effects*: liver, brain, and muscle damage; growth retardation; low resistance to disease	95

(*Continued*)

Table 9.3. (Continued)

Compound	Aquatic Organisms	Observed Effects	Ref.
TBTAc	*Crassostrea gigas*	NOEL[b] < 20 ng L^{-1} — At 50 ng L^{-1}: slow growth; high mortality after 10 days At 100 ng L^{-1}: slow growth; total mortality after 12 days At 200 ng L^{-1}: perturbation of food assimilation; total mortality after 12 days At 500 ng L^{-1}: larval anomalies; total mortality after 8 days At 1000 ng L^{-1}: abnormal veligers; malformation of trochophores	96
TBT	*Mytilus edulis*	At 1000 ng L^{-1}: inhibition of larval development	97
TBT	*Crassostrea gigas*	At 50 ng L^{-1}: anomalies in shell calcification mechanisms	98
TBTO	*Crassostrea gigas* (spat)	On algal diet, tested in the following: water + TBTO; low-contaminated sediments; sediments + TBTO At 150 ng L^{-1}: reduced growth; pronounced thickness of the upper shell valve At 1600 ng L^{-1}: severe inhibition of growth	15
Various organotins	Two marine diatoms	Growth and survival reduction	86
TBTO	*Mytilus edulis*	Juveniles (5 months old) tested at 100–10,000 ng L^{-1} — At 100 ng L^{-1}: no effects on growth in length up to 7 days At > 400 ng L^{-1}: significant toxic effects	17
TBTO	*Mytilus edulis*	Veliger larvae, at 100 ng L^{-1} — Mortality: 50% *Adverse effects*: significant reduction in growth rate	76

Compound	Species	Effect	Ref
TBTO	Orchestia traskiana	At 1000–6000 ng L^{-1} (9 day exposure period): evident viability reduction At > 10,000 ng L^{-1}: significant toxic effects and mortality	99
TBTO	Dunaliella tertiolata (microalga)	At 1000 ng L^{-1}: significant growth retardation At 100 ng L^{-1}: growth stagnation	85
TBTO	Daphnia magna	At 200 ng L^{-1}: 24 h EC$_{50}$^{b}; 50% immobilization	100
TBTF	Crassostrea gigas	At 200 ng L^{-1}: shell malformities (after 110 days)	14
TBT	Crassostrea gigas	At 1000 ng L^{-1}: decline of body weight	101
TBT	Crassostrea gigas	At > 100 ng L^{-1}: reduction of the condition index; significant reduction in the number of species and species diversity	102
TBT	Mercenaria mercenaria	At 50 ng L^{-1}: changes in growth rate	103
TBTO	Mytilus edulis	No evidence of genotoxicity	104
TBT	Nassarius obsoletus	At about 1 ng L^{-1}: induction of imposex (***the most sensitive effect recorded for TBT***)	105
TBT	Acanthorysis sculpta (crustacean)	At 190 ng L^{-1}, leached from panels coated with antifouling paint: NOEL^{b} 90 ng L^{-1} Estimate of final chronic value 140 ng L^{-1} Growth effects in juveniles 250 ng L^{-1} Reduction of release of viable juveniles	83
TBTO	Uca pugilator (fiddler crab)	At 500 ng L^{-1}: retardation of limb-generative processes; deformities	106
TBTO, TPhTO	Ophioderma brevispina (brittle star)	At 100 ng L^{-1}: inhibition in regeneration of the arms	107

(*Continued*)

Table 9.3. (*Continued*)

Compound	Aquatic Organisms	Observed Effects	Ref.
TBTF	*Crassostrea gigas*	At 2, 13, and 65 ng L^{-1}, four main target tissues: Digestive gland modifications Damage to epithelial cells of nephridial tubules; kidney hypertrophy Losses of glycogen in vesicular connective tissue Vacuolization, break off of cilia, and other damage to gills	108
TBTO	*Nucella lapillus*	1 ng injected: induction of imposex; significant increase of testosterone and to a lesser extent of progesterone and estradiol-17β	109
TBT	*Ocenebra erinacea*	Penis development and changes in other morphogenic factors	110
TBT	*Nucella lima*	At about 1000 ng L^{-1} — After 4 months, 27% died vs. 1% control, long-term LC$_{50}$ Feeding rates significantly lowered; no differences at about 60 ng L^{-1} Imposex induction increased in the treated groups Aqueous TBT seems to be more effective than the equivalent of TBT fed in the diet	111
TBT	*Scrobicularia plana*	At 100 ng L^{-1} (water) At 100 ng L^{-1} (water) + 1 ng g^{-1} spiked sediments Uncontaminated water + 1 ng g^{-1} spiked sediments Reproductivity reduction, almost dependent on sediment-bound TBT	112
TBT, DBTCl	*Poecelia reticulata*	At a few hundred nanograms per liter: chronic toxic effects on the immune system	113

Compound	Species	Effects	Ref.
TBTCl	*Phoxinus phoxinus* (embryos and larvae)	At < 1000 ng L^{-1}, 6-day exposure: 50% incidence of abnormal larvae and skeletal malformations, inhibition of coordinate movements, disruption of neuromuscular functions, etc., at 21 °C No effects at 16 °C	114
TBT	*Ilyanassa obsoleta*	At a few nanograms per liter: chronic toxic effects	115
TBT	*Morone saxatilis* (striped bass, larvae)	Decrease in growth Alterations in morphometry	116
TBT	*Leuresthes tenuis*	At 0–2000 ng L^{-1}, NOEL^b on: hatching success; notochord length	117
TBT	*Bythinia tentaculata*	At 1000 ng L^{-1}, 5-week exposure: low egg production, and eggs did not develop; growth reduction at concentration level > 200 ng L^{-1}	118
TBT	*Planorbis vortex*	At 1000 ng L^{-1}, 5-week exposure: no egg production; growth reduction at concentration level > 200 ng L^{-1}	118
TBT	*Salmo gairdneri*	At 11.7–5850 µg L^{-1}: 24 h EC$_{50}$ = 31 µg L^{-1}; damage to gill epithelium (suffocation) After 5-day exposure to 11.7 µg L^{-1}: flattening of bile duct columnar epithelial cells and destruction of corneal epithelium	119
^{14}C-TBTO	*Cyprinodon variegatus*	Life cycle test: toxic effects on progeny at concentration level > 0.24 µg L^{-1}; high mortality of the F1 generation progeny during 28-day posthatching period	71
TBT (from methacrylate leachates)	*Crassostrea gigas* *Mytilus edulis* *Venerupsis decussata* (clam, spat)	Significant reduction in growth at 0.24 µg L^{-1}	120

(*Continued*)

Table 9.3. (Continued)

Compound	Aquatic Organisms	Observed Effects	Ref.
TBT (leachates)	*Ostrea edulis* (European oyster, spat)	Predominance of maleness at 0.24 μg L^{-1} no differentiation at 2.6 μg L^{-1}; no larvae released	121
TBTO	*Ostrea edulis* (European oyster, spat)	Growth effect at 0.02 μg L^{-1} (marginal) At 0.06 μg L^{-1} growth rate was severely curtailed after 10 days	120
TBTO	*Nucella lapillus* (dogwhelk)	Imposex induced in the laboratory after exposure at 0.02 μg L^{-1}	122
TBT paint	*Nassarius obsoletus* (mud snail)	Anatomical abnormality; induction of imposex	123
TBTO	*Homarus americanus* (lobster, larvae)	90% decrease in growth at 1 μg L^{-1}	84
TBTO	*Daphnia magna*	At 0.5 μg L^{-1} daphnids displayed positive phototaxis, whereas nonexposed individuals showed a negative phototaxis	124
TBT	Indigenous algae sampled from Lake Ontario	Inhibitory effect on primary production at 0.1 μg L^{-1}	125

[a]Imposex is the superimposition of male characteristics on female gastropods.
[b]NOEL: no-observed-effect level. EC$_{50}$ is the effect concentration, i.e. the concentration of added Sn that reduces the algal growth by 50%.

induction) with steroid metabolism, possibly involving a cytochrome P-450 system (127).

Organo-Sn compounds can be taken up and metabolized by water plants and, very importantly from an environmental point of view, by microalgae that are able to produce hydroxyderivatives via a dioxygenase enzyme system (129–131). Numerous studies have demonstrated that fish are able to efficiently metabolize organo-Sn compounds (132–138). The liver appears to be the detoxifying organ, and the hydroxyderivatives are eliminated through transfer to the bile (in the gallbladder) (139). The bioconcentration factor is usually very high for organs with high lipid contents (brain, liver, or kidney), being lower for muscles.

Only a few reports have dealt with organo-Sn species metabolization and accumulation by aquatic invertebrates, which possess efficient mixed-function oxygenases in hepatopancreas F-cells (127, 139–141). Studies on kinetics and mechanism of accumulation have shown that marine bivalves rapidly and effectively accumulate organo-Sn compounds when exposed even to low ambient concentrations of dissolved material (142). In this respect, the behavior of organo-Sn compounds resembles that of a number of other nonpolar organic compounds such as petroleum hydrocarbons and synthetic organic molecules. Bivalves are capable of accumulating dissolved TBT directly from seawater, presumably directly into exposed tissues such as gills, followed by translocation to other tissues, or by ingesting tainted food. Very high concentrations can be reached in these organisms, because they are not capable of metabolizing many xenobiotics, including organo-Sn compounds, due to the low activity of their mixed-function oxidase system (43, 144).

The uptake of organo-Sn compounds from water and food (phyto- and zooplankton) has been investigated by several authors. The differential accumulation of TBT in various tissues has been studied as well (7, 19, 75, 145, 146). The accumulation of TBT in laboratory experiments has demonstrated high bioconcentration factors for oysters and mussels (15, 19, 75). However, when K_{ow} values are compared with experimental bioconcentration factors, there is approximately a 10-fold excess in TBT accumulation relative to that which would be predicted by the behavior of model hydrocarbons and synthetic organic compounds, suggesting that binding of organo-Sn compounds to macromolecules (mainly proteins) might play an important role in bioconcentration phenomena (19, 147). The reasons for the discrepancies between laboratory and field experiments are probably to be found in the difficulty of reproducing the true environmental conditions during microcosm experiments. In particular, stress conditions (heat, salinity, presence of other pollutants) can significantly alter growth and metabolic rates, and therefore quantification of the test animals' health is of capital importance in both laboratory and field studies (148, 149).

9.4. LEGAL PROVISIONS

At the beginning of the 1980s, it was clear that triorgano-Sn compounds, due to their toxicity and usage, could represent a risk to the marine environment. In order to minimize this risk, water quality standards were imposed in many countries. However, although it was fairly easy to control organo-Sn release from industrial water cooling systems and from mariculture sites, the control of such release from painted vessels presented great difficulties. Boats being so diffuse a pollution source, it was obviously impossible to control the emissions from each boat so as to keep, e.g., the TBT concentration under the limit imposed by the water quality standard. So, the only possibility was to restrict the use of these products.

In 1986, the Virginia Department of Agriculture and Consumer Services published data indicating that small boats contributed some 70% to TBT concentration in Virginia's waters and only 28 and 2% was from large commercial ships and from the U.S. Navy, respectively. It should be stressed, however, that these data probably did not take into account the fact that larger ships are located in areas characterized by greater water exchange and dilution than those areas where small boats are generally located.

Thus, for the foregoing reasons, most of the laws enacted by national legislatures have concerned the prohibition of the use of TBT paints on small boats. Recently, there have been recommendations to extend the restriction to all organo-Sn compound-based paints not only for boats but also for industrial water cooling systems, mariculture sites, etc. Many sources of legislative information on this topic can be found in the literature (6, 20, 150–154).

9.4.1. France

In January 1982 France prohibited for 3 years the application of TBT-based antifouling paints to hulls of boats shorter than 25 m. Boats with aluminum hulls were exempted because the use of copper-based antifouling paints, the only alternative to TBT paints present on the market, would result in heavy corrosion damage (7). The ban was initially applied to the French Atlantic coast and to the English Channel and only later (September 1982) to all French coasts.

In 1984 the ban was extended in its scope, prohibiting the use of antifouling paints containing more than 0.4% w/w of Sn. It should be stressed that the ban is on the use of such paints on certain types of boats and not on their production or sale, resulting in a very difficult control problem (surreptitious application is obviously possible). However, as mentioned earlier, a significant decrease of organo-Sn species concentration has been observed in Arcachon Bay (on the Atlantic coast) since 1982, together with an improvement in oyster production, suggesting significant public compliance.

9.4.2. United Kingdom

As of 1986 the use of TBT in antifouling paints has been restricted by the British government. The British approach was to favor products based on organometallic copolymer formulations over other products based on "free-association" formulations. In the first type of paint TBT is chemically bonded to a polymer, resulting in a relatively constant release rate depending on hydrolysis reactions, whereas in the second type TBT is free to diffuse into water, resulting in a higher release rate over shorter time periods. In particular, TBT was prohibited in copolymer paints at concentrations of Sn higher than 7.5% w/w and in conventional paints (the free-association type) at concentrations higher than 2.5% w/w (153). In 1987, the maximum allowable Sn concentration in copolymer paints was reduced to 5.5% w/w. The goal of these regulations was to keep dissolved TBT in British waters below 20 ng L^{-1}, considered as the water quality target. Further studies showed that this limit was not strict enough to avoid adverse effects on marine organisms. The water quality target was therefore lowered to 2 ng L^{-1} in 1988.

9.4.3. United States

The U.S. legislation is more complex due to the differences among individual states. In the early 1980s, the U.S. Navy began to study the potential environmental impact of TBT-based antifouling paints. The research included monitoring programs in several harbors, TBT release tests from the hulls of boats, acute and chronic toxicity tests, and studies on the degradation and ultimate fate of TBT in the marine environment. In 1985, the U.S. Navy published a report excluding significant impact in case of use of TBT paints on the fleet (155). In the same year, the Office of Pesticides of the Environmental Protection Agency (EPA), under the provisions of the Federal Insecticide, Fungicide and Rodenticide Act (FIFRA) (a law requiring the registration of all pesticides), initiated a special review on TBT. By 1986, the producers and manufacturers were asked to provide production, use, and sales information on TBT-based antifouling paints; thus, approximately 600 formulations were identified (156). Furthermore, many states began to take measures to restrict the use of TBT-based paints. North Carolina set a TBT water quality standard at 2 and 8 ng L^{-1} for salt- and freshwater, respectively (157). In 1987, legislation similar to the French law was enacted first in Virginia and then in other states. In 1988, the "Organo-Sn Antifouling Paint Control Act" restricted the use of these paints all over the United States. Paints leaching TBT at a rate higher than $4 \text{ }\mu\text{g cm}^{-2}$ per day were prohibited, and there was total prohibition of the use of TBT paints on boats shorter than 25 m. An exhaustive U.S. legislative history in this field has been recently published by Huggett et al. (152).

9.4.4. Italy

The only measure taken in Italy was promulgated by the Ministry of Health in 1985 (158). It prohibited the use of organo-Sn compound-based paints in industrial cooling water systems discharging into areas where aquaculture sites are located.

9.4.5. Other Countries

The use of organo-Sn compound-based antifouling paints has been restricted in many other countries, such as Canada, Japan, Switzerland, Germany, Norway, and Sweden. However, the regulations are usually less detailed than those in the countries examined in the previous subsections. As a conclusion, it is worth mentioning the United Nations Environment Programme (UNEP) recommendations for the Mediterranean Sea. The Commission for the Protection of the Mediterranean Sea, at its Sixth Ordinary Meeting held in Athens in 1989, recommended that the "contracting parties" (i.e., member states) take the following measures: (i) from July 1, 1991, not to allow the use of antifouling paint containing organo-Sn compounds on boats of less than 25 m (except boats with aluminum hulls) and on all structures, equipment, and apparatus, used in mariculture (those contracting parties not having access to substitute products by July 1, 1991, would be free to delay enforcement for 2 years, informing the Secretariat, which would contact the other contracting parties); (ii) each to report to the Secretariat on measures taken; and (iii) each to develop a code of practice to minimize the contamination (159).

The Commission recommended further that the contracting parties monitor the organo-Sn compound concentration in the marine environment after the adoption of the proposed measures.

9.5. REGIONAL CASES

Since the recognition of the potential environmental impact of the organo-Sn compounds, several studies and monitoring programs have been carried out. Most of this research has concerned sites where pollution sources were known (such as harbors and marinas), as well as sites where aquaculture activity was severely affected by organo-Sn compound contamination. Relatively little information is available about remote sites or protected areas, which could be affected by long-range transport of low levels of organo-Sn contaminants.

9.5.1. France

The first evidence of environmental damage attributable to the leaching of antifouling paints (and so to organo-Sn active ingredients) was detected in

Arcachon Bay, where 1000 ha oyster farms are located, producing about 10% [15,000 t (metric tons) in 1975] of the total French production. A dramatic decrease in productivity (down to 3000 t) was recorded in the period 1975–1983.

Arcachon Bay is characterized by heavy maritime traffic, mainly due to the presence of a large flotilla of pleasure boats. In order to investigate any link to antifouling paints, Alzieu *et al.* carried out field and laboratory studies (160). Thus, uncontaminated oysters were distributed in an organo-Sn compound polluted area (Boyardville marina), in an unpolluted area, and in tanks reproducing the conditions at the above locations. After 30 days of exposure only oysters affected by TBT contamination showed shell deformities, while the controls developed normally.

His and Robert demonstrated that TBT concentration levels higher than 50 ng L^{-1} seriously affected Pacific oyster *Crassostrea gigas* larvae growth, establishing a no-observed-effect level (NOEL) at 20 ng L^{-1} (96, 97). Environmental concentrations of total Sn were measured at 200–300 ng L^{-1} in Arcachon Bay (1982), where 70–100% of the 2-year-old oysters had shell deformities. A significant reduction of adverse effects was recorded after the January 1982 ban on organo-Sn compounds in antifouling paints. During the years 1982–1985 surveys were carried out three times a year, corresponding to those periods when wintering, cleaning/painting, and mooring of ships occurred. The surveying plan included two water-sampling stations and nine oyster-sampling stations (7). Oyster shell malformations were found to be substantially reduced, in good agreement with total Sn levels.

9.5.2. United Kingdom

Similar problems occurred in the U.K. oyster fisheries. The Pacific oyster *Crassostrea gigas* was introduced to the British east coast fisheries in the early 1970s, and field trials showed that in some areas oyster growth was poor and the shell valves were evidently malformed. These effects were later correlated with organo-Sn compound contamination by means of laboratory experiments. Waldock and Thain found organo-Sn compound levels higher than 200 ng L^{-1} in the Crouch area. Since this area was high in suspended sediments, they investigated the possibility that sediments could be the main cause of the adverse effects. They carried out oyster growth experiments in tanks with seawater and Crouch sediments, with and without addition of TBT. The results showed that concentrations of suspended Crouch sediments up to 75 mg L^{-1} did not produce any reduction in oyster growth rate, whereas addition of 0.15 μg L^{-1} of TBT produced an 80% decrease at the end of the 56-day experiment, confirming the strong correlation between growth reduction and TBT concentration (15).

Other aquatic organisms, such as dogwhelks (*Nucella lapillus*), were affected by organo-Sn compound contamination. Davies et al. found evidence of imposex (induction of male characters in female organisms) in Loch Crinan Argyll, Laxford, and Sween on the west coast of Scotland (161). Loch Crinan Argyll is subject to considerable small boat activity due to the passage to Crinan Canal. Loch Laxford (with not much boat activity) was studied because of the presence of two salmon farms that use TBT antifouling points on their cages. Loch Sween combines the two previously described situations: plenty of boating and salmon farming.

Results show that the degree of imposex is correlated with the main sources of TBT: a gradient effect was observed away from the fish farm cages and sites with intensive boating activity. An adequate dilution was generally observed in open coastal waters, resulting in unaffected populations of aquatic organisms (161).

Monitoring programs were carried out after the 1987 ban of TBT in the United Kingdom to ascertain the effectiveness of the protective measures in reducing environmental impact of this product. The TBT concentration levels in water and oyster flesh were reduced to some 1/3–1/4 of the preban levels. Adverse effects on aquatic organisms, even though reduced, were (and still are) present in the formerly most contaminated sites.

9.5.3. United States

Reports on the effects of Sn-based antifouling paints on aquatic populations in the United States are somewhat limited. Although it appears that adverse effects have indeed occurred, their extent is still unknown due to the limited monitoring of biological disturbances. Shell deformities in oysters from Coos Bay (on the Pacific coast), attributed to TBT residues from paint chips, were reported by the Oregon Department of Fish and Wildlife (22). Extensive work on organo-Sn compound concentration levels was carried out in San Diego Bay (Southern California) (162–165). Data were correlated with pleasure boats and commercial shipping activity as well as tidal, vertical, and spatial variability.

Stephenson et al. demonstrated that oysters (*Crassostrea gigas*) and mussels (*Mytilus edulis* and *Mytilus californianum*) exhibited shell thickening and growth alteration that were correlated with environmental exposure to TBT in San Diego Bay (166). Valkirs et al. monitored TBT concentration trends in water, sediments, and mussel tissue from selected areas of San Diego Bay following the imposition of legislative restrictions in 1988. A general decrease in water and mussel TBT concentration was observed: in the less contaminated area, tissue levels were 137 ng g^{-1} wet weight in February 1988 and 32 ng g^{-1} in January 1989; in the most heavily polluted region, values were respectively 2133 and 550 ng g^{-1} (167).

Twenty-six different environmental settings were tested by Stephenson demonstrating a clear correlation between growth anomalies and TBT tissue and water levels. The author concluded that a safe TBT water level should lie somewhere in the 10–50 ng L^{-1} range (168).

In 1986–87 mussel samples were collected from five bays on the Pacific coast. Reported TBT concentrations ranged from less than 5 ng g^{-1} to 1.1 μg g^{-1} wet weight (169).

Mytilus edulis was sampled from 15 Maine locations at various times between 1987 and 1989 by Page et al, who analyzed TBT and DBT tissue concentrations (170). Levels ranged from 40 ng g^{-1} in uncontaminated areas to 4300 ng g^{-1} at marina/commercial harbor sites. Monitoring results, corrected for seasonal variations, indicate that legislative regulations enacted in Maine in 1987 did not entirely eliminate TBT environmental concerns.

Oysters and mussels were collected at 36 U.S. coastal sites as a part of the National Oceanic and Atmospheric Administration's (NOAA) National Status and Trends Mussel Watch Program. The TBT, DBT (dibutyltin), and MBT (monobutyltin) concentrations in mussels were in the ranges 100–1540, < 5–870, and < 5–1240 ng g^{-1}, respectively; in oysters the ranges were < 5–1560, < 5–1280, and < 5–920 ng g^{-1}, respectively (147). Some of the samples were collected in remote zones in order to give baseline values that were useful for monitoring the restrictions' effectiveness.

From 1986 to 1990, the concentrations of butyl-Sn compounds in oysters were monitored at 17 sites in the Gulf of Mexico (171). No significant variations were observed during that period, suggesting that use of old paint stocks and/or release from sinks (e.g., sediments) may cause a slower than expected decrease of environmental concentrations of butyl-Sn species. A protocol for measurement of whelk imposex frequency was applied in 1987–1989 to monitor Puget Sound (northern Washington State) and Vancouver Island (British Columbia, Canada) coasts (172). Imposex levels were correlated with boat traffic, suggesting that TBT could be the main cause of contamination and that existing control procedures regulating only the use of antifouling paints by small boats may not be effective.

Butyl-Sn levels in sediments and bivalves from U.S. coastal areas were monitored by Wade et al. (173). Butyl-Sn compounds were detected in 75% of the sediment samples analyzed and concentrations ranged from < 5 to 282 ng g^{-1} (as Sn). TBT was the predominant butyl-Sn species, while DBT and MBT were detected in 30% of the sediment samples analyzed. Mean bivalve butyl-Sn concentrations were 18 times higher than mean sediments concentrations. The authors concluded that sediments are a possible source of bioavailable butyl-Sn compounds, but the lack of correlation between sediments and bivalve butyl-Sn concentration indicates that other sources may be predominant. Evidence supporting these conclusions is presented in Fig. 9.1, where

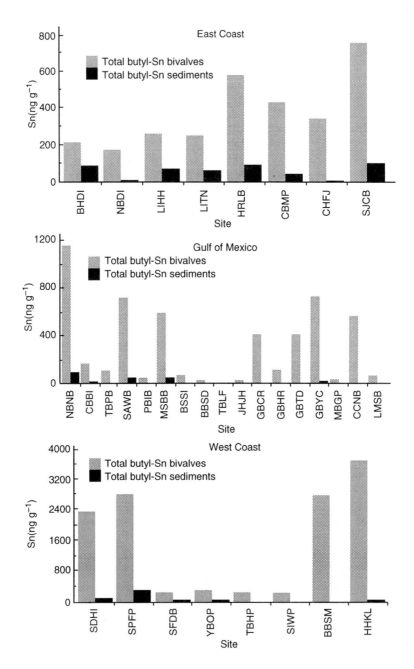

Figure 9.1. Comparison of total butyl-Sn concentrations in sediments and bivalves. See Fig. 9.3 for the location of sampling sites. Reprinted from Wade et al. (173) with permission of Pergamon Press Ltd, Oxford, United Kingdom.

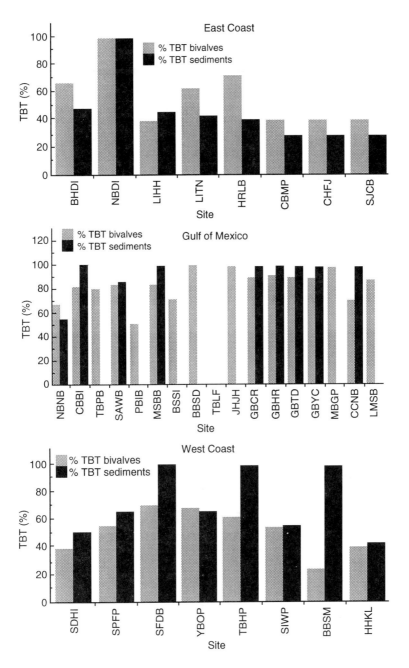

Figure 9.2. Comparison of TBT percentages in sediments and bivalves. See Fig. 9.3 for the location of sampling sites. Reprinted from Wade et al. (173), with permission of Pergamon Press Ltd, United Kingdom.

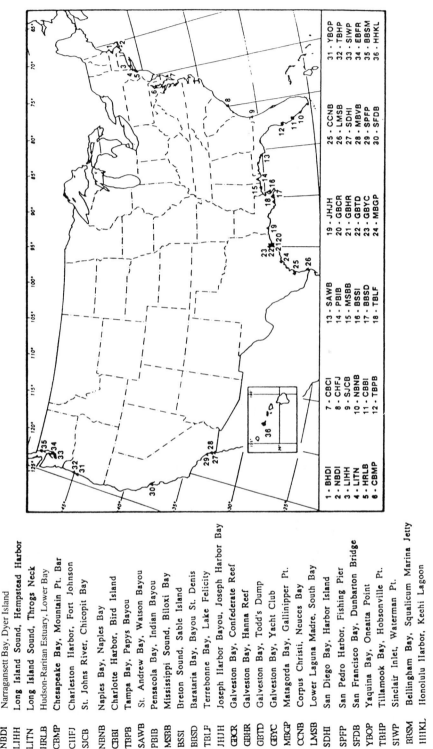

Figure 9.3. Sample site locations and designators. Reprinted from Wade et al. (173), with permission of Pergamon Press Ltd. United Kingdom.

total butyl-Sn concentrations in sediments and bivalves are compared, and in Fig. 9.2, where percentages of TBT are compared. Sampling site locations are shown in Fig. 9.3.

9.5.4. Italy

Data on organo-Sn compound environmental contamination in Italy are relatively recent and up to now rather scanty in view of extensive coastal development and aquaculture activities. Before 1991 only data on water concentration levels have been reported for Genoa, Leghorn, and La Spezia harbor areas (174–178). A general monitoring program was started by the present authors in 1989 to determine the organo-Sn compound levels in water, sediments, and aquatic organisms in various areas. These areas, mainly harbors, were selected on the basis of maritime traffic and aquaculture activities.

In Fig. 9.4 are shown the concentration levels of organo-Sn compounds found in mussel samples collected from seven different sites. Total organo-Sn compound concentrations range from a few hundred nanograms per gram (the Tiber estuary, Rome, and the Lagoon of Venice) up to more than $10\,\mu g\,g^{-1}$ (Genoa harbor). TBT was always the prevalent species, whereas phenyl-Sn compounds were detected only in samples with a total organo-Sn compound content higher than $1\,\mu g\,g^{-1}$ (179). Deeper investigations were carried out in La Spezia and Taranto harbors based on the following considerations:

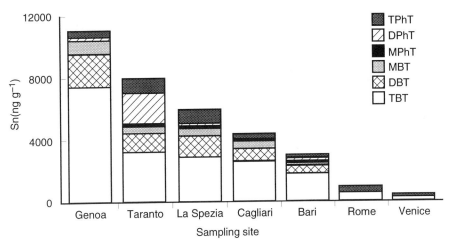

Figure 9.4. Concentration levels of organo-Sn compounds in mussels from five Italian harbors as well as the Lagoon of Venice and the Tiber estuary, Rome. Reprinted from Caricchia et al. (179), with permission of the Editor of *Analytical Sciences*, Japan Society for Analytical Chemistry, Tokyo, Japan.

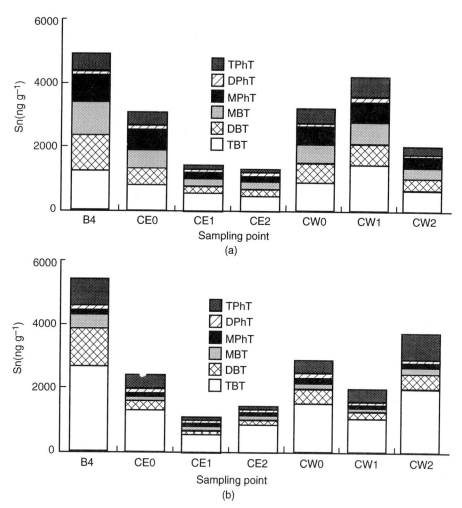

Figure 9.5. Concentration levels of organo-Sn compounds in mussels from the Gulf of La Spezia; (a) Summer sampling; (b) winter sampling. *Key*: C = outside of the harbor; E = East; W = West; B = mussel farm; 0, 1, 2, and 4 = numbering of sample locations. Reprinted from Caricchia et al. (179), with permission of the Editor of Analytical Sciences, Tokyo, Japan.

(i) data on mussel samples showed high organo-Sn compound levels; (ii) two of the most important Italian mussel farms were located in La Spezia and Taranto harbor areas.

The organo-Sn compounds concentration levels detected in August 1989 and in January 1990 in mussels from La Spezia are shown in Fig. 9.5 (179). As expected, the organo-Sn compound content was higher in mussels from the inner part of the gulf (B4 sampling point), where the mussel farm is located. The other samples were collected on the bottom of rocks outside the harbor.

The organo-Sn compound total content and the trend as functions of the sampling point are practically the same in 1989 and in 1990, but Sn speciation in mussels collected at the same sampling point is markedly different. In the 1989 samples the degradation products are more than 50%; on the other hand, in the 1990 samples the trisubstituted species are clearly prevalent. Analyses on sediments and water collected in the same area indicated that probably the trisubstituted compounds are the only species capable of accumulating at a significant rate and that di- and monosubstituted ones are formed by degradation processes (180). Moreover, the agreement between 1989 and 1990 data expressed as total content seems to show that a steady state is achieved in both cases.

Thus, the different ratios among species between summer and winter samplings might be attributable to different degradation rates; in particular, the degradation rate should be higher in summer than in winter. It is important to note that TBT concentration levels comparable with that found in the La Spezia mussels have been shown to produce some adverse effects on mussels (181). As yet, no significant adverse effects have been found in the La Spezia mussel farm (178). Comparison with results obtained on water and sediment samples (Table 9.4) showed that very high bioconcentration factors, up to 380,000, were reached (180).

Furthermore, concentration factors for both TBT and DBT are higher in mussels than in sediments, but, whereas the mussels/sediment concentration ratio ranges from 3:1 to 5:1 for TBT, it goes up to 23:1 for DBT. The explanation can be found in the fact that in mussels the primary degradation product is DBT, whereas in sediments it is MBT that, because of its hydrophilicity, may be released to the water column a short time after its formation, as reported in the literature (182). This suggests two different degradation mechanisms in mussels and sediments. Both these mechanisms are characterized by slow kinetics with respect to degradation mechanisms in water.

As regards the Taranto harbor, water, sediments, and mussel samples were collected in summer. The concentrations found in mussels were ca. 5-fold less than those at La Spezia and 8- to 10-fold less than that of a previous sample collected in Taranto 6 months before the monitoring program began. Actually, these concentrations could be affected by short exposure time, because only juvenile animals were available. Nevertheless, comparison with results obtained in water samples showed, in this case also, a high BCF in spite of the small size of the mussels collected. It is worth stressing that the BCF calculated from the La Spezia and Taranto results, as well as from other field determinations, are much higher than those that would be predicted by the behavior of model hydrocarbons and synthetic organic compounds (19).

Table 9.4. Water, Mussel, and Sediment Concentrations (as Sn) in the Gulf of La Spezia

Sampling Site[a,b]		Water: Surface (ng L^{-1})	Water: 5 m Depth (ng L^{-1})	Mussels (ng g^{-1} dry weight)	Bioconcentration Factor (BCF)	Sediments (ng g^{-1} dry weight)	Accumulation Factor
1 (B4)	TBT	15	22	2940	130,000–200,000	400	18,000–26,000
	DBT	85	14	1290	15,000–90,000	60	650–4200
	MBT	—	—	400	—	60	—
2 (A1)	TBT	11	13	—	—	550	42,000–50,000
	DBT	86	9	—	—	110	1300–12,500
	MBT	—	—	—	—	140	—
3 (A5)	TBT	19	12	—	—	350	18,400–29,000
	DBT	123	16	—	—	220	1800–14,700
	MBT	—	—	—	—	180	—
4 (CE2)	TBT	10	7	1010	100,000–150,000	580	58,000–83,000
	DBT	5 (CW2)	3	180	40,000–60,000	<50	—
	MBT	—	—	<50	—	<50	—
5 (CW2)	TBT	10	6	2270	230,000–380,000	750	75,000–125,000
	DBT	6	4	570	100,000–140,000	<50	—
	MBT	—	—	190	—	<50	—

Source: From Chiavarini et al. (180).

[a]Sites: **1.** mussel farm; **2.** dockyard activities; **3.** military harbor; **4 & 5**, open gulf outside the harbor.
[b]For key to letters and numbers in parentheses, see legend to Fig. 9.5 (page 316).

Recently, environmental studies in the Gulf of Olbia (Sardinia) have been begun. Their main goal is elucidation of accumulation phenomena of organo-Sn compounds in biota. The relationship between the organo-Sn concentration levels and the age of the organism is to be addressed, as well as identification of the organs where preferential accumulation occurs. Preliminary results indicate that cultured bass significantly accumulate organo-Sn compounds, and studies to assess the correlation with damage to hepatocytes are being planned.

9.5.5. Other Countries

Investigations on organo-Sn compound environmental distribution were also carried out in other countries. Data on natural populations of dogwhelks, pacific oysters, and mussels along the North Sea coast of the Netherlands have been reported by Laane et al. (183). Low imposex induction for dogwhelks and an appreciable valve thickening were observed. TBT concentrations in mussels ranged from 46 to 945 ng g^{-1} at four different sites.

A case study was conducted within the framework of the FAO/UNEP/IAEA/WHO Mediterrranean Organo-Sn Pilot Monitoring Programme. Mussel samples were collected on the coast from Genoa (Italy) to Sète (France) at 14 sampling sites. TBT and DBT concentrations ranged from 0.3 to 13.5 μg g^{-1} and from 0.5 to 14.5 μg^{-1}, respectively. High contamination levels (TBT concentrations higher than 10 μg g^{-1}) were found at Genoa, Toulon, and Marseille (184).

Data on 13 Irish coastal areas were reported by Minchin et al. (185). TBT concentrations in molluscan tissue (*Crassostrea gigas*, *Mytilus edulis*, *Lima hians*, *Pecten maximus*, and *Ostrea edulis*) were generally less than 0.3 μg g^{-1} and in six samples were under the detection limit of the employed technique (0.05 μg g^{-1}). In Mulray Bay bivalve beds either failed to be established or were reduced, presumably due to the use of organo-Sn compound net dips on salmon farms in the area (185).

The Environment Agency of Japan conducted organo-Sn compound pollution surveys on several coastal zones. The survey, which lasted for 4 years, revealed in 1986/87 high levels of TBT in fish tissue: for sea bass, the TBT concentrations were mainly in the range 0.1–0.3 μg g^{-1} fresh weight (186). The 1988 survey, after restrictions on TBT use were imposed in Japan, revealed that triphenyl-Sn was becoming a more serious pollutant than TBT. TPhT levels in fish were found to be two to three times those of TBT, ranging from 0.02 to 2.6 μg g^{-1} wet weight (187). After the 1989 restrictions on TPhT, its concentration in mussels was found to decrease rapidly, while the same did not happen in sediments (188).

The Sado estuary (Portugal), located south of Lisbon, suffered extensive death of oysters and other bivalves. Monitoring programs were carried out in the latter half of the 1980s and water, sediments, and mussels samples were collected. High contamination levels of butyl-Sn compounds in sediments were sometimes found, while in mussels data ranged from 10 to 300 ng g^{-1} dry weight on samples collected from January to May 1986 (18).

Samples of freshwater mussels (*Dreissena polymorpha*) and fish (*Leuciscus cephalus*) were taken in 1989 in marinas located in the oligotrophic Lake Lucerne in Switzerland, as were water and sediment samples. Both butyl-Sn compounds and phenyl-Sn compounds were recovered in all biological samples. TBT and TPhT were found to be the prevailing species at extremely high levels for mussels (up to 9.35 μg g^{-1} TBT and 3.88 μg^{-1} TPhT wet weight), far exceeding the levels usually found in marine mussels (189). A similar study has been performed in 1988 in Lake Geneva, both on the French and Swiss sides. Samples of freshwater mussels (*Dreissena polymorpha*), other bivalves (*Anodonta cignae*), and fish-eating birds (*Podiceps nigricollis*), together with water and sediments, were analyzed for butyl-Sn compounds and phenyl-Sn compounds. Mussel samples showed the highest TBT levels, especially in marinas on the Swiss side (up to 9 μg g^{-1} wet weight in a Geneva marina). TPhT was found only in mussel samples from Swiss marinas at levels in the range 0.6–3.3 μg g^{-1} wet weight. Bird livers contained mainly DBT at 0.35–0.57 μg g^{-1} levels. Mussels from a control point had TBT levels of 2.1 μg g^{-1}, demonstrating a generalized environmental contamination (190).

Organo-Sn compounds, along with a variety of other organic contaminants, were monitored in the marine environment around Barcelona, Spain. TBT, TPhT, and their degradation products were found in mussels, polychaetes, and fish, with bioconcentration factors of $3-6 \times 10^4$ for TBT (191–193).

9.6. CONCLUSIONS

Beginning in the early 1980s, mounting evidence that TBT antifouling paints has had deleterious effects on nontarget organisms, some of which are economically important, has generated more and more research activity seeking to assess the distribution and fate of TBT and related compounds and their lethal and sublethal effects on a wide variety of aquatic organisms.

That research focused on at least three priority fields: development of analytical methods; study of environmental distribution and fate; and study of toxicity and bioaccumulation.

Analytical methods were enhanced by the development and optimization of procedures able to determine TBT and its related compounds down to very low environmental concentrations (usually nanograms per liter in water and nanograms per grams in sediments). Furthermore, these methods have to be able to discriminate among different chemical forms due to the great difference in toxicity between TBT and its degradation products.

During the last decade many monitoring programs have also been carried out in several countries to assess the distribution of organo-Sn compounds throughout the environment in various biotic systems, as was the case for the regular sampling performed in Arcachon Bay in southwestern France. Data on water and sediments represent the main body of information, although during the last few years data on marine organisms have become more and more available.

Laboratory toxicity studies have been performed on a wide variety of organisms in order to ascertain toxicity parameters such as LD_{50}, mainly for TBT, but also for TPhT and other organo-Sn compounds. Moreover, various sublethal effects have been described, such as shell thickening, chambering, or imposex. At the same time several studies have sought to cast light on bioaccumulation phenomena. In this context, the BCF has been predicted for many organo-Sn compounds on the basis of their partition coefficients (K_{ow}) and laboratory experiments. However, calculated BCFs have shown poor correlations with those measured in the field, thus reducing the value of the predicted BCFs as reference parameters.

Legal restrictions on the use of TBT-based antifouling paints has been enacted by legislatures in several countries: France first introduced restrictive criteria, followed by the United Kingdom, the United States, Canada, and other countries. The effectiveness of such legal provisions is still open to debate, but it should be stressed that the whole matter is very problematic: the historical records of organo-Sn pollution levels are generally inadequate and data gathered have generally been obtained using different and not yet standardized analytical procedures, thereby affecting the homogeneity of results.

Recent trends in environmental legislation seem to indicate that an absolute ban of TBT-based antifouling paints is the ultimate goal. On the other hand, this trend is hindered by the absence of substitutes that have the same high degree of effectiveness against foulant organisms and lower environmental persistance and toxicity against nontarget organisms. At this point the need for expanded national and international monitoring programs validated by the use of standardized and certified operating procedures (sampling, storage, treatment, and analytical methodologies) poses a real challenge that will have to be met in the coming years.

REFERENCES

1. S. J. Blunden, L. A. Hobbs, and P. J. Smith, in *Environmental Chemistry* (H. J. M. Bowen ed), pp. 49–77. The Royal Society of Chemistry, London, 1984.

2. S. J. Blunden and A. H. Chapman, in *Organometallic Compounds in the Environment* (P. J. Craig, ed.), pp. 111–139. Longman, 1986.
3. C. J. Evans and R. Hill, *Rev. Silicon, Germanium, Tin, Lead Com.* **7**, 57–125 (1983).
4. B. Sugavanam, F. E. Smith, and S. Haynes, *I.T.R.I. Publ.* **607**, 1–8 (1981).
5. P. J. Smith, *I.T.R.I. Publ.* **538**, 1–20 (1978).
6. M. A. Champ, *Proc. Oceans Organotin Symp.*, Washington, DC, *1986*, Vol. 4, pp. 1–8 (1986).
7. C. Alzieu, J. Sanjuan, J. P. Deltrell, and M. Borel, *Mar. Pollut. Bull.* **17**, 494–498 (1986).
8. M. D. Müller, *Anal. Chem.* **59**, 617–623 (1987).
9. W. Langseth, *Talanta* **31**, 975–978 (1984).
10. R. J. Maguire, J. H. Carey, and E. J. Hale, *J. Agric. Food Chem.* **31**, 1060–1065 (1983).
11. R. J. Maguire, P. T. S. Wong, and J. S. Rhamey, *Can. J. Fish. Aquat. Sci.* **41**, 537–540 (1984).
12. J. A. J. Thompson, M. G. Sheffer, R. C. Pierce, Y. K. Chau, J. J. Cooney, W. R. Cullen, and R. J. Maguire, *Organotin Compounds in the Aquatic Environment*, p. 284. Nat. Res. Counc. Can., Ottawa, 1985.
13. R. J. Maguire, *Appl. Organomet. Chem.* **1**, 475–498 (1987).
14. C. Alzieu, M. Héral, Y. Thibaud, M. J. Dardignac, and M. Feuillet, *Rev. Trav. Inst. Péches Marit.* **45**, 101–116 (1982).
15. M. J. Waldock and J. E. Thain, *Mar. Pollut. Bull.* **14**, 411–415 (1983).
16. M. O. Andreae, J. T. Byrd, and P. N. Froelich, Jr., *Environ. Sci. Technol.* **17**, 731–737 (1983).
17. T. Stroemgren and T. Bongard, *Mar. Pollut. Bull.* **18**, 30–31 (1987).
18. Ph. Quevauviller, R. Lavigne, R. Pinel, and M. Astruc, *Environ. Pollut.* **57**, 149–166 (1989).
19. R. B. Laughlin, Jr., W. French, and H. E. Guard, *Environ. Sci. Technol.* **20**, 884–890 (1986).
20. C. Alzieu, *Proc. Int. Organotin Symp.*, 3rd, Monaco, *1990*, pp. 1–2 (1990).
21. Ph. Quevauviller and O. F. X. Donard, Chapter 10 of the present volume.
22. U. S. Environmental Protection Agency, *Tributyltin Technical Support Document*, U.S. EPA Position Doc. 2/3. Office of Pesticide Programs, Washington, DC, 1987.
23. P. J. Smith, *I.T.R.I. Publ.* **569**, 1–12 (1978).
24. A. Sylph, *I.T.R.I. Publ.* **LB12**, 1–14 (1984).
25. D. H. Calloway and J. J. McCullen, *Am. J. Chem. Nutr.* **18**, 1–12 (1966).
26. H. A. Schroeder, J. J. Balassa, and I. H. Tipton, *J. Chronic Dis.* **17**, 483–502 (1964).
27. P. J. Smith and L. Smith, *Chem. Br.* **11**, 208–212 (1975).
28. C. J. Evans and P. J. Smith, *J. Oil Colour Chem. Assoc.* **58**, 160–168 (1975).

29. R. B. Laughlin and O. Linden, *Ambio* **14**, 88–94 (1985).
30. E. J. Underwood, *Trace Elements in Human and Animal Nutrition*. Academic Press, New York, 1971.
31. J. E. Cremer, *Biochem. J.* **68**, 685–692 (1958).
32. H. Iwai and O. Wada, *Ind. Health* **19**, 247–253 (1981).
33. IRPTC, *Mediterranean Data Profiles: Organotins. Data Profiles for Chemicals for the Evaluation of Their Hazards to the Environment of the Mediterranean Sea*, Annex C, pp. 201–232. UNEP (United Nations Environment Programme), Geneva, 1978.
34. T. Alajouanine, L. Dérobert, and S. Thiéfry, *Rev. Neurol.* **98**, 85–96 (1958).
35. A. Sylph, *I.T.R.I. Publ.* **LB11** (1984).
36. S. J. Blunden, P. J. Smith, and B. Sugavanam, *Pestic. Sci.* **15**, 253–259 (1984).
37. M. J. Selwyn, *Adv. Chem. Ser.* **157**, 204–226 (1976).
38. J. S. Morrow and R. A. Anderson, *Lab. Invest.* **54**, 237–240 (1986).
39. W. N. Aldridge, *Biochem. J.* **69**, 367–376 (1958).
40. B. H. Gray, M. Porvaznik, C. Flemming, and L. H. Lee, *Proc. Oceans Organotin Symp.*, Washington, DC, *1986*, Vol. 4, pp. 1234–1245 (1986).
41. W. N. Aldridge, *Adv. Chem. Ser.* **157**, 186–196 (1976).
42. K. Cain, R. L. Hyams, and D. E. Griffiths, *FEBS Lett.* **82**, 23–28 (1977).
43. A. H. Henninks and W. Seinen, *Toxicol. Appl. Pharmacol.* **56**, 221–231 (1980).
44. A. G. Davies and P. J. Smith, in *Comprehensive Organometallic Chemistry: The Synthesis, Reactions and Structures of Organometallic Compounds* (G. Wilkinson, F. G. A. Stone, and E. W. Abel, eds.), pp. 519–627. Pergamon, Toronto, 1982.
45. A. H. Penninks, P. M. Verschuren, and W. Seinen, *Toxicol. Appl. Pharmacol.* **70**, 115–120 (1983).
46. M. Stockdale, A. P. Dawson, and M. J. Selwyn, *Eur. J. Biochem.* **15**, 342–351 (1970).
47. N. F. Cardarelli, *Controlled Release Molluscicides, Environ. Manage. Lab. Mon.*, Akron, OH, p. 34 (1977).
48. M. S. Rose and E. A. Lock, *Biochem. J.* **120**, 151–157 (1970).
49. F. Taketa, K. Siebenlist, J. Kasten-Jolly, and N. Palosaari, *Arch. Biochem. Biophys.* **203**, 466–472 (1980).
50. K. Cain and D. E. Griffiths, *Biochem. J.* **162**, 575–580 (1977).
51. B. M. Elliot, W. N. Aldridge, and J. W. Bridges, *Biochem. J.* **177**, 461–470 (1979).
52. M. T. Musmeci, G. Madonia, M. T. Lo Giudice, A. Silvestri, G. Ruisi, and R. Barbieri, *Appl. Organomet. Chem.* **6**, 127–138 (1992).
53. F. E. Brinckman, G. J. Olson, W. R. Blair, and E. J. Parks, *ASTM Spec. Tech. Publ.* **STP 971**, 219–232 (1988).
54. G. Eng, G. J. Tierney, J. M. Bellama, and F. E. Brinckman, *Appl. Organomet. Chem.* **2**, 171–175 (1988).
55. J. C. McGowan and A. Mellors, *Molecular Volumes in Chemistry and Biology: Applications Including Partitioning and Toxicity*. Wiley, New York, 1986.

56. G. Eng, E. J. Tierney, G. J. Olson, F. E. Brinkman, and J. M. Bellama, *Appl. Organomet. Chem.* **5**, 33–37 (1991).
57. G. Schuurman and G. Roderer, in *Heavy Metals in the Hydrological Cycle* (M. Astruc and J. N. Lester, eds.), pp. 433–440. Selper Ltd., London, 1988.
58. H. Floch, R. Deshiens, and T. Floch, *Bull. Soc. Pathol. Exot.* **57**, 454–465 (1964).
59. J. S. Alabaster, *Int. Pest Control* **11**, 29–35 (1969).
60. H. de Vries, A. H. Penninks, N. J. Snoeij, and W. Seinen, *Sci. Total Environ.* **103**, 229–243 (1991).
61. M. Polster and K. Halacka, *Ernaehrungsforschung* **16**, 527–535 (1971).
62. P. Matthiessen, *Proc. Int. Controlled Release Pestic.* Akron, OH, pp. 25.1–25.17 (1974).
63. M. Rexrode, *Proc. Oceans Organotin Symp.*, Halifax, *1987*, Vol. 4, pp. 1443–1455 (1987).
64. P. Schatzeberg and L. Harris, in *Report of the Organotin Workshop* (M. Good, ed.), pp. 95–107. University of New Orleans, New Orleans, LA, 1978.
65. W. Seinen, T. Helder, H. Vernij, A. Penninks, and P. Leeuwangh, *Sci. Total Environ.* **19**, 155–156 (1981).
66. R. Bock, *Res. Rev.* **79**, 1–262 (1981).
67. G. Gras and J. A. Rioux, *Arch. Inst. Pasteur Tunis* **42**, 9–22 (1965).
68. H. Floch and R. Deschiens, *Bull. Soc. Pathol. Exot.* **55**, 816–831 (1962).
69. T. E. Tooby, P. A. Hursey, and J. S. Alabaster, *Chem. Ind.* **12**, 523–526 (1975).
70. A. Linden, B. E. Bengtsson, O. Svanberg, and G. Sundström, *Chemosphere* **8**, 843–851 (1979).
71. G. S. Ward, G. C. Cramm, P. R. Parrish, H. Trachman, and A. Schlesinger, *ASTM Spec. Tech. Publ.* **STP 737**, 183–200 (1981).
72. S. J. Bushong, L. W. Hall, W. S. Hall, W. E. Johnson, and R. L. Herman, *Water Res.* **22**, 1027–1032 (1988).
73. M & T Chemicals Inc., Report L14-500, submitted by EG & G Bionomics to M & T Chemicals Inc., Rahway, NY, 1979.
74. J. E. Thain, *ICES Pap.* **1983/E:28** (1983).
75. J. W. Short and F. P. Thrower, *Proc. Oceans Organotin Symp.*, Washington, DC, *1986*, Vol. 4, pp. 1202–1205 (1986).
76. A. R. Beaumont and M. D. Budd, *Mar. Pollut. Bull.* **15**, 402–405 (1984).
77. L. S. Ritchie, V. A. Lopez, and J. M. Cora, in *Molluscicides in Schistosomiasis Control* (T. C. Cheng, ed.), pp. 77–88. Academic Press, New York, 1974.
78. U.S. Environmental Protection Agency, U.S. EPA Office of Pesticide Programs. USEPA, Washington, DC, 1985.
79. J. Dunkan, *Pharmacol. Ther.* **10**, 407–429 (1980).
80. S. C. U'ren, *Mar. Pollut. Bull.* **14**, 303–306 (1983).
81. M. Vighi and D. Calamari, *Chemosphere* **14**, 1925–1932 (1985).
82. S. J. Bushong, W. S. Hall, W. E. Johnson, and L. W. Hall, Jr., *Proc. Oceans Organotin Symp.*, Halifax, *1987*, Vol. 4, pp. 1499–1503 (1987).

83. B. M. Davidson, A. O. Valkirs, and P. F. Seligman, *Proc. Oceans Organotin Symp.*, Washington DC, *1986*, Vol. 4, pp. 1219–1225 (1986).
84. R. B. Laughlin, Jr. and W. J. French, *Bull. Environ. Contam. Toxicol.* **25**, 802–809 (1980).
85. A. R. Beaumont and P. B. Newman, *Mar. Pollut. Bull.* **17**, 457–461 (1986).
86. G. E. Walsh, L. L. McLaughlin, E. M. Lores, M. K. Louie, and C. H. Deans, *Chemosphere* **14**, pp. 383–392 (1985).
87. J. D. Paul and I. M. Davies, *Aquaculture* **54**, 191–203 (1986).
88. I. F. Lawler and J. C. Aldrich, *Mar. Pollut. Bull.* **18**, 274–278 (1987).
89. R. B. Laughlin, K. Norlund, and O. Linden, *Mar. Environ. Res.* **12**, 243–271 (1984).
90. R. B. Laughlin, W. J. French, and H. E. Guard, *Water, Air, Soil Pollut.* **20**, 69–79 (1983).
91. A. R. D. Stebbing, *Sci. Total Environ.* **22**, 213–234 (1982).
92. P. E. Gibbs and G. W. Bryan, *J. Mar. Biol. Assoc. U.K.* **66**, 767–777 (1986).
93. D. W. Bruno and A. E. Ellis, *Aquaculture* **72**, 15–20 (1988).
94. P. W. Balls, *Aquaculture* **65**, 227–237 (1987).
95. J. W. Short and F. P. Thrower, *Aquaculture* **61**, 193–200 (1987).
96. E. His and R. Robert, *Rev. Trav. Inst. Péches. Marit.* **47**, 63–88 (1983).
97. R. Robert and E. His, *ICES Pap. C. M.* **1981/F:42** (1981).
98. F. Gendron, Ph.D. Thesis, University of Aix–Marseille, France (1985).
99. R. B. Laughlin, O. Linden, and H. E. Guard, *Bull. Liaison Com. Int. Perm. Rech. Presérv. Matér. Milieu Marin (COIPM)*, No. 13, pp. 3–20 (1982).
100. H. Plum, *Inf. Chim.* **220**, 135–139 (1981).
101. A. O. Valkirs, B. M. Davidson, and P. F. Seligman, *Chemosphere* **16**, 201–220 (1987).
102. R. S. Henderson, *Proc. Oceans Organotin Symp.*, Washington, DC, *1986*, Vol. 4, pp. 1226–1233 (1986).
103. R. B. Laughlin, P. Pendoley, and R. G. Gustafson, *Proc. Oceans Organotin Symp.*, Halifax, *1987*, Vol. 4, pp. 1494–1498 (1987).
104. D. R. Dixon and H. Prosser, *Aquat. Toxicol.* **8**, 185–195 (1986).
105. B. S. Smith, *J. Appl. Toxicol.* **1**, 141–144 (1981).
106. G. S. Weis, J. Gottlieb, and J. Kwiatkowski, *Arch. Environ. Contam. Toxicol.* **16**, 321–326 (1987).
107. G. E. Walsh, L. L. McLaughlin, M. K. Louie, C. H. Deans, and E. M. Lores, *Ecotoxicol. Environ. Saf.* **12**, 95–100 (1986).
108. D. Chagot, *Proc. Int. Organotin Symp., 3rd*, Monaco, *1990*, p. 87 (1990).
109. N. Spooner and L. J. Goad, *Proc. Int. Organotin Symp., 3rd*, Monaco, *1990*, pp. 88–92 (1990).
110. C. Feral and S. Legall, in *Molluscan Neuroendocrinology* (J. Lever and H. H. Boer, eds.), pp. 173–175. North-Holland Publ. Amsterdam, 1983.

111. J. W. Short, J. L. Sharp-Dahl, W. B. Stickle, and S. D. Rice, *Proc. Int. Organotin Symp., 3rd*, Monaco, *1990*, pp. 105–109 (1990).
112. W. J. Langston, *Proc. Int. Organotin Symp., 3rd*, Monaco, *1990*, pp. 110–113 (1990).
113. P. W. Wester and J. H. Canton, *Aquat. Toxicol.* **10**, 143–165 (1987).
114. K. Fent, *Proc. Int. Organotin Symp., 3rd*, Monaco, *1990*, pp. 114–118 (1990).
115. G. W. Bryan, P. E. Gibbs, R. J. Huggett, L. A. Curtis, D. S. Bailey, and D. M. Dauer, *Mar. Pollut. Bull.* **20**, 458–462 (1989).
116. A. E. Pinkney, L. L. Matteson, and D. A. Wright, *Proc. Oceans Organotin Symp.*, Baltimore, *1988*, Vol. 4, pp. 987–991 (1988).
117. F. Newton, A. Thum, B. M. Davidson, A. O. Valkirs, and P. F. Seligman, Hormestic Effects on the Growth and Survival of Eggs and Embryos of the California Grunion (*Leuresthes tenuis*) Exposed to Trace Levels of Tributyltin, *Tech. Rep. 1040*. Naval Ocean Systems Center, San Diego, CA, 1985.
118. P. Kerrison and G. Phillips, *Proc. Int. Organotin Symp., 3rd*, Monaco, *1990*, pp. 119–126 (1990).
119. Y. P. Chliamovitch and C. Kuhn, *J. Fish. Biol.* **10**, 575–585 (1977).
120. J. E. Thain and M. J. Waldock, *ICES Pap.* **1985/E:13** (1985).
121. J. E. Thain, *Proc. Oceans Organotin Symp.*, Washington, DC, *1986*, Vol. 4, pp. 1306–1313 (1986).
122. G. W. Bryan, P. E. Gibbs, L. G. Hummerstone, and G. R. Burt, *J. Mar. Biol. Assoc. U.K.* **66**, 611–640 (1986).
123. B. S. Smith, *Veliger* **23**, 212–216 (1981).
124. J. P. Meador, *Proc. Oceans Organotin Symp.*, Washington, DC, *1986*, Vol 4, pp. 1213–1218 (1986).
125. P. T. S. Wong, Y. K. Chau, O. Kramar, and G. H. Bergert, *Can. J. Fish. Aquat. Sci.* **39**, 483–488 (1982).
126. D. W. Connell and G. J. Miller, *Chemistry and Ecotoxicology of Pollution*, pp. 31–37. Wiley, New York, 1984.
127. R. F. Lee, *Proc. Int. Organotin Symp., 3rd*, Monaco, *1990*, pp. 70–76 (1990).
128. R. H. Fish, E. C. Kimmell, and J. E. Casida, *J. Organomet. Chem.* **118**, 41–54 (1976).
129. R. François, F. T. Short, and J. H. Weber, *Environ. Sci. Technol.* **23**, 191–196 (1989).
130. R. F. Lee, A. O. Valkirs, and P. F. Seligman, *Environ. Sci. Technol.* **23**, 1515–1518 (1989).
131. D. Warshawsky, M. Radike, K. Jayasimhulu, and T. Cody, *Biochem. Biophys. Res. Commun.* **152**, 540–544 (1988).
132. T. Tsuda, H. Nakanishi, S. Aoki, and J. Takebayashi, *Toxicol. Environ. Chem.* **12**, 137–143 (1986).
133. T. Tsuda, H. Nakanishi, S. Aoki, and J. Takebayashi, *Water Res.* **21**, 949–953 (1987).

134. T. Tsuda, H. Nakanishi, S. Aoki, and J. Takebayashi, *Water Res.* **22**, 647–651 (1988).
135. R. C. Martin, D. G. Dixon, R. J. Maguire, P. V. Hodson, and R. J. Tkacz, *Aquat. Toxicol.* **15**, 37–52 (1989).
136. G. S. Ward, G. C. Cramm, P. R. Parrish, H. Trachman, and A. Sclesinger, *ASTM Spec. Tech. Publ.* **STP 737**, 183–200 (1981).
137. K. Fent, R. Lovas, and J. Hunn, *Naturwissenschaften* **78**, 125–127 (1991).
138. K. Fent, *Aquat. Toxicol.* **20**, 147–158 (1991).
139. R. F. Lee, *Proc. Oceans Organotin Symp.*, Washington, DC, *1986*, Vol. 4, pp. 1182–1188 (1986).
140. R. F. Lee, *Mar. Environ. Res.* **17**, 145–148 (1985).
141. R. D. Rice, J. W. Short, and W. B. Stickle, *Mar. Environ. Res.* **27**, 137–145 (1989).
142. C. Zuolian and A. Jensen, *Mar. Pollut. Bull.* **20**, 281–286 (1989).
143. R. E. Anderson, *Mar. Environ. Res.* **17**, 137–140 (1985).
144. R. F. Lee, *Mar. Biol. Lett.* **2**, 87–105 (1981).
145. M. J. Waldock and D. Miller, *ICES Pap. C. M.* **1983/E:12** (1983).
146. I. M. Davies and J. C. McKie, *Mar. Pollut. Bull.* **18**, 405–407 (1987).
147. T. L. Wade, B. García-Romero, and J. M. Brooks, *Environ. Sci. Technol.* **22**, 1488–1493 (1988).
148. H. H. White and M. A. Champ, *ASTM Spec. Tech. Publ.* **STP 805**, 299–312 (1983).
149. S. M. Salazar and M. H. Salazar, in *Proc. Int. Organotin Symp., 3rd*, Monaco, *1990*, pp. 79–83 (1990).
150. *UNEP/FAO/WHO/IAEA MAP Technical Report Series*, No. 33, p. 185. UNEP, Athens, 1989.
151. H. Vrijhof, *Sci. Total Environ.* **43**, 221–231 (1985).
152. R. J. Huggett, M. A. Unger, P. F. Seligman, and A. O. Valkirs, *Environ. Sci. Technol.* **26**, 232–237 (1992).
153. R. Abel, R. A. Hathaway, N. J. King, J. L. Vosser, and T. G. Wilkinson, in *Proc. Oceans Organotin Symp.*, Halifax, *1987*, Vol. 4, pp. 1314–1319 (1987).
154. R. B. Laughlin and O. Linden, *Ambio* **16**, 252–256 (1987).
155. U. S. Navy, *Fleetwide Use of Organotin Antifouling Paints: Environmental Assessment*. U.S. Navy, Washington, DC, 1984.
156. U.S. Environmental Protection Agency, *Data Call-in-Notice for Data on Tributyltins Used in Paint Antifoulants*, U.S. EPA Office of Pesticides and Toxic Substances, Washington, DC, 1986.
157. State of North Carolina, January 1985, Administrative Code Section 15 NCAC 2B 10100 and 15 NCAC 10200, Environmental Management Commission, Raleigh, NC, 1985.
158. *Gazzetta Ufficiale della Republica Italiana*, D.M. May 4, 1985 (Ministry of Health) published in G.U., May 27, 1985.

159. G.P. Gabrielides, *Proc. Int. Organotin Symp.*, *3rd*, Monaco, *1990*, pp. 3–4 (1990).
160. C. Alzieu, M. Heral, Y. Thibaud, M. J. Dardignac and M. Feuillet, *Rev. Trav. Inst. Pêches. Marit.* **44**, 301–348 (1982).
161. I. M. Davies, S. K. Bailey, and D. C. Moore, *Mar. Pollut. Bull.* **18**, 400–404 (1987).
162. P. F. Seligman, A. O. Valkirs, and R. F. Lee, *Environ. Sci. Technol.* **20**, 1229–1235 (1986).
163. P. F. Seligman, J. G. Grovhoug, and K. E. Richter, *Proc. Oceans Organotin Symp.*, Washington, DC, *1986*, Vol. 4, pp. 1289–1296 (1986).
164. P. F. Seligman, J. G. Grovhoug, A. O. Valkirs, P. M. Stang, R. Fransham, M. O. Stallard, B. Davidson, and R. F. Lee, *Appl. Organomet. Chem.* **3**, 31–47 (1989).
165. A. O. Valkirs, P. F. Seligman, P. M. Stang, V. Homer, S. H. Lieberman, G. Vafa, and A. C. Dooley, *Mar. Pollut. Bull.* **17**, 319–324 (1986).
166. M. Stephenson, D. R. Smith, J. Goetzl, G. Ichikawa, and M. Martin, in *Proc. Oceans Organotin Symp.*, Washington, DC, *1986*, Vol. 4, pp. 1246–1251 (1986).
167. A. O. Valkirs, B. Davidson, R. L. Fransham, J. G. Grovhoug, and P. F. Seligman, *Proc. Int. Organotin Symp.*, *3rd*, Monaco, *1990*, pp. 151–156 (1990).
168. M. Stephenson, *Proc. Int. Organotin Symp.*, *3rd*, Monaco, *1990*, pp. 157–159 (1990).
169. J. W. Short and J. L. Sharp, *Environ. Sci. Technol.* **23**, 740–743 (1989).
170. D. S. Page, T. Dassanayake, E. S. Gillfillan, and C. Kresja, *Proc. Int. Organotin Symp. 3rd*, Monaco, *1990*, pp. 170–175 (1990).
171. T. L. Wade and B. García Romero, *Proc. Int. Organotin Symp.*, *3rd*, Monaco, *1990*, pp. 202–204 (1990).
172. M. M. Saavedra Alvarez and D. V. Ellis, *Mar. Pollut. Bull.* **21**, 244–247 (1990).
173. T. L. Wade, B. García Romero, and J. M. Brooks, *Chemosphere* **20**, 647–662 (1990).
174. M. Di Cintio, *Inquinamento* **11**, 64–67 (1988).
175. E. Bacci and C. Gaggi, *Mar. Pollut. Bull.* **20**, 290–292 (1989).
176. E. Bacci and C. Gaggi, *Proc. Int. Organotin Symp.*, *3rd*, Monaco, *1990*, pp. 140–146 (1990).
177. S. Chiavarini, C. Cremisini, T. Ferri, R. Morabito, and A. Perini, *Sci. Total Environ.* **101**, 217–227 (1991).
178. A. M. Caricchia, S. Chiavarini, C. Cremisini, T. Ferri, and R. Morabito, *Sci. Total Environ* **121**, 133–144 (1992).
179. A. M. Caricchia, S. Chiavarini, C. Cremisini, R. Morabito, and R. Scerbo, *Anal. Sci.* **7** (Suppl.), 1193–1196 (1991).
180. S. Chiavarini, C. Cremisini, and R. Morabito, *UNEP/FAO/IAEA MAP Technical Report Series*, No. 59, pp. 179–187. UNEP, Athens, 1991.
181. D. S. Page and J. Widdows, *Proc. Int. Organotin Symp. 3rd*, Monaco, *1990*, pp. 127–131 (1990).

182. P. F. Seligman, R. F. Lee, A. O. Valkirs, and P. M. Stang, *Proc. Int. Organotin Symp., 3rd*, Monaco, *1990*, pp. 30–38 (1990).
183. R. W. P. M. Laane, R. Ritsema, and O. Donard, *Proc. Int. Organotin Symp., 3rd*, Monaco, *1990*, pp. 147–150 (1990).
184. D. S. Page, J. W. Readman, L. D. Mee, and M. McCarthy, *Proc. Int. Organotin Symp., 3rd*, Monaco, *1990*, pp. 179–184 (1990).
185. D. Minchin, C. B. Duggan, and W. King, *Mar. Pollut. Bull.* **18**, 604–608 (1987).
186. Environment Agency of Japan, *Sangyo Kougai* **24**, 56–59 (1988).
187. Environment Agency of Japan, *Sangyo Kougai* **25**, 715–717 (1989).
188. H. Shiraishi and M. Soma, *Chemosphere* **24**, 1103–1109 (1992).
189. K. Fent and J. Hunn, *Environ. Sci. Technol.* **25**, 956–963 (1991).
190. K. Becker, N. De Bertrand, and L. Merlini, *Rapport de Recherche, 3ème Cycle en Protection de l'Environment*. Ecole Polytechnique Fédérale de Lausanne, 1989.
191. J. M. Bayona, M. Valls, P. Fernandez, C. Porte, I. Tolosa, and J. Albaigés, *UNEP/FAO/IAEA MAP Technical Report Series* No. 59, pp. 85–97. UNEP, Athens, 1991.
192. M. Valls, J. M. Bayona, and J. Albaigés, *Int. J. Environ. Anal. Chem.* **39**, 329–348 (1990).
193. I. Tolosa, J. M. Bayona, J. Albaigés, L. Merlini, N. Bertrand, K. Becker, L. F. Alencastro, and J. Tarradellas, *Proc. Int. Organotin Symp., 3rd*, Monaco, *1990*, pp. 55–57 (1990).

CHAPTER

10

TIN SPECIATION MONITORING IN ESTUARINE AND COASTAL ENVIRONMENTS

PH. QUEVAUVILLER

European Commission, Standards, Measurements and Testing Programme,*
B-1040 Brussels, Belgium

O. F. X. DONARD

Laboratory of Photophysics and Molecular Photochemistry,
University of Bordeaux I, 3405 Talence, France

10.1. INTRODUCTION

The determination of organometallic species, i.e., research included under the rubric *speciation*, is necessary to improve our understanding of the direct input of trace metals into the environment and their subsequent integration into biogeochemical cycles. This is particularly so in the case of Sn, which is commonly used in organometallic compounds in a wide variety of agricultural and industrial applications including biocides. Expansion of the total amount of Sn utilized in such processes has brought about a serious increase in the quantity of this metal detected as an environmental contaminant over the last decade (1).

Organotin compounds are toxic at very low concentrations (in particular the trialkylated forms) in comparison with inorganic forms; this fact justifies the effort put into their determination. Further, these compounds are considered to be excellent indicators of anthropogenic activities; in fact, there are no "natural" butyl-Sn compounds. There have been no findings to date of any natural formation of butylated Sn species in the environment. On the other hand, while detection of methylated Sn species can seldon be attributed to wastewater treatment processes, their occurrence is known to be related

---

* Formerly BCR: Bureau Communautaire de Référence.

Element Speciation in Bioinorganic Chemistry, edited by Sergio Caroli.
Chemical Analysis Series, Vol. 135.
ISBN 0-471-57641-7 © 1996 John Wiley & Sons, Inc.

mainly to biogeochemical formation through methylation of inorganic Sn (1–3). It is thus recognized that Sn speciation in the environment still enables us, on the basis of the chemical forms identified, to distinguish between their natural vs. anthropogenic origin.

Estuaries are important geological transition zones in which hydrodynamic, biogeochemical, and ecological parameters, among others, fluctuate widely (4). These zones also constitute a key interface between the industrial world and the natural environment. In particular, direct inputs of contaminants resulting from harbor and/or leisure activities is bound to have some impact on fragile biological communities. Estuaries are indeed preferential areas for the spawning of many fish species as well as reproduction areas for many other organisms. The increased use of organo-Sn compounds inevitably theatens this fragile ecosystem by the release of a variety of organo-Sn species (e.g., tributyl-Sn used in antifouling paints and triphenyl-Sn used as a biocide in agriculture) (1, 5; see also Chapter 9 in this volume). Superimposed upon the cycle of these anthropogenic compounds a natural alkylation/dealkylation process generally occurs (e.g., methylation, degradation, and volatilization) that may be mediated by the high rate of biochemical activity in estuaries. These aspects have been widely studied in the past few years (3).

However, if the general biogeochemical pathways of organotins in the marine environment appear to be understood, well-defined monitoring strategies are needed to enable accurate data to be obtained that will in turn allow for worldwide comparison of results and modeling applications. The quality of measurements performed in monitoring programs relies on criteria such as the sample location, the type and treatment of samples, the collection protocol, the storage procedure, and quality control of the final determinations. It is of prime importance to limit all risk of contamination due to operations performed prior to the determination step. Indeed, although such errors may not be of great concern at the milligram per liter level, their effects can be dramatic when microgram per liter or nanogram per liter concentrations have to be detected. This matter is critical in the case of organotin compounds since levels detected in the environment are extremely low most of the time and yet can still significantly affect the biota. Careful sampling and storage procedures thus become critical steps in establishing the validity of the identification and quantification of Sn species.

An understanding of both the sources of variability and the main pathways in the marine environment is necessary to design an efficient sampling program and sample pretreatment strategy. In the following sections, we discuss the precautions to be used for monitoring butyl- and methyl-Sn compounds in estuarine and coastal waters, suspended matter, and sediments. Additional information on other compounds and related toxic effects can be found in the literature (1, 6, 7). As regards speciation of Sn in biological samples, the reader is referred to Chapter 9 in this volume (8).

10.2. THE DEVELOPMENT OF ENVIRONMENTAL CONCERNS REGARDING Sn COMPOUNDS

In the variety of contaminants monitored in estuarine environments, the case of organotin compounds is of particular importance, illustrating the extreme rapidity of the growth of international concern regarding environmental contamination. During the last few years, for this reason a large number of surveys have been carried out. The study of tributyl-Sn (TBT) toxicity and its impact on the biota has progressed rapidly. In 1983, His et al. were among the first authors to report the toxicity effects on oysters of TBT at microgram per liter levels (9). A few years later, Gibbs and Bryan demonstrated the deleterious effect on the small gastropod *Nucella lapillus* of TBT at the nanogram per liter level (10). As a result, the recommended acceptable TBT environmental quality target (EQT) value (Table 10.1), based on various criteria, rapidly decreased from 69 to $2 \, \text{ng L}^{-1}$ of TBT in saltwater (11–13).

When the critical values for TBT in water are evaluated using organisms as bioindicators, the following remarks should be taken into consideration. The impact of Sn compounds is usually based on chronic toxicity tests. If some organisms appear to be more responsive than usual to organotin compounds, one must bear in mind that organisms tend to integrate the great pressure of varied chemical stresses in the environment, which may induce synergistic

Table 10.1. Quality Level Recommendations for TBT Concentrations in Fresh- and Saltwaters

Concentrations [ng L^{-1} (as Sn)]	Sample	Speciation	Ref.
69	Saltwater	Based on chronic toxicity tests	11
303	Freshwater	Based on chronic toxicity tests	11
1.6	Saltwater	Not mentioned	12
26	Freshwater	Not mentioned	12
20	Estuarine water	Whole-water samples	13

Source: Adapted from ref. 60.

effects, resulting in an apparently amplified response to a specific chemical. Also, it is likely that the organisms tested are not the most responsive to TBT toxicity. Further, the source of entry and effects of the contaminant may vary according to the mode of ingestion of the chemicals by the animal. Some organisms are directly sensitive to the dissolved concentration of chemicals in the surrounding waters, whereas others (filter feeders) are more affected by the chemical load associated with suspended matter.

The use of bioassays to investigate the presence of contaminants in the environment is also growing rapidly and represents a more focused approach than the one based on monitoring mussel growth. The value of using bioassays to address the magnitude and impact of organo-Sn compounds in the environment has been discussed by Salazar (14). His main conclusion is that bioassays alone cannot give a clear view of the environmental impact of the presence of organo-Sn compounds. Hence, even if these screening tests may be effective, their utility should not be overestimated. Some of the main problems are related to the comparability of data between different areas. White and Champ (15) listed the following limiting factors for organo-Sn species surveys using bioassays: (i) there is considerable variation in results between laboratories using the same bioassay; (ii) bioassays do not accurately simulate the natural environment; and (iii) they are not good indicators of ecological consequences. The use and application of bioassays should, however, make important progress in the future and complement other types of surveys, such as chemical analysis.

These needs were addressed in 1986 by an interagency report in which recommendations for monitoring Sn species were given (16). Some of the discrepancies observed with regards to environmental quality levels reported in water may be partly due to the lack of comparability of methods used for assessing the organo-Sn compound contamination. Some considerations were based on theoretical calculations, whereas other were based on chronic toxicity tests or on direct measurements on water samples, either filtered or not filtered.

Also, some of these recommendations, with regard to sample collection, mainly originated from studies performed in areas with low hydrodynamic conditions such as prevail in enclosed bays or lakes. Most studies were also reported in microtidal estuaries associated with a very small suspended load, although the partitioning of organo-Sn compounds strongly varies in relation to the suspended matter content (as discussed below in Section 10.4). Nevertheless, all studies converged to establish the high toxicity of TBT. In view of the high biocidal action of TBT at extremely low concentrations, it is of prime importance to proceed a step further and to clearly delineate monitoring strategies as well as to establish sound environmental regulatory policies. Such recommendations ought to be adopted at an international level, and it should

10.3. GEOCHEMICAL PATHWAYS AND TRANSFORMATIONS

Very little is known about the global flux of Sn to the ocean. The major input from rivers have been estimated to be ca. 35,000 t (metric tons) per year. The inputs from the atmosphere is less than 1% of the global amount. Most of the inputs of Sn to the ocean have been estimated as occurring essentially in particulate form (7,17). Speciation via the formation of organo-Sn compounds (methylation) will completely modify its cycling in estuarine ecosystems (18). General pathways via methylation of inorganic Sn have been extensively described. However, these contributions have not received the same interest as those concerning the fate of toxic butyl-Sn compounds. It is nonetheless very likely that the environmental cycles of inorganic, methylated, and butylated Sn species are closely interrelated. The dealkylation of butyl-Sn compounds to finally produce inorganic Sn may possibly be reintegrated in the natural methylation cycle by natural biogeochemical methylation processes. It must be borne in mind that when determinations of Sn species in a sample are made, a series of alkylation/dealkylation reactions having different kinetics are competing with each other. A brief overview of the main biogeochemical pathways listed to date for Sn compounds is given in the following subsections. These pathways are closely linked to the chemical structure of the Sn compound, which controls its basic toxicity and also its fundamental biogeochemical properties with regard to partitioning, liquid–gas exchange or stability and persistence. A general overview of the diversity of these biochemical pathways is needed in order to design an appropriate sampling and monitoring strategy.

The variety of chemical species identified in natural and simulated estuarine environments illustrates the complexity of approaches required for the study of Sn and organo-Sn species cycle in such systems. Little is known about the different oxidation forms of Sn in the environment, this aspect still being controversial. Tin(IV) is assumed to be the predominant redox form in the environment (19). According to Pettine and coworkers, $Sn(OH)_2$ constitutes 94% of the Sn species in oxygenated seawater (20).

10.3.1. Methylation Reactions

Huey et al. first reported that methyl-Sn could be produced by the addition of inorganic Sn(II) or Sn(IV) to sediments (21). Since then the methylation of inorganic Sn was widely studied and mainly related to microorganism activity (22–26). Methylated species are generally thought to be derived from natural origins via a wide range of alkylation and dealkylation reactions, chemically or biologically mediated (27). Chemical methylating agents of Sn(II) and Sn(IV) have been investigated. Among them, methylcobalamin, the methyl coenzyme of vitamin B_{12} and CH_3I, produced by certain algae and seaweeds, have been demonstrated to be active (22, 28, 29). Manders et al. have also reported the oxidative methylation of SnS by methyl iodide in aqueous solution to form $MeSnI_3$ (30).

Methyl iodide is ubiquitous in marine environments and solubilizes bulk metals and refractory binary and ternary metal sulfides, producing methylated sulfur coproducts. Stannous sulfide and chloride react with MeI to produce methyl-Sn (IV) species. Recent studies have shown the occurrence of such pathways (31, 32). Similar pathways related to high biological turnover could explain the occurrence of high methyl-Sn contents observed in coral reef samples (33).

The feasibility of the methylation of Sn(II) compounds under environmentally simulated conditions has also been demonstrated (34, 35). Methylation reactions of inorganic Sn can be important in estuaries. The dissolved methylated form of Sn increases toward the seaward end of an estuary (17, 18). Methylated forms of Sn (mono-, di-, tri-, and tetramethyl-Sn) may represent between 60 and 90% of the total amount of Sn present in the dissolved phase (18, 36, 37). Little information is available on the particulate concentration of methylated Sn in estuaries. The concentrations reported to date suggest high concentrations in the nanogram per gram range (18, 38).

Gilmour et al. have shown that the Sn methylation rates were higher under anaerobic conditions; the production of methyl-Sn was significantly correlated with the numbers of both sulfate-reducing and sulfate-oxidizing bacteria (39). High levels of methylated Sn compounds have also been detected in sediments collected in anaerobic environments (33, 38).

Other possible methylating routes in the environment have been suggested. Transmethylation processes have been thought to occur between methylated Sn species and inorganic Hg to produce methylated Hg species (33). Finally, the possible methylation of butyl-Sn species to produce mixed methylbutyl-Sn compounds has been discussed, but this pathway was deemed to play only a minor role (40, 41). Moreover, the formal identification of the possible product (e.g., dimethylbutyl-Sn) is still controversial (42).

10.3.2. Volatilization Reactions

Other processes, such as volatilization leading to sea–air exchanges, have not been considered to a large extent in biogeochemical trace metal cycling studies. Until now, organo-Sn speciation studies have very seldom focused on the determination of volatile stannanes and tetraalkylated Sn species. Biotic and abiotic methylation and hydridization have been already recognized in experiments *in vitro* and under environmentally simulated conditions. However, the role of volatilization processes and their variability in the biogeochemical cycle of Sn in the aquatic environment is largely unknown (17).

Methylstannanes (Me_xSnH_{4-x}, where $x = 2$ or 3) were identified in cultures of *Pseudomonas* supplemented with inorganic Sn(IV) (43). In addition, the formation of methyl-Sn compounds and methylstannanes from $SnCl_2$ and $SnCl_4$ in sediments have been described (23, 24). More recently, in experiments *in vitro* Donard and Weber have demonstrated the formation of volatile stannanes (SnH_4) from decaying algal material exposed to inorganic and organic Sn (44). Volatilization of Sn as stannane and monomethyl-Sn trihydride has also been shown in sewage waters (2).

It has been suggested that 95% of the tetramethyl-Sn produced from trimethyl-Sn precursors in sediments occurs via a chemical rather than a biological pathway (34). It has also been found that following a treatment of naturally occurring sediments with CH_3I, dimethyl-Sn dihydride and dimethyl-Sn sulfide were formed almost immediately, whereas tetramethyl-Sn was detected after the sample was left standing overnight (45). Few experimental data are available as yet concerning the other alkylated volatile derivatives. In this regard, only Maguire has formally identified volatile Bu_3SnMe and Bu_2SnMe_2 in harbor sediments (40). The occurrence of Bu_3SnH has recently been reported in simulated batch reactors (46).

10.3.3. Degradation Reactions

Few results are available on the degradation of methyl-Sn compounds in the environment. Blunden showed that trimethyl-Sn is degraded photolytically (>235 nm) via dimethyl-Sn directly to inorganic Sn without formation of monomethyl-Sn (47). A half-life of trimethyl-Sn of 60 h was reported in water, whereas half-lives of 300 h and >300 h were estimated in separate experiments for di- and monomethyl-Sn, respectively.

The core of the studies performed deals with the persistence of toxic butylated species. Results are still controversial. TBT is known to degrade biologically and abiotically, yielding di- and monobutyl-Sn and inorganic Sn under natural conditions (30, 48). However, there is a broad range of half-lives reported in the literature (49). The TBT half-life in natural or spiked water samples was reported to vary between 6 days and 35 weeks depending on experimental conditions. Similarly, in sediments the reported half-lives ranged from 16 weeks to 15 years. Such a wide disparity certainly reflects the fact that samples and experimental conditions differ greatly from one author to another. It could also suggest that the accuracy of determinations was not equally mastered by all analysts over the past few years. In the marine environment,

TBT half-lives were shown to range between 7–15 days and several months in water samples (25, 50). However, strong degradation of TBT in water samples collected in European coastal environments was not clearly demonstrated since in most cases TBT was largely dominant in comparison with its degradation products monobutyltin (MBT) and dibutyltin (DBT) (51). Similar patterns were observed in waters from the Crouch estuary in the United Kingdom (13). This distribution is in the order TBT > DBT > MBT and may signify that a slight stepwise degradation may occur only in natural waters. This pattern could also mean, however, that constant inputs of TBT probably masked the real importance of the degradation. In the case of DBT, different groups of investigators reported half-lives of 5–15 days in fresh and seawater and 1–17 days in seawater (52–54). Thain et al. estimated a half-life of 90 days for DBT in seawater at 5 °C (55).

Reported half-lives for MBT are 2–14 days for seawater (53). However, Adelman et al. did not find any degradation of this compound in seawater over a 50 day period (54). In sediments, a distribution of butyl-Sn species in the order TBT > MBT > DBT was observed, which suggested that a direct degradation of TBT to MBT might be occurring or that DBT was less stable in

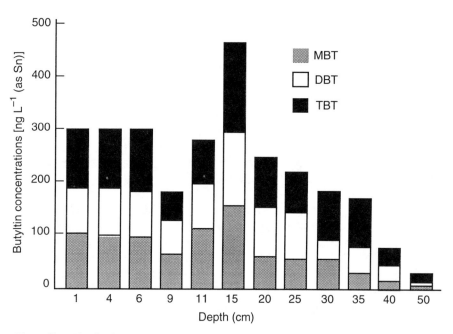

Figure 10.1. Distribution of butyl-Sn compounds in core sediments collected in Arcachon harbor. (Adapted from ref. 56.)

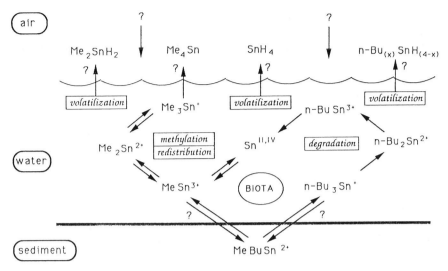

Figure 10.2. Biochemical pathways of Sn compounds in the environment. (Adapted from ref. 18.)

this medium and consequently desorbed (50, 51). The persistence of butyl-Sn in coastal sediments has been illustrated recently by the overall distribution of MBT, DBT, and TBT in a sediment core collected in Arcachon harbor (on the southwest coast of France), showing that the contents of the three species were displaying the same pattern, i.e., TBT degradation was apparently a minor phenomenon (Fig. 10.1). Hence the persistence of TBT in sediment presents a potential source of environmental contamination through harbor-dredging activities (56).

A summary of the main biochemical pathways and transformation of inorganic Sn and organo-Sn compound investigated to date is presented in Fig. 10.2.

10.4. SOURCES OF ENVIRONMENTAL VARIABILITY

The variety of chemical species identified in natural and simulated estuarine environments augments the complexity of the organo-Sn species cycle in such systems. Most reports cited in this chapter deal mainly with methylated or butylated organo-Sn compounds. Very few data are available as yet regarding phenyl- or octyl-Sn compounds in the environment. Many factors influence the persistence of organo-Sn chemicals in estuarine and coastal environments. Among the most important factors is variability, meaning that organo-Sn species inputs are discontinuous and related to localized point sources, primarily associated with antifouling paint leachates. Commercial ports or leisure harbors are necessarily the prime sources of direct inputs of organo-Sn compounds into the environment. Other biogeochemical factors controlling their dispersion and fate include the nature and concentration of dissolved and particulate material, the quality and concentration of microbial populations, the water temperature, the degree of insolation, etc. Maguire considered the main physical, chemical, and biological removal mechanisms of aquatic systems to be the following: (i) volatilization and adsorption to suspended solids and sediments; (ii) chemical and photochemical degradation or transformation; and (iii) uptake and transformation by microorganisms (1). Hydrodynamics in estuaries closely control the suspended load and hence the variability of organo-Sn compound contents; partitioning in water is likewise controlled by physical processes (dilution according to water discharge fluctuations and/or mixing of waters with different loads of suspended matter).

10.4.1. Temporal Variability

Short-term variability of organo-Sn species concentrations in water can often result from worst-case situations, such as increased concentrations during low tide or due to minimal rainfall, resuspension of contaminated bottom sediments during storms, or increased levels due to direct anthropogenic inputs (e.g., when vessels are being painted) (16). The variability of dissolved TBT over a tidal cycle has been demonstrated in San Diego Bay, where the maximum levels were observed at low tide (57). Important variability in the concentration of TBT (up to 100-fold) may be recorded on a yearly basis in enclosed marinas, as illustrated by the TBT values measured in a Dutch marina (Colijnsplaat) in 1989 (Fig. 10.3); highest concentrations detected in filtered water samples occurred in spring, associated with boat maintenance activities (either after setting the boats in water with antifouling paint freshly applied on their hulls or during scraping and repainting) (58). Long-term variations were also observed in some contaminated areas. However, significant variations in TBT concentrations in marine waters occurred between 1984 and 1986 along the southwest coast of England (13). Moreover, despite the fact that organo-Sn compound levels were expected to decrease after the ban of organo-Sn-containing antifouling paints in France in 1982, high levels were still found 10 years later in Arcachon Bay (56).

Seasonal variability is also an important parameter to be considered with respect to the quality of methylated species present in the water. The biological turnover is more important during summer and can completely modify the chemical distribution of methyl-Sn compounds through the natural methylation process of Sn. The relative amount of methyl-Sn species present in waters

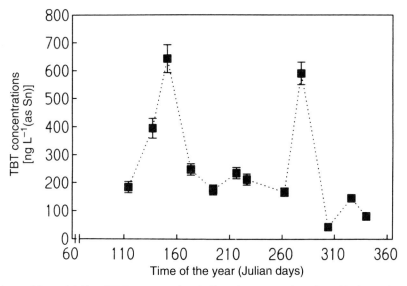

Figure 10.3. Variability of TBT concentrations in filtered water samples collected in the marina of Colijnsplaat (The Netherlands). High concentrations recorded occurred in May and at the beginning of October. (Adapted from ref. 58.)

may considerably increase in summer, as observed in the Great Bay estuary (New Jersey) (18). As a consequence of direct anthropogenic inputs, butyl-Sn compounds may be discharged in waters more frequently during spring (59). Important inputs have also been recorded in sediments of the Sado estuary (Portugal) near shipyards and in sediments at Pearl Harbor Naval Base (Hawaii) in summer and spring, probably due to discharges of TBT antifouling paint during these periods (38, 50).

10.4.2. Spatial Variability

Geographic variations of organo-Sn species concentrations are obviously another aspect of their variability in coastal environments. This topic has already been reviewed by Maguire (60) for TBT and butyl-, methyl-, and octyl-Sn compounds and by Clark et al. for TBT (61). Many surveys on the occurrence of organo-Sn compounds in the environment have been published in the proceedings of the Oceans Symposia from 1986 to 1990 (62–66).

On a large scale, primary factors controlling the variability of organo-Sn compounds are generally regulated by the location and intensity of the point sources. Dispersion is influenced by the hydrological regime of the area studied.

The diffusion of butyl-Sn compounds from contaminated areas is another source of high variability. Strong differences exist between TBT sources located in enclosed areas (e.g., marinas or harbors) and well-flushed channels nearby. Butyl-Sn levels differing by factors of 4–10 between well-flushed estuarine areas and areas of leisure craft activities were observed in U.K. coastal waters (13).

In the case of sediments, very high differences in butyl-Sn concentrations (in the below 60 μm fraction) were also shown between enclosed and well-flushed sampling areas (by a factor of 20) (51). This again indicated the effects of flushing limiting the accumulation of these compounds in sediments; this was also observed in sediments from San Diego Bay (California) and the Ria d'Aveiro (Portugal) (50,67). It is particularly significant that the accumulation of these compounds seems to be quite rapid in sediments from harbors and marinas. This might well explain why very low concentrations are often detected at only 500 m from TBT inputs. Thus it should not be surprising that low TBT levels are detected in sediments from areas previously suspected of being highly contaminated, if these areas are in well-flushed waterways.

The fact that these contaminants may remain in restricted areas does not solve the problem at all, for dredged sediments are often discharged in other places and may therefore contaminate clean areas, as was observed in Arcachon Bay (56, 68).

On a smaller scale, important variations must also be taken into account within short distances or within the water column. The variability of concentrations over a single site (samples within 50 m of each other) has been shown to be significant in the Sado estuary for both water and sediment (51). It is important to note that significant gradients occur at the same site upward in the water column. Clearly and Stebbing showed increasing gradients for TBT between bottom and surface waters and recorded the highest levels in the surface microlayer (69). Similar observations have been made in waters from the Great Bay estuary for methyl-Sn compounds (70). It was ascertained that the lipophilic nature of the surface microlayer waters was responsible for this higer organo-Sn compound concentration (69).

10.4.3. Partitioning in Marine Waters and Sediments

A first important step in the effort to understand the variability and fate of pollutants in the environment is to study its partitioning between waters and sediments. This information is necessary to enable a model of the dispersion to be formulated so that the fate of these chemicals may be predicted. Recent papers have dealt with theoretical aspects of the partitioning of butyl-Sn between the dissolved and particulate phases of the water sample particularly for TBT (71,72). However, few definitive conclusions could be drawn with regard to the possible distribution of the different organo-Sn species between water, suspended matter, and sediments according to these theoretical data. In addition, the partitioning of mono- and dibutyl-Sn and methyl-Sn compounds has presently not been studied to the same extent as has tributyl-Sn species.

TBT has a reported water solubility of ca. 1 mg L^{-1} and an octanol/water partition coefficient (K_{ow}) of 5500 in seawater (71). It was therefore expected that, under natural conditions, TBT would not remain truly dissolved, but rather would be strongly adsorbed to suspended materials. This hypothesis was supported by the calculation of log K_{ow} for MBT, DBT, and TBT (respectively 0.09, 0.05, and 2.2), which indicated that MBT and DBT would not be bound to the organic portion of sediments to the same extent as the more lipophilic TBT species (73).

The particulate/water partition coefficient (K_p) is another parameter often used. It has been estimated to be 3000 (μg kg^{-1})/(μg L^{-1}) for TBT, with a suspended particulate concentration of 10 mg L^{-1} (71). These results were in good agreement with field data for whole waters of San Diego Bay (72). Table 10.2 (74–78) presents the wide range of K_p values estimated by different authors under different experimental or field conditions (54,72,74,76–78). The wide range of values recorded may have various causes. Coefficients calculated on the basis of field data have serious limitations since these values may be modified by possible discharge of TBT in water during experiments. According to Maguire, discrepancies may be attributed to (i) differences in composition of the various solid phases (including the possible presence of paint chips); (ii) differences in the nature and concentrations of other solutes (including differences in salinity and dissolved matter contents); and (iii) a lack of equilibrium in harbor areas close to active sources of TBT (1). Such discrepancies were illustrated by high K_p seasonal variations observed for all the butyl-Sn species due to discharges of paint chips during periods of hull repainting (50). Giordano et al. ascertained that the mean sediment/water partition coefficient for DBT was about 100 times smaller than for TBT, this confirming the results obtained with K_{ow} studies (79).

The actual determination of organo-Sn compounds in the different phases (solid/dissolved) can also lead to discrepancies. Some experiments have suggested that TBT is mainly present in the dissolved phase of water samples. Valkirs et al. detected only 5% or less of measurable butyl-Sn species in particulate fractions of seawater samples after separation by filtration (72). Similar observations were made after DBT and TBT analysis in centrifuged vs. uncentrifuged water column samples (80). However, other studies have demonstrated the inverse tendency: TBT is weakly adsorbed onto kaolin but more strongly bound by humic acids (71). Butyl-Sn compounds were also shown to absorb onto suspended hydrous Fe oxides in the order MBT > TBT > DBT (78). Finally, in 1990, data obtained by us on the determination of organo-Sn

Table 10.2. Variability of the TBT Particle/Water Partition Coefficient (K_p)

Conditions	Salinity (g L^{-1})	Suspended Matter (mg L^{-1})	K_p	Ref.
Field	10–30	10–50	20	74
	Seawater	6	3.28	72
	Seawater	50	0.34	72
	Seawater	2	39.35	75
	Seawater	9	4.61	75
Mesocosm[a]	Seawater	0.7	71	54
Laboratory	32	60	1.4	76
	0	60	0.2	76
	0	3000–30,000	8	77
	24	3000–30,000	4.2	77
	5	1000	1.5	78
	35	10	1900	78
	27	15–57	0.12 ± 0.04	46

[a]The term *mesocosm* refers to tank experiments.

compounds in the water phase after filtering the sample and on the suspended matter after centrifugation of water indicated that TBT is strongly adsorbed onto particulate suspended matters whereas MBT and DBT remain mostly in the dissolved phase (51). These results are presented in Figs. 10.4 and 10.5.

In the case of methyl-Sn species in simulated environmental media, the partitioning between dissolved and particulate phases depends on their degree of substitution controlling the adsorption of the neutral methyl-Sn molecules. In this instance, partitioning would be explanable in terms of polarity of the organo-Sn molecules rather than on their ionic charges. Adsorption mechanisms have been conjectured to occur by one or more ion–dipole or dipole–dipole interactions, or by van der Waals forces between the organo-Sn molecules and the suspended matter. In estuarine water, monomethyl-Sn trichloride is the most polar molecule and was shown to be almost completely absorbed onto particulate material. Dimethyl-Sn dichloride was partly adsorbed according to the varied simulated estuarine conditions. Under these conditions, trimethyl-Sn trichloride was the less polar compound and predominated in the dissolved phase (81). Some of these mechanisms may partly explain the adsorptive behavior of butyl-Sn compounds. Results of simulated

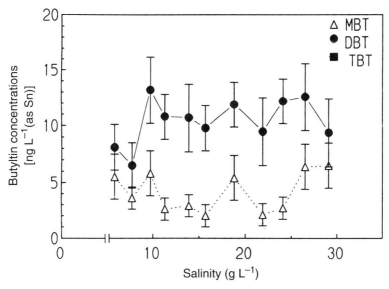

Figure 10.4. Distribution of butyl-Sn compounds in filtered water samples from the Western Scheldt (The Netherlands), October 1988. TBT was not present in any of the sample collected. (Adapted from ref. 95.)

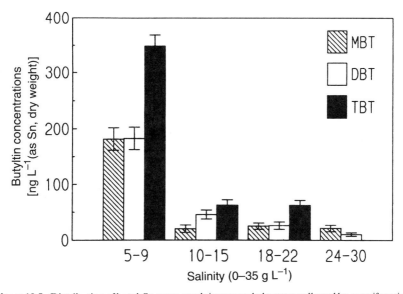

Figure 10.5. Distribution of butyl-Sn compounds in suspended matter collected by centrifugation in the Western Scheldt (The Netherlands), October 1988. Samples were collected by onboard continuous centrifugation. (Adapted from ref. 51.)

experiments indicate that mixed trends due to two opposing processes should be taken into account to explain the partition of butyl-Sn compounds. One process is likely to be similar to that described for methylated Sn, based on a dipole-type of interaction; it may be counterbalanced, however, by the second process, which involves the high hydrophobicity of butyl-Sn, overriding the dipolar mode of interaction and favoring hydrophobic interaction with the suspended matter (78).

Many of the experiments just described have severe limitations, such as the absence of control of suspended matter contents, use of an artificial sorbent, and heterogeneity of the quality of the suspended material used (78,81). No attempts have been made as yet to use a typical estuarine suspended material such as illite, a ubiquitous clay mineral present in aquatic systems. Moreover, no long-term adsorption kinetics has yet been studied for the different organo-Sn species.

In sediments, lower monomethyl-Sn compound contents were found in comparison to dimethyl- and trimethyl-Sn species, this being attributed to the lower lipophilicity of monomethyl-Sn (38). Other aspects of the sorption processes have also been studied in sediments for butyl-Sn compounds. A decrease in absorption capacity of butyl-Sn species was observed with increasing salinity, contrary to what was reported by Unger et al. (77) and Randall and Weber (78). Furthermore, the sorption–desorption processes were shown to be reversible (50). The adsorption of TBT on sediment particles

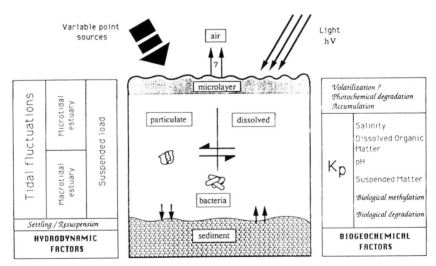

Figure 10.6. Factors and processes controlling the variability of TBT concentrations in estuarine and coastal environments.

is thought to be a rapid phenomenon, since 90% may be adsorbed in 30 min (82). Sorption processes may be different according to the butyl-Sn species studied. Seligman et al. confirmed the adsorption tendency for TBT and showed that the trend was, contrariwise, a desorption for DBT (50). These results are in agreement with the lower adsorption behavior observed for DBT on suspended matters (78). It is also necessary to stress that butyl-Sn compounds are likely to be preferentially adsorbed onto the finest fractions of sediments, as noted in the silt-clay fraction—as opposed to the bulk fraction—of some sediments of the Sado estuary (Portugal) (38). This is in agreement with similar observations on the partitioning of other micropollutants such as trace metals, organochlorine compounds, or polycyclic aromatic hydrocarbon compounds (83–85). The different factors likely to affect the variability of organo-Sn compounds are summarized in Fig. 10.6.

10.5. STRATEGIES FOR ORGANOTIN MONITORING

10.5.1. Sampling Program Design

The design of the sampling program should take into consideration the hydrodynamic characteristics of the area studied. Estuaries and enclosed bays and lakes should be the areas of choice for studying the organo-Sn species distribution. One should take into account the important sources of variability to obtain a meaningful set of data. In main water ways, particularly in estuarine environments, sample collection should be closely related to the presence of harbors or industrial activities. The upstream/downstream distribution should be based on the salinity gradient in order to estimate the organo-Sn compounds' behavior. The set of data should be sufficiently large to rule out extreme cases and give a clear view of the range of concentration (16). In enclosed areas such as bays or lakes, where the geographic diffusion of the organo-Sn compounds is limited, Lagrangian sampling strategies concerned essentially with processes-oriented mechanisms should be adopted. The Eulerian sampling distribution pattern should preferably be used on the basis of either a stratified random distribution approach or a radial distribution approach, rather than on the standard station grid design.

The organo-Sn species concentrations to be measured in waters and sediments are often less than $10\,\text{ng}\,\text{L}^{-1}$ or $50\,\text{ng}\,\text{g}^{-1}$, respectively. Thus it is imperative to minimize the risks of contamination during sampling, adsorption losses during storage, and biological and/or chemical redistributions that may occur. The precautions to be taken for monitoring organo-Sn species in estuarine and coastal environments have been reviewed recently (51).

10.5.2. Choice of Samples

As mentioned earlier in this chapter, it is presently almost impossible to evaluate even the most adequate sampling strategies from the data presented in the literature since methods are widely different. Many authors do not filter the water samples or use centrifuged waters, bulk sediments are analyzed in most cases, and suspended matter is generally not considered in organo-Sn species monitoring. For a thorough understanding of the organo-Sn species pathways, studies should involve analyses of microlayers, waters, suspended matter, sediments, and biota. Strategies with regard to the quality of the sample may vary according to the nature of the investigation. The analysis of filtered water samples (particle size below 0.45 μm) is justified for acute toxicity tests on marine organisms, which are often based on the butyl-Sn compounds contained in the dissolved phase. However, it is now suspected that the butyl-Sn compounds adsorbed on suspended matter may have an effect on filter- and deposit-feeding organisms (86). Therefore, it is recommended that organo-Sn species contents in bulk waters also be considered (51).

The number of sediment and biological tissue analyses performed every year is increasing as these matrices are recognized to be good indicators of organo-Sn species contaminations. Sediments act as a sink for, e.g., TBT, and mollusks accumulate a wide variety of organo-Sn compounds (13, 60, 87). Sampling strategies should therefore involve the collection and storage of sediment and biota samples.

10.5.3. Water Samples

Water samples are important for monitoring organo-Sn compounds in the environment. Two approaches may be considered, depending on the type of information needed and the hydrodynamic variability of the system studied.

The first approach should be used for sampling in highly variable environments such as estuaries. A time integrated, remotely moored, automated sampling and concentration system has been designed for this purpose (Seastar Instrumentation Ltd.) and is useful for the *in situ* evaluation of the concentration of trace constituents in an aquatic environment. This instrumentation is illustrated in Fig. 10.7. A programmable pump draws water through a resin extraction column that selectively adsorbs or chelates compounds of interest for later analysis in the laboratory (88). It has been applied to organo-Sn species monitoring in highly fluctuating tidal environments in order to integrate the Sn compound concentrations over several tidal cycles. Butyl-Sn compounds were extracted *in situ* on C_{18} covalently bound reversed-phase

Figure 10.7. *In situ* time-integrating water sampler from Seastar Instrumentation Ltd. This sampler enables the *in situ* preconcentration of trace contaminants (e.g., butyl-Sn compounds) on resins from seawater. The programmable pumping system allows the contaminant concentrations to be integrated and averaged over variable water masses on a single site.

silica on chromatographic sorbent. Recovery experiments showed that the C_{18}-silica absorbent was effective in quantitatively collecting and eluting di- and tributyl-Sn chlorides, but not monobutyl-Sn chloride (88). This type of application is very recent and still needs to be validated. Its application could favorably complete or support biomonitoring (e.g., by mussels). A comparison

Table 10.3. Comparison of the Possibilities of an *in Situ* Sampler and Sampling of Live Mussels for Monitoring Water Quality Levels

In Situ Water Sampler	Sampling of Live Mussels
It allows the determination of dissolved organo-Sn compounds in water (adsorbable on resins).	It provides an integrated view of the biological uptake of organo-Sn species from the water column.
It provides *in situ* concentration factors of, e.g., 1000–10,000, according to the organo-Sn species and type of resins.	It provides *in situ* concentration factors of about 100,000 (estimated based on data for DDT and other organics with similar octanol/water partition coefficients as TBT).
It integrates medium-term effects (1–3 days).	It integrates long-term effects (one to several months).
It samples effectively over a wide range of hydrodynamic and physical-chemical conditions.	It is limited to locations where the mussels can survive.
Known volumes of water are sampled; the sampling format is reproducible.	Biological variability is a problem: bioaccumulation varies with individuals, sex, reproduction state, season, body size, etc.
It provides separate measurements of dissolved and particulate phases (selectivity for organo-Sn species is provided by the analytical technique).	Responses to the total spectra of contaminants present in water are obtained; no differences can be discerned between dissolved and particulate phases.

Source: Adapted from ref. 18.

of the different respective capabilities with regard to organo-Sn species monitoring applications is presented in Table 10.3.

The second approach involves a more traditional way of collecting water samples. Samples are collected via discrete sampling methodology. In general, water samples should be collected 0.5 m below the surface to avoid contamination from the surface microlayer. A second sample should be taken deeper in order to avoid the potential degradation of organo-Sn compounds by photochemical processes. Special care should be taken when samples are collected close to sediments to avoid the possible resuspension of contaminated sediments. A large range of water samplers have already been designed and used. Samplers such as Niskin or Go-Flow systems may be suitable for water sample collection if they are polytetrafluoroethylene (PTFE, Teflon) coated. However, due to the low concentration of organo-Sn compounds in water (in the nanogram per liter range) and the quality of the material, particular care should be taken during the collection of water samples to avoid losses (e.g., adsorption) or contamination. The choice of the container material is of prime importance; for example, the use of polyvinyl chloride (PVC) must be avoided due to possible release of mono- and dibutyl-Sn after leaching, and some organo-Sn species adsorb onto PTFE and glass under certain conditions (89–92).

The use of "close–open–close" samplers for collecting coastal waters was found to be an efficient way to avoid cross-contamination with microlayer water (51). An example of such a sampler is given in Fig. 10.8. For the sampling itself, the use of a rubber boat was recommended for monitoring of trace metals and is justified in the case of organo-Sn species monitoring (especially when the hull of the ship is painted with organo-Sn-species- containing paint!) (93). The use of acid-prewashed Pyrex glass bottles to collect water in intertidal zones was shown to be suitable, provided that the bottles are opened and closed 50 cm below the surface. This type of container does not apparently display a strong adsorption capacity for butyl-Sn compounds.

Microlayers should also be sampled whenever possible, since organo-Sn species can accumulate preferentially in this stratum (69). Results from this type of sample may provide the widest distribution of organo-Sn compounds present in the environment studied, as all of them will accumulate at the air/seawater interface. Transient species such as volatile species (e.g., Me_4Sn) may be detected. The microlayer can be collected with a simple precombusted glass plate gently immersed and withdrawn from the water. Collection of the sample from the plate is performed with a PTFE spatula. Other techniques used for microlayer collection have been discussed (94).

Once the water sample has been collected, the sample preservation stage is critical. Acidification is commonly used to preserve the stability of various pollutants in water samples. It is also usual to filter the samples at the 0.45 μm level to compare the amounts of pollutants in the dissolved phase. In an instance of low suspended load, we observed no significant differences in butyl-Sn compound content between bulk and filtered water. However, as TBT was shown to strongly adsorb onto particulate suspended matter, the only determination of dissolved TBT concentration would underestimate the contamination if the suspended loads were very high (51). The release and adsorption of TBT depend on the amount and quality of suspended matter contained in the water samples. Therefore, desorption due to sample acidification prior to filtration may be more or less important according to whether the areas of collection do or do not display different parameters. The comparison of the dissolved phases obtained at different sites will lead in any case to discrepancies since TBT will certainly be released differently; this procedure is therefore not recommended.

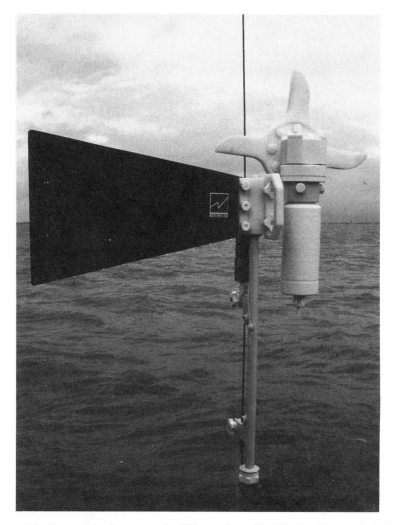

Figure 10.8. Ultratrace-level water sampler. This sampler enables 250 mL to 2 L water samples to be collected under ultraclean conditions. The sampling flask is the storage container. The sampler is opened and closed by a set of "messengers." The opening and closing can be performed at a selected depth to avoid contamination from the surface microlayers (sampler: B.U.T. Nereides Instrumentation).

As already mentioned, analyses of bulk turbid water revealed much higher TBT concentrations than in filtered samples. Suspended matter is ingested by organisms, and the TBT adsorbed onto particles clearly represents a risk of contamination that is not taken into account when only filtered samples are analyzed. This cause of underestimation can be avoided if the collected samples are left unacidified, filtered immediately in the field, and then both phases (dissolved and bulk) are acidified for storage. This procedure enables the butyl-Sn levels in the dissolved phase to be compared between different areas (same basis of comparison) and gives a more realistic view of the contamination underway considering the amounts in the bulk samples. Further, it is likely that turbid samples will be less stable over time. One may conclude that the choice of the quality of the water sample will depend on two different situations depending on the presence of the suspended load (95). First, when suspended concentrations are less than $10 \, mg \, L^{-1}$ or when organo-Sn species levels are very low, the direct analysis of whole-water samples will give a good estimation of the presence of the contaminants. Secondly, examination of both the dissolved and suspended phases should be applied to strongly hydrodynamic environments (mainly estuaries and bays) with a high and variable turbidity load.

10.5.4. Suspended Matter Samples

Very few data on organo-Sn compounds associated on suspended matter have been reported to date, probably because data obtained by filtration are not reproducible. The suspended load collected on the filter may indeed be highly variable even if similar water volumes are filtered. This error can be reduced by filtering large volumes, although filter clogging by colloids (especially in estuarine environments) may lead to discrepancies. Moreover, the amount of material collected on the filters is very small, and consequent misestimation of the mass of material collected, even if it is carefully weighted, generates a high probability of errors when the final concentration is reported on a mass basis. Finally, complete dissolution of the filter may release organo-Sn compounds used in the filter material. Further efforts in this area are needed to improve the quality of the measurements.

A recognized technique is the collection of suspended matter by centrifugation (51, 95). This approach requires onboard heavy equipment, but it is the only way to provide high-quality data on the organo-Sn species content associated with the suspended load. Gram quantities of suspended matter can be collected, allowing a higher accuracy and reproducibility to be achieved at the analytical stage. If for trace metal analysis both filtration and centrifugation have been shown to give satisfactory results with respect to total metal determination, the case of organo-Sn compounds certainly calls even more for onboard centrifugation (96). The detection of high butyl-Sn levels in suspensions collected by centrifugation have clearly confirmed that TBT is strongly absorbed to particles; therefore, analyses of water either filtered or containing less than $100 \, mg \, L^{-1}$ of suspended matter do not allow the occurrence of a high level of contamination to be detected if the evaluation is made with results of filtered-water analysis only. Centrifugation also allows large volumes of water to be integrated and minimizes the high variability of occurrence of organo-Sn species in the water column. Finally, it reduces the risk of losses (e.g., adsorption on the filter) and contamination when PTFE-lined centrifuging apparatus is used directly on board (51).

10.5.5. Sediment Samples

Sediment collection does not pose particular problems provided that the first few centimeter layers are sampled in order to avoid dilution with less contaminated layers. The sieving of sediment samples with a 60 μm mesh is a good procedure to enable the butyl-Sn levels to be compared in relation to the different areas of collection. The sieving procedure should be carefully designed to avoid cross-contamination (e.g., one polyethylene sieve per sample) and losses (limitation of organo-Sn species desorption from sedimentary particles by using water from the area of collection for the sieving). This procedure is also justified by the fact that butyl-Sn compounds accumulate preferably in the silt-clay fraction and therefore the total amount of butyl-Sn contained in the sediment may not give a realistic view of contamination processes (dilution with, e.g., uncontaminated sandy fractions) (87).

In the case of sand analysis, discrepancies may be caused by the presence of coarse light fragments, such as algae and leaves debris, which accumulate butyl-Sn compounds to a large extent. This fraction should be isolated (e.g., by flotation washing) in order to avoid overestimation of the butyl-Sn species levels in the sand samples and so to permit comparison of the concentrations to be done on the same basis.

After an extensive study of grain size effects it was concluded that the distribution of butyl-Sn species in the fine sediment fraction (particle size $<60\,\mu$m) provides a realistic means of detecting contaminated areas (87). Specifically, we suggest that one should at least consider the amount of the $<60\,\mu$m fraction (the percentage in relation to the total mass of sediment) and relate the butyl-Sn species levels in this fraction to the total mass of sediment so as to compare more accurately the distribution of the contamination over different sites (51).

10.5.6. Sample Storage

The instability of organometallic compounds during sample storage is another critical step for the full validation of the data, and storage procedures have therefore to be carefully designed in order to preserve the integrity of the samples as regards their content in organo-Sn species. Table 10.4 presents

Table 10.4. TBT Stability Under Storage Conditions in the Dark

Water Samples	Container	Treatment	Temperature (°C)	Time (weeks)	Loss	Ref.
Spiked natural	Polycarbonate	No	20	12	No	82
	Pyrex	No	20	12	No	82
	Polyethylene	No	20	12	Yes	82
Spiked natural	Teflon	No	22	20	No	98
	Polycarbonate	No	22	20	No	98
	Pyrex	No	22	20	No	98
	Polyethylene	No	22	20	Yes	98
	Polyethylene	pH 2	22	20	No	98
Natural	Not mentioned	pH 2	5	13	No	99
Extract	Not mentioned	No	5	13	No	99
Natural (filtered)	Pyrex	pH 2	4	12	No	100
	Pyrex	pH 2	22	12	No	100

Source: Adapted from ref. 51.

some of the main conclusions reached by various authors concerning water samples (82,97–99).

Some studies have been performed with synthetic solutions and demonstrated good stability of the compounds (82,97,100). However, the stability of natural samples is generally more difficult to achieve. Different methods were extensively tested for the storage of natural water and sediment samples. The present authors showed that the contents of butyl-Sn compounds are stable over 4 months in filtered coastal water, acidified at pH 2 and stored in the dark at 4 °C in prewashed Pyrex bottles (99). Under laboratory light, however, TBT was shown to degrade rapidly to yield MBT directly (99). Under similar storage conditions, half-lives of MBT, DBT, and TBT were below 1,1, and 4 days, respectively (101). Turbid water, both acidified at pH 2 and nonacidified, was more difficult to stabilize, as losses of butyl-Sn species were observed after 2 months of storage in the dark at 4 °C (99). Sediments were shown to be reasonably stable as regards their organo-Sn content, either wet-stored at 4 °C or frozen, and no major changes were observed using mild oven-drying conditions (50 °C) or a freeze-drying procedure (99). Further studies have shown that sediment samples may be heat-stabilized at 80 °C (during 10 h) and 120 °C (during 2 h) and remain stable for 1 year as regards their TBT content (102).

10.5.7. Determination

The techniques for determining butyl- and methyl-Sn species involve extraction steps (using organic solvents or acids), separation by gas chromatography (GC) or high-performance liquid chromatography (HPLC) and a wide variety of final quantification such as electrothermal atomization atomic absorption spectrometry (ETA–AAS), inductively coupled plasma mass spectrometry (ICP–MS), and flame photometric detection (FPD). Extensive reviews of the techniques used for Sn speciation may be found in the literature (103, 104) and it should only be stressed here that the methods should be applied under good quality control conditions involving the use of reference materials certified for their contents in organo-Sn species (105). These already exist for butyl-Sn species in polluted marine sediment (106), TBT and DBT in coastal sediment (107), butyl-Sn species in fish tissue (108); a mussel material is in preparation for the certification of butyl- and phenyl-Sn compounds (109); finally, a reference material (not certified) of harbor sediment is also available for the quality control of TBT determination (110). The aspects of quality control for speciation analyses are dealt with in more detail in Chapter 6 of this volume.

10.6. CONCLUSIONS

Nowadays, the determination of chemical species in environmental matrices is becoming increasingly necessary for the study of environmental pathways and assessment of their toxic impact leading later to the establishment, monitoring, and enforcement of regulations. In the past few years, Sn speciation has been of growing interest owing to the wide variety of species existing and the different problems to be met, i.e., studies of the toxic impact of trialkyl-Sn compounds (e.g., TBT or TPhT) on marine biota, monitoring of environmental levels to check on compliance with regulations (e.g., for TBT), and studies of biogeochemical pathways (e.g., methylation, degradation, of volatilization). Results discussed in this chapter clearly indicate that considerable harmonization of the findings is needed. A future approach to this necessary harmonization will certainly lead to enhanced efforts of the international community to develop standard procedures for sampling, storage, and analytical quality control involving the use of certified reference materials. None of the different approaches described in this chapter can be considered sufficient to yield a full understanding of the fate and impact of organo-Sn compounds in the environment. Harmonization, and hence comparability of data, will also lead to better utilization of bioassays, mussel water, *in situ* monitoring, and classical sampling strategies.

All these approaches should be validated by establishing well-defined monitoring strategies in order to make results comparable throughout the world and to enhance modeling applications and theory development. Several points should be borne in mind when one is monitoring Sn species in estuarine and coastal waters and sediments, as follows:

a. Precautions should be taken to avoid contamination or losses during sampling, e.g., for waters the use of "close–open–close" samplers is recommended to limit contamination from surface microlayers. For sediments, the first few centimeters should be sampled using, e.g., polyethylene or PTFE material.

b. The pretreatment step should also be considered with great care. The use of precleaned filtration units is a good way to avoid contamination or adsorption. It is recommended that suspended matter be analyzed separately by centrifuging with a PTFE-lined apparatus. The separate analysis of filtered water and suspended matter is thought to be more accurate than the analysis of turbid waters. In the case of sediments, wet sieving using water from the area of collection is recommended to avoid physical-chemical changes (change of species present; desorption).

c. Sample storage should be considered with special care. Waters acidified at pH 2 at 4 °C in the dark have been shown to be stable with regard to butyl-Sn species. Freezing and freeze-drying or oven-drying (50–80 °C) has also been shown to preserve the sample integrity.

d. Finally, a good analytical quality control system is a prerequisite for accurate analysis. The analytical method should be validated, and no systematic error should be left undetected. One of the methods of choice for this validation is the use of reference materials certified for their organo-Sn species contents. If such materials do not exist, the analyst should compare the results obtained by a given method with those obtained by another basically different method.

A clear monitoring strategy, accepted and used by the scientific community, is a prerequisite if we are to gain an understanding of the complex environmental pathways of Sn species. The lack of comparability of data and pervasive doubts as to a variety of newly developed analytical techniques have blocked the way to full comprehension of the biogeochemical cycle of Sn. In other words, methylation, degradation, and volatilization pathways are still very controversial. Future efforts should thus focus on the achievement of good quality assurance of data produced.

REFERENCES

1. R. J. Maguire, *Water Pollut. Res. J. Can.* **26**, 243 (1991).
2. P. Quevauviller, O. F. X. Donard, and A. Bruchet, *Water Res.* **27**, 1085 (1993).
3. P. J. Craig, in *Organometallic Compounds in the Environment* (P. J. Craig, ed.), pp. 1–58. Longman, Harlow, 1986.
4. D. A. Wolfe and B. Kjerfve, in *Estuarine Variability* (A. Wolfe, ed.), pp. 3–17, Academic Press, 1986.

5. S. J. Blunden and A. Chapman, in *Organometallic Compounds in the Environment* (P. J. Craig, ed.), pp. 111–150. Longman, Harlow, 1986.
6. J. A. J. Thompson, M. G. Sheffer, R. C. Pierce, Y. K. Chau, J. J. Cooney, W. R. Cullen, and R. J. Maguire, *Nat. Res. Counc. Can. Rep.*, ISSN 0316-0114, NRCC 22494, p. 284 (1985).
7. C. Alzieu, *Rapp. Sci. Tech., IFREMER* **17**, 93 (1989).
8. S. Chiavarini, C. Cremisini, and R. Morabito, Chapter 9 of the present volume.
9. E. His, D. Maurer, and R. Robert, *J. Moll. Stud.* **124**, 60 (1983).
10. P. E. Gibbs and G. W. Bryan, *J. Mar. Biol. Assoc. U.K.* **66**, 767 (1986).
11. R. D. Cardwell and A. W. Sheldon, *Proc. Oceans Organotin Symp.*, Washington, DC, *1986*, p. 1117 (1986).
12. M. A. Champ and W. L. Pugh, *Proc. Ocean, Int. Workplace*, Halifax, *1987*, p. 1296 (1987).
13. M. J. Waldock, J. E. Thain, and M. E. Waite, *Appl. Organomet. Chem.* **1**, 287 (1987).
14. M. H. Salazar, *Proc. Oceans Organotin Symp.*, Washington, DC, *1986*, p. 1240 (1986).
15. H. H. White and M. A. Champ. *ASTM, Spec. Tech. Publ.* **STP 805**, 299 (1983).
16. D. R. Young, P. Schatzberg, F. E. Brinckman, M. E. Champ, S. E. Holm, and R. B. Landy, *Proc. Oceans Organotin Symp.*, Washington, DC, *1986*, p. 1135 (1986).
17. J. T. Byrd and M. O. Andreae, *Science* **218**, 565 (1982).
18. O. F. X. Donard, *Oceanis* **13**, 381 (1987).
19. M. Astruc and R. Pinel, *Report CNRS–PIREN*. ASP Déchets et Environnement, Paris (1982).
20. M. Pettine, F. J. Millero, and G. Macchi, *Anal. Chem.* **53**, 1039 (1981).
21. C. Huey, F. E. Brinckman, S. Grim, and W. P. Iverson, *Proc. Int. Conf. Transp. Persistent Chem. Aquat. Exosyst. 1974*, Vol. II, p. 73 (1974).
22. Y. K. Chau, P. T. S. Wong, O. Kramar, and G. A. Bengert, *Proceedings of the International Conference on Heavy Metals in the Environment* (W. H. Ernst, ed.). Amsterdam, 1980.
23. H. E. Guard, A. B. Cobet, and W. M. Coleman, *Science* **213**, 770 (1981).
24. J. R. W. Hallas, J. C. Means, and J. J. Cooney, *Science* **215**, 1505 (1982).
25. R. J. Maguire and R. J. Tkacz, *J. Agric. Food Chem.* **33**, 947 (1985).
26. J. H. Weber and J. J. Alberts, *Environ. Technol.* **11**, 3 (1990).
27. S. Rapsomanikis and J. H. Weber, in *Organometallic Compounds in the Environment* (P. J. Craig, ed.), p. 279. Longman, Harlow, 1986.
28. S. Rapsomanikis and J. H. Weber, *Environ. Sci. Technol.* **19**, 352 (1985).
29. O. F. X. Donard, F. T. Short, and J. H. Weber, *Can. J. Fish. Aquat. Sci.* **44**, 140 (1987).
30. W. F. Manders, G. J. Olsen, F. E. Brinckman, and F. E. Bellama, *J. Chem. Soc. Chem. Commun.*, 538 (1984).

31. R. M. Ring and J. H. Weber, *Sci. Total Environ.* **68**, 225 (1988).
32. J. R. Ashby and P. J. Craig, *Sci. Total Environ.* **100**, 337 (1991).
33. Ph. Quevauviler, O. F. X. Donard, J. C. Wasserman, F. Martin, and J. Schneider, *Appl. Organomet. Chem.* **6**, 221 (1992).
34. P. J. Craig and S. Rapsomanikis, *Environ. Technol. Lett.* **5**, 407 (1984).
35. S. Rapsomanikis, O. F. X. Donard, and J. H. Weber, *Appl. Organomet. Chem.* **1**, 115 (1987).
36. R. S. Braman and M. A. Tompkins, *Anal. Chem.* **51**, 12 (1979).
37. V. F. Hodge, S. L. Seidel, and E. D. Goldberg, *Anal. Chem.* **51**, 1256 (1979).
38. Ph. Quevauviller, R. Lavigne, R. Pinel, and M. Astruc, *Environ. Pollut.* **59**, 267 (1989).
39. C. C. Gilmour, J. H. Tuttle, and J. C. Means, in *Marine and Estuarine Geochemistry* (A. C. Sigleo and A. Hattori, eds.), pp. 239–258. Lewis Publishers, Chelsea, MI, 1985.
40. R. J. Maguire, *Environ. Sci. Technol.* **18**(4), 291 (1984).
41. S. Rapsomanikis and R. Harrison, *Appl. Organomet. Chem.* **2**, 25 (1988).
42. Ph. Quevauviller, R. Ritsema, R. Morabito, W. M. R. Dirkx, S. Chiavarini, J. M. Bayona, and O. F. X. Donard, *Appl. Organomet. Chem.* **8**, 541 (1994).
43. J. A. Jackson, W. R. Blair, F. E. Brinckman, and W. P. Iverson, *Environ. Sci. Technol.* **16**, 110 (1982).
44. O. F. X. Donard and J. H. Weber, *Nature (London)* **332**, 339 (1988).
45. J. S. Thayer, G. J. Olsen, and F. E. Brinckman, *Environ. Sci. Technol.* **18**, 726 (1984).
46. F. Leguille, Ph.D. Thesis, University of Pau (1992).
47. S. J. Blunden, *J. Organomet. Chem.* **248**, 149 (1983).
48. D. Barug, *Chemosphere* **10**, 1145 (1981).
49. C. Steward and S. J. De Mora, *Environ. Sci. Technol.* **11**, 565 (1990).
50. P. F. Seligman, A. O. Valkirs, and R. F. Lee, *Proc. Oceans Organotin Symp.*, Washington, DC, *1986*, p. 1189 (1986).
51. Ph. Quevauviller and O. F. X. Donard, *Appl. Organomet. Chem.* **4**, 353 (1990).
52. Y. Hattori, A. Kobayaski, K. Nonaka, A. Sugimae, and M. Nakamoto, *Water Sci. Technol.* **20**, 71 (1988).
53. R. François, F. T. Short, and J. H. Weber, *Environ. Sci. Technol.* **23**, 191 (1989).
54. D. Adelman, K. R. Hinga, and M. E. Q. Pilson, *Environ. Sci. Technol.* **24**, 1027 (1990).
55. J. E. Thain, M. J. Waldock, and M. E. Waite, *Proc. Ocean, Int. Workplace*, Halifax, *1987*, p. 1398 (1987).
56. Ph. Quevauviller, H. Etcheber, and O. F. X. Donard, *Environ. Pollut.* **84**, 89 (1994).
57. C. Clavell, P. F. Seligman, and P. M. Stang, *Proc. Oceans Organotin Symp.*, Washington, DC, *1986*, p. 1152 (1986).
58. R. Ritsema, R. W. P. M. Laane, and O. F. X. Donard, *Mar. Environ. Res.* **32**, 243 (1991).

59. P. F. Seligman, C. M. Adema, P. M. Stang, A. O. Valkirs, and J. G. Grovhoug, *Proc. Ocean, Int. Workplace*, Halifax, *1987*, p. 475 (1987).
60. R. J. Maguire, *Appl. Organomet. Chem.* **1**, 475 (1987).
61. E. A. Clark, R. M. Sterritt, and J. N. Lester, *Environ. Sci. Technol.* **22**, 600 (1988).
62. *Proceedings of the Organotin Symposium of the Oceans 1986 Conference*, Washington, DC, Vol. 4, IEEE Service Center, Piscataway, NJ, 1986.
63. *Proceedings of the Organotin Symposium of the Oceans 1987 Conference, Halifax*, Vol. 4, IEEE Service Center, Piscataway, NJ, 1987.
64. *Proceedings of the Organotin Symposium of the Oceans 1988 Conference, Baltimore*, Vol. 4, IEEE Service Center, Piscataway, NJ, 1988.
65. *Proceedings of the Organotin Symposium of the Oceans 1989 Conference, Seattle*, Vol. 2, IEEE Service Center, Piscataway, NJ, 1989.
66. *Proceedings of the Third International Organotin Symposium, Monaco, 1990*, (1990).
67. L. Cortez, Ph. Quevauviller, F. Martin, and O. X. F. Donard, *Environ. Pollut.* **82**, 57 (1993).
68. P. M. Sarradin, A. Astruc, V. Desauziers, R. Pinel, and M. Astruc, *Environ. Technol.* **12**, 537 (1991).
69. J. J. Cleary and A. R. D. Stebbing, *Mar. Pollut. Bull.* **18**, 238 (1987).
70. O. F. X. Donard, S. Rapsomanikis, and J. H. Weber, *Anal. Chem.* **58**, 772 (1986).
71. R. B. Laughlin, H. E. Guard, and W. M. Coleman, *Environ. Sci. Technol.* **20**, 201 (1986).
72. A. O. Valkirs, P. F. Seligman, and R. F. Lee, *Proc. Oceans Organotin Symp.*, Washington, DC, *1986*, p. 1165 (1986).
73. T. Tsuda, H. Nakanishi, S. Aoki, and J. Takebayashi, *Water Res.* **22**, 647 (1988).
74. O. F. X. Donard, Ph. Quevauviller, M. Ewald, R. W. P. M. Laane, J. M. Marquenie, and R. Ritsema, *Proc. Heavy Met. Environ.*, Geneva, *1989*, p. 526 (1989).
75. A. O. Valkirs, M. O. Stallard, and P. F. Seligman, *Proc. Ocean, Int. Workplace*, Halifax, *1987*, p. 1375 (1987).
76. J. R. W. Harris and J. J. Cleary, *Proc. Ocean, Int. Workplace*, Halifax, *1987*, p. 1370 (1987).
77. M. A. Unger, W. G. McIntyre, and R. J. Huggett, *Proc. Ocean, Int. Workplace*, Halifax, *1987*, p. 1381 (1987).
78. L. Randall and J. H. Weber, *Environ. Sci. Technol.* **57**, 191 (1986).
79. R. Giordano, L. Ciaralli, G. Gattorta, M. Ciprotti, and S. Costantini, *Microchem. J.* **49**, 69–77 (1994).
80. W. E. Johnson, L. W. Hall, S. J. Bushong, and W. S. Hall, *Proc. Ocean, Int. Workplace*, Halifax, *1987*, p. 1364 (1987).
81. O. F. X. Donard and J. H. Weber, *Environ. Sci. Technol.* **19**, 1104 (1985).
82. C. A. Dooley and V. Homer, *Nav. Ocean Syst. Tech. Rep.* **197**, 19 (1983).

83. W. Salomons and U. Förstner, *Environ. Technol. Lett.* **1**, 506 (1980).
84. Ph. Quevauviller, S. Laforge, D. Lecuona, J. Faure, and J. G. Faugère, *J. Rech. Océanogr.* **13**, 125 (1988).
85. C. Raoux, *DEA Report*, No. 348. University of Bordeaux, CNRS UA 1987.
86. W. J. Langston and G. R. Burt, *Mar. Environ. Res.* **32**, 61 (1991).
87. Ph. Quevauviller, H. Etcheber, C. Raoux, and O. F. X. Donard, *Oceanol. Acta* **SP 11**, 247 (1992).
88. P. Schatzberg, C. M. Adema, W. M., Thomas, Jr., and S. R. Mangum, *Proc. Oceans Organotin Symp.*, Washington, DC, *1986*, p. 1155 (1986).
89. E. A. Boettner, G. L. Ball, Z. Hollingsworth, and R. Aquino, *U. S. Environ. Prot. Agency Rep.* **EPA-600/1-81-062**, 102 (1981).
90. Ph. Quevauviller, O. F. X. Donard, and A. Bruchet, *Appl. Organomet. Chem.* **5**, 125 (1991).
91. H. A. Meinema, T. Burger-Wiersma, G. Versluis-de Haan, and E. C. Gevers, *Environ. Sci. Technol.* **12**, 288 (1978).
92. R. J. Maguire, J. H. Carey, and E. J. Hale, *J. Agric. Food Chem.* **31**, 1060 (1983).
93. L. Mart, *Fresenius' Z. Anal. Chem.* **27**, 55 (1986).
94. P. Liss, in *Chemical Oceanography* (J. P. Riley and G. Skirrow, eds.), p. 243. Academic Press, New York, 1975.
95. O. F. X. Donard, Ph. Quevauviller, R. Ritsema, and M. Ewald, in *Heavy Metals in the Hydrological Cycle* (M. Astruc and J. N. Lester, eds.), p. 401, Selper Ltd., London, p. 401. 1988.
96. H. Etcheber and J. M. Jouanneau, *Estuarine Coastal Mar. Sci.* **11**, 701 (1980).
97. W. R. Blair, G. H. Olson, F. E. Brinckman, R. C. Paule, and D. A. Becker, *Nat. Bur. Stand. Rep.*, Gaithersburg, p. 55 (1986).
98. R. J. Huggett, M. A. Unger, and D. J. Westbrook, *Proc. Oceans Organotin Symp.*, 86, Washington, DC *1986*, Vol. 4, p. 1262 (1986).
99. Ph. Quevauviller and O. F. X. Donard, *Fresenius' J. Anal. Chem.* **339**, 599 (1991).
100. Ph. Quevauviller, M. Astruc, L. Ebdon, H. Muntau, W. Cofino, R. Morabito, and B. Griepink, *Mikrochim. Acta* in press (1995).
101. D. T. Burns, M. Harriott, and F. Glockling, *Fresenius' Z. Anal. Chem.* **327**, 701 (1987).
102. V. Desauziers, Thesis, No. **105**, p. 159 University of Pau (1991).
103. O. F. X. Donard and R. Pinel, in *Environmental Analysis Using Chromatography Interfaced with Atomic Spectroscopy* (R. Harrison and S. Rapsomanikis, eds.), 189. Ellis Horwood, Chichester, UK 1989.
104. M. Leroy, Ph. Quevauviller, O. F. X. Donard, and M. Astruc, *Appl. Chem.* in press (1995).
105. Ph. Quevauviller, *Analyst*, **120**, 597 (1995).
106. National Research Council of Canada, Marine Sediment PACS-1 (1991).

107. Ph. Quevauviller, M. Astruc, L. Ebdon, V. Desauziers, P. M. Sarradin, A. Astruc, G. N. Kramer and B. Griepink, *Appl. Organomet. Chem.* **8**, 629 (1994).
108. National Institute for Environmental Studies, Fish Tissue No. 11 (1990).
109. Ph. Quevauviller, S. Chiavarini, C. Cremisini, R. Morabito, M. Bianchi, and H. Muntau, *Mikrochim. Acta* in press (1995).
110. Ph. Quevauviller, M. Astruc, L. Ebdon, G. N. Kramer, and B. Griepink, *Appl. Organomet. Chem.* **8**, 639 (1994).

CHAPTER

11

THE ANODIC STRIPPING VOLTAMMETRIC TITRATION PROCEDURE FOR STUDY OF TRACE METAL COMPLEXATION IN SEAWATER

G. SCARPONI,[1] G. CAPODAGLIO,[1,2] C. BARBANTE,[1,2] AND P. CESCON[1,2]

[1] *Department of Environmental Sciences, Ca' Foscari University of Venice, 30123 Venice, Italy*
[2] *Center for Studies on Environmental Chemistry and Technology, CNR, 30123 Venice, Italy*

11.1. INTRODUCTION

Remarkable and continuous advances and improvements in the field of environmental analytical chemistry over the past two decades have brought about a real revolution in marine science, particularly with respect to our knowledge of the distribution patterns and chemical behavior of trace elements in seawater (1, 2).

Awareness of the risks of contamination of samples during the entire analytical process (collection, storage, treatment, and analysis) has prompted the development of better analytical methodologies mainly with the aim of performing measurements under careful contamination control (3–13). In turn, novel investigations conducted under stringent, noncontaminating conditions have led to the discovery that concentrations of many trace elements are in fact lower than those previously accepted by more than 1 order of magnitude (10–1000 times) and to the assessment of vertical profiles of trace elements in deep waters that show patterns which are consistent with known biogeochemical and/or physical processes active in the oceans (2, 3).

On the other hand, as instrumental analytical methods have been refined, the physical-chemical forms of trace elements have gradually been identified and/or classes of metal species have been discriminated (individual species, oxidation states, association with inorganic and organic ligands). Speciation

---

Element Speciation in Bioinorganic Chemistry, edited by Sergio Caroli.
Chemical Analysis Series, Vol. 135.
ISBN 0-471-57641-7 © 1996 John Wiley & Sons, Inc.

research now offers a new tool to be used when assessing the environmental impact of metals since physical, chemical, and biological behaviors (such as adsorption on particles, biological effects, and toxicity to organisms) depend on their chemical forms (14–17).

Initial investigations into the complexing capacity of aquatic systems, carried out by sample titration with metal (principally Cu) (18–26), were followed during the 1980s by the development of two voltammetric titration methodologies for studying trace metal complexation with natural organic ligands in seawater (27–38). In both cases the procedure involves the titration of ligands present in the sample with the metal under study. Differential pulse anodic stripping voltammetry (DPASV) can be used to determine the electroactive fraction of metal (DPASV-labile metal, or "free" metal, i.e., ionic plus inorganically complexed metal) directly during the titration (27–33). As an alternative, a competitive equilibrium with another ligand or surface site (adsorption on MnO_2) can be set up, after which titration is followed by cathodic stripping voltammetry (CSV) or DPASV (34–38). In both cases an independent measurement of total metal content is required; then a suitable treatment of titration data allows one to determine the fraction of metal that is not complexed by organic ligands, the total ligand concentration, and the related conditional stability constant.

Here discussion is confined to the direct anodic stripping voltammetry (ASV) titration procedure for determining metal complexation in seawater. The method involves adding aliquots of a standard metal solution to the untreated sample and measuring the DPASV peak height for the metal in the sample and after each addition. A suitable equilibration period is allowed to pass after each addition before the instrumental measurement is carried out.

A typical titration curve obtaind for Antarctic coastal seawater titrated with Pb is shown in Fig. 11.1 (39). The first part of the curve shows a low response due to complexation of added metal with free ligand present in the sample. Beyond the end point, when sufficient titrant excess is present, the ligand becomes unavailable to complex the added metal, which thus remains free and gives a normal DPASV response. Consequently, a straight line is finally obtained in the upper region of the curve.

The concentration of "free" metal present in the original sample and after any addition is computed as the ratio of the relevant peak current to the slope of the straight line observed to the right of the end point in the titration curve. Finally, the complexation parameters are computed through a linear transformed plot of titration data such as the one displayed in Fig. 11.2 (39).

In this chapter a review of ASV titration procedures for the study of trace metal complexation in seawater is presented. After a thorough discussion of theoretical aspects, a brief outline of the instrumentation involved, the experimental procedure, and the contamination control is given and finally results

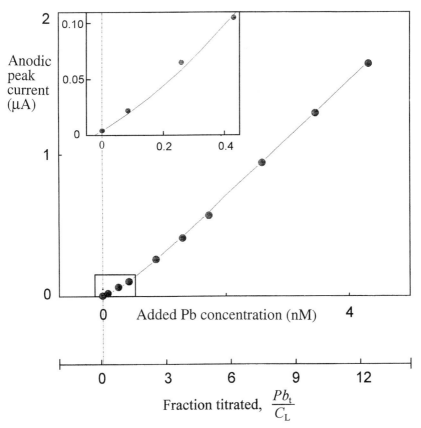

Figure 11.1. Titration curve obtained in the study of Pb complexation in Antarctic coastal seawater (39).

pertaining to Cd, Co, Cu, Pb, and Zn obtained from various geographic areas are reviewed.

11.2. METHODOLOGY: THEORETICAL ASPECTS

Several authors have contributed to the development of basic theory and experimental procedures for the study of trace metal complexation in natural waters based on the voltammetric titration of ligands by metal additions (27–48).

Theoretical aspects have gradually been refined over the past 10 years and are scattered in several journals and in different contexts. This chapter

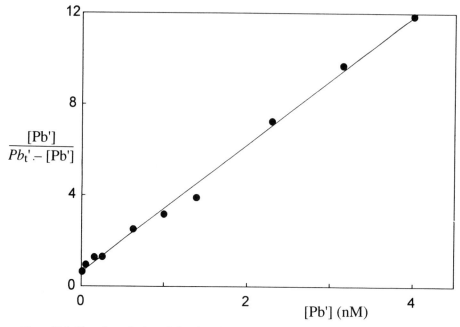

Figure 11.2. Transformed plot of titration data reported in Fig. 11.1, showing the ratio of electroactive to nonelectroactive Pb concentration against the electroactive concentration (39).

presents a unified, rigorous theory of the titrimetric method of metal speciation. Attention is mainly focused on the formation of complexes with 1:1 stoichiometry, the presence of two or more ligands, the competition between different metals, and the general formation of complexes with other than 1:1 stoichiometry.

For the sake of convenience, charges of ions, ligands, and complexes are omitted in the following discussion except in a few examples dealing with specific ions.

11.2.1. Single 1:1 Complex Formation

Consider the simple 1:1 complexation reaction between the metal, M, and the ligand, L, to form the complex ML. If one assumes that M participates in side-reactions with inorganic ligands (or weak organic ligands) X_i (Cl^-, OH^-, CO_3^{2-}, etc.) to form complexes MX_i, that L participates in side-reactions with proton H^+ and major cations, M_j (Ca^{2+}, Mg^{2+}, etc.), and that ML does not participate in side-reactions, the involved equilibria can be represented as follows:

$$\begin{array}{ccc} \text{MX}_i & & \text{H}_n\text{L} \\ +X_i \updownarrow & & +n\text{H} \updownarrow \\ \text{M} & + & \text{L} \rightleftharpoons \text{ML} \\ & & +M_j \updownarrow \\ & & \text{M}_j\text{L} \end{array} \qquad (1)$$

The equilibrium of the complexation reaction (1) is defined, in terms of the concentration stability constant K, by the expression

$$K = \frac{[\text{ML}]}{[\text{M}][\text{L}]} \qquad (2)$$

where K is related to the true or thermodynamic stability constant K^* by the following relationship:

$$K = K^* \frac{\gamma_M \gamma_L}{\gamma_{ML}} \qquad (3)$$

where γ_i is the activity coefficient of species i.

The conditional concentration of metal $[M']$ is defined as the concentration of the metal not complexed with L, i.e. (49),

$$[M'] = [M] + \sum_i [MX_i] \qquad (4)$$

and the conditional concentration of ligand $[L']$ as the concentration of the ligand not complexed with M, i.e.,

$$[L'] = [L] + \sum_n [H_n L] + \sum_j [M_j L] \qquad (5)$$

Then the side-reaction coefficients for metal, α_M, and for ligand, α_L, are given by

$$\alpha_M = [M']/[M] \qquad (6)$$

and

$$\alpha_L = [L']/[L] \qquad (7)$$

respectively, and the equation for the concentration constant, substituted for

Equations 6 and 7, becomes

$$K = \frac{[ML]}{[M'][L']} \alpha_M \alpha_L \tag{8}$$

where the conditional concentration ratio is defined as the conditional stability constant, K':

$$K' = \frac{[ML]}{[M'][L']} = \frac{K}{\alpha_M \alpha_L} \tag{9}$$

The voltammetric titration methodology for the study of trace metal complexation in seawater is based on the assumption that, under carefully controlled experimental conditions, the DPASV technique allows the conditional concentration of a metal, i.e., [M'], to be measured.

We shall now consider the titration procedure and the theoretical equation for the titration curve. A solution containing a mixture of the metal and ligand is titrated by successive additions of the metal. The titrant reacts with the excess of ligand according to reaction (1). In this context C_M denotes the initial analytical concentration of metal M; C_L, the analytical concentration of ligand L; M_a, the total added metal concentration after each addition; and M_t, the overall (initial plus added) analytical concentration of metal after each addition. It is assumed that the dilution effect is negligible.

The equation system for the solution of the equilibrium problem is the following:

$$K'[M'][L'] = [ML] \tag{10}$$

$$M_t = C_M + M_a = [M'] + [ML] \tag{11}$$

$$C_L = [L'] + [ML] \tag{12}$$

This includes expressions for the conditional stability constant and the mass balances for metal and ligand, respectively.

The system of Equations 10–12 can be solved for [M'] as described hereafter. Canceling out [L'] from Equations 10 and 12 and solving for [ML], one obtains

$$[ML] = \frac{K'C_L[M']}{1 + K'[M']} \tag{13}$$

Substituting Equation 11 for Equation 13 gives

$$M_t = [M'] + \frac{K'C_L[M']}{1 + K'[M']} \tag{14}$$

which represents the theoretical equation of the titration curve. Equation 14 can be rearranged into the quadratic expression (27, 44, 45, 47):

$$K'[M']^2 + [1 + K'(C_L - M_t)][M'] - M_t = 0 \qquad (15)$$

and solved for $[M']$ as a function of M_t and therefore of M_a.

To evaluate the combined influence of the conditional stability constant and of the ligand concentration on the shape of the titration curve and, in turn, to assess the possibility of determining the complexation parameters (K' and C_L) from titration data, a convenient rearrangement of Equation 15 in terms of the fraction titrated, i.e., the ratio M_t/C_L, can be of help. Divided by $K'C_L^2$, Equation 15 becomes

$$\left(\frac{[M']}{C_L}\right)^2 + \left[\frac{1}{K'C_L} + 1 - \frac{M_t}{C_L}\right]\frac{[M']}{C_L} - \frac{1}{K'C_L}\frac{M_t}{C_L} = 0 \qquad (16)$$

This represents a generalized theoretical equation of the titration curve and shows that the extent of metal complexation, and thus the shape of the curve, is affected by the dimensionless product $K'C_L$.

Figure 11.3 shows titration curves (as $[M']/C_L$ vs. M_t/C_L) calculated by Equation 16 for different values of the product $K'C_L$ varying from 10^{-3} to 10^3. In each case the metal is assumed to be absent before titration, i.e., $C_M = 0$ and then $M_a = M_t$. It can be observed that there exist practical lower and upper limits of $K'C_L$ values ($K'C_L = 10^{-2}$ and 10^2, respectively) within which the complexation reaction can profitably be studied (detection window) (31). For $K'C_L \leqslant 10^{-3}$, the ligand shows no appreciable interaction with metal and the titration curve practically coincides with the straight line obtained for blank solution in the absence of ligand, i.e., the calibration curve. For $K'C_L = 10^{-2}$, the initial part of the titration curve is practically superimposed on the calibration curve, until the total amount of metal is doubled or even tripled with respect to the ligand quantity. After this a slight differentiation can be observed progressively with further excess of titrant. On the contrary, when $K'C_L \geqslant 10^3$ very strong complexes are formed, the complexation reaction is quantitative in each step, and no detectable metal is revealed until the ligand is almost completely titrated. After this the added metal remains totally free in solution and a straight line with a slope identical to that of the calibration curve (see above) is displayed. For $K'C_L = 10^2$ a slight differentiation from the previous curve is observed in the vicinity of the equivalence point. Finally, in cases in which values of the $K'C_L$ product are included in the range $10^{-2} < K'C_L < 10^2$, titration curves show intermediate shape between these two extremes.

These results highlight the influence of the ligand concentration on the interval of stability constant values that can be determined by the titration

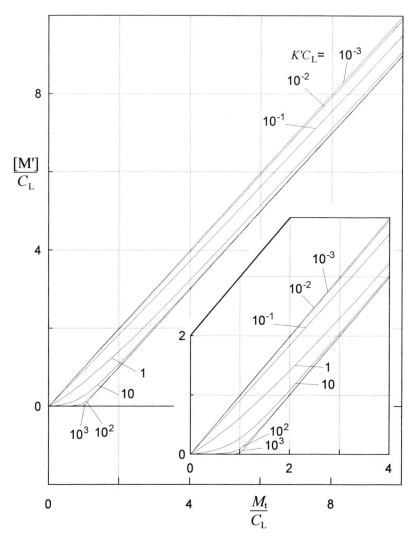

Figure 11.3. Generalized theoretical titration curves for 1:1 complex formation; $K'C_L$ values vary from 10^{-3} to 10^3.

approach (the procedure for simultaneous evaluation of K' and C_L is outlined below). In particular, for a ligand concentration at the nanomolar level, the range of stability constants that can be explored is between about 10^7 and 10^{11} M^{-1}. In fact, in the case of stability constants higher than 10^{11} M^{-1} only one of the two complexation parameters, the ligand concentration, can be evaluated by direct titration (end-point detection), the constant remaining undefined (but anyway $> 10^{11}\,\mathrm{M}^{-1}$). In the case of $K' < 10^7\,\mathrm{M}^{-1}$ no complexation at all can be observed (when reasonable quantities of titrant are added) and neither C_L nor K' are obtainable. An increase in ligand concentration to 10 nM or a decrease to 0.1 nM would lead to a shift of the interval of the observable stability constants to the range of 10^6–10^{10} and 10^8–$10^{12}\,\mathrm{M}^{-1}$, respectively.

A second observation of some importance, based on the theoretical titration curves plotted in Fig. 11.3, concerns the slope of the final part of the curve. This slope reaches a value indistinguishable from that obtained in the absence of the ligand (the value required to measure sensitivity in DPASV and to make the required computations) solely after an addition of remarkable excess of metal, strongly dependent on the product (complexation parameter) $K'C_L$. In fact, while the condition is already reached after the addition of an excess of 2–3 times the equivalent amount when $K'C_L = 10^2$, the same situation is obtained with excesses of 3–5, 6–7, 8–9, and even > 10 times when $K'C_L$ values are 10^1, 1, 10^{-1}, and 10^{-2}, respectively. In conclusion, it is worth noting that a substantial excess of titrant is normally employed and that care is required to be sure that the final, linear part of the curve has really been reached.

In the titration of a real sample ($C_M \neq 0$) the initial part of the curve is to the right of the origin in Fig. 11.3 (the higher the metal concentration in the sample, the further it lies to the right, as shown in Fig. 11.4). This fact precludes the possibility of the first part of the titration curve being entirely observed

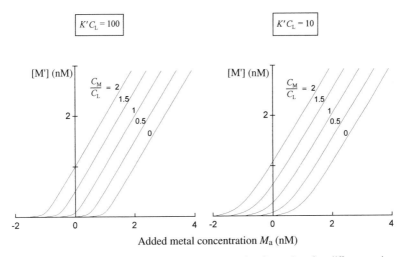

Figure 11.4. Theoretical titration curves for 1:1 complex formation for different values of C_M/C_L.

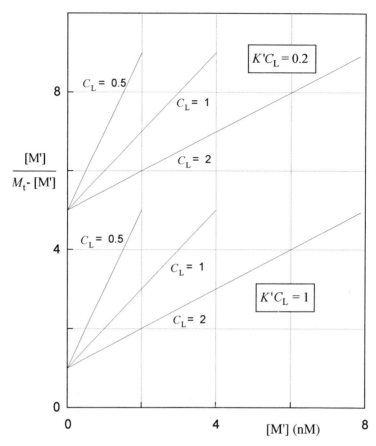

Figure 11.5. Plots of $[M']/(M_t - [M'])$ vs. $[M']$ for titration curves with different values of $K'C_L$ and C_L (values in nM units).

and, as the ratio C_M/C_L increases, the curved part of the titration line tends to be completely obscured. Finally, for C_M/C_L higher than about 1 (depending on the complex strength) the observable curve becomes practically indistinguishable from a straight line. In the latter situation the ligand is already practically totally saturated at the beginning of the titration and the conditional stability constant can no longer be determined with confidence. As shown in Fig. 11.4, this condition is obtained for C_M/C_L values which are higher the weaker the complex.

The evaluation of complexation parameters K' and C_L from the titration experiment can easily be obtained by rearranging Equation 14 in the form

(27, 36, 41–44, 46)

$$\frac{[M']}{M_t - [M']} = \frac{[M']}{C_L} + \frac{1}{K'C_L} \tag{17}$$

i.e., the ratio of free-to-bound metal concentration is linearly related to the free metal concentration. From titration data and the independent measurement of the total metal concentration in the sample C_M, one can compute $[M']/(M_t - [M'])$ and plot it against $[M']$. In this way a transformed linear plot of titration data is obtained, and from this it is possible to compute both the ligand concentration, as the reciprocal of the slope, and the conditional stability constant from the intercept, taking account of the C_L value just obtained from the slope.

Figure 11.5 shows two sets of curves on the transformed diagram, corresponding to $K'C_L$ values of 0.2 and 1, respectively. In each set, curves for C_L equal to 0.5, 1, and 2 nM are plotted. It can be observed that the effect of increasing ligand concentration is a decrease in the slope of the curve, whereas the effect of increasing the constant (with identical C_L) is a decrease in the intercept and consequently the whole curve.

In the case of high C_M/C_L values, for which only the straight-line part of the titration curve is observed and the ligand is practically totally saturated (i.e., $[ML] \approx C_L$; see Equation 11), the ligand concentration can again be computed as $C_L \approx [ML] \approx C_M - [M']_{M_a} = 0$, but the K' cannot be determined with confidence using Equation 17 ($M_t - [M'] \approx C_L$ and the degree of uncertainty in the evaluation of the intercept in the transformed plot is too high).

11.2.2. Ligand Competition: Two 1:1 Complexes of One Metal with Two Ligands

If two ligands (L_1 and L_2) compete in complexing one metal, forming two 1:1 complexes (ML_1 and ML_2) with diffferent stabilities, the equilibrium problem can be described by the following system:

$$K'_1[M'][L'_1] = [ML_1] \tag{18}$$

$$K'_2[M'][L'_2] = [ML_2] \tag{19}$$

$$M_t = C_M + M_a = [M'] + [ML_1] + [ML_2] \tag{20}$$

$$C_{L_1} = [L'_1] + [ML_1] \tag{21}$$

$$C_{L_2} = [L'_2] + [ML_2] \tag{22}$$

where K_1' and K_2' are the conditional stability constants for the two complexes; [M'] is the conditional concentration of metal (with respect to inorganic side-reactions); $[L_1']$, $[L_2']$ and C_{L_1}, C_{L_2} are the conditional and the total concentrations of the two ligands, respectively; $[ML_1]$ and $[ML_2]$ are the equilibrium concentrations of the two complexes; and M_t, C_M, and M_a have the same meaning as defined earlier.

Canceling out $[L_1']$ from Equations 18 and 21 and solving for $[ML_1]$ as a function of [M'], one obtains

$$[ML_1] = \frac{K_1' C_{L_1} [M']}{1 + K_1'[M']} \tag{23}$$

Analogously, from Equations 19 and 22 one can solve for $[ML_2]$:

$$[ML_2] = \frac{K_2' C_{L_2} [M']}{1 + K_2'[M']} \tag{24}$$

Combining Equations 23 and 24 with Equation 20, one obtains the theoretical equation of the titration curve:

$$M_t = [M'] + \frac{K_1' C_{L_1} [M']}{1 + K_1'[M']} + \frac{K_2' C_{L_2} [M']}{1 + K_2'[M']} \tag{25}$$

or

$$K_1' K_2' [M']^3 + [K_1' + K_2' + K_1' K_2'(C_{L_1} + C_{L_2}) - K_1' K_2' M_t][M']^2$$
$$+ [1 + K_1' C_{L_1} + K_2' C_{L_2} - (K_1' + K_2') M_t][M'] - M_t = 0 \tag{26}$$

The same equation was reported, in other forms, by Ruzic (27) and van den Berg (34,36).

It one defines C_{L_t} as

$$C_{L_t} = C_{L_1} + C_{L_2} \tag{27}$$

Equation 26 can be rearranged in terms of the fraction titrated, i.e., M_t/C_{L_t}, by dividing it by $K_1' K_2' C_{L_t}^3$ and replacing C_{L_2} with $C_{L_t} - C_{L_1}$:

$$\left(\frac{[M']}{C_{L_t}}\right)^3 + \left(1 + \frac{1}{K_1' C_{L_t}} + \frac{1}{K_2' C_{L_t}} - \frac{M_t}{C_{L_t}}\right)\left(\frac{[M']}{C_{L_t}}\right)^2$$
$$+ \left[\frac{1}{K_1' C_{L_t}}\left(1 + \frac{1}{K_2' C_{L_t}}\right) + \frac{C_{L_1}}{C_{L_t}}\left(\frac{1}{K_2' C_{L_t}} - \frac{1}{K_1' C_{L_t}}\right)\right.$$
$$\left. - \left(\frac{1}{K_1' C_{L_t}} + \frac{1}{K_2' C_{L_t}}\right)\frac{M_t}{C_{L_t}}\right]\frac{[M']}{C_{L_t}} - \frac{1}{K_1' C_{L_t}} \frac{1}{K_2' C_{L_t}} \frac{M_t}{C_{L_t}} = 0 \tag{28}$$

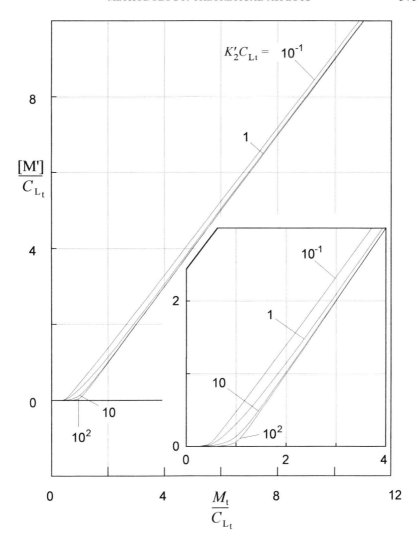

Figure 11.6. Generalized theoretical titration curves for the formation of 1:1 complexes between one metal and two ligands: $K'_1 C_{L_t} = 100$; $C_{L_1}/C_{L_t} = 0.5$, $K'_2 C_{L_t}$ varies from 0.1 to 100.

Equation 28 makes it possible to observe titration curves for different values of K'_1 and K'_2 by fixing parameters $K'_1 C_{L_t}$, $K'_2 C_{L_t}$, and C_{L_1}/C_{L_t}. Figure 11.6 shows titration curves obtained for $K'_1 C_{L_t} = 10^2$ and $C_{L_1}/C_{L_t} = 0.5$ and for several values of $K'_2 C_{L_t}$, i.e., 10^{-1}, 1, 10, and 10^2.

It can be observed that the titration curve of two ligands tends to be superimposed on that of the single stronger one in the very first part of the curve as well as that of the single weaker one in the final part (titrant excess). In particular, the separation in the two previously defined curves is quite evident when the difference in the K' values is of 3 orders of magnitude or more. In fact, in such cases the titration curve is practically coincident with that of the single

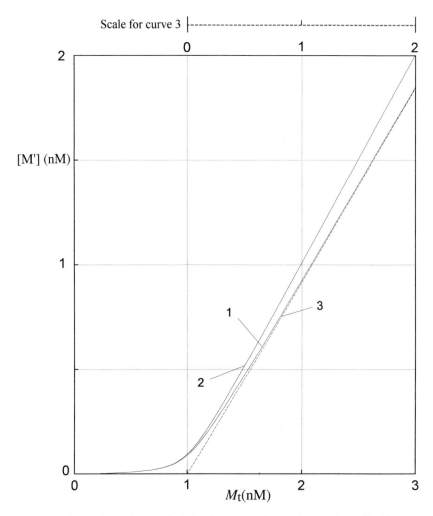

Figure 11.7. Comparison of theoretical titration curves for a mixture of two ligands (curve 1: $K'_1 = 10^{11}\,\mathrm{M}^{-1}$, $K'_2 = 10^8\,\mathrm{M}^{-1}$, $C_{L_1} = C_{L_2} = 1\,\mathrm{nM}$) and those of the single ligands (curve 2: stronger, $K' = 10^{11}\,\mathrm{M}^{-1}$, $C_L = 1\,\mathrm{nM}$; curve 3: weaker, $K' = 10^8\,\mathrm{M}^{-1}$, $C_L = 1\,\mathrm{nM}$).

stronger ligand until the latter is completely titrated. After this the curve tends progressively to be superimposed on that of the single weaker ligand (see Fig. 11.7). In contrast, when constants are of the same order of magnitude, no difference between the titration curve of two ligands and that of one ligand is observed (compare curve for $K'_1C_{L_1} = K'_2C_{L_1} = 100$ in Fig. 11.6 with that of $K'C_L = 100$ in Fig. 11.3). Behaviors between the foregoing two extremes are found with intermediate situations.

In the case of two ligands the transformed plot of titration data as proposed in the previous subsection, i.e., $[M']/(M_t - [M'])$ vs. $[M']$, is no longer linear (see Fig. 11.8). In fact, the theoretical equation for the transformed plot, obtained after rearrangement of Equation 25, becomes (27)

$$\frac{[M']}{M_t - [M']} = \frac{1}{\dfrac{K'_1 C_{L_1}}{1 + K'_1[M']} + \dfrac{K'_2 C_{L_2}}{1 + K'_2[M']}} \tag{29}$$

or (34, 36)

$$\frac{[M']}{M_t - [M']} = \frac{[M']}{C_{L_t}} + \frac{[ML_1]}{[ML_1] + [ML_2]} \frac{1}{K'_1 C_{L_t}}$$

$$+ \frac{[ML_2]}{[ML_1] + [ML_2]} \frac{1}{K'_2 C_{L_t}} \tag{30}$$

where $[ML_1]$ and $[ML_2]$ can be replaced with Equations 23 and 24, respectively.

In Fig. 11.8 it can be noted that curves tend to asymptotic straight lines when the titrant excess is sufficiently high (high values of $[M']$) depending on the difference between the two constants. The greater the difference between constants, the later the asymptotic line is reached. Conversely, in the case of differences lower than 1 order of magnitude, the result is practically an entirely straight line, which is indistinguishable from that obtained for the presence of a single ligand having a concentration equal to the sum of those of the two ligands and approximately an average value for the logarithm of the constant (e.g., for $K' = 2 \times 10^{10} \, M^{-1}$ and $C_L = 2 \times 10^{-9} \, M$ the slope is $0.5 \, nM^{-1}$ and the intercept is 0.025).

In the situation of ligand competition, the determination of complexation parameters ($C_{L_1}, C_{L_2}, K'_1, K'_2$) is no longer easily accomplished directly from the transformed plot, as in Fig. 11.8, and different procedures for the solution of the problem have been developed independently by van den Berg (34, 36) and Ruzic (27).

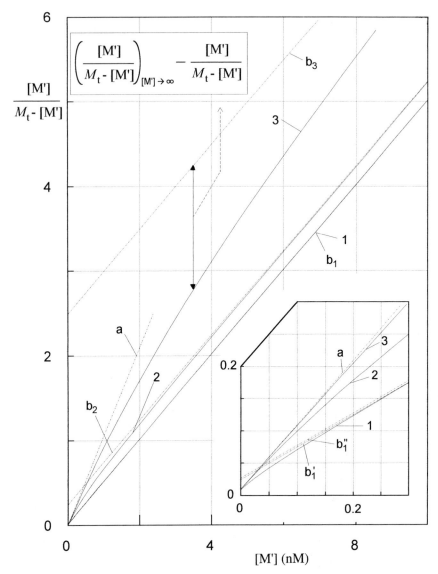

Figure 11.8. Plots of $[M']/(M_t - [M'])$ vs. $[M']$ for titrations of two ligands with one metal (1:1 complexes). $C_{L_1} = C_{L_2} = 10^{-9}$ M; $K'_1 = 10^{11}$ M^{-1}. K'_2 (M^{-1}): curve 1, 10^{10}; curve 2, 10^9; curve 3, 10^8. Lines a and b: asymptotic straight lines.

11.2.2.1. The van den Berg Approach

If differences between the two constants are of at least 1 order of magnitude, approximations of Equation 30 can be considered at the very beginning of the titration (low [M'] values) and in the presence of titrant excess (high [M'] values) (34, 36).

In the first part of the titration curve, the formation of the weaker complex is negligible and the problem becomes that in which a single ligand is present (see Fig. 11.8, straight line a). The transformed plot of this part of the curve can be used to compute parameters of the stronger complex (K_1' and C_{L_1}). Data reported in Fig. 11.8 show that in practice accurate estimates of the asymptotic line a and consequently of the complexation parameters can be obtained from experimental measurements through this simple procedure only when the difference between constants is higher than 2 orders of magnitude.

In the presence of titrant excess and if $K_2' \ll K_1'$, both ligands can be considered totally saturated (i.e., $[ML_1] = C_{L_1}$ and $[ML_2] = C_{L_2}$) and the second term of the right member of Equation 30 can be neglected, giving the following approximated equation for the final part of the transformed plot:

$$\left(\frac{[M']}{M_t - [M']}\right)_{[M'] \to \infty} = \frac{[M']}{C_{L_1} + C_{L_2}} + \frac{C_{L_2}}{K_2'(C_{L_1} + C_{L_2})^2} \quad (31)$$

from which the remaining parameters C_{L_2} and K_2' can be computed from the slope and from the intercept, respectively (Fig. 11.8, straight lines b and b_1').

In cases where the values of K_1' and K_2' are not very different, the values just determined can be considered as a first approximation and better results can be computed through an iterative approach. Essentially, the procedure involves the development of two separate transformed plots related to the two complexes (basically on the first part of the titration for the stronger and on the second part for the weaker), taking account of the reciprocal influence due to the formation of one complex in the range of study of the other. From the approximate values of K_1' and C_{L_1} obtained in the first run, one can compute the corresponding value of $[ML_1]$ (Equation 23) for each [M'] value of the experimental data points; this value is substituted for the second term of the right member of Equation 25, and the following new equation is used for the transformed plot (principally at high values of [M']):

$$\frac{[M']}{M_t - [ML_1] - [M']} = \frac{[M']}{C_{L_2}} + \frac{1}{K_2' C_{L_2}} \quad (32)$$

which allows K_2' and C_{L_2} to be calculated to a first approximation.

From the approximated values of K_2' and C_{L_2} the foregoing procedure can be repeated with respect to the stronger complex. Thus, $[ML_2]$ values can be computed (Equation 24) for each data point and substituted for the third term of the right member of Equation 25, yielding the following equation:

$$\frac{[M']}{M_t - [ML_1] - [M']} = \frac{[M']}{C_{L_1}} + \frac{1}{K_1' C_{L_1}} \tag{33}$$

The new linear expression can be used to plot a second transformed diagram (principally at low values of $[M']$), which allows better results for K_1' and C_{L_1} to be computed.

Finally, the procedure is repeated until two subsequent sets of results do not differ significantly with respect to the experimental error.

11.2.2.2. The Ruzic Approach

Unlike the procedure described in the previous subsection, the concept here is to use all the experimental data points of the titration curve and not only those that fit into the two linear parts of the transformed plot (27).

The only assumption is that when titrant excess is sufficient, i.e., $[M'] \to \infty$, both ligands are totally saturated. Then $[ML_1] = C_{L_1}$ and $[ML_2] = C_{L_2}$ can be substituted for the mass balances of ligands (Equations 21 and 22), and the final equation for the ratio $[M']/(M_t - [M'])$ becomes (see Equation 30)

$$\left(\frac{[M']}{M_t - [M']}\right)_{[M'] \to \infty} = \frac{[M']}{C_{L_1} + C_{L_2}} + \frac{C_{L_1}}{K_1'(C_{L_1} + C_{L_2})^2} + \frac{C_{L_2}}{K_2'(C_{L_1} + C_{L_2})^2} \tag{34}$$

Note that the Ruzic approximation for titrant excess, Equation 34, is similar to that of van den Berg, Equation 31. Here, however, no zero assumption is made regarding the contribution of the stronger complex.

The asymptotic lines obtained for the transformed plot at titrant excess are shown in Fig. 11.8, curves b_1, b_2, b_3, and b_1''.

In this context it is also worth considering a new transformed function to be used for linearization of the whole titration curve, i.e., the reciprocal of the difference between two values of the ratio $[M']/(M_t - [M'])$, the extrapolated value given by the asymptotic line of Equation 34 (obtained from the approximation for titrant excess) and the value on the real curve given by Equation 29 (and obtained without any approximation). It is easy to show that the new straight line obtained by the above procedure obeys the following linear expression:

$$\frac{1}{\left(\dfrac{[M']}{M_t - [M']}\right)_{[M']\to\infty} - \left(\dfrac{[M']}{M_t - [M']}\right)} = \frac{(K'_1 K'_2)^2 C^3_{L_t}[M']}{K'_1 K'_2 (C^2_{L_1} + C^2_{L_2}) + C_{L_1} C_{L_2}(K'^2_1 + K'^2_2) - K'_1 K'_2 C^2_{L_t}} + \frac{K'_1 K'_2 C^2_{L_t}(K'_1 C_{L_1} + K'_2 C_{L_2})}{K'_1 K'_2 (C^2_{L_1} + C^2_{L_2}) + C_{L_1} C_{L_2}(K'^2_1 + K'^2_2) - K'_1 K'_2 C^2_{L_t}} \quad (35)$$

Figure 11.9 is a graphic representation of the transformed plot of $1/\{[[M']/(M_t - [M'])]_{[M']\to\infty} - [M']/(M_t - [M'])\}$ vs. $[M']$ obtained for a few representative cases.

In conclusion, one now has two experimental straight lines (according to Equations 34 and 35) with two intercepts and two slopes that can be utilized in calculating the four requested parameters. If one rewrites Equations 34 and 35 as

$$\left(\frac{[M']}{M_t - [M']}\right)_{[M']\to\infty} = a[M'] + b \quad (36)$$

and

$$\frac{1}{\left(\dfrac{[M']}{M_t - [M']}\right)_{[M']\to\infty} - \dfrac{[M']}{M_t - [M']}} = \alpha[M'] + \beta \quad (37)$$

respectively, with

$$a = \frac{1}{C_{L_t}} \quad (38)$$

$$b = \left(\frac{C_{L_1}}{K'_1} + \frac{C_{L_2}}{K'_2}\right) \Big/ C^2_{L_t} \quad (39)$$

$$\alpha = \frac{(K'_1 K'_2)^2 C^3_{L_t}}{K'_1 K'_2(C^2_{L_1} + C^2_{L_2}) + C_{L_1} C_{L_2}(K'^2_1 K'^2_2) - K'_1 K'_2 C^2_{L_t}} \quad (40)$$

$$\beta = \frac{K'_1 K'_2 C^2_{L_t}(K'_1 C_{L_1} + K'_2 C_{L_2})}{K'_1 K'_2(C^2_{L_1} + C^2_{L_2}) + C_{L_1} C_{L_2}(K'^2_1 + K'^2_2) - K'_1 K'_2 C^2_{L_t}} \quad (41)$$

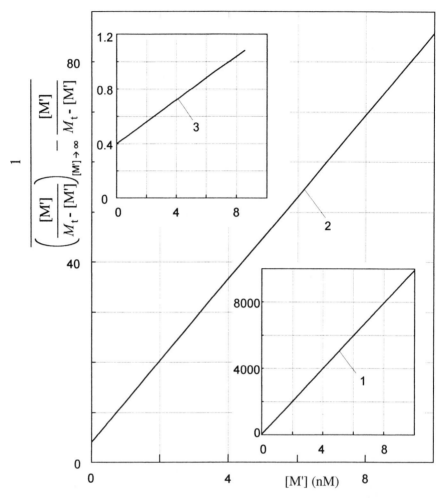

Figure 11.9. Plot of $1/\{[[M']/(M_t - [M'])]_{[M'] \to \infty} - [M']/(M_t - [M'])\}$ vs. $[M']$ for titrations of two ligands with one metal (1:1 complexes). Obtained from data plotted in Fig. 11.8, as explained in the text.

and then solves, progressively, for K'_2, K'_1, C_{L_1}, and C_{L_2}, one obtains

$$K'_2 = \frac{\alpha b + a\beta + [(\alpha b + a\beta)^2 - 4a\alpha(b\beta - 1)]^{1/2}}{2(b\beta - 1)} \quad (42)$$

$$K'_1 = \frac{a\alpha}{K'_2(b\beta - 1)} \quad (43)$$

$$C_{L_1} = \frac{K'_1(K'_2 b - a)}{a^2(K'_2 - K'_1)} \tag{44}$$

$$C_{L_2} = \frac{1}{a} - C_{L_1} \tag{45}$$

A somewhat different procedure for the solution was discussed by Ruzic (27).

As regards both the van den Berg and the Ruzic approaches, any assumption that the linear asymptotic part of the curve $[M']/(M_t - [M'])$ vs. $[M']$ has really been reached should be treated with caution because the required excess of titrant can be extremely high (practically impossible to obtain) in cases of differences between constants higher than 2 orders of magnitude (see Fig. 11.8, curve 3).

11.2.2.3. Ligand Competition: Generalization

In the presence of more than two ligands it is easy to show that the following general equation is valid for titration data (cf. Equation 29):

$$\frac{[M']}{M_t - [M']} = \frac{1}{\sum_i K'_i C_{L_i}/(1 + K'_i[M'])} \tag{46}$$

In such a situation a nonlinear least-square curve fitting procedure can be applied to determine the number of ligands (classes of ligands) to be introduced into the model, together with the relevant complexing parameters (34, 48).

11.2.3. Metal Competition: Two 1:1 Complexes of One Ligand with Two Metals

In the case of competition of a second metal N with the studied metal M in forming 1:1 complexes with the same ligand L, i.e., ML and NL, the equilibrium problem can be described by the following system:

$$K'_M[M'][L'] = [ML] \tag{47}$$

$$K'_N[N'][L'] = [NL] \tag{48}$$

$$M_t = C_M + M_a = [M'] + [ML] \tag{49}$$

$$C_N = [N'] + [NL] \tag{50}$$

$$C_L = [L'] + [ML] + [NL] \tag{51}$$

where K'_M and K'_N are the conditional stability constants for the two complexes; [N'] and C_N are the conditional concentration (with respect to inorganic side-reactions) and the total concentration, respectively, of the competing metal N; [NL] is the equilibrium concentration of complex NL; and the remaining quantities have the same meaning as defined earlier.

Solving for [ML], [L'], and [NL] as a function of [M'] from Equations 47–50, one obtains

$$[ML] = M_t - [M'] \tag{52}$$

$$[L'] = \frac{M_t - [M']}{K'_M[M']} \tag{53}$$

$$[NL] = \frac{K'_N C_N \dfrac{M_t - [M']}{K'_M[M']}}{1 + K'_N \dfrac{M_t - [M']}{K'_M[M']}} \tag{54}$$

These expressions, when substituted in Equation 51, lead to the theoretical equation of the titration curve:

$$C_L = \frac{M_t - [M']}{[M']} \left([M'] + \frac{1}{K'_M} + \frac{K'_N C_N}{K'_M \left(1 + K'_N \dfrac{M_t - [M']}{K'_M[M']}\right)} \right) \tag{55}$$

or

$$K'_M(K'_M - K'_N)[M']^3 + \{(K'_M - K'_N)[1 + K'_M(C_L - M_t)] + K'_M K'_N(M_t + C_N)\}[M']^2 \\ + [K'_M K'_N(C_L - M_t - C_N) + 2K'_N - K'_M]M_t[M'] - K'_N M_t^2 = 0 \tag{56}$$

The same equation was given, in another form, by Ruzic (27).

Equation 56 can be rearranged in terms of the fraction of ligand titrated by metal M, i.e., M_t/C_L, dividing it by $K'_M(K'_M - K'_N)C_L^3$:

$$\left(\frac{[M']}{C_L}\right)^3 + \left[1 + \frac{1}{K'_M C_L} - \frac{M_t}{C_L} + \frac{1}{(K'_M/K'_N) - 1}\left(\frac{M_t}{C_L} + \frac{C_N}{C_L}\right)\right]\left(\frac{[M']}{C_L}\right)^2 \\ + \left[\frac{1}{(K'_M/K'_N) - 1}\left(1 + \frac{1}{K'_M C_L} - \frac{M_t}{C_L} - \frac{C_N}{C_L}\right) - \frac{1}{K'_M C_L}\right]\frac{M_t}{C_L}\frac{[M']}{C_L} \\ - \frac{1}{K'_M C_L}\frac{1}{(K'_M/K'_N) - 1}\left(\frac{M_t}{C_L}\right)^2 = 0 \tag{57}$$

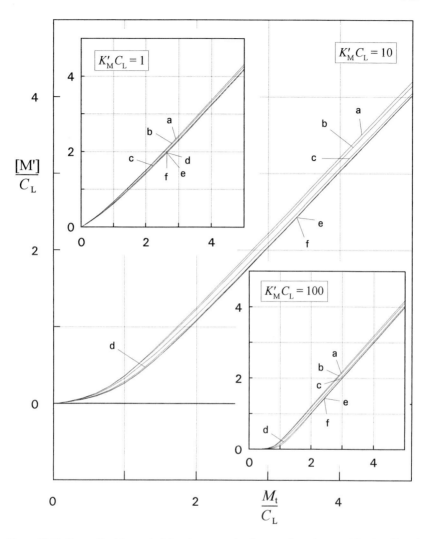

Figure 11.10 Generalized theoretical titration curves for the case of metal competition (one ligand with two metals forming 1:1 complexes). $C_N/C_L = 0.2$. K'_M/K'_N: curves a, 0.01; b, 0.1; c, 1; d, 10; e, 100; f, presence of single metal M.

It should be noted that when $K'_M = K'_N$ the theoretical equation of the titration curve can be simplified as follows:

$$\left(\frac{M_t}{C_L} + \frac{C_N}{C_L}\right)\left(\frac{[M']}{C_L}\right)^2 + \left[1 - \frac{M_t}{C_L} - \frac{C_N}{C_L} + \frac{1}{K'_M C_L}\right]\frac{M_t}{C_L}\frac{[M']}{C_L} - \frac{1}{K'_M C_L}\left(\frac{M_t}{C_L}\right)^2 = 0 \tag{58}$$

Figure 11.10 shows generalized titration curves obtained from Equation 57 and for $K'_M C_L$ equal to 1, 10, and 100, respectively. In each case the ratio C_N/C_L is taken equal to 0.2 and curves are reported for values of K'_M/K'_N between 0.01 and 100. For comparative purposes the curve obtained in the case of the presence of only one metal is also reported. It can be observed that curves range between that related to the single metal M, as if metal N were absent (the case of an ML complex much stronger than NL, i.e., $K'_M/K'_N > 10$) and that of the same single metal M, but with part of the ligand (in this case 20% because $C_N/C_L = 0.2$) masked by metal N, i.e., the curve is displaced by -0.2 units along the abscissa with respect to the previous one throughout the titration experiment except at the beginning (the case of the ML complex that is much weaker than the complex containing the competing metal N, i.e., $K'_M/K'_N < 0.1$). In the latter case only part of the complexation capacity of the system can be determined. Finally, it should be noted that the general shape of the curves in the first part of the titration depends strictly on the value of $K'_M C_L$, as already observed for the presence of a single metal (see Fig. 11.3).

A special case arises when both complexes are very strong ($K'_M C_L > 100$; $K'_N C_L > 100$). Now the titration curve tends to be reduced to two straight lines (typical shape of reversed L) and constants are no longer detectable, while ligand content can still be determined. Two limiting situations can be recognized: $K'_M \gg K'_N$ or $K'_N \gg K'_M$. For $K'_M \gg K'_N$ the total ligand content can be determined, whereas for $K'_N \gg K'_M$ only part of the ligand, i.e., $C_L - C_N$, can be detected because the remaining amount is masked by the formation of the stronger complex NL.

The transformed plot of titration data as $[M']/(M_t - [M'])$ vs. $[M']$ in the case of metal competition is no longer linear [Fig. 11.11(a,b) shows the generalized plot obtained as a function of $[M']/C_L$ for the same cases as displayed in Fig. 11.10]. The theoretical equation, obtained after rearrangement of Equation 55, becomes

$$\frac{[M']}{M_t - [M']} = \frac{[M']}{C_L} + \frac{1}{K'_M C_L} + \frac{K'_N C_N}{K'_M C_L \left(1 + K'_N \dfrac{M_t - [M']}{K'_M [M']}\right)} \quad (59)$$

It should be noted, however, that curves approach asymptotic straight lines (see below for the theoretical equations) for high values of $[M']/C_L$, i.e., when sufficient titrant excess is added, although in cases of low values of K'_M/K'_N the required excess can be very high, practically beyond the range that is reasonably observable experimentally. Again, a behavior almost coincident with the one obtained for the single metal M (practically the same straight line, except at the very beginning of the curve) is observed when the complex with competing metal N is much weaker than that with metal M (i.e., for $K'_M/K'_N > 10$). Deviations from the latter model are higher as the NL complex is stronger with respect to ML. In particular, for $K'_M/K'_N \leq 0.01$ (see Fig. 11.11, curve a) the behavior practically corresponds to that of single metal M with part of the ligand (equivalent to C_N) masked by N. In fact, as can be noted from curve a in Fig. 11.11, the slope of the transformed plot approximates 1.25, as predicted by the first term to the right of Equation 17, which in this case is $[M']/(C_L - C_N)$, i.e., equal to $1.25 [M']/C_L$.

The solution of the problem, which in this case pertains to the determination of C_N, K'_N, C_L, and K'_M, was developed by Ruzic, who considered that for titrant excess, and if NL complex is not by far stronger than ML, equilibrium is totally displaced toward ML formation (27). This means that for $[M'] \to \infty$ Equations 50 and 54 can be approximated as

$$C_N = [N'] \qquad [M'] \to \infty \qquad (50')$$

and

$$[NL] = K'_N C_N \frac{M_t - [M']}{K'_M [M']} \qquad [M'] \to \infty \qquad (54')$$

In such a situation the solution of the equilibrium problem leads to the following linear equation for the transformed curve:

$$\left(\frac{[M']}{M_t - [M']}\right)_{[M'] \to \infty} = \frac{[M']}{C_L} + \frac{1}{K'_M C_L} + \frac{K'_N C_N}{K'_M C_L} \qquad (60)$$

Figure 11.11 shows the asymptotic lines obtained by Equation 60 for each case displayed (dashed lines a', b', etc.).

Now, if one considers the transformed function of titration data as introduced in the previous subsection and given by

$$1/\{[[M']/(M_t - [M'])]_{[M'] \to \infty} - [M']/(M_t - [M'])\}$$

and combines Equations 59 and 60, the following equation is obtained:

$$\frac{1}{\left(\dfrac{[M']}{M_t - [M']}\right)_{[M'] \to \infty} - \left(\dfrac{[M']}{M_t - [M']}\right)} = \frac{K'^2_M C_L}{K'^2_N C_N} \frac{[M']}{M_t - [M']} + \frac{K'_M C_L}{K'_N C_N} \qquad (61)$$

which is linear with respect to $[M']/(M_t - [M'])$. The straight line obtained is valid for the whole titration curve.

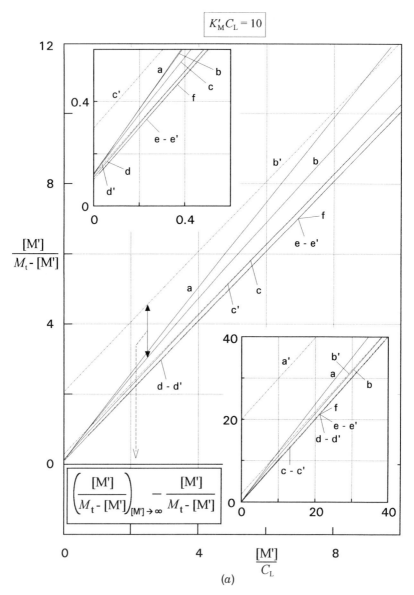

Figure 11.11. Plot of $[M']/(M_t - [M'])$ vs. $[M']/C_L$ for titrations in the case of metal competition (one ligand with two metals forming 1:1 complexes): $C_N/C_L = 0.2$. Parameters and symbols are as in Fig. 11.10. Lines a' to e': asymptotic straight lines. (a) Case $K'_M C_L = 10$. (b) Cases $K'_M C_L = 1$ and 100.

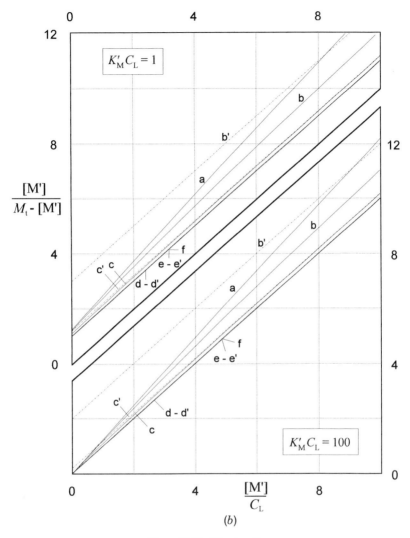

Figure 11.11. (*Continued*)

Plots of titration data transformed according to this procedure are shown in Fig. 11.12 for the cases displayed in Figs. 11.10 and 11.11. In this kind of plot the behavior is conditioned by the relative strength of the two complexes (i.e., the ratio K'_M/K'_N). Thus, as predicted by Equation 61, straight lines related to cases with the same value of K'_M/K'_N coincide independently of the $K'_M C_L$ value.

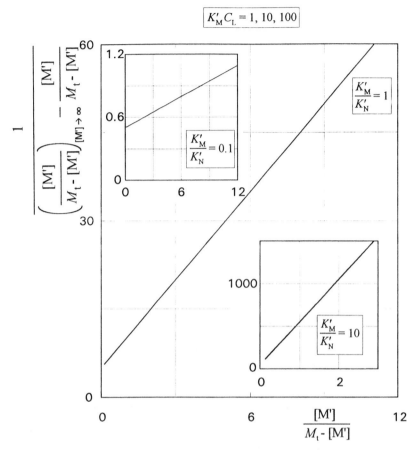

Figure 11.12. Plot of $1/\{[[M']/(M_t - [M'])]_{[M'] \to \infty} - [M']/(M_t - [M'])\}$ vs. $[M']/(M_t - [M'])$ for titrations in the case of metal competition (one ligand with two metals forming 1:1 complexes): $C_N/C_L = 0.2$. Obtained from data plotted in Fig. 11.11, as explained in the text.

Finally, two straight lines can be obtained from titration data according to the transformations expressed by Equations 60 and 61. In this way two intercepts and two slopes can be experimentally determined to calculate the four unknown quantities. If one rewrites Equations 60 and 61 as

$$\left(\frac{[M']}{M_t - [M']}\right)_{[M'] \to \infty} = c[M'] + d \qquad (62)$$

and

$$\frac{1}{\left(\frac{[M']}{M_t - [M']}\right)_{[M'] \to \infty} - \frac{[M']}{M_t - [M']}} = \gamma \frac{[M']}{M_t - [M']} + \delta \qquad (63)$$

respectively, with

$$c = \frac{1}{C_L} \qquad (64)$$

$$d = \frac{1 + K'_N C_N}{K'_M C_L} \qquad (65)$$

$$\gamma = \frac{K'^2_M C_L}{K'^2_N C_N} \qquad (66)$$

$$\delta = \frac{K'_M C_L}{K'_N C_N} \qquad (67)$$

and solves, progressively, for C_L, C_N, K'_N, and K'_M, one obtains

$$C_L = \frac{1}{c} \qquad (68)$$

$$C_N = \frac{C_L \gamma}{\delta^2} \qquad (69)$$

$$K'_N = \frac{\delta}{C_L d\delta - C_N \delta} \qquad (70)$$

$$K'_M = \frac{K'_N \gamma}{\delta} \qquad (71)$$

A slightly different procedure for the final computation was presented by Ruzic (27).

Discrimination between the two cases of ligand competition and metal competition, both giving a slight departure from linearity in the low region of the transformed diagram $[M']/(M_t - [M'])$ vs. $[M']$, can be obtained by plotting the experimental data on diagrams of Figs. 11.9 and 11.12, i.e., plotting the function

$$1/\{[[M']/(M_t - [M'])]_{[M'] \to \infty} - [M']/(M_t - [M'])\}$$

vs. $[M']$ or vs. $[M']/(M_t - [M'])$, respectively, and observing the model which shows the best fit (discrimination between Equation 35 and Equation 61).

11.2.4. Other than 1:1 Complex Formation

The general complexation reaction between metal and ligand, in any stoichiometry, to form the complex M_aL_b, is represented by the equilibrium:

$$a\mathrm{M} + b\mathrm{L} \rightleftharpoons \mathrm{M}_a\mathrm{L}_b \tag{72}$$

and the system which characterizes it is the following:

$$K'[M']^a[L']^b = [M_aL_b] \tag{73}$$

$$M_t = C_M + M_a = [M'] + a[M_aL_b] \tag{74}$$

$$C_L = [L'] + b[M_aL_b] \tag{75}$$

Canceling out $[L']$ from Equations 73 and 75 and replacing $[M_aL_b]$ as obtained from Equation 74, one obtains the following general equation for the theoretical titration curve:

$$b(M_t - [M']) + a^{1-(1/b)}K'^{-(1/b)}[M']^{-a/b}(M_t - [M'])^{1/b} - aC_L = 0 \tag{76}$$

or, in terms of the transformed function as proposed earlier in the case of a single 1:1 complex formation, one obtains

$$\frac{[M']}{M_t - [M']} = \frac{[M']^{1-a}}{aK'}\left[C_L - \frac{b}{a}(M_t - [M'])\right]^{-b} \tag{77}$$

or

$$\frac{[M']}{M_t - [M']} = \frac{b}{aC_L}\left[[M'] + \frac{a^{1-(1/b)}}{bK'^{1/b}}[M']^{1-(a/b)}(M_t - [M'])^{(1/b)-1}\right] \tag{78}$$

and the nonlinearity can easily be distinguished from those of both ligand competition and metal competition in cases of 1:1 stoichiometry by fitting data into Equations 35 and 61, respectively.

In this situation, two different diagrams can be used to obtain complexation parameters. In fact, from Equations 77 and 78 one can obtain

$$\frac{(M_t - [M'])^{1/b}}{[M']^{a/b}} = (aK)^{(1/b)-1} C_L - a^{1/b-1} K^{1/b} b (M_t - [M']) \qquad (79)$$

and, respectively,

$$\frac{[M']^{a/b}}{(M_t - [M'])^{1/b}} = \frac{b}{aC_L} \frac{[M']^{a/b}}{(M_t - [M'])^{(1/b)-1}} + \frac{1}{(aK')^{1/b} C_L} \qquad (80)$$

from which linear plots can be obtained and, if a and b are known, K' and C_L values can be determined.

It may be observed that results obtained here (Equations 78 and 80) cannot confirm those previously reported by Ruzic for the same general case of complexation [compare Equations 78 and 80 with Equations 28 and 29 given in Ruzic (27)].

11.3. METHODOLOGY: THE INSTRUMENTAL APPROACH

DPASV, used together with the thin-mercury-film electrode (TMFE) plated onto a rotating glassy carbon disk electrode (RGCDE), is one of the most sensitive and powerful techniques available at present for the determination of ultratrace metals in real samples (50, 51). Detection power is about $0.5 \, \text{ng L}^{-1}$ (8, 51). Cadmium, copper, lead, and zinc are among the metals most frequently determined at subnanomolar or even picomolar concentration in seawater, without any preconcentration step (8, 52, 53)

The fundamental characteristic of DPASV with respect to the speciation problem is that when it is applied to the untreated (undigested) sample, it is not sensitive to the total amount of metal present but only to the so-called electroactive fraction, which, under carefully controlled conditions (especially the deposition potential and the rotation speed of the electrode), is almost entirely composed of the ionic and the inorganically complexed fractions of the element itself. The metal aliquot strongly bound to organic ligands remains undetected (50, 54, 55).

The titrimetric methodology for the study of metal speciation (complexation) in seawater includes both the determination of the total content, using an aliquot of the sample preliminarily subjected to acid digestion [and also to ultraviolet (UV) irradiation in the case of high organic material content] and the voltammetric titration of organic ligands with the metal of interest, using an untreated aliquot of the sample (55).

In this section the instrumental approach and the experimental procedure used in the present authors' laboratory for Cd, Cu, and Pb speciation measurements is described in some detail. Several reviews regarding the general theoretical and experimental aspects of ASV and its applications in seawater (or natural water) analysis can be found in the literature (6–8, 17, 50–57).

11.3.1. DPASV Instrumentation

The instrumental assembly used in our laboratory consists of an electrochemical cell especially developed for ultratrace metal determination in seawater by Nürnberg and Mart (8, 30) (EG&G, Model Rotel 2, Munich, Germany) and of a voltammetric device with anodic stripping capability in the differential pulse mode (EG&G, Polarographic Analyzer, Model 384B, Princeton, New Jersey). The electrochemical cell is equipped with an RGCDE and an Ag/AgCl, KCl (sat.) reference electrode to which all potentials are referred. Potentiostatic control is accomplished using a Pt wire as the auxiliary electrode. Reference and auxiliary electrodes are inserted inside small fluorinated ethylene propylene (FEP) tubes, filled with saturated KCl/AgCl solution, and fitted with porous Vycor tips. A Teflon (PTFE) cell cup and a Teflon cap are used, and the cell compartment is separated from the cell controller and the electrode motor by a Plexiglas box. The electrochemical cell assembly and a detail of the measurement cell are illustrated in Fig. 11.13(a,b).

The entire titrimetric procedure for the speciation measurements (requiring some 15–20 standard additions and a time of about 20 h; see Section 11.3.4, below) is completely automated by interfacing the electrochemical equipment with a Perkin-Elmer robot (MasterLab System, Model 9000, Norwalk, Connecticut) and a personal computer. The robot operates the polarographic analyzer, makes the necessary additions of the metal standard solution, and acquires and stores all the voltammetric data for subsequent processing.

The equipment is installed in a clean chemistry laboratory, inside a Class 100 laminar flow area, or, during oceanographic cruises, under a laminar flow hood in a clean chemistry laboratory container (ISO20) available on board.

11.3.2. Preparation of the Thin-Mercury-Film Electrode

The TMFE (prepared daily immediately before analysis) is electrolytically deposited on the surface of the RGCDE just before the beginning of the measurement. The glassy carbon disk electrode is rotated at 1000 rpm, and the smooth surface is polished with wetted alumina powder (0.075 mm grain size or lower) using a filter paper. Afterward the electrode is rinsed for 5 min with 1:200 diluted ultrapure HCl acid and then two or three times with ultrapure water. The Hg film is prepared by controlled potential electrolysis of a Hg(II) nitrate solution. The electrolytic solution is made 2.5×10^{-2} M in KCl (ultrapure) and 10^{-4} M in Hg(NO$_3$)$_2$ and purged with N$_2$ for at least 15 min; Hg(NO$_3$)$_2$ is prepared by oxidizing hexadistilled Hg with HNO$_3$. The film deposition is carried out at a potential of -1.00 V for 20 min while the electrode is rotating at 4000 rpm. After film deposition, a quiescent period of 30 s is allowed to pass and then a differential pulse potential scan is carried out in the positive direction until a potential of -0.18 V is reached with a scan rate of $10\,\text{mV s}^{-1}$, a pulse height of 50 mV, and a pulse frequency of $5\,\text{s}^{-1}$.

If the voltammogram obtained does not show any irregular peak and the base current is sufficiently low (300–400 nA), then it is possible to carry on the analysis of the samples; otherwise the Hg film is destroyed and prepared again. To avoid alteration of the complexation equilibria due to residual ionic Hg and to clean the electrode assembly, the latter is rinsed with a purged aliquot of the sample before placing the cell cup with the sample in the measurement position.

11.3.3. Total Metal Concentration

The measurement of total dissolved metal concentration is carried out on filtered acid-digested samples. Normally, 100 µL of HCl [30% Suprapur (Merck, Darmstadt, Germany), or 32% Ultrapure, NIST (National Institute of Standards and Technology, U.S.)(59)] are added to approximately 50 mL of seawater directly into the same PTFE vessel subsequently used as the cup of the electrochemical cell (the exact volume of the sample is determined at the end of the measurement; see below). A pH of about 2 is obtained, and digestion is carried out at room temperature for at least 2 days.

In the case of samples with large amounts of dissolved organic matter, such as those collected in coastal areas and/or near river mouths, acid digestion alone may be not sufficient to completely release metals (especially in the case of Cu) due to very strong binding by macromolecular ligands. In this event UV irradiation of acidified samples by a high-power Hg lamp (1.2 kW) for at least 3 h (Cu normally requires longer times) is carried out. Measurements performed on a sample collected in the open sea in an unpolluted area show the same results for the total amount of Cd, Cu, and Pb independent of whether the sample was subjected to acid digestion or to UV irradiation (33, 39, 48).

After digestion as appropriate, measuremens of Cd, Cu, and Pb, are carried out separately according to the procedure outlined below.

The sample, already in the electrochemical vessel (in the case of acid digestion) or transferred to it (in the case of UV irradiation), is purged for 15–20 min with N$_2$ in the outgassing position of the cell assembly (see Fig.11.13). The vessel with the sample is then rapidly transferred and screwed

396   TRACE METAL COMPLEXATION IN SEAWATER

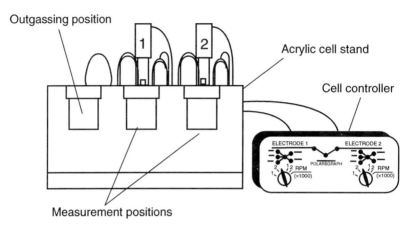

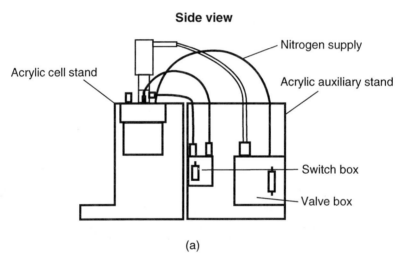

(a)

Figure 11.13. Electrochemical cell. Complete assembly (a) and detail of the measurement cell (b).

under the cell head of the electrochemical device where the TMFE has already been prepared and tested. The metal deposition into the TMFE (electrolytic accumulation) is then carried out by constant potential electrolysis. The deposition potential is set at -0.95 V for Cd and Pb and at -0.85 V for Cu. The deposition time, depending on the metal concentration, is usually about 20 min, although sometimes a shorter time is sufficient for Cu. During the

METHODOLOGY: THE INSTRUMENTAL APPROACH 397

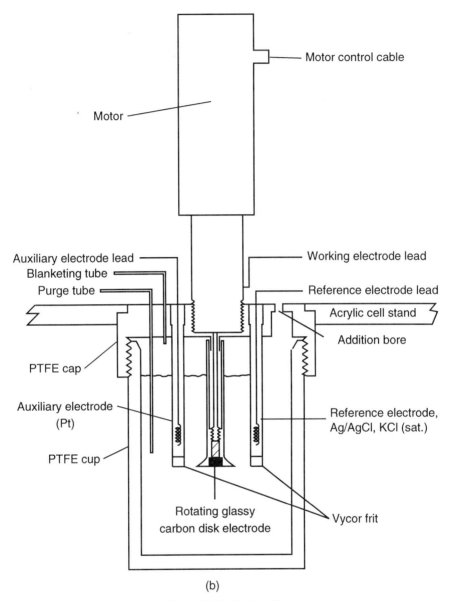

(b)

Figure 11.13. (*Continued*)

deposition period the working electrode rotates at a constant rate of 4000 rpm. This is followed by a quiescent period of 30 s and then by the stripping step, during which a pulsed potential scan is applied in the positive direction from the deposition potential to the final potential, fixed at -0.18 V for Cd and Pb and at -0.15 V for Cu. The scan rate is set at $10\,\text{mV s}^{-1}$, the pulse height at 50 mV, and the pulse frequency at $5\,\text{s}^{-1}$, while the voltammogram is being recorded. At the end of the scan the potential is held at -0.20 V for 5 min, while the electrode is rotating, to allow the amalgamated metals to be removed completely from the working electrode. To test electrode stability and repeatability of the voltammogram, a second measurement is performed on the sample solution. The multiple standard addition method is used for quantification. Three or four metal additions are made, and after each addition the voltammetric measurement is repeated. The volume of the sample is finally determined using a graduated cylinder.

The blank of the acid employed in the digestion step is evaluated by repeating the acid treatment on a 3×10^{-2} M KCl solution prepared with ultrapure water and KCl. The voltammetric measurement is carried out before and after the acid treatment, and the concentration increment observed gives the blank value to be used for correction. When the Merck Suprapur HCl is used, the Cd and Pb blanks are found to be 17 and 19 pM, respectively (33, 39), while the blank for Cu is below the detection limit (48). When the NIST Ultrapure acid is used, all the blanks are lower than the detection limit.

11.3.4. Metal Speciation (Voltammetric Titration of Ligands)

The study of complexation of each metal is carried out individually on separate aliquots of the sample, freshly defrozen and untreated (samples destined for speciation measurements are not subjected to acidification or other treatment, except filtration, when prepared for storage; see Section 11.4.4, below). The voltammetric sequence is mostly the same as the one just described in Section 11.3.3, and only differences are outlined hereafter.

The deposition potential is selected from pseudopolarograms obtained by plotting the anodic stripping current against the applied potential during the deposition step (e.g., see Fig. 11.14). In this way, values of -0.95, -0.80 to -0.85, and -0.75 to -0.80 V are selected for Cd, Pb, and Cu, respectively.

To follow the titration with sufficient precision and to reach the linear part of the titration curve with certainty (as required by the theory outlined in Section 11.2.1) numerous metal standard additions (at least 10, but preferably between 15 and 20) are made, while the amount of added metal increases throughout the titration experiment (see Fig. 11.1). After each standard addition and before the voltammetric measurement, a period of 15 min for Cd and Cu and 25–30 min for Pb is allowed to pass to reach the chemical equilibrium (44).

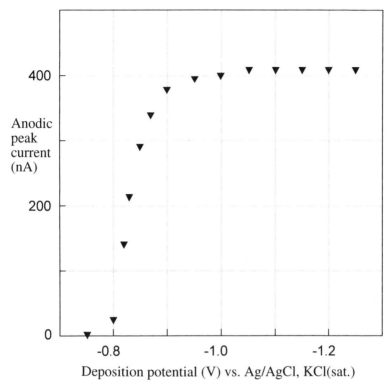

Figure 11.14. Pseudopolarogram of Cd obtained in a seawater sample collected in Antarctica (the Ross Sea).

In the case of very low metal concentration (especially of Pb), measurements on the unspiked sample are carried out with a higher deposition time (normally 30–40 min) to enhance sensitivity and are repeated two or three times to ensure conditioning of the working electrode and repeatability (8, 32). Currents are then normalized to the deposition time of the rest of the titration, given the linear relationship observed between the two quantities (59).

11.3.5. Defining the DPASV-Measured Species in a Seawater Sample

The fundamental requirement for the application of the titrimetric methodology to the study of metal complexation in natural waters is that an instrumental technique must be available to measure a signal related to the inorganic fraction of the metal, i.e., the free ion and its inorganic complexes, and then to determine the conditional concentration of metal $[M']$.

The DPASV has the sensitivity and selectivity required to directly detect the requisite fraction and is therefore one of the most frequently applied techniques. Other methodologies include the following: (i) the detection of free or labile forms of metals by equilibration with MnO_2 (38, 41), liquid–liquid extraction (60), adsorption on ion-exchange resins (61–64), cathodic stripping voltammetry (34, 65–67), the use of bioassays, or ion-selective electrodes (68–70); (ii) the detection of the complexed fraction by adsorption on resins at low polarity (Sep-Pack) (71, 72) and on ion-exchange resin (73); and (iii) the detection of uncomplexed fluorescent ligands by fluorescence quenching when metals are added to the sample (74).

The capacity of ASV to discriminate accurately between the inorganic and organic fractions of metal has been questioned by some authors (17, 75, 76). Criticisms concentrate mainly on the possible overestimation of the inorganic fraction due to both direct electrochemical reduction of organic complexes and reduction caused by rapid dissociation of complexes within the electrode diffusion layer (kinetic contribution to the stripping current). The problem of adsorption of organics on the electrode surface, thereby decreasing sensitivity, has also been raised.

A discussion of all these criticisms together with suggestions as to possible actions to eliminate or minimize the problems has already been given by Bruland (31) in considering the application of the voltammetric methodology in seawater using the TMFE–RGCDE. In short, accurate measurements can be obtained if the experimental conditions are selected with great care. In particular, the type of electrode, the deposition potential, and the hydrodynamic conditions at the electrode surface (which influence the diffusion layer thickness) play a fundamental role.

The use of TMFE in which adsorption of organics has not been observed in analyzing seawater is recommended. The deposition potential should be selected from a pseudopolarographic experiment (see Fig. 11.14) in order to perform the plating step at potentials negative enough to reduce only the inorganic forms of the metal. Optimal hydrodynamic conditions at the electrode surface during the plating step [stirring of the solution in a hanging mercury drop electrode (HMDE) or rotating rate in a rotating disk electrode (RDE)] are very important if the kinetic contribution to the stripping current is to be eliminated or minimized. The aim is to minimize the thickness of the diffusion layer in order to keep the residence time of a complex in it well below the time required for the complex to dissociate. The best conditions are obtained by using a rotating electrode at the highest rotation rate.

The theory regarding kinetic currents arising from dissociation of labile complexes within the diffusion layer is well established and throughly reviewed (17, 77, 78). Semiquantitative experimental evidence concerning the kinetic lability of organic complexes detected in seawater by the DPASV titration procedure and following the aforementioned recommendations is given below.

In the case of a homogeneous chemical reaction (e.g., complex dissociation) preceding the electron transfer step (here the reduction of metal during the plating step of DPASV), the degree of perturbation of the electron transfer itself due to the chemical reaction can be judged by comparing the residence time of the compound (complex) in the diffusion layer (defined as the timescale of the electrochemical experiment, τ) with the lifetime of the compound (complex) due to the chemical reaction, t_l, which for a first-order reaction (complex dissociation) is given by $t_l = 1/k_d$ and represents the time required for the concentration to drop to 37% of its initial value. If the residence time is small compared with the lifetime, then the electron transfer reaction will be largely unperturbed by the preceding chemical reaction (79).

The range of available experimental values of τ (time window) can vary drastically as a function of the particular electrochemical experiment, depending on the kind of electrode and technique applied. In particular, the time window determines the range of rate constants of the preceding chemical reaction that is accessible to each electrochemical technique (79).

In the preelectrolysis step of DPASV, when an HMDE and a magnetic stirrer are used the time window ranges between 0.01 and 1 s, whereas when an RDE is used the time window is 1–100 ms (17, 79). The kinetics of the complexation reaction of Pb and Cu with natural ligands present in seawater was studied by Capodaglio et al. (32) and Coale (80). They found that the formation rate constant k_f is of the order of $10^4 \, \text{s}^{-1} \, \text{M}^{-1}$. As a consequence, if one considers that the detection interval of the titration methodology for the conditional stability constants, as explained above, is between 10^7 and $10^{11} \, \text{M}^{-1}$ (for 1 nM ligand concentration), a value of k_d is attained, which can vary between 10^{-3} and $10^{-7} \, \text{s}^{-1}$ corresponding to lifetimes in the range of 10^3–10^7 s.

In conclusion, typical residence times for both HMDE and RDE in the plating step of ASV appear by orders of magnitude lower than the lifetime of metal complexation reactions with natural ligands, at least for Pb and Cu in seawater. Less favorable situations may occur for different metals or different natural waters. In these last events adjustment of the stirring rate to the maximum allowable value (HMDE) or the increment of the electrode rotation rate as much as possible (RDE) may be necessary to eliminate or at least minimize the effect of kinetic lability.

In view of the doubts and criticisms directed at operational speciation procedures because of the potential perturbation of the equilibrium of the system analyzed, it was recommended that two or more independent techniques should be used in parallel. A few investigators evaluated ASV measurements by comparing results with those of at least another procedure of speciation.

Del Castilho compared results obtained by the ASV procedure on Cu complexation in seawater with data obtained by other fractionation methods based on ion exchange on a NH_4–chelex column and on the retention on a Sep-Pak C_{18} column (81). Analyzing filtered soil solutions, this author found good agreement between the labile concentration measured by ASV at pH 4.5 and the ion-exchangeable Cu from an NH_4–chelex column. In seawater, reasonably comparable concentrations were obtained between the nonlabile metal fraction detected by ASV and the fraction adsorbed onto the C_{18} column.

Boussemart et al. used DPASV and fluorescence quenching to determine the complexing capacity of fulvic acid solutions and interstitial water extracted from marine sediment (82). They found good agreement between results obtained by the two methods when the complexation from fulvic acids was examined, but systematic higher values were obtained by DPASV when interstitial water was analyzed. This observation suggested that interstitial water contains two classes of complexing organic compounds, one fluorescent and the other nonfluorescent. Voltammetric measurements seem to detect the total complexing capacity, whereas the nonfluorescent but complexing organic matter is not revealed by fluorescence measurements.

Donat and Bruland compared results obtained by DPASV and DPCSV to detect the complexation of Zn in seawater from an open-ocean station (surface euphotic zone) (83). Both methodologies showed that $\geqslant 95\%$ of the dissolved Zn is organically complexed with one class of naturally occurring ligands. Excellent agreement between the values of ligand concentration was reported. The values of the conditional stability constant obtained by the two techniques differed by about 1 order of magnitude. The authors assumed that this difference may be a result of a different analytical detection window, but additional data are required to reach a definitive conclusion.

11.4. CONTAMINATION CONTROL

Because of the very low concentrations at which heavy metals are present in seawater, it is of primary importance to keep all the steps of speciation studies under rigorous contamination control, from preparation of the sampling equipment to instrumental analysis. Moreover, in the case of speciation measurements, the goal of contamination control is pertinent not only to the change in metal concentration but also to metal distribution between different species and particularly the extent of metal complexation. Thus, it is important to prevent contamination from external ligands and, in general, equilibrium displacements. Careful consideration with respect to any potential contribution to contamination is to be given to general laboratory equipment, chemical reagents, materials used for sampling and storage bottles and their cleaning, sampling equipment, onboard laboratories and other facilities in the field, filtration devices and procedures, storage conditions, pretreatment of samples, and final instrumental analysis. The contamination control procedures routinely adopted in our laboratory are briefly outlined below (33, 39, 48); a more detailed description can be found in the literature (3–13).

11.4.1. Laboratory and Chemicals

As regards airborne contamination due mainly to suspended dust, a first step in contamination control consists of the improvement of existing laboratories, obtained by installing laminar flow hoods, which enable samples to be treated and measurements to be carried out in a satisfactorily dust-free area.

Further improvement entails the design of clean rooms in which air is pumped through a dust-stop prefilter and then forced by a blower through an absolute filter inside the working area. These devices are installed in the ceiling of the clean laboratory and allow measurements to be done under a laminar flow pattern in which contamination problems from the atmosphere are greatly reduced (a Class 100 laboratory). The filtered blown air produces a slight positive pressure that prevents dust from entering through chinks in doors, windows, and the ceiling. Furthermore, the walls of the laboratory are lined with plastic coatings, so that dust that contains trace elements cannot contaminate the atmosphere. The preroom floor is covered with a sticky plastic mat to keep dust attached to the bottom of operators' and visitors' shoes outside the clean area. Obviously, all metallic furniture must be avoided, or, if this is not possible, it must be carefully coated with metal-free plastic paint. If one wants to protect samples from problems arising from contamination, all handling of sampling equipment and instrumentation devices must be performed in this kind of environment.

Reagents and water employed to rinse all items and to perform analyses must be extremely pure. Only reagents with a very low heavy-metal content can be used (58). Ultrapure water is produced by quartz distillation apparatus or with special ion-exchange devices (6, 84, 85). These latter can produce a satisfactory amount of ultrapure water in a reasonable time. Obviously, the blanks of pure water and reagents must be checked periodically in order to enable an estimate to be made of the metal content added to the samples during analysis.

The overall quality of laboratory measurements should be checked every few weeks by determining the metal content of seawater reference materials, now easily available (86,87). These measurements are of particular importance to test the metal standard solutions employed and the overall calibration procedure. The control chart in Fig. 11.15 shows recent results obtained by our laboratory on NASS [National Research Council of Canada, Ottawa, Canada (86)] certified reference materials.

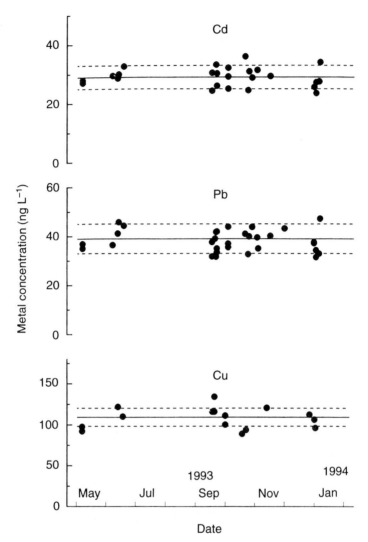

Figure 11.15. Control chart from our recent measurements on NASS-3 (86) seawater certified reference material. *Key:* (———) mean; (----) tolerance interval.

A further step to ensure that a laboratory is working with the required accuracy is participation in international intercomparison campaigns (13, 87–90). In these exercises one can compare results obtained on an unknown sample with values reported by many other expert laboratories that have analyzed the same sample using different instrumental approaches. Intercomparison campaigns are now being organized in which the overall procedure from sampling to chemical analysis, carried out independently by each of the participating laboratories, is being tested (90–92). Surprisingly, past experience in these intercalibration exercises has showed that relatively few laboratories involved in trace metal determination take into account the real problems of contamination of seawater samples during sampling, storage, and analysis and consequently adopt the stringent clean procedures required to obtain uncontaminated samples and accurate results.

11.4.2. Sampling

In on-site activity it is of primary importance to avoid possible contamination coming from the ship itself, from the equipment used for gathering samples, or from the operators involved in the operations. In this respect the availability on board of a clean area (Class 100) for the chemistry laboratory to handle all the sampling material and to carry out the necessary sample treatments is essential. Our research team employs a mobile laboratory (ISO20 container) specifically designed to provide suitable conditions for the activities that have to be carried out on site. This is equipped with laminar flow hoods and an ultrapure water production system and prepared and cleaned with care before each oceanographic cruise (92).

Two different approaches are used in collecting seawater samples from surface, shallow, and deep water. In the collection of surface water it is preferable that operators move away from the ship in a rubber dinghy and gather samples using a Teflon pump (possibly submersible) for temporary storage in a high-capacity polyethylene container (20–50 L). The procedure consists of reaching the sampling area by dinghy, stopping the engine, and rowing upwind while sampling uncontaminated water from the prow. Before collecting the sample, the polyethylene tank (cleaned as reported below for sample containers) is conditioned two or three times with seawater. In some cases, when it is possible to avoid filtration (particularly when working far from the coast), it is suggested that samples be collected directly into the storage bottles in order to avoid a subsequent transference step. The operation can be carried out manually and with the use of long polyethylene gloves in the case of surface waters, or by a plastic telescope bar with the sampling bottle inserted at its end (7).

Deep-water samples (more than 20 m from the surface) can be collected directly from the ship, using a sampling bottle immersed to the desired depth by a nonmetallic (Kevlar) hydrowire. A "close–open–close" bottle, internally lined with Teflon and with pressurization capability, can be recommended. This system allows the bottle to enter water in the closed position, open itself automatically at a depth of about 10 m, and close at the desired depth by the "messenger" system.

In our laboratory Go-Flo bottles (General Oceanics, Miami, Florida) with capacities of 20 or 30 L are routinely used. The bottles are made of polyvinyl chloride (PVC) tubing coated internally with Teflon and plugged at the extremities by an opening–closing system. In the wall of the cylinder there are a tap to connect the bottle to the filtration device, a safety valve to avoid high-pressure differences between the inner part and the external part, and a pressure opening device, which allows the bottle to open automatically at a depth of about 10 m. High-capacity bottles (20 or 30 L) are preferred in order to achieve the highest possible value of the volume/surface ratio and thus minimize the effects of possible small releases from the walls into the sample. The very heavy ballast required to dip the bottle, which in the closed position tends to float, should be covered with plastic and attached to the wire at least 20 m below the bottle (when possible). The pressurization capability of these samplers allows them to be used as the reservoir of the filtration device, thus avoiding transference of the sample to another container (see below).

The cleaning procedure for sampling bottles is slightly complicated by their large size. However, a common way of proceeding when it is necessary to employ teflon-coated samplers (available with a high degree of cleanness from the manufacturer) for the first time is to rinse them repeatedly with ultrapure water to remove dust particles. The bottles are then transferred to the clean laboratory where the cleaning operations can continue with operators wearing polyethylene gloves and clean garments. The procedure adopted consists of repeated washings with ultrapure HCl (Merck Suprapur, Darmstadt, Germany) diluted 1:100 with ultrapure water. At the end of each washing the bottles are filled with the same solution and allowed to stand for a few days before repeating the treatment. After repeating treatment three or four times the sampling bottles are rinsed abundantly and repeatedly with ultrapure water. Bottles are stored dried and sealed in multiple clean polyethylene bags to protect them from dust particles during storage and transport. Before use the samplers are conditioned by filling them several times with seawater collected offshore.

As regards the submersible or other Teflon pumps, the parts in contact with the sample are cleaned as described for storage containers (see Section 11.4.4, below). On site the pumping system is conditioned with seawater by pumping and discarding at least the first 10 L.

11.4.3. Filtration

Filtration of samples through a conventional filter of 0.45 μm pore size must be carried out as soon as possible after sampling and inside the clean laboratory that ought to be available on board. A Teflon apparatus developed for on-line pressure filtration (e.g., Sartorius SM16540, Göttingen, Germany) and a cellulose nitrate membrane filter (e.g., Sartorius SM11306, 142 mm diameter) can be used for pressure filtration, with maximum pressure at 0.5 bar. Technically the filtration is performed by exerting a slight pressure inside the 50 L polyethylene tank used for collection, or the Go-Flo bottle, by means of pure N_2 gas. The sample is forced to pass through the filtration apparatus and, after discarding a first aliquot of about 3–4 L, the filtrate is collected directly in the storage bottle.

Filters must be cleaned before use with 1:10 diluted HCl, ultrapure grade, rinsed with ultrapure water, and stored in 1:100 diluted ultrapure acid until use. The filtration apparatus follows a cleaning procedure similar to that described below for storage containers. Some investigators prefer to collect the samples directly into the storage bottles avoiding, when possible, the filtration step (especially under open-sea conditions or for short-term storage) (6, 7).

11.4.4. Storage

Prolonged preservation of the integrity of samples with respect to metal speciation is a very difficult task and, in principle, it would be better to carry out the analyses on board immediately after collection and filtration. In fact, the DPASV instrumentation can easily be installed and operated on a ship provided that a clean laboratory is available. Measurements obtained in polar regions under ice-breaking conditions are also reported in the literature (33, 39, 93). However, due to the very long time required for the analyses, measurements on board must be restricted to relatively few samples, mainly to make tests of contamination and assess reference values to compare with results obtained later in the laboratory ashore. As a consequence, storage of samples for a few months can become mandatory.

As regards containers for storage, it is preferable to employ FEP or Teflon bottles rather than those made of polyethylene. Over long periods of storage polyethylene can, in fact, release plasticizers (above all phthalates or amines), which behave as ligands and modify the complexation equilibria of the solution. Purer but more expensive materials, such as FEP or Teflon, do not present this kind of problem, but it is in any case preferable that the storage time be as short as possible.

To preserve integrity as much as possible samples are stored frozen at the temperature of $-20\,°C$ without adding mineral acids that certainly would alter the complexation equilibria of metals in seawater. There is some experimental evidence that freezing of samples does not significantly alter the results of the complexation studies carried out on heavy metals according to the methodology discussed here (92).

For new bottles the cleaning procedure starts with a first wash with tap water, followed by rinsing and cleaning with acetone. Then bottles are left in a heated (50 °C) detergent solution for 10 days to maximize the degreasing stage. During these operations long polyethylene gloves are worn to prevent the bottles from coming into contact with hands. After rinsing with ultrapure water, bottles are immersed in a heated (50 °C) 10% solution of HNO_3 (analytical reagent grade) where they are kept for 2 weeks in order to allow heavy metals to be released from the walls of the containers. After further rinsing, bottles are enclosed in polyethylene bags and transferred to the clean laboratory for final treatment. They are filled with a 10-fold diluted solution of ultrapure HCl (Merck Suprapur). After a week this step is repeated for another week. After a further rinsing with ultrapure water, bottles are filled with a 1:100 diluted solution of ultrapure HCl and sealed in double polyethylene bags until sampling. Used bottles are subjected to repeated washings with small aliquots of concentrated HNO_3 (Merck Suprapur). After a rinsing with ultrapure water, the final treatment described above is carried out.

11.5. REVIEW OF THE LITERATURE

In chemical oceanography one of the most ambitious ongoing research projects over the last 20 years has been the investigation of the relationships between the speciation of metals and their bioavailability, toxicity, or general biogeochemical behavior. In this field the study of metal complexation by the ASV titration procedure represents one promising methodology in the advanced analytical chemistry of seawater. In this section a brief survey of the results published on this subject matter is presented, as well as a comparison of these data with information obtained using different procedures, particularly as regards Cd, Co, Cu, Pb, and Zn.

11.5.1. Copper

Copper is an element whose complexation with organic ligands present in natural waters is frequently studied because it is a biologically essential metal and forms stable complexes that heavily affect its bioavailability. Traditionally, the ligand concentration detected by using Cu as the titrant was referred to as the complexing capacity of seawater or other natural water.

There is considerable evidence to demonstrate the existence of a correlation between Cu(II) activity in seawater and the toxic effect of Cu on phytoplankton (94, 95) and on higher organisms (94, 96, 97). Srna et al. studied the ability of complexing compounds present in lagoon waters of southern California to mitigate Cu toxicity toward phytoplankton (98). They found a good correlation between complexing capacity detected by ASV and effect concentration EC_{50} (i.e., the concentration of added Cu that reduced the algal growth by

50%) measured on *Thalassiosira pseudonana*. Analogous results were obtained by Camusso et al., who investigated the toxicity of Cu to *Selenastrum capricornutum* in river waters in northern Italy (99). They observed a significant correlation between complexing capacity and algal growth. Complexing capacities were determined by using either DPASV or the procedure previously proposed in the literature (100) based on solubilization of Cu(OH)$_2$ at pH 10. The ligand concentrations obtained by the two procedures were in good agreement and, according to the observations of Buffle et al. (101), the complexing capacity was found to be strongly correlated with dissolved organic carbon (DOC).

Studies of Cu complexation in waters from oceanic and coastal marine areas by voltammetric titration techniques showed that data fit well into two-ligand models and that models with more than two ligands do not produce a significant improvement in fitting (102). The first ligand (identified here as L_1) is present at low concentrations, normally at the nanomolar level, and has a conditional stability constant higher than 10^{11} M^{-1}. The second ligand (L_2) is present at a concentration 1 or 2 orders of magnitude higher and has a conditional stability constant of about 10^8 M^{-1}.

Investigations carried out by Coale and Bruland (30) in the North Pacific and by Moffett et al. (103) in the Atlantic Ocean (Sargasso Sea) showed that the stronger class of ligands is localized in the upper part of the water column in correspondence with higher values of both primary production and chlorophyll content. The ligand concentration reaches its maximum value of 2.4 nM at a depth of 60 m in the Pacific Ocean (30). In the Sargasso Sea the maximum fraction of complexed Cu is present at a depth of between 140 and 160 m; at the same depth the amount of L_1 is 2.0 nM (103). These results suggest that substances complexing Cu are produced, for the most part, by phytoplankton. Laboratory culture experiments also indicate that phytoplankton can produce extracellular substances that have an affinity for Cu and other metals (104–106).

Midorikawa et al. studied the complexation of metals by natural organic ligands in seawater samples collected in the North Pacific and Japan Sea using the ion selective electrode (ISE) approach (107). They concentrated and demetalized the organic ligands by freeze-drying and dialysis to avoid the effects of side-reactions, which can alter the conditional stability constants and prevent any comparison of results obtained on different environmental samples with different analytical procedures. Measurements carried out on Cu showed that two classes of ligands or two complexing sites are detected also by the ISE method. It was found that the concentration of the stronger ligand varies between 1.1 and 3.9 nM and the logarithm of the conditional stability constant (log K'_1) varies between 8.41 and 9.60. The weaker ligand had a concentration of between 4.0 and 11 nM and a log K'_2 between 7.09 and 7.94.

Coale and Bruland observed the variability of L_1 in a seasonal investigation of the North Pacific gyre (a near-oligotrophic area) and along a transect between this site and the Alaskan gyre (a eutrophic area) (108). They found a variation of the L_1 concentration by a factor of 2 in the mixed layer and some differences in the vertical distribution between the near-oligotrophic area and the eutrophic area. The presence of the stronger ligand was observed down to a depth of 200 m in the oligotrophic area and in the top 40 m in the eutrophic area. From the vertical profiles of ligands the authors concluded that a consumption or degradation of L_1 must occur on a timescale shorter than 1 year and that a depth-specific L_1 production, characteristic of organisms capable of maintaining their position in the water column, should occur.

Capodaglio et al. examined the Cu complexation in Terra Nova Bay (Antarctica) in three successive years (austral summer seasons from 1987–88 to 1989–90) and observed an effect of seasonal evolution on the L_1 concentration (48). The mean values in samples collected during the 1987–88 and 1988–89 expeditions, i.e. quite normal seasons, were 1.2 and 1.3 nM, respectively, whereas in samples of the 1989–90 expedition, in which an abnormal biological development was observed, the value was 2.9 nM. Conversely, the L_2 concentration did not show any significant variation in the three expeditions, even when samples collected under the pack ice were analyzed (mean concentrations of 32, 30, and 29 nM were detected, respectively). Similar concentration values of L_2 and homogeneous spatial distributions (both vertical and horizontal) have been observed in different oceanic areas by other researchers (29, 35, 103). This finding on the weaker class of ligands suggests that it includes refractory compounds with residence times in the oceanic environment on the order of hundreds of years or greater.

The relationship between Cu complexation by natural ligands and phytoplankton growth was examined in an extensive biological investigation by Robinson and Brown (109). They carried out a time-course study of Cu complexation by the ASV technique during the bloom in Esquimolt Lagoon (near Vancouver Island, British Columbia, Canada). A linear relationship between complexing capacity and the number of cells of *Gymnodinium sanguineum* was observed. The hypothesis that ligands might have been present in the sample due to cell rupture was discarded because phytoplankton was removed by prefiltration under gravity through a 30 μm net before filtration with a membrane filter of 0.45 μm and and no grazing by zooplankton in the studied area was observed. The authors, pointing out that no significant variation was noted in complexing capacity with depth or at the end of the bloom (both situations lead to an increase in the lysis of cells), attributed the detected ligand to phytoplankton exudates. They also showed that a negative correlation exists between complexing capacity and log K'; the same relationship has been found by other investigators using different techniques (37, 110–

119), indicating that it is not an artifact of the ASV method. However, Robinson and Brown noted that different values are obtained for the constant when different techniques are applied. In particular, measurements carried out by ASV lead to values that are 1 or 2 orders of magnitude lower than those obtained with other methods. They concluded, in agreement with Buffle et al. (101), that this discrepancy can be ascribed to the possible overestimation of free metal by ASV due to direct reduction of labile complexes and to saturation of ligand in the electrode surface during the stripping step. A possible explanation for this observation [deduced from Coale and Bruland (30)] deals with the forms of Cu that are detected by different analytical techniques. Measurements carried out by methodologies such as bioassay, ISEs, and adsorptive CSV generally determine the ionic concentration (or activity) of metals, whereas ASV or amperometric techniques detect labile species of metal, i.e., [M'], constituted by the ionic form of metals, inorganic complexes, and some labile organic complexes. This implies that ASV detects a conditional constant that takes into account the side-reactions of both metal and ligand, $K_{M',L'}$ [according to the usual notation (49)], whereas the other aforementioned techniques detect a different conditional constant that accounts for the side-reactions of the ligand only, $K_{M,L'}$. The discrepancy observed by Robinson and Brown can be theoretically explained and calculated if one considers that $K_{M,L'} = \alpha_M K_{M',L'}$, and knows that α_M in the case of Cu in seawater has been estimated as 24 (120). This leads to the conclusion that $\log K_{Cu,L'} - \log K_{Cu',L'} = 1.38$.

Kozarac et al. applied DPASV to study the Cu complexing capacity of exudates of axenic cultures of *Dunaliella tertiolecta* when micronutrient metals (Co, Cu, Fe, Mn, Mo, and Zn) were added to the culture media (121). They showed that Cu was complexed by ligands released from phytoplankton and that the complexing capacity computed per cell decreased when the essential metals were added to the culture media, indicating that under stress conditions phytoplankton cells excrete a completely different material from that produced under normal conditions (no limitation of essential trace metals).

11.5.2. Zinc and Cobalt

ASV has been applied to study the complexation of other biologically essential elements in seawater and estuarine waters. Muller and Kester studied the Zn–organic ligands interaction in Narragansett Bay (Rhode Island) and in the Slope Water region off the U.S. mid-Atlantic bight (122). They observed distinct situations for the two areas studied. In Narragansett Bay the analysis of filtered samples showed that only one class of ligands is present in the soluble phase ($\log K' = 7.7$; concentration 51 nM), whereas a second class, corresponding to surface complexation by particles, was observed when unfiltered seawater was considered ($\log \beta \approx 7.0$; concentration of particle-bound ligand 220 nM). Samples collected near the western edge of the Gulf Stream showed two classes of dissolved ligands with $\log K_1' = 9.3$ and $\log K_2' = 7.6$ and concentration values of 5 and 35 nM, respectively. The authors pointed out that, owing to the complexity of the systems under study and due to the observed kinetic lability of thermodynamically stable complexes, optimization of experimental parameters (pH and electrode rotation rate) is required to yield accurate and precise calculations of stability constants and ligand concentrations.

Bruland determined the Zn complexation by natural organic ligands using DPASV in the upper 600 m of the North Pacific (31). He observed the presence of a strong ligand (mean value of $\log K' = 11$) at a low concentration (about 1.2 nM) and with a relatively uniform distribution along the vertical profile. He observed also that this class of ligands complexes more than 99% of the Zn present in the upper 200 m and that the complexation could influence the geochemical behavior of the metal and determine the species distributions in phytoplankton populations.

Zhang et al. studied the Co complexation in the River Scheldt and the Irish Sea by adsorptive CSV in which Co(II) is collected by adsorption of its complexes with Nioxime* (N) and dimethylglyoxime (DMG) (67). Conditional stability constants for complexes Co–N and Co–DMG were evaluated by calibration with ethylenediaminetetraacetic acid (EDTA). On the basis of the ligand competition between the added chelator used for the preconcentration step and the natural organic ligands the parameters of complexation in seawater were obtained. The authors found that a fraction of 45–100% of Co is complexed by strong organic ligands present at a concentration of between 0.29 and 1.13 nM, while $\log K'$ fell within the range 15.6–17.5.

11.5.3. Lead and Cadmium

Speciation of elements that are not essential from the biological point of view but have toxic effects, such as Pb and Cd, has also been studied in marine waters. Capodaglio et al. (32, 39) and Scarponi et al. (123) showed that also for Pb a considerable fraction of the metal is complexed by organic ligands and that the complexing material seems related to biological activity. The ligand concentration detected in surface seawater ranges between 0.2 and 0.5 nM in open ocean and increases by 1 order of magnitude in coastal and shallow waters, while the conditional stability constant (average $\log K'$ is 9.7) does not seem affected by the area studied. The fraction of complexed Pb depends on

---

* Trade name of 1,2-cyclohexanedione dioxime.

the characteristic of the studied areas and varies between about 50% in open ocean and about 80% in coastal areas.

As regards Cd, a study carried out in a mesocosm experiment by Slauenwhite and Wangersky showed that the formation of biogenic particles associated with phytoplankton bloom is an important factor controlling the dissolved concentration and speciation of the element (104). Capodaglio et al. (33) and Scarponi et al. (123) studied the interaction of Cd with naturally occurring organic material in seawater in different geographic areas, revealing considerable complexation of the metal (50–70%), although less than that of Cu. Bhat and Weber (124) and Plavsic et al. (125) showed that natural organic matter, such as fulvic acids or humic acids, has a complexing capacity for Cd and that complexation probably affects metal reactivity and its interaction with particulate matter that determine its rapid depletion (33, 126) in areas involved in phytoplankton bloom.

In their study cited earlier, Midorikawa et al. used the ISE approach to show the presence of one class of ligands for Cd with a concentration ranging between 1.2 and 4.5 nM and the logarithm of the conditional stability constant between 6.74 and 7.21 (107).

11.6. CONCLUSIONS

ASV titration procedures applied to the study of trace metal complexation in seawater appear sufficiently reliable and precise. Best results are obtained using the TMFE plated on a rotating glassy carbon support and applying the differential pulse mode in the stripping scan. Typical values of measurement repeatability [as relative standard deviation (RSD)] at the very low metal concentration levels involved in seawater (10^{-9}–10^{-11} M) are 10–15% for total content, 20–25% for labile amount, 10–15% for the ligand concentration, and 20–30% for the conditional stability constants.

Contamination control of the overall analytical procedure, from sample collection to instrumental analysis, is a very critical step. Contamination, when present, is likely to alter results by orders of magnitude. Moreover, the preservation of sample integrity with respect to speciation requires that both metal and ligand contamination be avoided.

Previous studies carried out using the ASV methodology have suggested that natural organic ligands play an important role in the complexation of metals in marine waters. The organically complexed fraction of metals always represents a significant portion of the total concentration and, in some cases, values of up to 95–100% have been reported. The implications for the development of theoretical, equilibrium-based speciation models and for the understanding of the biogeochemical cycles of metals in seawater are evident.

Studies on the relationship between metal complexation and bioavailability and toxicity are also reported, but new developments in this field are expected in the near future.

ACKNOWLEDGMENTS

The authors gratefully acknowledge useful discussions with G. Toscano and the technical assistance of M. Cecchini, C. Cozzi, and C. Turetta.

REFERENCES

1. E. D. Goldberg, *Mar. Pollut. Bull.* **12**, 225 (1981).
2. K. W. Bruland, in *Chemical Oceanography* (J. P. Riley and R. Chester, eds.), Vol. 8, Chapter 45. Academic Press, London, 1983.
3. C. C. Patterson and D. Settle, *NBS Spec. Publ. (U.S.)* **422**, 321 (1976).
4. K. W. Bruland, R. P. Franks, G. A. Knauer, and J. H. Martin, *Anal. Chim. Acta* **105**, 233 (1979).
5. J. R. Moody and E. S. Beary, *Talanta* **29**, 1003 (1982).
6. L. Mart, *Fresenius' Z. Anal. Chem.* **296**, 350 (1979).
7. L. Mart, *Fresenius' Z. Anal. Chem.* **299**, 97 (1979).
8. L. Mart, H. W. Nürnberg, and P. Valenta, *Fresenius' Z. Anal. Chem.* **300**, 350 (1980).
9. B. K. Schaule and C. C. Patterson, *Earth Planet. Sci. Lett.* **54**, 97 (1981).
10. E. A. Boyle, S. S. Huested, and S. P. Jones, *J. Geophys. Res.* **86**, 8048 (1981).
11. J. M. Bewers and H. L. Windom, *Mar. Chem.* **11**, 71 (1982).
12. P. M. Harper, *Mar. Chem.* **21**, 183 (1987).
13. K. W. Bruland, K. H. Coale, and L. Mart, *Mar. Chem.* **17**, 285 (1985).
14. D. R. Turner, M. Whitfield, and A. G. Dickson, *Geochim. Cosmochim. Acta* **45**, 855 (1981).
15. F. M. M. Morel, *Principles of Aquatic Chemistry*, pp. 237–310. Wiley, New York, (1983).
16. T. A. Neubecker and H. E. Allen, *Water Res.* **17**, 1 (1983).
17. J. Buffle, *Complexation Reactions in Aquatic Systems*. Ellis Horwood, Chichester, UK, 1988.
18. Y. K. Chau, R. Gächter, and K. Lum-Shue-Chan, *J. Fish. Res. Board Can.* **31**, 1515 (1974).
19. R. Ernst, H. E. Allen, and K. H. Mancy, *Water Res.* **9**, 969 (1975).
20. T. A. O'Shea and K. H. Mancy, *Anal. Chem.* **48**, 1603 (1976).
21. T. M. Florence and G. E. Batley, *Mar. Chem.* **4**, 347 (1976).

22. T. M. Florence and G. E. Batley, *J. Electroanal. Chem.* **75**, 791 (1977).
23. J. C. Duinker and C. J. M. Kramer, *Mar. Chem.* **5**, 207 (1977).
24. M. S. Shuman and G. P. Woodward, Jr., *Environ. Sci. Technol.* **11**, 809 (1977).
25. J. S. Young, J. M. Gurtisen, C. W. Apts, and E. A. Crecelius, *Mar. Environ. Res.* **2**, 265 (1979).
26. M. Plavsic, S. Kozar, D. Krznaric, H. Bilinski, and M. Branica, *Mar. Chem.* **9**, 175 (1980).
27. I. Ruzic, *Anal. Chim. Acta* **140**, 99 (1982).
28. M. Plavsic, D. Krznaric, and M. Branica, *Mar. Chem.* **11**, 17 (1982).
29. C. J. M. Kramer, *Mar. Chem.* **18**, 335 (1986).
30. K. H. Coale and K. W. Bruland, *Limnol. Oceanogr.* **33**, 1084 (1988).
31. K. W. Bruland, *Limnol. Oceanogr.* **34**, 269 (1989).
32. G. Capodaglio, K. W. Coale, and K. W. Bruland, *Mar. Chem.* **29**, 221 (1990).
33. G. Capodaglio, G. Scarponi, G. Toscano, and P. Cescon, *Ann. Chim. (Rome)* **81**, 279 (1991).
34. C. M. G. van den Berg, *Mar. Chem.* **15**, 1 (1984).
35. P. J. M. Buckley and C. M. G. van den Berg, *Mar. Chem.* **19**, 281 (1986).
36. C. M. G. van den Berg, *Mar. Chem.* **11**, 307 (1982); correction: **13**, 83 (1983).
37. C. M. G. van den Berg, *Mar. Chem.* **11**, 323 (1982).
38. C. M. G. van den Berg, P. M. Buckley, and S. Dharmavanij, in *Complexation of Trace Metals in Natural Waters* (C. J. M. Kramer and J. C. Duinker, eds.), pp. 213–216. Nijhoff/Junk, The Hague, The Netherlands 1984.
39. G. Capodaglio, G. Toscano, G. Scarponi, and P. Cescon, *Ann. Chim. (Rome)* **79**, 543 (1989).
40. C. M. G. van den Berg, Ph.D. Thesis, McMaster University, Ontario (1979).
41. C. M. G. van den Berg and J. R. Kramer, *Anal. Chim. Acta* **106**, 113 (1979).
42. C. M. G. van den Berg and J. R. Kramer, in *Chemical Modeling in Aqueous Systems* (E. A. Jenne, ed.), pp. 115–132. American Chemical Society, Washington, DC, 1979.
43. I. Ruzic, *Thalassia Jugosl.* **16**, 325 (1980).
44. I. Ruzic and S. Nikolic, *Anal. Chim. Acta* **140**, 331 (1982).
45. I. Ruzic, in *Complexation of Trace Metals in Natural Waters* (C. J. M. Kramer and J. C. Duinker, eds.), pp. 131–147. Nijhoff/Junk, The Hague, The Netherlands, 1984.
46. J. Lee, *Water Res.* **17**, 501 (1983).
47. S. C. Apte, M. J. Gardner, and J. E. Ravenscroft, *Anal. Chim. Acta* **212**, 1 (1988).
48. G. Capodaglio, G. Toscano, G. Scarponi, and P. Cescon, *Int. J. Environ. Anal. Chem.* **55**, 129 (1994).
49. A. Ringbom, *Complexation in Analytical Chemistry*. Wiley, New York, 1979.
50. J. Wang, *Stripping Analysis*. VCH, Deerfield Beach, FL 1985.

51. M. Kopanica and F. Opekar, in *Electrochemical Methods in Chemical and Environmental Analysis* (R. Kalvoda, ed.), pp. 58–84. Pergamon, New York, 1987.
52. A. Zirino, in *Marine Electrochemistry* (M. Whitfield and D. Jagner, eds.), Chapter 10, pp. 421–503. Wiley, New York, 1981.
53. C. M. G. van den Berg, in *Chemical Oceanography*, (J. P. Riley and R. Chester, eds.), Vol. 9, Chapter 51, pp. 197–245. Academic Press, London, 1988.
54. H. W. Nürnberg and P. Valenta, in *Trace Metals in Sea Water* (C. S. Wong, E. Boyle, K. W. Bruland, J. D. Burton and E. D. Goldberg, eds.), pp. 671–697. Plenum, New York, 1983.
55. W. Lund, in *The Importance of Chemical "Speciation" in Environmental Processes* (M. Bernhard, F. E. Brinkman, and P. J. Sadler, eds.), pp. 533–561. Springer-Verlag, Berlin, 1986.
56. H. W. Nürnberg and P. Valenta, in *The Nature of Sea Water*, (E. D. Goldberg, ed.), pp. 87–136. Dahlem Konferenzen, Berlin, 1975.
57. W. Davison and M. Whitfield, *J. Electroanal. Chem.* **75**, 763 (1977).
58. P. J. Paulsen, E. S. Beary, D. S. Bushee, and J. R. Moody, *Anal. Chem.* **60**, 971 (1988).
59. H. W. Nürnberg, P. Valenta, L. Mart, B. Raspor, and L. Sipos, *Fresenius' Z. Anal. Chem.* **282**, 357 (1976).
60. J. W. Moffett and R. G. Zika, *Mar. Chem.* **21**, 301 (1987).
61. T. M. Florence and G. E. Batley, *Talanta* **23**, 179 (1976).
62. J. R. Hasle and M. I. Abdullah, *Mar. Chem.* **10**, 487 (1981).
63. K. Hirose, Y. Dokiya, and Y. Sugimura, *Mar. Chem.* **11**, 343 (1982).
64. M. Zhang, and T. M. Florence, *Anal. Chim. Acta* **197**, 137 (1987).
65. C. M. G. van den Berg, *Mar. Chem.* **16**, 121, (1985).
66. C. M. G. van den Berg and M. Nimmo, *Sci. Total Environ.* **60**, 185 (1987).
67. H. Zhang, C. M. G. van den Berg, and R. Wollast, *Mar. Chem.* **28**, 285 (1990).
68. W. G. Sunda and R. L. Ferguson, in *Trace Metals in Sea Water* (C. S. Wong, E. Boyle, K. W. Bruland, J. D. Burton, and E. D. Goldberg, eds.), pp. 871–890. Plenum, New York, 1983.
69. W. G. Sunda, D. Klaveness, and A. V. Palumbo, in *Complexation of Trace Metals in Natural Waters* C. J. M. Kramer and J. C. Duinker, eds.), pp. 399–409. Nijhoff/Junk, The Hague, The Netherlands, 1984.
70. J. G. Hering, W. G. Sunda, R. L. Ferguson, and F. F. M. Morel, *Mar. Chem.* **20**, 299 (1987).
71. G. L. Mills, A. K. Hanson, Jr., J. G. Quinn, W. R. Lammela, and N. D. Chasteen, *Mar. Chem.* **11**, 355 (1982).
72. D. J. Mackey, *Aust. J. Mar. Freshwater Res.* **37**, 437 (1986).
73. K. Kremling, A. Wenck, and C. Osterroht, *Mar. Chem.* **10**, 209 (1981).
74. P. Berger, M. Ewald, D. Liu, and J. H. Weber, *Mar. Chem.* **14**, 289 (1984).
75. J. R. Tushall and P. L. Brezonik, *Anal. Chem.* **53**, 1986 (1981).

76. M. Betti and P. Papoff, *CRC Crit. Rev. Anal. Chem.* **19**, 271 (1988).
77. W. Davison, *J. Electroanal. Chem.* **87**, 395 (1978).
78. H. P. van Leeuwen, *Sci. Total Environ.* **60**, 45 (1987).
79. A. J. Bard and L. R. Faulkner, *Electrochemical Methods*. Wiley, New York, 1980.
80. K. H. Coale, *Copper Complexation North Pac. Ocean, Diss. Symp. Chem. Oceanogr. (DISCO)*, Honolulu, Hawaii, *1988*, 55 (1989).
81. P. del Castillo, *Anal. Proc.* **28**, 253 (1991).
82. M. Boussemart, C. Benamou, M. Richou, and J. Y. Benaim, *Mar. Chem.* **28**, 27 (1989).
83. J. R. Donat and K. W. Bruland, *Mar. Chem.* **28**, 301 (1990).
84. E. W. Wolff and D. A. Peel, *Ann. Glaciol.* **7**, 61 (1985).
85. C. F. Boutron, *Fresenius' Z. Anal. Chem.* **337**, 482 (1990).
86. NASS-3, *Open Ocean Sea Water Reference Material for Trace Metals*, Nat. Res. Coun. Can. Div. Chem., MACSP, Ottawa, Canada, 1990.
87. Ph. Quevauviller, K. J. M. Kramer, E. M. van der Vlies, K. Vercoutere, and B. Griepink, *Mar. Pollut. Bull.* **24**, 33 (1992).
88. S. S. Berman and V. J. Boyko, *ICES 16th Round Intercalibration Trace Met. Sea Water*, JMG 6/TM/SW (1987).
89. J. M. Bewers, J. Dalziel, P. A. Yeats, and J. L. Barron, *Mar. Chem.* **10**, 173 (1981).
90. C. S. Wong, K. Kremling, J. P. Riley, W. K. Johson, V. Stukas, P. G. Berrang, P. Erickson, D. Thomas, H. Petersen, and B. Imber, in *Trace Metals in Sea Water* (C. S. Wong, E. Boyle, K. W. Bruland, J. D. Burton, and E. D. Goldberg, eds.), pp. 175–195. Plenum, New York, 1983.
91. J. M. Bewers and H. L. Windom, in *Trace Metals in Sea Water* (C. S. Wong, E. Boyle, K. W. Bruland, J. D. Burton, and E. D. Goldberg, eds.), pp. 143–154 Plenum, New York, 1983.
92. G. Capodaglio, G. Scarponi, G. Toscano, C. Barbante, and P. Cescon, *Fresenius' J. Anal. Chem.* **351**, 386 (1995)
93. L. Mart, H. W. Nürnberg, and D. Dyrssen, in *Trace Metals in Sea Water* (C. S. Wong, E. Boyle, K. W. Bruland, J. D. Burton, and E. D. Goldberg, eds.), pp. 113–130. Plenum, New York, 1983.
94. W. G. Sunda and R. R. L. Guillard, *J. Mar. Res.* **34**, 511 (1976).
95. D. M. Anderson and F. M. M. Morel, *Limnol. Oceanogr.* **23**, 283 (1978).
96. D. M. McKnight and F. M. M. Morel, *Limnol. Oceanogr.* **24**, 823 (1978).
97. A. G. Lewis, A. Ramnarine, and M. S. Evans, *Mar. Biol.* **11**, 1 (1971).
98. R. F. Srna, K. S. Garrett, S. M. Miller, and A. B. Thum, *Environ. Sci. Technol.* **14**, 1482 (1980).
99. M. Camusso, G. F. Gaggino, R. Marchetti, A. Provini, and G. Tartari, *Acqua & Aria* **1**, 31 (1992).
100. R. Kunkel and S. E. Manahan, *Anal. Chem.* **45**, 1465 (1973).

101. J. Buffle, A. Tessier, and W. Haerdi, in *Complexation of Trace Metals in Natural Waters* (C. J. M. Kramer and J. C. Duinker, eds.), pp. 301–316. Nijhoff/Junk, The Hague, The Netherlands, 1984.
102. S. E. Cabaniss, M. S. Shuman, and B. J. Collins, in *Complexation of Trace Metals in Natural Waters* (C. J. M. Kramer and J. C. Duinker, eds.), pp. 165–179. Nijhoff/Junk, The Hague, The Netherlands, 1984.
103. J. W. Moffett, R. G. Zika, and L. E. Brand, *Deep-Sea Res.* **37**, 27 (1990).
104. D. E. Slauenwhite and P. J. Wangersky, *Mar. Chem.* **32**, 37 (1991).
105. N. S. Fisher and J. G. Fabris, *Mar. Chem.* **11**, 245 (1982).
106. B. Imber, M. G. Robinson, and F. Pollehne, in *Complexation of Trace Metals in Natural Waters* (C. J. M. Kramer and J. C. Duinker, eds.), pp. 429–440. Nijhoff/Junk, The Hague, The Netherlands, 1984.
107. T. Midorikawa, E. Tanoue, and Y. Sugimura, *Anal. Chem.* **62**, 1737 (1990).
108. K. H. Coale and K. W. Bruland, *Deep-Sea Res.* **37**, 317 (1990).
109. M. G. Robinson and L. N. Brown, *Mar. Chem.* **33**, 105 (1991).
110. C. M. G. van den Berg, P. T. S. Wong, and Y. K. Chau, *J. Fish. Res. Bd. Can.* **36**, 901 (1979).
111. D. M. Anderson, J. S. Lively, and R. F. Vaccaro, *J. Mar. Res.* **42**, 677 (1984).
112. A. Nelson, *Anal. Chim. Acta* **169**, 287 (1985).
113. T. D. Waite and F. M. M. Morel, *Anal. Chem.* **55**, 1268 (1983).
114. S. Sueur, C. M. G. van den Berg, and J. P. Riley, *Limnol. Oceanogr.* **27**, 536 (1982).
115. B. E. Imber and M. G. Robinson, *Mar. Chem.* **14**, 31 (1983).
116. A. Seritti, D. Pellegrini, E. Morelli, C. Barghigiani, and R. Ferrara, *Mar. Chem.* **18**, 351 (1986).
117. B. E. Imber, M. G. Robinson, A. M. Ortega, and J. D. Burton, *Mar. Chem.* **16**, 131 (1985).
118. A. Seritti, D. Pellegrini, R. Ferrara, and C. Barghigiani, in *Proceedings of the Conference on Heavy Metals in the Environment, Heidelberg*, pp. 1120–1123. Vol. 6, CEP Consultants, Edinburgh, 1983.
119. C. J. M. Kramer and J. C. Duinker, in *Complexation of Trace Metals in Natural Waters* (C. J. M. Kramer and J. C. Duinker, eds.), pp. 217–228. Nijhoff/Junk, The Hague, The Netherlands 1984.
120. R. H. Byrne and W. L. Miller, *Geochim. Cosmochim. Acta* **49**, 1837 (1985).
121. Z. Kozarac, M. Plavsic, and B. Cosovic, *Mar. Chem.* **26**, 313 (1989).
122. F. L. L. Muller and D. R. Kester, *Mar. Chem.* **33**, 71 (1991).
123. G. Scarponi, G. Capodaglio, G. Toscano, C. Barbante, and P. Cescon, *Microchem. J.* **51**, 214 (1994).
124. G. A. Bhat and J. H. Weber, *Anal. Chim. Acta* **141**, 95 (1982).
125. M. Plavsic, B. Cosovic, and S. Miletic, *Anal. Chim. Acta* **255**, 15 (1991).
126. G. Scarponi, G. Capodaglio, C. Turetta, C. Barbante, G. Toscano, and P. Cescon, *Int. J. Environ. Anal. Chem.* in press (1996).

CHAPTER

12

PROBLEMS OF SPECIATION OF ELEMENTS IN NATURAL WATERS: THE CASE OF CHROMIUM AND SELENIUM

L. CAMPANELLA

Department of Chemistry, University of Rome "La Sapienza," 00185 Rome, Italy

12.1. INTRODUCTION

Chemical speciation may be defined as the determination of the individual concentrations of the various chemical forms of an element that together make up the total concentration of that element in a given matrix. Although the total concentrations of a dissolved element may be similar in two distinct aqueous systems, the chemical forms of that element may be quite different.

The toxicity, transport, or—more generally—the biogeochemical behavior of inorganic micropollutants (especially heavy metals and metalloids) is strongly dependent on their chemical species. The methods available for separating the dissolved forms of chemical elements have been reviewed in the last several years by authors who covered ion-exchange, ultraviolet (UV) irradiation, resin adsorption, solvent extraction, filtration, centrifugation, dialysis, ultrafiltration, and gel filtration chromatography [see, e.g., Morrison (1) and Buffle (2)]. On the other hand, suitable detection techniques are anodic stripping voltammetry (ASV), ion-selective electrodes (ISEs), atomic absorption spectrometry (AAS), Inductively coupled plasma atomic emission spectrometry (ICP–AES), and inductively coupled plasma mass spectrometry (ICP–MS).

The fact that all these techniques are now available can be seen as both cause and effect of the growing interest in this type of research. Moreover, this interest can be expected to expand further in the near future, since it is becoming increasingly apparent that the toxicity and bioavailability of an element are strictly related to the form in which it occurs. For example, many essential elements are present in a variety of forms with widely differing

Element Speciation in Bioinorganic Chemistry, edited by Sergio Caroli.
Chemical Analysis Series, Vol. 135.
ISBN 0-471-57641-7 © 1996 John Wiley & Sons, Inc.

degrees of bioavailability. The most bioavailable form of e.g., Fe for humans is its combination with heme, whereas the inorganic salts and simple organic complexes are far less bioavailable. Similarly, the only assimilable form of Co is vitamin B_{12}, whereas complexes of Cr and Zn with amino acids are far more bioavailable than the corresponding inorganic compound.

With regard to the toxicity of metals, it is generally accepted that the free ion (aquated ion) form is the most toxic for aquatic life, whereas the highly complexed forms and those associated with colloidal material are far less toxic. There are however exceptions: in the case of Hg, for example, the most toxic form is methyl-Hg.

Clearly, a knowledge of the total essential concentration is fundamental for this type of study, but it is not sufficient, as it gives no information on the bioavailability or toxicity of the element itself.

12.2. CHEMICAL AND BIOLOGICAL ASPECTS

Table 12.1 shows factors influencing the toxicity of a heavy metal in water, and Table 12.2 highlights the various forms in which an element may occur naturally in water. An initial broad distinction between these various forms may be made on the basis of distribution between the dissolved and adsorbed states, defining as dissolved anything passing through a 0.45 μm filter. However, the speciation of elements can be approached using other parameters including, first of all, the dimensions of particles.

The dissolved state is the phase in which the transport of heavy metals and interaction with suspended material (sediment or organisms) most frequently occur. Thus, speciation of elements must be approached in terms of the basic natural water parameters, such as salinity, pH, oxygen concentration, depth, and concentration and composition of the dissolved organic material. The total concentration of an element dissolved in natural waters is generally made up of labile complex MX_n species and nonlabile, stable complex ML_j species. The determination of these various forms is a very hard task. Even obtaining data on the total concentration alone is often difficult given the low concentration at which many elements occur in natural waters. At such levels of concentration it is obviously extremely arduous to determine the individual forms contributing to the total concentration.

Moreover, given such considerations, theoretically the scientist ought to carry out determinations without altering the system since any disturbance could modify the equilibrium among the chemical forms of the element under examination. In short, speciation of an element in natural waters, whether polluted or not, cannot be carried out without knowledge of the following characteristics: (i) total quantity of carbon present, as this provides informa-

Table 12.1. Factors Influencing the Toxicity of a Heavy Metal in Water

(A) *Metal form in water*: inorganic, organic soluble (ion, complex ion, molecule), particulate (colloid, precipitate, adsorbed)
(B) *Presence of other metals or toxic substances*: combined action, no interaction, antagonism
(C) *Factors influencing both physiology of organisms and form of metals in water*: temperature, pH, dissolved oxygen, light, salinity
(D) *Conditions of organisms*: stages of reproduction and/or growth cycle, age and size, sex, feeding conditions, additional protection (shell, etc.), adaptation to metals

Table 12.2. Possible Forms of an Element in Water

Chemical Forms	Possible Examples	Diameter of Species (nm)
Particulate	Blocked by 0.45 μm filters	>450
Hydrated element	$Cd(H_2O)^{2+}$	0.8
Simple inorganic complexes	$Pb(H_2O)_4Cl_2$	1
Simple organic complexes	Cu–glycinate	1–2
Stable inorganic complexes	$PbS \cdot ZnCO_3$	1–2
Stable organic complexes	Cu–fulvate	2–4
Adsorbed on inorganic colloids	$Cu^{2+}-Fe_2O_3$	10–500
Adsorbed on organic colloids	Cu^{2+}–humic acids	10–500
Adsorbed on organic/inorganic colloids	Cu^{2+}–humic acids/Fe_2O_3	10–500

tion on the possible complexation of the element by organic substances; (ii) pH, as most reactions are influenced by the acidity of water; (iii) pε (formally defined as $-\log e$, where e is the electronic concentration), indicating the possibility of redox reactions in the environment investigated; and (iv) salinity, temperature, and pε, as all these parameters directly influence the speciation equilibria of the elements. With this in mind various methods can be utilized to obtain a composite picture comprising the total concentration of the element and the fraction adsorbed by solids or complexed in stable and labile forms and, finally, the free element occurring as an aquated ion.

Speciation studies of this type require specific, extremely sensitive, and accurate methods ranging from voltammetric techniques to ISEs or AAS. With voltammetric techniques it is possible to study many heavy metals as well as nonmetals of particular ecotoxicological interest, thus accessing many types of data. The natural pH value of dissolved substances corresponds to the total concentration of labile complexes with inorganic anions, such as chloride and sulfate, and some organic substances, including amino acids. The pH 2 value corresponds, e.g., to the amount of Cd released by dissolved organic material. Determinations carried out after UV irradiation yield data on complexes with anthropogenic species so stable that they decompose only under these drastic conditions.

Titrations with a solution of easily complexed inorganic ion yield data on the complexing capacity of water. Although still in the experimental phase, many studies carried out with such methods have elucidated the speciation state of several elements and point to a number of general considerations, two of which are of particular interest: (i) the comparatively minor role played by amino acids and (with the exception of freshwater) by humic acids in speciation studies due to the low values of the stability constants of the relevant complexes and of the dissolved aliquots; (ii) the demonstration that not only labile complex species (including aquated ions) can be adsorbed by plankton, but also stable species. Potentiometry with ISEs is one of the simplest, fastest and most economic techniques. With a Nernst-type equation it is possible to calculate the activity (or the concentration, if an ionic strength buffer is applied) of the free ion in solution from the values of the potential. An important feature of these electrodes is the membrane structure, which can lower the interferences, while the technique also has the great advantage of not disturbing the system under investigation (very often the ionic strength buffer is already present in the sample, as in the case of seawater). This approach is also theoretically useful for any real matrix, since the signal refers only to the ion as a function of the selectivity of the electrode. In practice, however, these electrodes may display two types of interference, i.e., that of the method itself and that due to the electrode. The former leads to errors by default caused by the subtractive effect on the determined ion, whereas the latter results in bias, since the signal from the primary ion is accompanied by signals from other interfering ions. The intensity of these signals is a function, on the one hand, of the concentration ratio between primary ion and interfering ions and, on the other hand, of the selectivity constant. This second type of interferences is particularly serious in speciation studies.

AAS is one of the most sensitive analytical tools. In fact, it makes use of atoms in the fundamental state and not the small fraction of atoms in the excited state. Broadly speaking, the technique consists of aspiration of the sample into an atomization cell where the temperature converts molecules into atoms in a vaporized state; this can absorb radiation of certain wavelengths emitted by a lamp whose cathode consists of the same element as that which is under examination. The quantity of radiation absorbed is proportional to the concentration of the element in the sample. Since any element will only absorb radiation of a particular wavelength, this technique is very selective and, theoretically at least, devoid of interferences. Thus, it may be applied to analysis of real matrices. Actually, it is well known that this technique is subject to interferences and that some precautions must therefore be taken.

The only information obtainable with this technique is the total concentration of an element. However, an operation mode, i.e., AAS with hydride generation, possesses speciation capabilities for those elements to which it is applicable. Not only does it reduce interferences and bring the limits of detection to levels of micrograms per liter, but it also discriminates among the various forms of certain elements, such as Se.

The most important parameters to be determined when dealing with dissolved species are labile and nonlabile associations with dissolved organic matter. For this purpose, voltammetry is still the best tool available. The effect of organic matter on solubility, transport, and bioavailability of inorganic micropollutants is one of the areas where information, both from laboratory experiments and real aquatic systems, is still scarce.

The two cases discussed below, namely, Cr and Se, are particularly representative in this context because they demonstrate that speciation studies need a critical consideration of the system analyzed in order to perform the test most suited to reach useful and correct information. The many varied techniques used nowadays are not yet fully satisfactory for all needs, indicating that further development of new methods and approaches is still essential.

12.3. THE CASE OF CHROMIUM

The concern as regards Cr speciation in natural waters arises from the different levels of toxicity of the species Cr(III) and Cr(VI). In fact, the low uptake by biota of the metal as Cr(III) and the relatively greater bioavailability of Cr(VI) as chromate or dichromate due to the presence of specific membrane carriers threaten living organisms. Recently, the importance of Cr(III) complexes in the mechanism of toxicity has also been considered (3–7).

The use of Cr in many industrial activities, such as tanning, plating, and stainless steel welding, leads to substantial disposal of sludges containing this metal, which can then contaminate soil and deep waters, thereby subsequently entering the food chain. In an assessment of the actual toxicity of Cr-containing matrices, speciation turns out to be of great significance, particularly on account of the possible redox reactions liable to occur in the system. It is well known that both Cr(III) and Cr(VI) can exist in natural waters; theoretical models suggest that in well-oxygenated water Cr(III) should be oxidized to Cr(VI). Unfortunately, this is by no means certain, as there are a number of contradictory experimental reports in the literature (8–10).

The Cr(III)–Cr(VI) equilibrium is influenced in real systems by a number of side-reactions as well as by factors, such as dissolved oxygen, organic matter, particulate matter, and kinetic parameters. In order to further clarify this issue an investigation was undertaken in the present author's laboratory as described hereafter.

Calculations were performed by the computer program MINEQL (11). The input data were as follows. Cr(III), 1×10^{-6} mol L^{-1}; oxalic acid (OXA), 4.81×10^{-4} mol L^{-1}; ethylenediamine (EN), 4.81×10^{-4} mol L^{-1}; ethylenediaminetetraacetic acid (EDTA), 4.81×10^{-4} mol L^{-1}; glycine (GLY), 4×10^{-4} mol L^{-1}; Ca(III), 2.5×10^{-4} mol L^{-1}; carbonates, 2.5×10^{-4} mol L^{-1}; dissolved oxygen = 0.15, 0.30, 0.61, 1.25, 2.5, and 5.0×10^{-4} mol L^{-1}; pH = 6–8. Conditional stability constants were chosen in accord with Martell and Smith (12).

Stock solutions of Cr(III) and Cr(VI) were prepared that contained respectively 1 g L^{-1} Cr(NO$_3$)$_3$ (Spectrosol, BDH, Milan, Italy) and 1 g L^{-1} K$_2$Cr$_2$O$_7$ [RPE* (electronic purity reagent) grade]. To avoid losses of the metal through wall adsorption, all glassware was preliminarily rinsed with a Cr solution of the same concentration as used in the tests. Element concentration for the Cr(III) oxidation tests was 3.8×10^{-4} mol L^{-1}. Tests were performed at pH 7: (i) under stirring conditions for 7 days; (ii) under UV irradiation for 4h; (iii) under an airstream for 7 days. All tests were carried out both in the presence and in the absence of EDTA (3.8×10^{-2} mol L^{-1}). Concentration of Cr(VI) for the reduction tests was 2 mg L^{-1}. The tests were performed by stirring the solutions for 7 days under different chemical oxygen demand (COD) conditions (5, 10, and 20 mg L^{-1} glucose), in the presence (3.8×10^{-2} mol L^{-1}) or in the absence of EDTA.

In the exchange tests, Cr(III) concentration was 1×10^{-6} mol L^{-1} and the concentration of complexing agents (RPE grade) was 4.81×10^{-4} mol L^{-1}. Chelex-100 resin (Bio-Rad Laboratories, Richmond, California) was in the Ca form according to Campbell et al. (13). Tests were performed in the batch mode by putting 125 mg of resin in 50 mL of test solution. At a chosen time an aliquot was filtered and analyzed for Cr as described below.

Chromium was analyzed by electrothermal atomization or flame AAS using the spectral line at 357.9 nm. Chromium (VI) was determined after extraction by methyl isobutyl ketone (MIBK) of the complex formed between Cr(VI) and ammonium pyrrolidinedithiocarbamate (APDC) (14). As stability

---

* RPE is the Italian acronym.

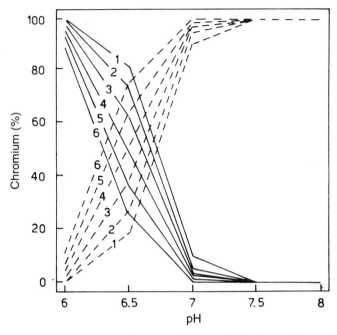

Figure 12.1. Theoretical calculation of Cr(III) and Cr(VI) distribution vs. pH as a function of dissolved O_2. Key: (----) Cr(VI); (——) Cr(III). Dissolved O_2 (mol $L^{-1} \times 10^{-4}$): (1) 0.15; (2) 0.30; (3) 0.61; (4) 1.25; (5) 2.50; (6) 5.00.

constants for Cr complexes with substances occurring in nature (as fulvic or humic acids) are not available, in the case at hand four representative agents with different complexing behavior such as glycyne, oxalate, ethylenediamine, and EDTA have been chosen. The concentration variations of different Cr species in relation to pH, pε, and the nature of the complexing agents have been considered.

From theoretical calculations, Cr(III) proves to be the only form up to pH 6 in the absence of complexing agents, as shown in Fig. 12.1. On the other hand, in the presence of a large excess of EDTA, Cr(VI) is the main form up to pH 7, whereas in the pH range from 6 to 7, the Cr(III)/Cr(VI) ratio is strictly dependent on dissolved O_2 concentration and on pH. The other complexing agents considered do not appreciably affect the equilibrium ratio.

It is important to note that the theoretical model is thermodynamic in nature and does not take into account the kinetic aspects, even though these are probably important in the case of Cr equilibria. In order to estimate the behavior of this metal in a simulated natural system the oxidation of Cr(III) has been tested in synthetic solutions at pH 7 under several strongly oxidizing

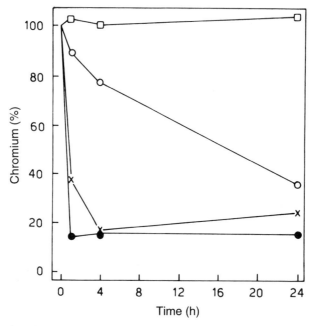

Figure 12.2. Chromium exchange behavior with Chelex-100 resin in the presence of complexing agents: residual Cr percentage in solution vs. contact time. *Key*: ●, EDTA; ○, oxalate; ×, ethylendiamine; □, absence of complexing agents. Reproducibility: ±6%.

conditions, as mentioned earlier. In the presence or in the absence of organic matter all these solutions show less than 1% oxidation of Cr(III). In the case of Cr(VI) reduction tests, less than 1% of the metal turns out to be reduced under each of the selected conditions. With regard to the well-known stability and inertness of Cr(III) complexes, the exchange behavior with Chelex-100 resin of Cr(III) has been studied in the presence of the same complexing agents chosen for the theoretical exercise of speciation.

Because of the affinity of this resin for Cr(III), it is possible to distinguish among different modes for the exchange of complexed Cr species (15). Figure 12.2 shows three different patterns with reference to the hydrocomplex $(Cr_xOH_y^{+(3x-y)})$ (i.e., free metal). The exchange behavior of Cr(III)–EN complexes is very close to that of free Cr. In fact, after 4h of contact with the resin, more than 80% of Cr is exchanged. The Cr–oxalate complexes show very little exchange even after 24h of contact. No important difference was noted if tests were performed in solutions after a pause of 13 days.

The same behavior can be found in natural systems; however, because of the lack of a successful model and of data in the literature, there is no information

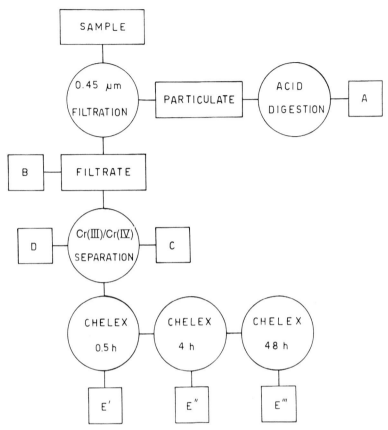

Figure 12.3. Separation of Cr in various species: A = total particulate metal [Cr(III) + Cr(VI)]; B = total dissolved metal [Cr(III) = Cr(VI)]; C = total dissolved Cr(III); D = total dissolved Cr(VI); E' = nonexchangeable Cr(III) after 0.5 h; E" = nonexchangeable Cr(III) after 4 h; E''' = nonexchangeable Cr(III) after 48 h. Total exchangeable Cr(III) is given by (C − E').

on the effect of particulate matter other than the well-known fact that Cr(III) has a pronounced ability to be absorbed on solids.

Finally, Fig. 12.3 proposes a scheme for differentiating the various classes of species. From this model it is possible to distinguish Cr species on the basis of their cationic or anionic nature and of different exchange behavior depending on stability and inertness on Chelex-100 resin.

At the moment such a model is being tested on real systems and correlation should be worked out between speciation data and toxicity results. An interesting application of this study concerns the Cr recovery from tannery sludge by incineration and acid extraction. In fact, it frequently happens that wastewaters from the various tanning operations are not kept separate. The treatment of this mixed wastewater results in very large quantities of sludge in which Cr(III) is present in concentrations ranging from 1–5% on a dry solid basis (16).

As a general rule the alternative based on the reuse of the substances contained in the sludges is to be preferred since it is consistent with the general trend of maximum recovery of resources. Utilization for agricultural purposes of the organic and nutrient matter contained in the sludges (e.g., spreading on farmland, composting with urban solid waste, or incineration combined with flash drying and subsequent use of the dried material) is not feasible owing to the presence of Cr, which can affect plant metabolism and can also have cumulative toxic effects. Chromium also hinders the reuse of tannery sludges in animal feedstuff manufacturing. Pyrolysis (i.e., high-temperature cracking of the organic matter contained in the sludges in the presence of an excess of oxygen) has only been implemented in a small number of pilot plants treating combined municipal refuse and sewage sludge; therefore, no reliable data are available on which to base technical-economic prospects for tannery wastes.

Of great interest is the alternative based on the acid treatment of sludges followed by mechanical separation of the acid extract (subsequently subjected to selective Cr separation for recycling in the tanning process) from solid matter (for use in agriculture or in animal feedstuff manufacturing). In principle, numerous separation methods are available, although none has yet been exhaustively tested (e.g., electrolytic oxidation, selective precipitation, and membrane processes). Promising results have been obtained by using cupferron as a precipitating agent for Al and Fe interfering in the tanning process (17). However, an economic limitation to this method is its inability to completely recover cupferron, a very expensive reagent, from the precipitate. Further studies to improve the reagent recovery yield (e.g., through reagent immobilization on inert matrices) have been planned.

Nevertheless, disposal methods without recovery, such as mechanical removal of water and transportation to a controlled landfill (with or without intermediate incineration), are still widely used. Possible options in this direction have been taken into account in a study by Beccari et al., who used well-defined criteria for assessing the design of treatment units and evaluating capital and operating costs (18). This analysis showed that if high postcombustion temperatures are not required, incineration of the sludges (dried up to 35–40% of the solid content) before transportation to disposal sites is the most economical solution. Furthermore, there is a much smaller quantity of material to be disposed of. This fact, as well as the frequent difficulties in locating a suitable disposal site, can in some cases make incineration a compulsory choice.

However, landfilling with ashes from tannery sludge incineration can pose an environmental hazard, as possible oxidation of Cr(III) to Cr(VI) during the incineration process can render the total load of metal more toxic and leachable (19). Some patented methods for preventing Cr oxidation or minimizing leaching by heating in a reducing atmosphere or by adding chemicals have also been proposed (20-23).

Conversely, owing to the increased leachability of Cr in its hexavalent form, an easier and more selective separation of the metal (for recycling in the tanning process) can be performed by extraction from the ashes rather than from sludges, provided that during incineration most of the Cr(III) content is oxidized to Cr(VI). In this regard, the high Ca content of sludges is a positive factor for increasing the yield of Cr oxidation during incineration. However, more knowledge of Cr behavior during the incineration of tannery sludges is needed if the environmental impact of the process and the possible metal recovery from the ashes are to be assessed. Tables 12.3 and 12.4 list extraction data from the ashes as a function of pH of sludges, weight ratio of ashes, final amount of extracting solution, and incineration temperature.

Table 12.5 shows that Cr(VI) is absent in the leachate; under such conditions, however, the extraction yield is slightly lower than in the absence of simultaneous reducing treatment and the total Cr content exceeds the allowable values. The evaluation of the state of the art of treatment and disposal of tannery sludges has shown that a promising method for selectively recovering Cr from the sludges for subsequent recycling in the tannery consists of oxidative incineration followed by Cr(VI) extraction, reduction to Cr(III), residue stabilization, and disposal of the inert ashes in a conventionally controlled discharge. Ad hoc experiments have been carried out to evaluate process feasibility. Incineration tests have shown that in the 600–800 °C temperature range, Cr(III) oxidation takes place with the highest yields (about 80%) without the need for additional reagents (provided that the Cr/Ca weight ratio in the sludges to be treated is ≤ 0.30–0.35, as in most tannery sludges) or with large oxygen excess. A contact time of 1 h is sufficient, at pH 3–6, to extract in the form of Cr(VI) a very large aliquot of the total Cr originally contained in the sludge, even for comparatively high ash/extracting solution ratios (up to 6–7%). The corresponding acid consumption has been found to be quite low (5–10 kg of 100% H_2SO_4 per kilogram of Cr actually extracted).

Results obtained in preliminary extraction tests combined with simultaneous reduction of the extracted Cr(VI) are promising in light of the environmental hazards associated with the leachability of Cr(VI) from the residual ashes. On the basis of the favorable results described herein, an adequately parametric model can be developed for the purpose of predicting performances as a function of a given set of adopted conditions. This will therefore allow

Table 12.3. Chromium, Calcium, and Iron Extraction and Acid Consumption vs. the Ash/Solvent Weight Ratio at Various pH's (Incineration Temperature, 600 °C)

Ash/Solvent Ratio (w/w)	pH	Recovery				Extracted			Concentration in the Extract			Acid Consumption	
		Cr (%)	Cr(VI) (%)	Cr(VI)/Cr (%)		Ca (%)	Cr (mg L^{-1})		Ca (mg L^{-1})	Fe (mg L^{-1})		Acid/Ash Ratio (w/w)	Acid/Cr Recovered Ratio (w/w)
0.3	3	80.3	79.1	96.58		91.8	155.9		307.6	0.18		0.42	8.8
1.3		78.9	75.5	95.71		62.9	593.1		810.5	0.25		0.36	7.8
4.0		77.1	76.5	99.29		51.2	1810.1		2063.3	0.71		0.37	8.3
6.6		75.7	75.5	99.72		22.2	2952.8		1481.6	0.71		0.38	8.6
8.4		61.4	66.2	107.85		17.9	3036.6		1818.0	0.94		0.38	9.6
0.4	6	76.7	75.4	98.24		64.8	158.9		303.3	0.10		0.29	6.5
1.5		75.0	74.5	99.32		61.0	652.0		910.8	0.23		0.19	4.3
4.5		75.4	74.9	99.44		64.9	1992.8		2946.1	0.80		0.22	4.8
8.4		72.6	72.9	100.36		32.3	3591.6		2739.1	0.98		0.19	4.5
8.4		65.1	71.9	110.47		33.2	3221.1		2812.9	1		0.17	4.4
11.6		59.6	62.6	105.12		23.0	4082.9		2705.7	1		0.17	4.4
0.4	8	76.1	73.6	96.58		56.8	157.5		201.7	0.10		0.24	5.3
13.3		57.8	61.8	106.97		26.9	4543.4		3628.6	0.97		0.11	3.0
0.4	>10	70.9	68.4	96.42		45.9	155.3		172.3	0.10		0.10	2.8
8.3	—[a]	52.5	54.0	102.76		24.9	2592.1		2108.6	0.78		0.05	1.5

[a] Natural pH of sludges.

Table 12.4. Chromium, Calcium, and Iron Extraction and Acid Consumption vs. the Ash/Solvent Weight Ratio at Various pH's (Incineration Temperature, 800 °C)

Ash/Solvent Ratio (w/w)	pH	Recovery				Extracted		Concentration in the Extract			Acid Consumption	
		Cr (%)	Cr(VI) (%)	Cr(VI)/Cr (%)	Ca (%)	Cr (mg L^{-1})	Ca (mg L^{-1})	Fe (mg L^{-1})		Acid/Ash Ratio (w/w)	Acid/Cr Recovered Ratio (w/w)
0.3	3	86.1	81.4	94.5	81.7	178.6	273.6	0.28		0.38	7.3
1.2		90.8	84.2	92.7	74.8	675.2	911.6	0.45		0.33	9.1
3.9		91.5	82.5	90.2	61.7	2271.2	2467.4	0.63		0.33	5.8
6.5		85.6	81.5	95.2	33.5	3504.8	2212.1	0.71		0.33	6.2
8.3		74.1	69.0	93.0	15.5	3935.8	1297.9	0.93		0.37	8.1
0.3	6	81.9	83.5	101.9	68.8	172.3	237.3	0		0.43	8.6
1.3		86.5	84.0	97.1	69.1	685.8	987.1	0.49		0.32	5.9
4.1		77.2	80.7	104.4	46.4	2010.1	1946.7	0.25		0.31	6.5
7.4		82.3	79.4	96.5	30.7	3857.0	2319.9	0.42		0.29	5.7
9.2		65.2	66.3	101.8	23.2	3874.1	2175.3	0.87		0.38	9.1
0.3	8	64.2	59.1	92.1	55.0	134.7	184.1	0.13		0.37	9.2
11.8		52.6	54.5	103.5	26.2	3986.5	3124.4	1.03		0.33	9.7
0.3	>10	53.6	57.7	107.5	25.5	112.9	87.9	0		0.12	3.8
1.7	—[a]	58.5	54.7	93.5	28.9	608.5	493.1	0		0.08	2.1
8.3	—[a]	35.1	38.4	109.5	19.5	1974.9	1645.8	0.61		0.06	2.8

[a] Natural pH of ashes.

Table 12.5. Extraction–Reduction Test and Leaching Test on Residual Solids

Incineration Temperature (°C)	Recovered Cr		Leached Cr		Leached Cr	
	Cr (%)	Cr(VI) (%)	Cr (mg L^{-1})	Cr(VI) (mg L^{-1})	Cr (%)	Cr(VI) (%)
800	66.8	0.11	35.6	0	0.67	0
600	46.6	0.28	48.9	0	1.01	0

a cost–benefit analysis to be made of the process under optimum operating conditions, taking environmental constraints into account as well.

12.4. THE CASE OF SELENIUM

It is well known that many elements are both toxic and essential. Among these, Se plays an important role for two reasons: it is the most toxic and the least abundant among the essential elements. The concentration range within which Se is essential is, in fact, very narrow (24–31). It is only within the past few years that Se essentiality to humans has been ascertained. In particular, it has been ascertained that its compounds can protect cell membranes from oxidative damage, catalyze intermediate metabolic reactions, and inhibit the toxicity of some heavy metals.

There is abundant evidence regarding the effects of Se deficiency, especially for animals. Very low Se concentrations can be related with, e.g., necrotic degeneration of the liver, pancreas, heart, and kidneys. On the other hand, high Se concentrations can trigger toxicity phenomena in animals by inflammation of the feet and softening and loss of hooves and horns, and in humans by loss of hair and nails and irritation of skin and eyes. Generally, Se toxicity in animals appears to be dependent on its chemical forms, the animal species, and other factors; as of now, only scanty information exists concerning Se toxicity in humans (32, 33).

Because of its physical-chemical properties, this element is closely associated with S (it is mainly found in metal sulfide deposits). The Se average crustal abundance is about 0.09 mg kg^{-1}, but this value can greatly vary (the total levels can amount up to 80 mg kg^{-1} in seleniferous soils). As regards living organisms, Se can sum up to several thousands of milligrams per kilogram in primary bioaccumulators, whereas in secondary bioaccumulators and in nonbioaccumulators its levels are about 200–300 and 0.2–20 mg kg^{-1}, respectively (26, 32, 33).

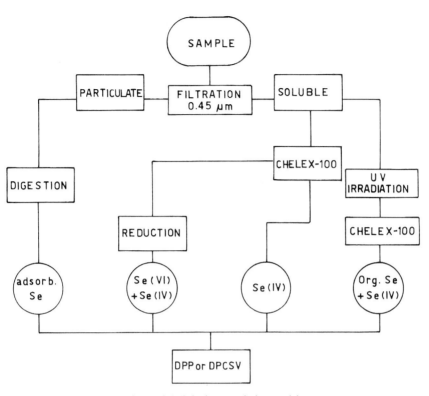

Figure 12.4. Selenium speciation model.

In seawater the Se concentration is rather constant (about $0.09\,\mu g\,L^{-1}$), whereas in freshwater it is usually much lower. In water, Se can exist in the oxidation states $(-\mathrm{II})$, (0), (IV), and (VI). The prevalence of one or another of these oxidation states depends on many factors, e.g., salinity, redox potential, pH, and kinetic parameters.

Within this framework a study was carried out in the present author's laboratory to develop an Se speciation model and to verify its applicability to real matrices. The Se speciation model proposed for studies in natural waters is shown in Fig. 12.4.

The first step is the filtration of the sample on $0.45\,\mu m$ filters in order to distinguish between undissolved (particulate and/or adsorbed on solid materials) and soluble Se. The former is determined on the collected solid after wet digestion. The inorganic Se(IV) is directly determined on the filtered solution, whereas the inorganic Se(VI) and organic Se are determined by differentiation after reduction of the filtered solution by 6 M HCl and destruction of organic compounds. A proper choice of experimental conditions for these two processes can avoid errors in the evaluation of Se contents. Poor efficiency of the reduction process can lead to a too low determination of Se(VI), while an erroneous choice of the organic compound destruction method causes errors in the organic and adsorbed Se determination because of, e.g., incomplete mineralization.

Furthermore, Se losses can occur during the storage of samples because of element adsorption on the container walls. The techniques employed for the experimental measurements are differential pulse polarography (DPP) and differential pulse cathodic stripping voltammetry (DPCSV), both applied to the electrochemical determination of Se(IV). Two mechanisms have been proposed to explain the Se(VI) behavior at the mercury electrode: the first assumes the formation on the drop surface of elemental Se, which is subsequently reduced to SeH_2; the other, commonly accepted as more reliable, is based instead on the following reactions:

Film—

$$H_2SeO_3 + 6e + 6H^+ \rightleftarrows H_2Se + 3H_2O$$

Formation—

$$H_2Se + Hg \rightleftarrows HgSe + 2e + 2H^+$$

Total—

$$H_2SeO_3 + 4e + 4H^+ \rightleftarrows HgSe + 3H_2O$$

Film dissolution—

$$HgSe + 2e + xH^+ \rightleftarrows Hg + H_x + H_xSe^{(x-2)}$$

The instrumentation and the operative conditions used in this investigation include a Model 472 multipolarograph Amel (Milan, Italy) equipped with a three-electrode potentiostatic control, a dropping mercury working electrode, a satured calomel reference electrode, and a platinum counterelectrode. Metrohom polarographic cells were also used. Reagents used were as follows: Biorad Chelex-100 treated with 1 M CH_3COONa; Se(IV) as SeO_2 (Aldrich "Gold Label", Milwaukee, Wisconsin); Se(VI) as Na_2SeO_4 (Sigma Chemicals, Saint Louis, Missouri), and selenourea (Sigma Chemicals). In order to characterize the technique's linearity range, sensitivity, detection limits, peak potential value, accuracy, and reproducibility were determined in different supporting electrolytes. The results are shown in Tables 12.6 and 12.7 for DPCSV and DPP, respectively, while Tables 12.8 and 12.9 list the experimental conditions adopted.

CONCLUSIONS

Table 12.6. DPCSV Performance in Different Media

	$HClO_4$	HNO_3
Linearity range	$5.1 \times 10^{-10} - 3.6 \times 10^{-8}$	$5.1 \times 10^{-9} \times 4.8 \times 10^{-8}$
Sensitivity[a]	14.0	2.1
Detection limit	5.1×10^{-10}	5.1×10^{-9}
E_p (mV)[b]	-460	-450
Accuracy	$+6.2\%$	-8.1%
Reproducibility	$\pm 3.6\%$	$\pm 4.1\%$

[a] Units: nanoamperes/(nanomoles/liter).
[b] E_p = Potential energy.

Table 12.7. DPP Performance in Different Media

	HCl	H_2SO_4	$HClO_4$	HNO_3
Linearity range	$7.6 \times 10^{-8} -$ 2.0×10^{-6}	$1.3 \times 10^{-7} -$ 1.9×10^{-6}	$1.3 \times 10^{-7} -$ 2.0×10^{-6}	$2.5 \times 10^{-7} -$ 2.0×10^{-6}
Sensitivity[a]	1.22	1.28	1.17	0.45
Detection limit	7.6×10^{-8}	1.3×10^{-7}	1.3×10^{-7}	2.5×10^{-7}
E_p (mV)	-430	-420	-470	-450
Accuracy	-3.6%	-4.6%	-4.2%	-6.1%
Reproducibility	$\pm 3.1\%$	$\pm 3.7\%$	$\pm 2.9\%$	$\pm 5.8\%$

[a] Units: nanoamperes/(nanomoles/liter).

Table 12.8. Experimental Conditions for the DPCSV Experiments

Electrolysis potential	-250 mV
Electrolysis time	300 s
Stirring rate	400 rpm
Potential scanning rate	10 mV s^{-1}
Pulse height	-50 mV
Drop surface	$2.145 \pm 0.042 \times 10^{-3}$ cm^2

Table 12.9. Experimental Conditions for the DPP Experiments[a]

Sample volume	25 mL
Hg column height	66 cm
Dropping time	2 s
Pulse height	−50 mV
Scale	500 nA
Potential scanning rate	2 mV s^{-1}

[a] Applied method: standard additions (daily prepared solutions).

12.4.1. Procedures

Selenium(VI) reduction was achieved in 100 μL of a 20 mg L^{-1} Se(VI) solution, after 2.7 mL of 30% HCl was mixed with 1.5 mL of deionized water. The solution was then cooled and quantitatively transferred into a polarographic cell for the determination, after adjusting the final volume to 25 mL. The reduction times investigated were 5, 7.5, 10, 15, 20, 25 and 30 min at the boiling temperature under refluxing conditions.

To carry out digestion, aqueous samples containing Se were added with 10 mL of either a single acid or a mixture of acids. The sample was digested by refluxing at 80 °C for 5 h. The solution was first concentrated to a final volume of about 2 mL, subsequently diluted to 25 mL with deionized water so that the resulting Se concentration was 80 μg L^{-1}, and finally transferred into a polarographic cell for the determination. The resulting acid concentration was about 1 M. The same procedure was applied to selenourea solution.

12.4.2. Application of the Speciation Model

To evaluate the ability of the model to mimic real systems, Se(IV) was added to water sampled from the Tiber River (1:1) so that the final concentration was ca. 80 μg L^{-1} at the moment of measurement. The sample was divided into five aliquots; in each aliquot Se (−II), Se(IV), Se(VI), and adsorbed Se were determined according to the speciation model. Analyses were performed after 1, 2, 3, 6, and 10 days of storage in Teflon containers at room temperature and at natural pH. The same procedure was employed both in the selenourea addition and in the simultaneous addition of Se(IV), Se(VI), and selenourea. As regards sample storage, some authors investigated the Se losses during this phase (32–35). Unfortunately, in some cases, the results of these efforts, although useful, are inadequate mainly because the Se concentration value

CONCLUSIONS

used (1 mg L^{-1}) is not representative of the element concentration in environmental samples (microgram per liter level).

Among the factors that can affect the Se adsorption on the container walls, the following should be taken into account: (i) container materials; (ii) storage time; (iii) Se concentration; (iv) storage medium; and (v) storage temperature.

As regards the container materials, Pyrex, polyethylene, and Teflon were employed. The study was carried out with Se concentrations of 0.4 and 80 μg L^{-1}. Element determinations were performed after 1, 2, 3, 6, and 10 days. In both cases the Se recovery was about 100% when Teflon beakers were used. In terms of storage time, Se adsorption on the container walls depends on the duration of this phase. In fact, the Se loss increases with time up to a constant value; probably this is due to saturation of the adsorbing walls and/or to previous container treatment.

The adsorption percentage is shown in Fig. 12.5 as a function of the sample Se concentration. The experiment was carried out with Se concentrations from 0.4 to 100 μg L^{-1} stored in polyethylene bottles. The Se determinations were performed after 6 days. In order to test the influence of the storage medium, the Se samples (containing 0.4 and 80 μg L^{-1}) were stored both in acid media (1 M HNO$_3$ and 1 M HClO$_4$) and at the natural pH using Pyrex, polyethylene, and Teflon storage containers. In both cases no influence due to the storage medium on the Se adsorption was noted.

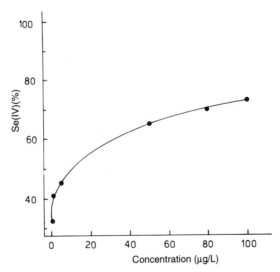

Figure 12.5. Selenium adsorption percentage as a function of Se(IV) concentration in samples stored in polyethylene bottles.

Finally, as regards storage temperature, samples containing 0.4 and 80 µg L^{-1} Se were stored in the three types of containers already mentioned at 4 °C and at room temperature. Results showed that when Teflon containers were used percentage recovery was the same for both temperatures, whereas for Pyrex and polyethylene containers the recovery was higher at 4 °C. Selenium losses from samples containing selenourea and Se were investigated. In both cases no element loss was observed.

Another problem in sample storage is the analyte desorption from the container walls. In fact, the container walls can yield erroneously high blank values if Se adsorbed during a previous storage is successively desorbed. It is thus necessary to perform a pretreatment of the containers by soaking in HNO$_3$ and rinsing with double-distilled and deionized water. All the factors investigated, except the storage medium, affect the Se losses. The aforementioned results show that in order to avoid or minimize element adsorption, use of Teflon containers is recommended.

The experimental procedure just described for deionized water was applied to tap water. It was modified only in that a Chelex-100 treatment was added. This treatment was needed because tap water blank polarograms showed a very intense peak in the Se electroactivity potential range. Twenty minutes of treatment with Chelex-100 in batch conditions were sufficient to obtain

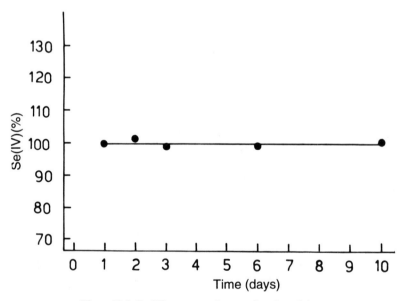

Figure 12.6. Se (IV) concentration as a function of time.

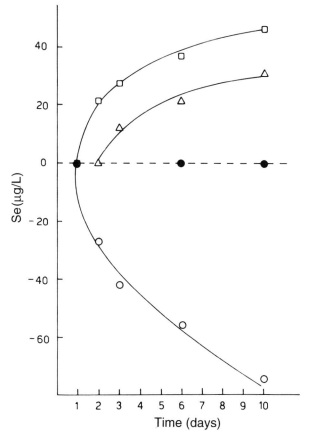

Figure 12.7. Variation with time of the Se species concentrations after selenourea addition to water from the Tiber River. *Key*: ●, Se(IV); △, Se(VI); ○, selenourea; □, adsorbed Se.

adequate blanks. It was also verified that no adsorption occurs on Chelex-100 of inorganic Se (selenite and selenate). In contrast, it was found that selenourea was adsorbed on Chelex-100, so that UV irradiation had to be performed before the treatment. If recovery data in deionized water and in tap water are considered when the treatment with Chelex-100 precedes UV irradiation, comparable results are then observed.

Much along the same line, as already mentioned, the method was applied to the determination of Se in water from the Tiber River. The Se content in this

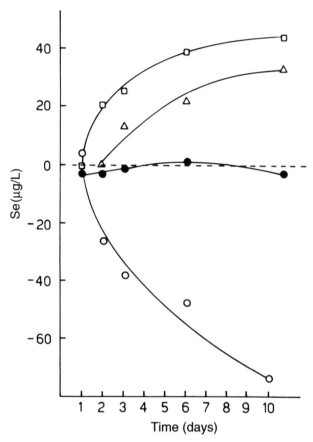

Figure 12.8. Variation with time of the Se species concentrations after Se(VI) and selenourea addition to water from the Tiber River. *Key:* ●, Se(IV); △, Se(VI); ○, selenourea; □, adsorbed Se.

case was found to be lower than the DPCSV detection limits. The water samples were then spiked with a known amount of Se(IV), Se(VI), and selenourea. In Figs. 12.6 and 12.7 the Se(IV) and the selenourea concentrations are shown as a function of time, respectively. As can be noted, Se(IV) remained "frozen" in its oxidation state, whereas selenourea over 10 days was first oxidized to elemental Se and later to Se(IV). These results were confirmed by adding to the Tiber water Se(IV), Se(VI), and selenourea simultaneously. In Fig. 12.8 the percentage variation of the different Se species is shown as a function of time.

Thus, the characteristics of the proposed speciation model are as follows: (i) the use of voltammetric techniques that allow, in combination with filtering and chemical processes (redox mechanism, digestion), the Se species to be determined; (ii) the possible minimization of errors due to both Se adsorption on the container walls and Se volatilization during the digestion process by optimization of experimental conditions; and (iii) the introduction of the Chelex-100 treatment in order to reduce interferences from other elements. As the Se concentration in Tiber water was lower than the detection limits of the described techniques, the addition of the element to the natural matrix was necessary to test the model. However, the Se concentration range in which the speciation model can be applied includes both concentration levels resulting from the usual natural conditions and concentration levels caused by anthropogenic sources.

A possible limitation of this approach may lie in the choice of selenourea as a substance representative of all the organo-Se compounds. In fact, there could be some organo-Se compounds for which UV irradiation does not supply the best mineralization efficiency. In this case wet digestion methods might be adopted according to the procedure described above.

12.5. CONCLUSIONS

The problem of element speciation in water is of increasing interest and importance. This is due, on the one hand, to health and environmental concerns and, on the other hand, to economic reasons. In the first case it must be emphasized that the differences in toxicity of the various chemical species of the same element underscore the need to ascertain its speciation state if human health and essential ecosystems are to be preserved. In the latter case it must be observed that the recovery and recycling approach can be based only on a cost–benefit analysis for the assessment of which fundamental information is needed on speciation.

Many problems concerning this kind of investigation are still unresolved, although significant steps have already been taken. Electrochemical methods and ad hoc computer programs can yield particularly useful contributions, but further progress is expected from phase distribution studies (i.e., exchange, extraction) and membrane methods. Nor should it be overlooked that applications to real matrices require particular attention to be paid to concentration ranges and experimental conditions.

Finally, special care should be devoted to the proper use of reference compounds and solutions and their storage in order to avoid material losses and misleading results.

REFERENCES

1. G. M. P. Morrison, in *Speciation of Metals in Water, Sediment and Soil Systems* (L. Lander, ed.), Lect. Notes Earth Sci., No. 11, pp. 55–73. Springer-Verlag, Berlin, 1987.
2. J. Buffle, *Complexation Reactions in Aquatic Systems: An Analytical Approach* (Ellis Horwood Series in Analytical Chemistry), pp. 384–426. Ellis Horwood, Chichester, UK, 1988.
3. National Research Council Canada, *Effect of Chromium in the Canadian Environment*, NRCC Rep. No. 15017, p. 168. NRCC, Ottawa, Canada, 1976.
4. U.S. Environmental Protection Agency, *Ambient Water Quality Criteria from Chromium*, U.S. EPA Rep. No. 440/5-80-035. USEPA, Washington, DC, 1980.
5. S. Langard, *Metals in the Environment*, pp. 111–132. Academic Press, New York, 1980.
6. R. A. Anderson, *Sci. Total Environ.* **17**, 13 (1981).
7. B. L. Odell and B. J. Campbell, *Compr. Biochem.* **21**, 179–259 (1970).
8. H. Elderfield, *Earth Planet. Sci. Lett.* **9**, 10 (1970).
9. G. E. Batley and J. P. Matousek, *Anal. Chem.* **52**, 1570 (1980).
10. J. F. Pankov, D. P. Leta, J. W. Lin, Ohl, S. E., W. P. Shum, and G. E. Janauer, *Sci. Total Environ.* **7**, 17–26 (1977).
11. J. C. Westall, J. L. Zachary, and F. M. L. Morel, Tech. Note No. 18. MINEQL Computer Program, Department of Civil Engineering, Massachussets Institute of Technology, Cambridge, MA, 1976.
12. A. R. Martell and R. M. Smith, *Critical Stability Constants*. Plenum, New York, 1982.
13. G. C. Campbell, M. Bisson, R. Bougiè, A. Tessier, and J. P. Villeneuve, *Anal. Chem.* **55**, 2246 (1983).
14. *Metodi Analitici per le Acque*, Vol. 1. Consiglio Nazionale delle Ricerche, Instituto di Ricerca delle Acque, Rome, Italy, 1972.
15. Bio-Rad Laboratories, *Chromatography, Electrophoresis and Membrane Technology*, Prod. Bull. No. 11. Bio-Rad Lab., Richmond, VA, 1975.
16. A. Simoncini, *Acqua & Aria* **1**, 55–57 (1982).
17. M. Majone, *Environ. Technol. Lett.* **7**, 531–538 (1986).
18. M. Beccari, M. Majone, R. Passino, and E. Rolle, *Ing. Sanit.* **34**, 137–145 (1987).
19. V. A. Sprinffield, (Ed.), in *Environmental Effects of Pollutants III*, pp. 6–8. U.S. Environ. Prot. Agency, Washington, DC, 1978.
20. Y. Matsumoto, Jpn, Kokai Tokkyo Koho 79/128,158 (Cl.C02C3/00) (1979).
21. H. Aritoshi, E. Tamura, H. Nagamatsu, and S. Usada, Jpn. Koksi Tokkyo Koho 79/139,261 (Cl.C02C3/00) (1979).
22. T. Sugawara, Jpn. Kokai Tokkyo Koho 81/15,883 (Cl.B09B3/00) (1981).
23. Ebara-Infilco Co. Ltd Jpn. Kokai Tokkyo Koho 82/22,000 (Cl.C02F11/14) (1981).
24. H. W. Nürnberg, *Fresenius' Z. Anal. Chem.* **316**, 557 (1983).

25. T. M. Florence and G. E. Batley, *CRC Crit. Rev. Anal. Chem.* **27**, 219 (1980).
26. T. M. Florence, *CRC Crit. Rev. Anal. Chem.* **29**, 345 (1982).
27. T. M. Florence and G. E. Batley, *CRC Crit. Rev. Anal. Chem.* **23**, 179 (1976).
28. W. C. Cooper and J. R. Glover, in *The Toxicity of Selenium and its Compounds* (R. A. Zingaro and W. C. Cooper, eds.), p. 32ff. Van Nostrand-Reinhold, New York, 1974.
29. V. M. Cowgill, *Biol. Trace Elem. Res.* **5**, 345 (1983).
30. V. M. Demayo, M. C. Taylor, and J. W. Reeder, in *Direction Générale des Eaux Intérieures: Lignes Directrices Concernant la Qualité des Eaux de Surface*, Vol. 1, pp. 1–21. Ottawa, Canada, 1976.
31. K. W. Frenke and E. P. Painter, *J. Nutr.* **10**, 599 (1935).
32. R. J. Jarr, *Hazard. Toxic Subst.* **2**, 393 (1978).
33. H. A. Schroeder, V. F. Douglas, and J. J. Balassa, *J. Chronic Dis.* **23**, 227 (1970).
34. H. Robberecht and R. Van Grieken, *Talanta* **29**, 823 (1982).
35. D. C. Reamer, C. Veillon, and T. Tokousbalides, *Anal. Chem.* **53**, 245 (1981).

CHAPTER

13

ARSENIC SPECIATION AND HEALTH ASPECTS

S. CAROLI, F. LA TORRE, F. PETRUCCI, and N. VIOLANTE

*Istituto Superiore di Sanità,
00161 Rome, Italy*

13.1. INTRODUCTION

Arsenic is an element that raises much concern from both the environmental and human health standpoints. This element has long been associated with criminal activity and still is an emotionally highly charged topic, as large doses can cause death. Ingestion via food or water is the main pathway of this metalloid into the organism, where absorption takes place in the stomach and intestines, followed by release into the bloodstream. Arsenic is then converted by the liver to a less toxic form, which is eventually largely excreted in the urine. Only very high exposure can in fact lead to appreciable accumulation in the body. Minor alternative pathways of entry are through inhalation and dermal exposure.

Health effects, both acute and systemic, are known in detail and include nausea, vomiting, diarrhea, decreased production of red and white blood cells, cardiac anomalies, blood vessel damage, impairment of liver, kidneys and nerve functions, and skin abnormalities. Arsenic is ubiquitous in nature, where it is present in a variety of minerals. In fact, about 70% of the element builds up in the environment in the form of inorganic salts, mostly $NaAsO_2$, $Pb_3(AsO_4)_2$, arsenides, and sulfide and oxide compounds. The remaining 30% is accounted for by organoarsenicals, such as monomethylarsonic acid (MMAA) and dimethylarsinic acid (DMAA) (1). Arsenic compounds are widely used for agricultural purposes as herbicides, pesticides, and wood preservatives (1–5).

Minor applications are in the production of elemental As, special alloys, catalysts, pigments, and glasses. For these reasons a large amount of As-containing compounds is released annually into the environment, where a considerable interconversion of As compounds by biological and chemical action can occur (6).

Nowadays it is generally acknowledged that different species of an element can exert diverse toxicological and biological effects in animal and human systems (7). This obviously also applies to compounds whose toxicity greatly varies, with inorganic forms of As exhibiting the highest toxicity levels (8). Organoarsenicals are usually less toxic than the inorganic As species. Indeed, some organic As compounds, such as arsenobetaine (AsBet) and arsenocholine (AsChol), are well tolerated by living organisms (9,10). From this point of view it is becoming increasingly important that the various forms of As be qualitatively and quantitatively determined in biological fluids and tissues as well as in matrices of nutritional and environmental relevance, especially in marine ecosystems (11, 12). This will allow for a much better assessment of the risk associated with exposure to As compounds.

Legal provisions are at present almost exclusively concerned with the total amount of the element in foodstuffs and drinking water. According to the World Health Organization (WHO), e.g., the provisional total daily intake should not exceed 2 μg of inorganic As per kilogram of body weight. Marine organisms are considered to be among the greatest bioaccumulators of As, given the tendency shown by this element to replace N or P in several compounds, thus producing e.g., AsBet, AsChol, and algal arsenosugars.

13.2. EXPERIMENTAL APPROACHES

In order to arrive at reliable results and correct evaluations, numerous analytical methods of separation and detection have been proposed in recent years to identify and measure different As species. Gas chromatography (GC) and high-performance liquid chromatography (HPLC) are especially attractive to analysts seeking to achieve speciation. Particularly HPLC displays great flexibility and can be used to separate nonvolatile and thermally labile compounds. Also ion liquid chromatography (ILC) is being employed to an increasing extent to separate both ionic and nonionic components of either organic or inorganic nature. This technique has a great resolution capability, i.e., it is applicable to a broad range of species. As regards detection, element-specific or multielement detectors based, e.g., on microwave emission spectrometry (MES), electrothermal atomization–atomic absorption spectrometry (ETA–AAS), flame atomic absorption spectrometry (FAAS), inductively coupled plasma atomic emission spectrometry (ICP–AES), and inductively coupled plasma mass spectrometry (ICP–MS) can be recommended as analytical techniques of choice.

Among the most promising combined approaches, the direct coupling of HPLC to ICP–AES has been attempted in several laboratories, as this offers the advantages of a powerful fractionation technique characterized by a high degree of selectivity together with the simultaneous multielemental detection ability, relative freedom from matrix interferences, element specificity, and wide dynamic range typical of ICP–AES (13–15). This combined approach is also possible thanks to the compatibility of the typical solution uptake rate of ICP–AES nebulizers with the flow rate range generally used in HPLC work as well as in chromatographic ion exchange systems where an aqueous mobile phase is used.

On the other hand, a wealth of other strategies have been worked out, as is amply reported in the relevant literature. Monomethyl-, dimethyl-, ethyl-, n-propyl-, and n-butylarsonic acids were, e.g., determined in commercial pesticides after reduction of As compounds to arsines by means of $NaBH_4$ and separation through a GC system. By use of a highly selective and sensitive MES device a detection limit of 0.25 μg L^{-1} was reached. The reduction of inorganic and organic As species to generate the corresponding volatile arsines generally provides an excellent method to isolate As from the matrix and to improve detection limits, as discussed by Talmi and Bostick (16).

Other authors have reported that MMA and DMA can be separated and quantified after conversion into stable derivatives, i.e., methyl esters of thioglycolic acids, by GC using flame ionization detection or the more selective thermoionic specific detection system. The detection limit was found to be ca. 10 ng mL^{-1} in both urine and blood (17). As regards the species arsenite, arsenate, and MMA, these were separated and determined by GC in aqueous samples after reduction of the element to the trivalent state and subsequent reaction with 2,3-dimercaptopropanol. Absolute amounts of 0.02 ng of As for arsenite and arsenate and of 0.04 ng of As for MMA were determined by using a flame-photometric detection system (18).

The combination of GC with multiple ion detection MS enabled inorganic and monomethyl-, dimethyl-, and trimethylarsenic compounds in the urine of rats, mice, and hamsters to be determined with a limit of detection of 0.2–0.4 ng L^{-1}. In this case the arsenical compounds were detected as arsines (19). According to another study, arsenite, arsenate, MMA, and DMA were separated as the corresponding methylthioglycolate derivatives by means of capillary gas–liquid chromatography (CGLC) on wide-bore borosilicate glass and fused-silica columns under programmed-temperature conditions (20). With this derivatization technique, however, arsenite and arsenate could not be resolved from each other. As a rule, detection limits for the various As compounds are ca. 0.1 ng L^{-1} with electron-capture detection and 5 ng L^{-1} with flame ionization detection.

As regards equipment development aimed at providing better tools for As speciation, a low-cost automated interface for a Zeeman GFAAS apparatus was devised that makes possible the use of this instrument as an element-specific detector for HPLC. With the aid of a reversed-phase C_{18} column AsChol, AsBet, and inorganic As were separated and then sent on to the aforementioned ETA–AAS system for detection (21).

Other investigations are also worthy of note, as follows. Arsenical residues from soil were extracted and separated by HPLC on an anion-exchange column. Arsenite, arsenate, DMA, and MMA were quantified on-line at a level of 0.5 mg L^{-1} by means of ETA–AAS (22). The separation of arsenite, arsenate, MMA, and DMA was attempted on a strong anion exchanger, but only the fractionation of arsenite, DMA, and arsenate was achieved. However, all four As compounds were separated on a C_{18} column. The quantification of arsenical compounds at the level of 10–20 μg L^{-1} was independently accomplished by using two automated detectors consisting of (i) an ETA–AAS instrument with an especially adapted autosampler and (ii) a Zeeman ETA–AAS apparatus (23).

Both approaches were used to speciate organo-As compounds in soil extracts and drinking waters. Five arsenical species, namely, arsenite, arsenate, MMA, DMA, and p-aminophenylarsonic acid, were separated utilizing an ion-exchange column by optimization of the ionic strength of the mobile phase (24). The detection apparatus consisted of a continuous arsine-generation system connected with a heated quartz furnace atomizer and an AAS apparatus. For each form a detection limit of less than 10 ng mL^{-1} was obtained.

Moreover, an analytical procedure was developed for the determination of As in blood and urine and for the speciation of its metabolites in the latter fluid. This method was successfully applied to assess the As levels in subjects prone to environmental or occupational exposure or liable to oral ingestion (25). The separation of inorganic As, MMA, and DMA was obtained by ion-exchange chromatography (IEC) coupled directly with AAS (25). With reference to the separation of arsenite, arsenate, MMA, and DMA in water samples, these forms were fractionated on a strong anion-exchange resin. Tye et al. attached constant-flow hydride generation (HG) system to the outlet of the chromatographic system, which in turn was connected to an AAS apparatus (26). In order to improve the capability of the method, the authors employed a preconcentration technique on thin-layer anion-exchange material. The absolute detection limit was approximately 2 ng.

An inexpensive system consisting of an HPLC connected with HG–FAAS was devised for continuous monitoring of arsenical compounds (27). This allowed arsenite, arsenate, MMA, and DMA to be separated on an ion-exchange column. The procedure was applied to samples of seaweed extracts and liver supernatant fractions obtained from mice that had been given arsenite orally.

In an attempt to assess the performance of different approaches, a comparative study was carried out using a variety of combinations of chromatographic and atomic emission techniques (28). Separation of the compounds was obtained by either GC or HPLC, while detection was carried out by FAAS, flame atomic fluorescence spectrometry (FAFS), and ICP–AES. In the GC separation the arsenical compounds were analyzed either as the corresponding hydrides or as the methylthioglycolate derivatives. Inorganic As, MMA, and DMA were thus determined at the sub-microgram per liter levels with an HG–cryogenic trapping–GC–FAAS system. In this context, it is not out of place to recall that quantification and speciation of As in plants and marine organisms has always been a challenge. Arsenite, arsenate, MMA, DMA, and AsBet were separated using a C_{18} column connected in series either to a SAX column or to an ion-exchange resin (29). An AAS detector was used for the initial optimization, and quantitative analyses were carried out by using an ICP–MS instrument.

A novel HPLC–HG–AAS interface has been proposed for the determination of the biogenic arsonium species AsBet, AsChol, and tetramethyl cations (30). This on-line interface is based on thermospray nebulization of the HPLC methanolic eluent, pyrolysis of the analyte in a methanol/oxygen flame, gas-phase thermochemical hydride generation (THG) using excess hydrogen, and cool diffusion flame atomization of the resulting arsine in a quartz cell mounted in the optical pathway of an AAS device. The analytes were separated isocratically in a cyanopropyl-bonded phase HPLC column. This procedure led to absolute detection limits for AsBet, AsChol and tetramethylarsonium (TMA) cations of 13, 14, and 8 ng, respectively.

Inorganic As, MMA, and DMA were separated from untreated urine of volunteers on a combined cation-exchange/anion-exchange column and determined by HG–AAS (31). A water-soluble organic As compound, isolated from the muscle tissues of shark, was identified as AsBet by comparison of retention times on either an anion- or cation-exchange column (32). The analytical signal was quantified by ICP–AES. Arsinic and arsonic acid, MMA, DMA, and AsBet were determined in seaweed samples by separation on a strong anion exchanger and quantification by ICP–AES (33). The absolute detection limits were estimated to be 30, 19, 41, and 30 ng, respectively, for As(III), MAA, DMAA, and As(V) [based on a 2 SD (standard deviation) criterion]. Sodium arsenite, arsenate, and dimethylarsenate in samples of well water were separated with the aid of a paired-ion reversed-phase HPLC (34). After derivatization to arsines with sodium borohydride, these species were determined at a level of 50 μg L^{-1} by ICP–AES.

Water-soluble As compounds, namely, MMA, DMA, AsBet, and AsChol, were separated and quantified in three crab species by means of a macroporous divinylbenzene–styrene copolymer column and an ICP–AES detector

(35). Amounts as low as 3 ng for MMA, 5 ng for DMA, 14 ng for AsChol, and 165 ng for AsBet were ascertained. Arsenite, arsenate, MMA, DMA, and AsBet were separated by switching between a C_{18} column and an anion-exchange column (36). A mobile phase consisting of a 3 mM ammonium dihydrogen orthophosphate solution was used, while the detection system was an ICP–AES apparatus.

Arsenite, arsenate, MMA, and DMA were separated in 15 min by means of an anion-exchange column using a linear gradient from water to 0.5 M ammonium carbonate (37). Arsenic species detection was accomplished by the direct coupling of the column effluent to an ICP–AES equipment. This study was developed for application to the determination of As metabolites in cultured cell suspensions. The absolute detection limits (as micrograms of As injected) were 0.39 for arsenite, 0.060 for DMA, 0.057 for MMA, and 0.126 for arsenate.

With the aid of an anion-exchange column and a phosphate buffer solution at pH 6.75 as the mobile phase, arsenite, arsenate, MMA, and DMA were separated and determined (38). The column was connected to an HG system coupled with an ICP–AES detector. Concentrations as low as 3.5, 3.8, 21.3, and 9.2 µg L^{-1} for arsenite, MMA, DMA, and arsenate, respectively, were quantified.

Six environmentally significant arsenical compounds (arsenite, arsenate, MMA, DMA, AsBet, and AsChol) were separated on a strong anion-exchange column connected in series with a NH_2-based column. A phosphate buffer was used as the mobile phase. Due to the resistance of AsBet and AsChol to conversion into hydrides, these species were preliminarily irradiated by UV. By using an ICP–AES system as the detector, absolute detection limits of 1 ng for arsenite, 5 ng for MMA, arsenate, and AsBet, and 10 ng for DMA and AsChol were achieved (39). With a similar approach, six arsenical species (arsenite, arsenate, MMA, DMA, AsBet, and AsChol) were fractionated on an anion-polymer-based column and determined by means of a UV–HG–ICP–AES system (40).

Arsenobetaine was identified as the principal As species (84%) in a dogfish muscle reference material (41). The quantification of this product was accomplished by interfacing a C_{18} column to an ICP–MS detector using a 10 mM sodium dodecyl sulfate solution containing 5% methanol and 2.5% glacial acetic acid as the mobile phase. The absolute detection limit of 300 pg of As was lowered to 30 pg when the column was replaced by a flow injection analysis (FIA) system.

Urine excretion is the major pathway for the elimination of As compounds from the organism. Arsenite, arsenate, MMA, and DMA in this fluid were separated via anion-exchange HPLC and detected on-line by using ICP–MS (42). The absolute detection limits were 38 pg for arsenite, 20 pg for DMA,

44 pg for MMA and 91 pg for arsenate. Combining the separation capability of HPLC with the selectivity and sensitivity of ICP–MS enabled Sheppard et al. to separate and detect arsenite, arsenate, and DMA in urine, club soda, and wine (43, 44). To eliminate or reduce the mass spectral interference due to the presence of chloride (namely, the molecular ion $ArCl^+$) the authors performed preliminary IC separation.

A thermospray nebulizer was used as an interface between LC and ICP–AES for separation of arsenite, arsenate, DMA, and phenylarsonic acid (45). A Dionex high-performance ion chromatography (HPIC) AS4A column was used, with a detection power of 234, 3.4, 31, and 2.4 ng for the four species, respectively. Finally, four arsenical species, i.e., arsenite, arsenate, MMA, and DMA, were separated by HPLC by using an anion-exchange column (38). After hydride conversion concentrations as low as 3.5, 9.2, 3.8, and 21.3 μg L^{-1} for the four species were detected, respectively.

Arsenite, arsenate, MMA, and DMA were separated by micellar liquid chromatography (MLC) on a polymeric reversed phase column coupled with an ICP–MS. An aqueous mobile phase containing a surfactant was more compatible with the spectrometric system than were organic solvents (46).

Hovanec and Northington separated arsenite and arsenate in drinking water samples by HG–ICP–MS using a glass insertion probe. The detection limit of arsenite was 0.01 μg L^{-1} (47).

Samples of shrimp, crab, fish, fish liver, shellfish, and lobster digestive gland, as well as five certified reference materials, were investigated for their content of As compounds. The speciation and detection of six arsenical species were achieved by Larsen et al. by means of a cation-exchange column coupled with an ICP–MS system. Trimethylselenium ion, detected at an m/z of 82, was used as the internal chromatographic standard (48).

Arsenic species have also been separated by using capillary zone electrophoresis (CZE) with ultraviolet (UV) detection. Various approaches to interfacing CZE with ICP–MS are currently being explored (49).

13.3. ONGOING ACTIVITIES

The scenario depicted in the previous section is probative of the vast interest raised by As speciation in the scientific community. On the other hand, this wealth of information clearly testifies to the enormous investigative potential of the particularly promising combination of HPLC with ICP–AES and ICP–MS. As regards this last aspect, systematic studies have been undertaken in the present authors' laboratory in order to elucidate the actual capabilities of these hyphenated techniques and, at the same time, to identify those features that can be further developed to increase their applicability to fieldwork.

As already briefly mentioned, the separation of six arsenical species, namely, arsenite, arsenate, MMA, DMA, AsChol, and AsBet, and the simultaneous determination of their total As content was achieved by coupling an HPLC system with ICP–AES and/or ICP–MS detectors for the quantification of the element.

In a first attempt, a series of reversed-phase columns were used to separate the aforementioned compounds. A Nucleosil C_{18} column (Macherey-Nagel, Düren, Germany) was unsuccessfully employed with a phosphate-based aqueous mobile phase (pH = 6.0) at a flow rate of 1 mL min^{-1}. With this approach, in fact, AsChol is blocked on the stationary phase. Similarly, the arsenical compounds could not be separated using either a Carbohydrate Analysis (Waters, Milford, Massachusetts) or a Lichrosorb NH_2 (Merck, Darmstadt, Germany) column and a phosphate buffer solution at pH 6.0 as the mobile phase. Coelution of arsenite and AsBet was observed with the Waters column; use of the NH_2-based column resulted in the same problem with AsChol and AsBet as well as with arsenite and DMA.

Since the arsenical compounds exist in a charged form in aqueous solution, a Supelcosil LC-SAX anion-exchange column (Supelco, Bellefonte, Pennsylvania) was thought to be most suited to the purpose. The ionic strength and the pH values of the electrolyte in the mobile phase were optimized in an effort to obtain the best possible separation and reproducibility. At lower ionic strength a good separation of the six arsenical species was achieved, but reproducibility was not satisfactory. Also, coelution of MMA with arsenate was observed. Decreasing the pH value resulted in the coelution of AsBet with DMA, while the separation between MMA and arsenate improved. An opposite trend was observed when the pH value was increased. The use of two mobile phases at different pH values did not give rise to any improvement of the chromatographic pattern, probably due to an overly long equilibration time of the column. A more satisfactory chromatographic profile was obtained with a mobile phase consisting of a 0.005 M ammonium phosphate buffer (pH = 4.75). A drawback of the system consisted in the coelution of MMA with arsenate after a certain number of chromatographic runs. A NH_2-based column connected in series to the Supelco column allowed more chromatographic runs (at least 40) to be carried out without problems. However, the As minimum amount detectable with the HPLC–ICP–AES system was not adequate to the expected levels of As in matrices of environmental interest. Thus, the detection power had to be dramatically improved by adding an HG setup to the HPLC–ICP–AES combination.

The incompatibility of the HPLC flow rate (1–2 mL min^{-1}) with that used in conventional HG systems (8 mL min^{-1} for the sample and 4 mL min^{-1} for

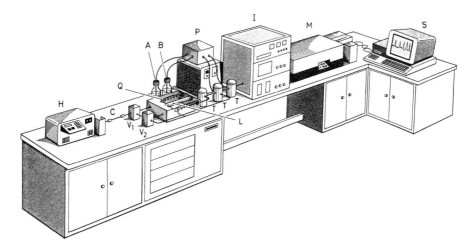

Figure 13.1. Overall view of the HPLC–UV–HG–ICP–AES system: A & B, reagents for hydride generation; C, column; I, ICP torch; H, HPLC pump; L, UV lamp; M, monochromator; P, peristaltic pump; Q, quartz coil; S, integration software; T, tee connections; V_1 & V_2, two-way valves.

$NaBH_4$) suggested the use of a Babington-type nebulizer with a Scott chamber as the gas–liquid separator. In this case, in fact, a 3 mL min^{-1} flow rate was achieved. To promote the decomposition of AsBet and AsChol before their conversion into hydrides, these two compounds were loaded into two separate quartz coils soon after the chromatographic separation. The coils were then irradiated in an air-cooled box with a 500 W UV lamp. An overall view of the HPLC–UV–HG–ICP–AES system is shown in Fig. 13.1. An irradiation time of 30 min was sufficient to produce an 80% decomposition. The absolute detection limits thus obtained were 1 ng for arsenite, 5 ng for arsenate, MMA, and AsBet, and 10 ng for DMA and AsChol.

An $N(CH_3)_2$ silica-based column (Macherey-Nagel) was later employed in order to improve the reproducibility of the chromatographic system. A buffer solution consisting of 0.005 M ammonium phosphate was used as the mobile phase and phenylarsonic acid (PAA) was added as the internal standard. At pH 3.5 a strong retention of MMA, arsenate, and PAA took place. Increasing the pH caused a decrease of the retention time of these compounds. A partial solution to this problem was provided by the use of a pH gradient, although the chromatographic pattern obtained through the experimental conditions reported in Table 13.1 showed a broadening of the PAA peak. An increase in the ionic strength of eluent B (0.01 M) allowed this drawback to be reduced.

Table 13.1. Chromatographic Instrumentation and Operative Conditions for the Macherey-Nagel $N(CH_3)_2$-Based Column

Pump	Biocompatible 250 (Perkin-Elmer, Norwalk, Connecticut)
Solvents	(A) 0.005 M $NH_4H_2PO_4$, pH 3.5
	(B) 0.01 M $NH_4H_2PO_4$, pH 6
Flow rate	1.0 mL min^{-1}

Gradient Program

	Solvents	
Time (min)	A (%)	B (%)
0	100	0
5	100	0
15	0	100[a]
20	0	100
30	100	0

[a] Convex increase curve.

The chromatographic behavior of the aqueous mixture of the seven As species separated by the Macherey-Nagel column, as shown in Fig. 13.2, was similar to that of the Supelco column. The detection limits reached when an HG setup was used were of the same order of magnitude as those obtained with the Supelco column, while reproducibility was also satisfactory.

In order to improve the shape and the resolution of the peaks, a polymer-based Dionex AS7 column was used. This column can be operated at a pH range of 2–11. As the mobile phase, a 0.03 M $NaHCO_3$ solution under conditions of pH and flow rate gradients was employed. The operative conditions are described in Table 13.2.

An aqueous standard solution containing inorganic arsenite, arsenate, MMA, DMA, PAA, AsChol, and AsBet was tested. The chromatographic pattern (presented in Fig. 13.3) shows an improvement of the chromatographic parameters as compared with those of the columns previously used. The reproducibility of this column, under the experimental conditions just described, was tested on different days and turned out to be excellent.

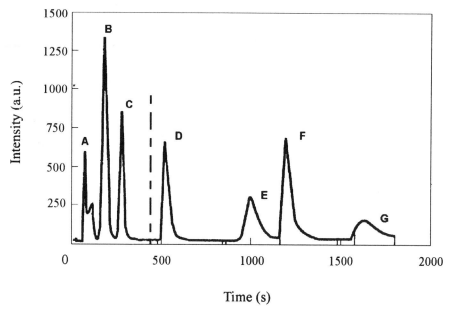

Figure 13.2. Chromatographic separation obtained with the $N(CH_3)_2$-based column (Macherey-Nagel) as revealed by ICP–AES detection. Aqueous solution of seven As compounds (5 mg L^{-1} each): A, AsChol; B, arsenite; C, AsBet; D, DMA; E, MMA; F, arsenate; G, PAA.

To improve the detection limits, as in the case of the Supelco and Macherey-Nagel columns, an HG system was used. The main disadvantage of this system is a dramatic increase in the baseline during the separation. This phenomenon is due to the evolution of carbon dioxide in consequence of the reaction between the alkaline mobile phase and the acidic solution used for the hydride generation. To overcome this problem and obtain better detection limits, the HPLC apparatus was connected directly with an ICP–MS detector, thus replacing the entire UV–HG–ICP–AES system (Fig. 13.4). The ICP–MS is a relatively new technique for trace element analysis. The method offers the possibility of multielement analysis with excellent detection limits (sub-nanogram per milliliter), a broad linear range (at least 4 orders of magnitude for concentration), and rather simple spectra when compared to those of atomic emission spectrometry. Furthermore, it is characterized by good reproducibility over either the short or the long term. The ICP–MS apparatus

Table 13.2. Chromatographic Instrumentation and Operative Conditions for the Dionex AS7 Column

Pump	Biocompatible 250 (Perkin-Elmer, Norwalk, Connecticut)
Precolumn and column	IonPac AG7 and IonPac AS7 (Dionex, Sunnyvale, California)
Solvents	(A) Deionized water
	(B) 0.03 M NaHCO$_3$, pH 9

Gradient Program

Time (min)	Flow Rate (mL min^{-1})	Solvents A (%)	B (%)
0	0.7	100	0
3	0.7	100	0
18	1.0	0	100[a]
33	1.5	0	100
38	0.7	100	0

[a]Convex increase curve.

and working conditions as adopted for the As speciation study are listed in Table 13.3. In order to better assess the detection limits of HPLC–ICP–MS combinations, a standard solution containing seven arsenical species at the level of nanograms per liter (arsenite, arsenate, MMA, DMA, PAA, AsChol, and AsBet) was subjected to separation and quantification, producing the chromatogram shown in Fig. 13.5.

Within the framework of a pilot study organized by the Standards, Measurements and Testing (SMT, formerly BCR) Programme of the European Union, fish and mussel extracts containing six As species (arsenite, arsenate, MMA, DMA, AsChol, and AsBet) were analyzed with the system described above, in order to better test its capabilities (50). PAA was added to the extract solutions as the internal standard. The extracts were prepared by the Service Central d'Analyse (Vernaison, France) according to the following procedure: fish and mussel powders, obtained after lyophilization, were extracted five times with 10 mL of a 1:1 methyl alcohol–water mixture in an ultrasonic bath. The liquid, after filtration, was carefully evaporated to dryness and then reextracted with water (five times with 10 mL). The aqueous solution was extracted with ethyl ether (five times with 10 mL). At this stage no

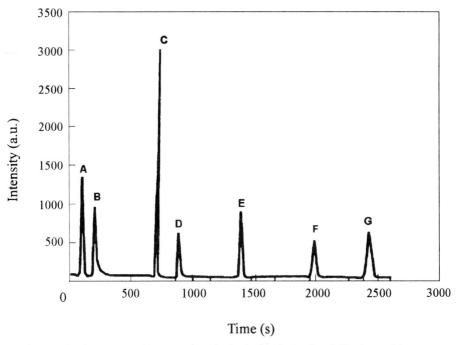

Figure 13.3. Chromatographic separation obtained with the IonPac AS7 column (Dionex) as revealed by ICP–AES detection. Aqueous solution of seven As compounds (5 mg L^{-1} each): A, AsBet; B, arsenite; C, DMA; D, AsChol; E, MMA; F, PAA; G, arsenate.

detectable levels of As were found in the organic phase. Finally, the extracts were tested for their NaCl content, which turned out to be 0.15 and 1.0% in fish and mussel, respectively. The extracts were injected after 1:20 dilution with double-distilled water without any further cleanup procedure. Calibrants were checked in order to exclude the presence of any other As-containing impurity detectable by the analytical system.

The chromatographic pattern of the fish extract did not reveal the presence of peaks other than those corresponding to the calibrant mixture (Fig. 13.6). Instead, the presence of two unidentified peaks was observed in the mussel extracts (Fig. 13.7). The peak showing the highest retention time was ascribed to the presence of chloride, detected by the mass spectrometer as $^{40}Ar^{35}Cl$.

Some difficulties were encountered in the separation of AsBet from arsenite, as well as in the quantitative determination of the latter species. In fact,

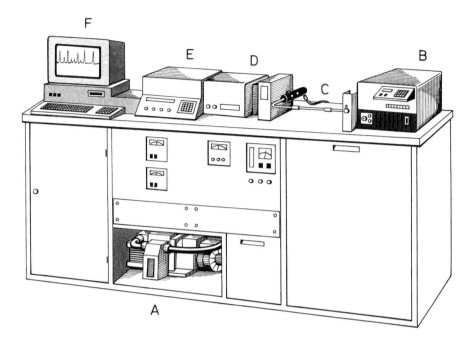

Figure 13.4. Overall view of the HPLC–ICP–MS setup: A, vacuum pumps; B, HPLC pump; C, column; D, ICP torch; E, quadrupole; F, Integration software.

Table 13.3. ICP–MS Instrumentation and Operative Conditions

Spectrometer	Sciex Elan 5000 (Perkin-Elmer, Norwalk, Connecticut)
Nebulizer	Cross-flow type, with Ryton condensation chamber
Power (kW)	1000
Argon flow (L min^{-1})	Plasma gas, 16; auxiliary gas, 0.9; aerosol gas, 1.0.
Analysis mode	Transient signal, graphic mode
Scan conditions	Dwell time, 800 ms; sweeps/reading, 1; readings/replicate, 1; number of replicate, 2700
Analytical mass	^{75}As

the said compounds elute with similar retention times. In addition to this, fish samples usually contain the two As species at a ratio of 1:100. As a consequence, under these conditions, the quantitative determination of arsenite turned out to be rather inaccurate because of the large tailing of the AsBet peak.

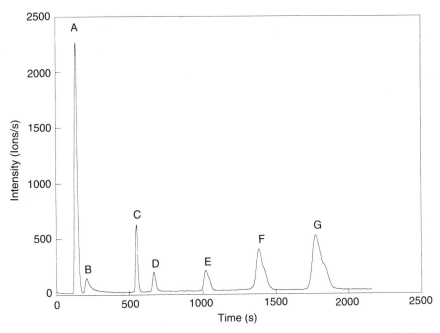

Figure 13.5. Chromatographic separation obtained with the IonPac AS7 column (Dionex) as revealed by ICP–MS detection. Aqueous solution of seven As compounds: A, AsBet, 50 μg L^{-1}; B, arsenite, 5 μg L^{-1}; C, DMA, 10 μg L^{-1}; D, AsChol, 10 μg L^{-1}; E, MMA, 10 μg L^{-1}; F, PAA, 50 μg L^{-1}; G, arsenate, 50 μg L^{-1}.

13.4. CONCLUSIONS

The vast quantities of As circulating in the environment and the differences in noxious potential of its various chemical forms have fostered intense research activities to develop novel and reliable analytical systems. Diversity in matrices and levels to be expected for the As species of interest have prompted many investigators to design more and more sophisticated methodologies.

GC and HPLC techniques are being widely used for the speciation of As compounds, the second approach being by far the most promising. Particularly, the use of aqueous mobile phases in IEC and MLC have improved the compatibility of the HPLC system with that of ICP–AES or ICP–MS. The detection power characterizing ICP–AES appears to be inadequate for the levels expected in matrices of environmental interest. Coupling it with a UV–HG attachment made the detection power of the system suitable for the

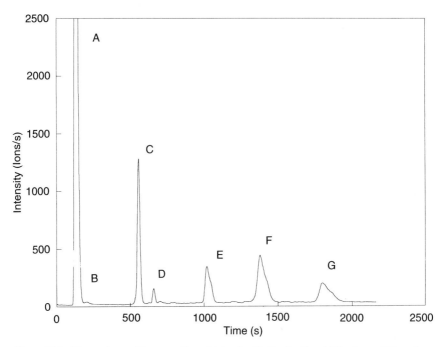

Figure 13.6. Chromatographic separation obtained with the IonPac AS7 column (Dionex) as revealed by ICP–MS detection. Fish extract (1:20 dilution with deionized water): A, AsBet; B, arsenite; C, DMA; D, AsChol; E, MMA; F, PAA; G, arsenate.

purpose, although the overall setup was quite complicated. These problems were solved by the use of ICP–MS. This technique, in fact, turned out to be a powerful investigative tool from both a qualitative and quantitative point of view, with detection limits quite suitable for determining the arsenical compounds in matrices such as fish and mussel extracts.

CZE, in turn, is today one of the more promising separative techniques for As species, although the UV detector normally used in this approach appears to be scarcely specific. ICP–MS, in contrast, is specific and sensitive enough to detect analytes emerging from CZE. Owing to the different flow rates characterizing the two systems, however, a suitable interface between CZE and the ICP–MS detector needs to be developed, this posing a challenging task for the future.

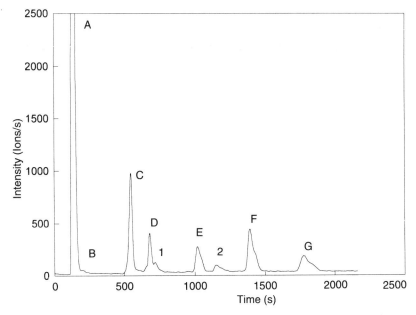

Figure 13.7. Chromatographic separation obtained with the IonPac AS7 column (Dionex) as revealed by ICP–MS detection. Mussel extract (1:20 dilution with deionized water): A, AsBet; B, arsenite; C, DMA; D, AsChol; E, MMA; F, PAA; G, arsenate; 1, unknown peak; 2, Cl^- ion (detected as $^{40}Ar^{35}Cl$).

REFERENCES

1. C. J. Sonderquist, D. G. Crosby, and J. B. Bowers, *Anal. Chem.* **46**, 155 (1974).
2. N. T. Crosby, *Analyst (London)* **102**, 225 (1977).
3. J. S. Edmonds and K. A. Francesconi, *Nature (London)* **265**, 436 (1977).
4. R. S. Braman, in *Arsenical Pesticides* (E. A. Woolson, ed.), pp. 108–123. Am. Chem. Soc., Washington, DC, 1975.
5. R. S. Braman and C. C. Foreback, *Science* **182**, 1247 (1973).
6. Y. K. Chauand and P. T. S. Wong, *ASC Symp. Ser.* **82**, 65–68 (1978).
7. International Agency for Research on Cancer, *IARC Monograph on the Evaluation of the Carcinogenic Risk of Chemicals to Humans*, Vol. 23. IARC, Lyon, France, 1980.
8. Food and Agriculture Organization (FAO), World Health Organization (WHO), *WHO Food Addit. Ser.* **18** (1983).
9. C. R. Penrose, *CRC Crit. Rev. Environ. Control* **4**, 465 (1974).
10. Food and Agriculture Organization (FAO) and World Health Organization

(WHO), *Toxicological Evaluation of Certain Additives and Contaminants*, 33rd Meet. Jt. FAO/WHO Expert Comm. Food Addit., Geneva, 1989.
11. W. Maher and E. Butler, *Appl. Organomet. Chem.* **2**, 191 (1988).
12. W. R. Cullen and K. J. Reimer, *Chem. Rev.* **89**, 713 (1989).
13. L. Ebdon, S. Hill, and R. W. Ward, *Analyst (London)* **112**, 1 (1987).
14. I. S. Krull, *Trends Anal. Chem.* **3**, 76 (1984).
15. J. R. Dean, S. Munro, L. Ebdon, M. H. Crews, and C. R. Massey, *J. Anal. At. Spectrom.* **2**, 607 (1987).
16. Y. Talmi and D. T. Bostick, *Anal. Chem.* **47**, 2145 (1975).
17. B. Beckermann, *Anal. Chim. Acta* **135**, 77 (1982).
18. S. Fukui, T. Hirayawa, M. Nohara, and Y. Sakagami, *Talanta* **30**, 89 (1983).
19. Y. Odanaka, N. Tsuchiya, O. Matano, and S. Goto, *Anal. Chem.* **55**, 929 (1983).
20. K. Dix, C. J. Cappon, and T. Y. Toribara, *J. Chromatogr. Sci.* **25**, 164 (1987).
21. R. A. Stockton and K. J. Irgolić, *Int. J. Environ. Anal. Chem.* **6**, 313 (1979).
22. R. Iadevaia, N. Aharonson, and E. A. Woolson, *J. Assoc. Off. Anal. Chem.* **63**, 742 (1980).
23. F. E. Brinckman, K. L. Jewett, W. P. Iverson, K. J. Irgolić, K. C. Ehrard, and R. A. Stockton, *J. Chromatogr.* **191**, 31 (1980).
24. G. R. Ricci, L. S. Shepard, G. Colovos, and N. E. Hester, *Anal. Chem.* **53**, 610 (1981).
25. V. Foà, A. Colombi, M. Maroni, M. Buratti, and G. Calzaferri, *Sci. Total Environ.* **34**, 241 (1984).
26. C. T. Tye, S. J. Haswell, P. O'Neill, and K. C. C. Bancroft, *Anal. Chim. Acta* **169**, 195 (1985).
27. T. Mayani, S. Uchiyama, and Y. Saito, *J. Chromatogr.* **391**, 161 (1987).
28. L. E. Ebdon, S. Hill, A. P. Walton, and R. W. Ward, *Analyst (London)* **113**, 1159 (1983).
29. S. Branch, K. C. C. Bancroft, L. Ebdon, and P. O'Neill, *Anal. Proc.* **26**, 73 (1989).
30. J. S. Blais, G. M. Monplaisir, and W. D. Marshall, *Anal. Chem.* **62**, 1161 (1990).
31. L. R. Johnson and J. G. Farmer, *Bull. Environ. Contam. Toxicol.* **46**, 53 (1991).
32. S. Kurosawa, K. Yasuda, M. Taguchi, S. Yamazaki, S. Toda, M. Morita, T. Uehiro, and K. Fuwa, *Agric. Biol. Chem.* **44**, 1993 (1980).
33. M. Morita, T. Uehiro, and K. Fuwa, *Anal. Chem.* **53**, 1806 (1981).
34. D. S. Bushee, I. S. Krull, P. R. Demko, and S. B. Smith, *Liq. Chromatogr.* **7**, 861 (1984).
35. K. A. Francesconi, P. Micks, and K. J. Irgolić, *Chemosphere* **14**, 1443 (1985).
36. G. K. C. Low, G. E. Batley, and S. J. Buchanan, *J. Chromatogr.* **368**, 423 (1986).
37. W. D. Spall, J. G. Spall, J. G. Lynn, J. L. Andersen, J. G. Valdez, and L. R. Gurley, *Anal. Chem.* **58**, 1340 (1986).
38. G. Rauret, R. Rubio, and A. Padró, *Fresenius' J. Anal. Chem.* **340**, 157 (1991).
39. N. Violante, F. Petrucci, F. La Torre, and S. Caroli, *Spectroscopy* **7**, 36 (1992).

40. R. Rubio, A. Padró, J. Albertí, and G. Rauret, *Anal. Chim. Acta* **283**, 160 (1993).
41. D. Beauchemin, M. E. Bednas, S. S. Berman, J. W. McLaren, K. W. M. Siu, and R. E. Sturgeon, *Anal. Chem.* **60**, 2209 (1983).
42. D. Heitkemper, J. Creed, J. A. Caruso, and F. L. Fricke, *J. Anal. At. Spectrom.* **4**, 279 (1983).
43. B. S. Sheppard, W. L. Shen, J. A. Caruso, D. T. Heitkemper, and F. L. Fricke, *J. Anal. At. Spectrom.* **5**, 431 (1990).
44. B. S. Sheppard, J. A. Caruso, D. T. Heitkemper, and K. A. Wolnik, *Analyst (London)* **117**, 971 (1992).
45. S. B. Roychowdhury and J. A. Koropchak, *Anal. Chem.* **62**, 434 (1990).
46. H. Ding, J. Wang, J. D. Dorsey, and J. A. Caruso, *Speciation of Arsenic in Biological Samples by Micellar Chromatography with Inductively Coupled Mass Spectrometric Detection*, Proc. Winter Conf., San Diego, CA, p. 102 (1994).
47. B. M. Hovanec and D. J. Northington, *Inorganic Arsenic Speciation in Drinking Waters Samples by Continuous Hydride Generation Inductively Coupled Plasma Mass Spectrometry Using a Glass Insertion Probe Separator*, Proc. Winter Conf., San Diego, CA, p. 103 (1994).
48. E. H. Larsen, G. Pritzl, and S. H. Larsen, *J. Anal. At. Spectrom.* **8**, 1075 (1993).
49. L. Lin, J. Wang, and J. A. Caruso, *Speciation of Arsenic in Capillary Zone Electrophoresis with Inductively Coupled Plasma Mass Spectrometry*, Proc. Winter Conf., San Diego, CA, p. 69 (1994).
50. S. Caroli, F. La Torre, F. Petrucci, and N. Violante, *Environ. Sci. Pollut. Res. Int.* **1**, 205 (1994).

INDEX

Accumulation process, 297, 305
Acidification process, 351
Activity coefficients, 367
Adenosine monophosphate, determination of, 147
Adenosine phosphate, determination of, 147
Adenosine triphosphate, determination of, 147
Adsorbent resins, 33, 42, 63–65
Adsorption process, 49–51, 56, 57, 63, 122, 211 344, 347, 351, 354, 357, 419
Aerosols, 132
Agglomeration, 122
Aggregates, fibrillary, 244
Air, analysis of, 9
Albumin, determination of, 235–237, 268, 271 272, 281
Algae, analysis of, 296, 297
Alkylation process, 335, 336
Alkyllead, determination of, 206
Alkylmercury chloride, determination of, 107
Aluminosilicates, determination of, 223, 225, 243, 244, 246
Aluminosilicate species, formation of, 244
Aluminum:
  accumulation in ageing, 246
  in blood serum, 228
    ultrafiltrable in, 234
  body, accumulation in, 223, 224, 239, 246, 247
    removal from, 239
  brain, transport to, 244
  detection limits of, 228
  determination of, 47, 60, 227, 228, 258, 265, 281
  detoxification mechanism of, 240
  speciation of, 4, 9, 12, 13, 15, 205, 232, 234–241, 245, 247
  supplementation of, 246
  toxicity of, 245, 247

Aluminum citrate complexes, determination of, 238, 239
Aluminum fluoride, separation of, 239
Aluminum oxalate, determination of, 239
Aluminum–silicon, colocalization, 243
Aluminum toxins, 247
Alzheimer's disease, 243–245
American Society for Testing and Materials, 202
Aminophenylarsonic acid, determination of, 448, 461
Amperometry, 127
Amyotrophic lateral sclerosis, 243
Analytical curves, 137
Analytical techniques, time scale of, 30, 31, 39 42, 63, 65
Anodic stripping voltammetry, 3, 47, 48–51, 53, 54, 63–65, 67, 69, 70, 81, 82, 259, 363, 364, 411, 419
  instrumentation for, 393, 394
Anoxic media, 23, 24, 66, 68
Anthropogenic activities, 331, 332, 340, 341
Antifouling paints, 287, 306–310, 321
Antimony:
  determination of, 178, 184
  speciation of, 58, 67, 106, 108, 189
Arsenate:
  conversion into, 10
  detection limits of, 448, 450, 453
  determination of, 110, 206, 448–452, 454, 456, 460, 461
Arsenic:
  absorption of, 445
  health effects of, 445
  natural presence of, 445
  speciation of, 4, 9–12, 16, 24, 57–59, 67, 103, 104, 108–110, 130, 150, 190, 196, 197, 204–206, 208, 211, 212, 214, 215, 448–452
  toxicity of, 446

INDEX

Arsenic compounds, levels of, 448
Arsenite:
  detection limits of, 448, 450, 453
  determination of, 110, 206, 448–452, 454, 456–458, 460
  separation of, 448, 449, 454, 456
Arsenobetaine:
  detection limits of, 450, 453
  determination of, 206, 208, 214, 448–450, 452, 456–458, 460, 461
  ultraviolet decomposition of, 453
Arsenocholine:
  detection limits of, 450, 453
  determination of, 206, 208, 214, 448–450, 452, 454, 456, 461
  ultraviolet decomposition of, 450, 453
Arsine, determination of, 206
Ashes, analysis of, 431
Atomic absorption spectrometry, 3, 33, 56, 58, 60, 64, 65, 67, 68, 97, 98, 127, 129, 201, 203, 204, 207–209, 211, 215, 216, 259, 419, 422, 424
Atomic emission spectrometry, 121, 127
Atomic fluorescence spectrometry, 9, 97, 98
Atomic populations, 132
Atomization mechanism, 100

Barium, determination of, 258, 266, 281
Baseline drift, 132
Batch techniques, 31, 42, 64, 65
Bicarbonate–transferrin complexes, 238
Binding capacity, 150
Bioassays, 3, 28, 32, 33, 61, 291, 334, 411
Biochemical pathways, 338
Bioconcentration factors, of organic compounds, 317, 318, 320
Biogeochemical cycles, 413
Biogeochemical pathways, 219, 332, 335, 336, 356
Bioindicators, 333
Biological fluids, analysis of, 1, 17, 157, 231
Biological models, 28
Biological samples, analysis of, 216
Biological tissues, analysis of, 1, 17, 127, 141, 157, 164, 219, 333
Biomedical analysis, 195
Biota, analysis of, 196, 204, 205, 216, 332, 333, 348, 356
Bivalves, analysis of, 305, 312–314

Blood serum:
  analysis of, 17, 230, 238, 247
  human, analysis of, 181, 184, 187
Bogs, analysis of, 12
Brain, analysis of, 17
Brain tissues, aluminum in, 246
Bromine:
  determination of, 184
  speciation of, 189, 191
Butylarsonic acid, determination of, 447
Butylation process, 205
Butyltin, determination of, 196, 197, 199, 200, 204, 207, 209, 212, 216

Cadmium:
  determination of, 29, 47, 258, 262, 266, 281, 395, 398, 412
  speciation of, 4, 10, 13–15, 17, 31, 32, 43, 53, 54, 56, 63, 65, 398, 399, 412, 421
Calcium, determination of, 37, 281
Calibration, determination of, 197, 198, 201, 202, 209–211, 219
Calorimetry, 126
Cancer treatment, 144
Capillary gas liquid chromatography, 447
Capillary zone electrophoresis, 208, 451, 460
Carboplatin, determination of, 144
α-Caseins, determination of, 267, 268, 271
β-Caseins, determination of, 267, 268, 271
κ-Caseins, determination of, 267, 268, 271, 272, 281
Cathodic stripping voltammetry, 47, 52, 69, 70, 364, 411
Cell membrane damage, 290
Cellulose acetate-coated mercury film electrode, 51
Centrifugation, 33, 34, 99, 343–345, 348, 353, 354, 357, 419
Certification process, 217
Certified reference materials, 261, 262, 356, 403
Cesium, speciation of, 189, 191
Chemical models, 28
Chlorine:
  determination of, 184
  speciation of, 189
Chloroform, determination of, 204
Chromatographic techniques, 231, 234, 236
Chromium:
  determination of, 258, 262, 266, 281

leachability of, 428, 429
speciation of, 4, 9–12, 14, 24, 54, 58, 100, 102, 103, 105, 114, 138, 140–142, 150, 190, 196, 197, 205, 211, 419, 423, 427–429
Cisplatin, determination of, 144
Clean room facilities, 394, 403
Cleanup process, 202, 205, 207, 211, 214
Coastal water, analysis of, 331, 339, 342
Cobalt:
determination of, 178, 184, 258, 262, 266, 411, 412,
speciation of, 15, 17, 24, 148, 189, 190, 412
Cold trapping, 207–209
Cold vapor generation, 216
Colloids, 27, 31, 39
Colorimetric tests, 259
Colorimetry, 3, 127
Colostrum, analysis of, 271, 272, 282
Comité Européenne de Normalisation, 219
Complexation:
metals competition for, 383
theory, 365
Complexation process, 126
Complexes:
organic, kinetic lability, 400
stability of, 25, 45, 51–53, 65, 66, 77, 82, 83, 86
Complexes formation, side-reactions, 366, 367
Complexing capacity, 45, 53, 65, 74, 78, 364, 408
Complexing parameters, 25, 45, 51–53, 65, 66, 74, 77, 78, 83–86
Conditional ligand concentration, 367
Conditional metal concentration, 367, 368, 374
Conditional stability constants, 364, 368, 369, 374
Conductivity, detectors for, 134
Contamination control, 364, 402, 413
Contamination process, 197, 201, 204, 206, 211, 215, 217, 219, 225, 226, 233, 247, 332–334, 339, 347, 348, 350–354, 356, 357, 363
Control charts, 199, 200, 202
Copper:
determination of, 29, 47, 52, 184, 258, 262, 265, 266, 281, 395, 398, 408, 409
speciation of, 4, 10, 13, 14, 17, 23, 24, 26, 29–31, 42, 43, 47, 53, 54, 60, 61, 63, 65, 66, 189, 408–410, 421

toxicity of, 408, 410, 411
Coulometry, 127
Crab, analysis of, 451
Crustaceans, analysis of, 295, 296
Curve fitting, 383
Cyclic voltammetry, 81

Dealkylation process, 335, 336
Deferoxamine, chelating agent, 240, 241, 243, 247
De Ford–Hume equation, 75
Degradation process, 203, 211, 288, 332, 337–340, 350, 355–357
Deposition potential, 396, 398
Deposition time, 396, 399
Derivatization process, 105, 148, 149, 197, 198, 201, 203, 205, 206, 208, 210, 215, 216
Desorption process, 354
Detection process, 197, 199, 203, 208–211, 215, 216, 218
Detection window, 369, 401, 402
Dialysis, 33, 39, 64, 99, 232, 234, 240, 258
Dialysis encephalopathy, 243
Dichloromethane, determination of, 203, 204
Differential equilibrium parameters, 84, 85
Differential pulse anodic stripping voltammetry, 259, 364, 368, 371, 393, 394, 400
Differential pulse cathodic stripping voltammetry, 51, 434, 435, 440
Differential pulse polarography, 46, 47, 49, 54, 66, 67, 70, 80, 215, 433
Differential scanning calorimetry, 144
Diffusion process, 36, 37, 39, 40, 44, 45, 50, 63, 68, 74, 78
Dilution process, 354
Dimethylarsinic acid:
detection limits of, 447, 448, 450, 453
determination of, 109, 110, 212, 214, 445, 447–452, 454, 460, 461
Direct current plasma atomic emission spectrometry, 9, 97, 129, 130, 145, 151, 229
Discrete models, 52, 84, 87
Dissociation process, 33, 42
Dissolution process, 124
Distillation, 99
Dog fish, analysis of, 450

468 INDEX

Down syndrome, 243
Dropping mercury electrode, 45, 68
Dry ashing method, 260, 261
Drying process, 203

Electrochemical detection, 208
Electrodeposition, 99
Electron capture detection, 203, 209, 210, 216
Electron paramagnetic fluorescence, 60
Electron paramagnetic resonance, 60
Electron spin resonance, 60
Electrophoresis, 99, 114
Electrothermal atomization atomic absorption spectrometry, 9, 98, 110, 126, 209, 215, 227–231, 236, 239, 356, 424, 446, 448
Element-selective detectors, 129
Elements:
 bioavailability of, 9, 21–25, 52, 63, 65, 414, 419
 detection limits of, 45, 47, 54, 56, 58–60, 69, 133, 172, 393
 determination limits of, 199
 essential, 157, 160, 282
 leachability of, 1, 123, 146
 nonessential, 157, 160
 toxicity of, 21–25, 29, 41, 44, 47, 51–53, 59, 61–63, 121, 265, 282, 333–335, 421
 uptake of, 22, 23, 29–31, 59, 62, 63
Environmental analysis, 11, 195, 197, 199, 216, 363
Environmental contamination, 219
Enzymatic reactions, 3
Equilibration time, 398
Errors, sources of, 202–204, 205, 208, 210, 211, 215, 216
Estuarine water, analysis of, 331–333, 335, 336, 338, 339, 341, 342, 347
Ethylarsonic acid, determination of, 447
Ethylation process, 203, 205, 206, 216
Ethylmercury chloride, determination of, 107
European Union, 196
Experimental data, comparability of, 334, 356, 357
Extraction process, 99, 102, 126, 147, 197, 201–206, 210, 211, 213, 215, 218, 348, 355

Ferritin, analysis of, 17
Ferritin drugs, analysis of, 18
Filter feeders, 334
Filtration process, 33, 35, 64, 65, 67, 99, 112, 343, 344, 352, 353, 357, 406, 419
Fish:
 analysis of, 219, 292–295, 456, 460
 arsenic compound extraction from, 457
Fish liver, analysis of, 451
Flame atomic absorption spectrometry, 209, 446, 449
Flame atomic fluorescence spectrometry, 449
Flame ionization detection, 203, 209, 215
Flame photometry, 203, 209, 215, 216, 356
Flow injection analysis, 61, 126, 236
Fluorescence spectrometry, 61, 65, 68
Fluorimetric detection, 236
Foam flotation, 99
Food, analysis of, 104, 195–197, 216
Fractionation process, 123
France legislation on organotin compounds, 306, 308, 309
Freeze-drying process, 203, 357
Freshwater, analysis of, 15, 292, 333
Fulvic acids, complexation by, 13, 43, 61, 413
Fundamental units, 201

Gamma-ray spectrometry, 163
Gas chromatography, 33, 57, 99, 201, 204, 206–209, 211, 215, 216, 259, 355, 446, 447, 449, 456, 459
Gas liquid chromatography, 208
Gel chromatography, 241, 258, 269
Gel electrophoresis, 241, 258
Gel filtration, 235
Gel filtration chromatography, 3, 33, 40, 175
Geochemical pathways, 335
Gold, determination of, 144, 145
Gravitational settling, 122
Grignard reagents, 205

Hanging mercury drop electrode, 47, 53, 401
Heart, analysis of, 17
Heavy metals:
 determination of, 13, 147, 421
 speciation of, 13
Hexane, determination of, 204

High performance liquid chromatography, 9, 33, 59, 126, 129, 147, 201, 204, 208, 209, 215, 216, 235, 236, 239, 241, 255, 258, 259, 267, 356, 446, 448–456, 458, 459
Histochemical techniques, 231
Homeostasis process, 265, 283
Humic acids, complexation by, 13, 23, 26, 27, 34, 37, 43, 52, 60, 61, 67, 84, 413
Hydride generation, 33, 57, 67, 108, 109, 203, 205, 206, 210, 215, 216, 423, 447, 448, 450, 452–454, 456, 459
Hydridization process, 337
Hydrolysis process, 135
8-Hydroxyquinolinesulfonic acid, complexing agent, 236
Hyphenated techniques, 99, 148, 151, 258, 264, 447, 448, 449, 451, 453, 455, 458

Identification procedures, 332
Immunoassay tests, 3
Immunocytochemical techniques, 231
Immunoglobin, determination of, 235
Immunoglobulins, determination of, 268, 270, 272, 281
Imposex, 301, 304, 310, 311, 319, 321
Inductively coupled plasma atomic emission spectrometry, 9, 97, 129, 203, 208, 209, 211, 215, 227, 229, 231, 255, 257, 259, 261, 263–265, 419, 446, 449–453, 458, 459
Inductively coupled plasma mass spectrometry, 3, 97, 201, 208, 209, 211, 215, 216, 356, 419, 446, 449, 451, 452, 455, 456, 459, 460
Inert complexes, 30, 33, 45, 48, 52, 53, 64, 65, 71–73, 82, 84
Instrumentation, development of, 9
Intercomparison exercises, 404, 196, 202, 210, 211, 214–216, 218, 219
Interferences, analytical, 197, 204–206, 209, 210, 213, 215, 216
International Atomic Energy Agency, 265
International Standardization Organization, 219
Iodine:
 determination of, 178, 184
 speciation of, 189, 191
Ion chromatography, 11, 126, 147, 151

Ion exchange chromatography, 205, 259
Ion exchange process, 99, 126, 419
Ion exchange resins, 33, 42, 63, 64, 68
Ion liquid chromatography, 446, 448
Ion-selective electrodes, 54–57, 63, 409, 411, 419
 modified, 50
Iproplatin, determination of, 144
Iron:
 determination of, 178, 184, 258, 262, 266, 281
 speciation of, 4, 11, 12, 15, 18, 23, 24, 26, 42, 44, 61, 66, 67, 128, 136–138, 143, 148, 150, 189, 190
Iron chlorocomplexes, determination of, 137
Isotope dilution, 209
Isotope dilution mass spectrometry, 213
Isotopic composition, 162
Italy legislation on organotin compounds, 308, 315, 316

Kidneys, analysis of, 17

Labile complexes, 30–33, 43–45, 48, 52, 64, 65, 67, 71, 74–79, 81, 82, 84
α-Lactalbumins, determination of, 269, 271–273, 281
β-Lactoglobulin, determination of, 269–272, 281
Lagoon water, analysis of, 408, 410
Lake water, analysis of, 14
Laser microprobe mass spectrometry, 231
Lead:
 determination of, 47, 258, 266, 281, 365, 366, 395, 398, 412
 speciation of, 4, 9, 11, 13–16, 31, 51, 53, 54, 56–59, 63, 65, 101, 196, 208, 211, 364, 398, 421
Legal provisions, 306, 307, 308, 321
Ligands:
 competition of, 373, 383, 391
 conditional concentration of, 367
 organic, 393
Liquid chromatography, 99, 121, 209, 211, 452
Liquid extraction, 33, 41
Lithium, determination of, 281
Liver diseases, 224
Liver, analysis of, 17
Lobster digestive gland, analysis of, 451

470 INDEX

Local binding parameters, 84, 86
Long-lived radionuclides, 169, 182, 185
Loss process, 197, 201, 204–206, 211
Lungs, analysis of, 17

Macromolecules, 27, 30
Magnesium:
  determination of, 258, 262, 265, 266, 281
  speciation of, 189
Magnetic resonance spectrometry, 231
Manganese:
  determination of, 178, 184, 258, 262, 266, 281
  speciation of, 11, 12, 15, 23, 24, 26, 66, 67, 148, 150, 189, 190
Marine matrices, analysis of, 196
Mass balance, 136
Mass spectrometry, 203, 215, 216
Matrix composition, 199
Matrix effects, 135, 210
Matrix-matched solutions, 197
Matrix modifiers, 209
Mean diffusion, 45, 74
Measurement sensitivity, 134
Measurements:
  accuracy of, 2, 195, 197, 201, 202, 204, 209, 211, 219, 353
  comparability of, 195
  precision of, 2, 199, 204
  reliability of, 159
  reproducibility of, 195, 207, 213, 217, 353, 454, 455
  traceability of, 201
  uncertainty of, 20
Membrane filtering, 142
Mercury film electrode, 47, 53, 69
Mercury:
  determination of, 107
  speciation of, 4, 10, 12, 14, 15, 57, 58, 105, 196, 204–207, 211, 214
Metabolism, effects of chemicals on, 297, 305
Metallothioneins complexation by, 17
Metals:
  electroactive fraction of, 364, 393
  labile fraction of, 364
  total, determination of, 395
  trace, complexation of, 365, 398, 408, 409, 411, 412
Methoxyethylmercury chloride, determination of, 107

Methyl iodide, determination of, 336
Methylation process, 332, 335–337, 340, 356, 357
Methylcobalamin, determination of, 336
Methylmercury, determination of, 107, 196, 197, 207, 216, 217, 219
Methylmercury chloride, determination of, 106
Micellar liquid chromatography, 451, 459
Microelectrodes, 51, 54
Microwave atomic emission spectrometry, 446, 447
Microwave-induced plasma atomic emission spectrometry, 97, 98
Milk:
  analysis of, 17, 255, 256, 265–267, 273
  cow, analysis of, 230, 256, 258–260, 263, 265–267, 272, 274–280
    pasteurized, analysis of, 259, 260, 266, 267, 269, 274–281
  formula, analysis of, 256, 258–260, 263, 266, 273, 274–282
  human, analysis of, 256, 258–260, 263, 265–267, 271, 272, 274–282
  ultrahigh temperature, sterilized, analysis of, 259, 260, 265, 266, 274–281
Molybdenum, determination of, 47
Monitoring programs (on) organotin compounds, 308–310, 315, 319, 321
Monomethylarsonic acid:
  detection limits of, 447, 448, 450, 453
  determination of, 109, 110, 212, 214, 445, 447, 449, 460, 461
  separation of, 448–451
Muscle, analysis of, 17,
Mussels:
  analysis of, 216, 217, 294, 295, 350, 356, 456, 460
  arsenic compounds extraction from, 457

Nafion-coated thin mercury film, 50
National Institute for Environmental Studies, 197
National Institute of Standards and Technology, 141, 197
National legislations on organotin compounds, 308, 319
National Research Council of Canada, 197
Natural samples, heterogeneity of, 48, 49

Natural water:
  analysis of, 9, 13, 26, 35, 37, 51, 365
  binding capacity of, 149
Nebulizers, 133
Neurofibrillary tangles, 243
Neurological disorders, 224
Neuropathological disorders, 247
Neutron activation analysis, 9, 33, 155, 156, 161, 164, 182, 208, 209, 213, 215, 216, 227
Nickel:
  determination of, 24, 47, 54, 258, 262, 266, 281
  speciation of, 15
Nitrilotriacetic acid, complexation by, 15
Nitrogen, speciation of, 12
Normal pulse polarography, 46, 49, 50, 69, 70, 80
Nuclear magnetic resonance, 243
Nuclear reactors, 166

Operationally defined procedures, 196
Organic exudates, 22, 25, 53
Organic traces, determination of, 195
Organisms:
  aquatic, analysis of, 298
  marine, analysis of, 289, 292
Organochlorine compounds, determination of, 347
Organotin compounds:
  determination of, 196, 204–206, 208 211, 214–216, 219, 331–335, 337, 339–356
  legislation on, 356
  lipophilicity of, 346
  monitoring, 331, 332, 334, 335, 347, 348, 350, 351, 356, 357
  source variability of, 332, 340, 342, 344, 346–348, 350
Organs, analysis of, 1
Orthophosphate, determination of, 147
Oyster, analysis of, 199, 216

Pancreas, analysis of, 17
Parkinson disease, 243
Particle size distribution, 123
Particle size measurement, 122
Particulate matter, analysis of, 122

Partition process, 334, 335, 340, 342–344, 346
Patients:
  hyperaluminemic, 247
  thalassemic, 240
Peak broadening, 132
Peak shape, 134
Peak symmetry, 134
Pentylation process, 206
Peptones, determination of, 281
Pesticides, determination of, 287
Phenylarsonic acid, determination of, 451, 453, 454, 456
Phenylmercury chloride, determination of, 106, 107
Phosphorus, speciation of, 147
Photometry, 127
Plankton, analysis of, 14
Plasma, analysis of, 230
Plasma sources, 126
Platinum:
  determination of, 144
  speciation of, 11, 144, 150
Polarography, 46, 68, 69, 70
Pollutants, stability of, 335, 351, 355
Polycyclic aromatic hydrocarbons, determination of, 347
Polymers, stabilizers for, 288
Pond water, analysis of, 14
Pore size measurement, 122
Postcolumn derivatization, 127
Potassium, speciation of, 189
Potentiometry, 33, 54, 127, 422
Precipitation, 99
Preconcentration, 133, 349
Preconcentration process, 202
Preconcentration techniques, 448
Propylarsonic acid, determination of, 447
Proteins:
  analysis of, 179, 181, 187
  metal carrying, in blood serum
  separation of, 174, 179
Proteoses, determination of, 281
Pseudopolarograms, 398, 399
Pyrophosphate, determination of, 147

Quality assurance, 197–199, 202, 209, 247, 357
Quality control, 195–197, 199, 200, 202, 204, 208, 214–216, 219, 220, 332, 334, 340, 346, 348, 350, 351, 353, 356, 357

Quantification procedures, 332
Quartz furnace atomic absorption spectrometry, 209, 215
Quasi-labile complexes, 30, 45, 48, 79, 81
Quiescent period, 398

Radiographic techniques, 244
Rain water, analysis of, 15, 26
Random errors, 197
Reaction kinetics, 22, 25, 29–33, 42–45, 48, 52, 53, 60, 64, 65, 67, 71–84
Recovery, 198, 205, 211, 213, 349
Reduction process, 135
Reference materials, 197, 199, 202, 213, 218, 247, 357, 441, 450, 451
  certified, 12, 18, 197, 199, 201, 202, 219, 220
Retention time, 134, 213
River water, analysis of, 14, 26, 439
Road runoff, analysis of, 13
Rotating disk electrode, 401
Rotating glassy carbon disk electrode, 393, 394
Rubidium:
  determination of, 178, 184
  speciation of, 189
Ruzic, approach of, 380

Saliva, analysis of, 230
Saltwater, analysis of, 293, 333
Samples:
  homogeneity of, 199, 211
  introduction of, 122, 133, 135
  pretreatment of, 332, 357
  processing of, 135
  storage of, 203, 332, 347, 348, 352, 354–357, 407, 438
  treatment of, 126, 173, 201
Sampling procedures, 332, 333, 335, 342, 347, 350–352, 354, 356
Sampling strategy, 172, 332, 349, 351, 405
Scanning electron microscopy, 231
Seawater:
  analysis of, 14, 15, 26, 318, 335, 338, 343, 344, 346, 347, 363–365, 368, 408–411, 413, 433
  antarctic, analysis of, 399, 410
Secondary ion mass spectrometry, 244

Sediment:
  analysis of, 15, 123, 124, 149, 199, 200, 204, 209, 210, 212, 215–219, 312–314, 318, 332, 334, 335, 337–343, 346–348, 350, 355–357
  bacteria in, 336
  grain size of, 354
Sediment core, analysis of, 338, 339
Selenium:
  determination of, 47, 57, 58, 258, 266, 439, 432–434
  speciation of, 9, 11, 57, 131, 178, 189, 191, 196, 197, 208, 211, 214, 419, 433, 439
Seminal fluid, analysis of, 17
Senile plaques, 243, 244
Separation process, 126, 130, 197, 201, 202, 203, 207–209, 211, 213, 215, 343, 355
Separation techniques, 446
Sequential extraction process, 196, 217, 218
Settling effect, 123
Shellfish, analysis of, 451
Short-lived radionuclides, 168, 182, 183
Shrimp, analysis of, 451
Sieving process, 354, 357
Silicon:
  blood serum levels, 225
  brain, transport to, 244
  detection limits of, 231
  determination of, 229
  in body fluids, 231
Silicon–aluminum, colocalization, 243
Silicon supplementation, 246
Silver, speciation of, 23, 51
Single extraction process, 196, 217, 218, 219
Size exclusion chromatography, 99, 264
Sludges, analysis of, 430
Small intestine, analysis of, 17
Sodium:
  determination of, 178, 184
  speciation of, 189
Soil, analysis of, 9, 15, 123, 210, 214, 217, 218, 448
Solid, suspended, analysis of, 332, 334, 343–348, 351–353, 357
Solubility process, 343
Speciation literature, 4
Speciation methods, 1, 3, 23, 29, 33, 63, 121
Speciation models, 427, 433, 436
Speciation studies, procedural aspects of, 402

Spectrofluorimetry, 215, 227
Spectrophotometry, 60, 65–67
Spleen, analysis of, 17
Square wave voltammetry, 53
Stability, 199, 211, 212
Stability constants:
  concentration, 367
  thermodynamic, 367
Standard, internal, 207, 209, 213, 214
Standard additions, 197, 210
Standard additions method, multiple, 398
Standard methods, 201
Standards, Measurement and Testing Programme, 8, 197
Statistical control, 207
Statistical errors, 173
Strontium, determination of, 258, 266, 281
Sublethal effects, 291, 298
Sulfur:
  determination of, 47
  speciation of, 12, 67
Supercritical fluid extraction, 3
Surface microlayer, analysis of, 342, 348, 350–352, 357
Systematic errors, 174, 211, 215, 216

Tetraethyllead, determination of, 101, 114
Tetraethyltin, determination of, 112, 113
Tetramethyllead, determination of, 101
Tetramethyltin, determination of, 112, 113
Tetraplatin, analysis of, 144, 146, 147
Thin layer chromatography, 258, 259
Thin mercury film electrode, 393, 394, 396
Tin:
  organic species, determination of, 287–298 301, 302, 304–311, 315, 317–321
  speciation of, 9, 12, 56, 57, 59, 110, 196, 211, 335–337
Titanium, speciation of, 15
Titration curves:
  voltammetric, 364, 365, 368–371, 374, 375, 384, 385, 392
  voltammetric, generalized, 369, 370, 375, 385, 392
  voltammetric, shape of, 369
  voltammetric, transformed, 364, 366, 373, 377, 379, 381, 386
Titration procedures, voltammetric, 363, 364, 393
Titration process, 52

Titration, voltammetric, theoretical equations for, 368–370, 374, 377, 384, 385, 387
Toulene, determination of, 203, 204
Toxicity (of) organotin compounds, 289, 291–296
Trace elements:
  bioavailability of, 257, 267
  determination of, 195–199, 201–209, 211, 215–217, 219, 225, 227, 229
  extractable, determination of, 196, 210, 217, 218
  speciation of, 255–258, 263, 265–267, 273, 282, 283, 331–333, 335, 336, 347, 351, 356, 363–365, 368, 393, 398, 419
Trace metals, determination of, 218
Transferrin, determination of, 235–237
Transferrin fraction, determination of, 235
Transferrin receptors, 244
Transition metals, 127
Tributyltin, determination of, 332–334, 338, 340, 341, 343–345, 347–349, 352, 355
Triphenyltin, determination of, 208
Tuna, analysis of, 217

Ultracentrifugation, 99
Ultrafiltration process, 33, 37, 64, 99, 232, 233, 258
Ultrahigh temperature processes, 259, 260, 267, 270
Ultramicrofiltration process, 234, 237
Ultraviolet detection, 145
Ultraviolet irradiation, 419, 441
United Kingdom legislation on organotin compounds, 307, 309
United States legislation on organotin compounds, 307, 310
Universal detection, 134
Urine, analysis of, 230, 450, 451

Validation process, 196, 206, 210–212, 354, 357
van den Berg, approach of, 379
Vanadium:
  determination of, 184
  speciation of, 11, 138, 150, 189, 190
Vegetables, analysis of, 1
Volatilization process, 110, 203, 332, 336, 337, 340, 356, 357
Voltammetry, 33, 34, 44–54, 63–77, 79, 80–82, 209

Wastewater, analysis of, 14, 142
Water:
    analysis of, 122, 209, 215, 333, 334, 338, 341, 343, 345, 347, 348, 350, 351–355, 448
    elements speciation in, 11
Water sampling, 351, 355
    containers for, 351, 352, 355
Wet ashing methods, 260, 261
Whey proteins, determination of, 272
World Health Organization, 265

X-ray fluorescence spectrometry, 33, 155, 208, 215, 229
X-ray microanalysis, energy dispersive, 243

Zinc:
    determination of, 29, 47, 52, 178, 184, 258, 262, 265, 266, 411, 412
    speciation of, 4, 10, 15, 17, 31, 43, 44, 53, 54, 56, 65, 189, 190, 411, 412